D0215889

# THE HISTORY OF THE BRITISH COAL INDUSTRY

VOLUME 5

1946–1982: THE NATIONALIZED INDUSTRY

# THE HISTORY OF THE BRITISH COAL INDUSTRY

VOLUME 5

1946–1982: The Nationalized Industry

BY

WILLIAM ASHWORTH

*with the assistance of*

MARK PEGG

CLARENDON PRESS · OXFORD

1986

*Oxford University Press, Walton Street, Oxford OX2 6DP*

*Oxford New York Toronto*
*Delhi Bombay Calcutta Madras Karachi*
*Kuala Lumpur Singapore Hong Kong Tokyo*
*Nairobi Dar es Salaam Cape Town*
*Melbourne Auckland*

*and associated companies in*
*Beirut Berlin Ibadan Nicosia*

*Oxford is a trade mark of Oxford University Press*

*Published in the United States*
*by Oxford University Press, New York*



*British Library Cataloguing in Publication Data*

*The History of the British coal industry.*
*Vol. 5; 1946-1982: the nationalized industry.*
*1. Coal mines and mining–Great Britain–History*
*I. Ashworth, William II. Pegg, Mark*
*338.2′724′0941 HD9551.5*
*ISBN 0-19-828295-8*

*Library of Congress Cataloging in Publication Data*

*The History of the British coal industry.*
*Includes bibliographies and indexes.*
*Contents: –v. 2. 1700-1830, the Industrial Revolution / by Michael W. Flinn with the assistance of David S. Stoker– –v. 5. 1946-1982, the nationalized industry / by William Ashworth with the assistance of Mark Pegg.*
*1. Coal trade–Great Britain–History.*
*2. Coal mines and mining–Great Britain–History.*
*3. Coal trade–Government ownership–Great Britain–History.*
*1. Flinn, Michael W. (Michael Walter), 1917-*
*II. Ashworth, William. III. Pegg, Mark.*
*HD9551.5.H57 1984 338.2′724′0941 83-4194*
*ISBN 0-19-828295-8*

*Set by Joshua Associates Ltd*
*Printed and bound in Great Britain by*
*Biddles Ltd, Guildford and King's Lynn*

# Foreword

BY SIR NORMAN SIDDALL

THIS volume is the third to be published (but the last chronologically) of a major undertaking, the five volume *History of the British Coal Industry* initiated and supported by the National Coal Board. It covers the period of the nationalized coal industry from 1946 up to mid-1982, coincidentally the period of most of my own career in management up to the time of my appointment as Chairman, and a time of significant and sometimes convulsive change in the industry.

It is extremely difficult to write contemporary or near-contemporary history which is objective, lucid, and fluent but captures the feel of the period. Professor Ashworth achieves this task admirably, doing full justice to the complexities of mining operations and of the political and commercial context in which the coal industry operates. This volume, then, continues the very high standard already established for the series and I am glad to commend it.

# Preface

This book has a dual character. It is the final volume of a series covering the whole history of the British coal industry and it should be read in association with the earlier volumes, in whose subject matter many of the strongest roots of conditions and behaviour in the coal industry of the later twentieth century are to be found. But, as it deals only with the period since the industry was nationalized, it also incorporates the history of a very large and often controversial institution, the National Coal Board. Much of it has also been written very close to the events which it examines. When, in the summer of 1978, I first discussed with the NCB chairman, Sir Derek Ezra (as he then was), the possibility of writing the book, he suggested conversationally that 1980 would make a good round terminal date. Eventually, in 1982, it appeared to me that that year would provide a more natural break, as it marked the end of both a long chairmanship of the NCB and a long presidency of the NUM, and the cumulative effects of recession had then clearly set in train a serious decline in the financial condition of the coal industry. Thus I found myself dealing historically with events that were still in the future when I began my research, and using policy papers that were almost contemporary. This would not have been possible without an exceptional degree of co-operation from the National Coal Board. It is rare for any business undertaking voluntarily to let loose an outsider on material that is so nearly contemporary, and I was able to write a good deal of the history from sources in current administration rather than archives. As the latest events of the history came in the middle of a financial year, some of the annual statistics have been carried through to the end of the year 1982–3.

The book is a subject of contemporary influence in another way. Although the necessary research was then almost complete, nearly all the book was written between the middle of 1983 and the end of 1984. For most of that time the coal industry was in turmoil, with first an overtime ban and then its longest strike, and was the daily subject of incessant public comment. I have tried to ensure that the objectivity of the book was not reduced by these circumstances. Obviously, its whole subject matter has an important bearing on the state of the coal industry since

1982, but, despite its date of writing, the book was not designed with immediately following events in mind as an offstage climax. Its terminal date was already decided and, while it was being written, there were influential developments that are outside its scope.

It would be impossible to write such a book as this without a lot of help and co-operation from others. Though they are too numerous for everyone to be acknowledged individually, they all have my gratitude. In particular I benefited enormously from my visits to various collieries, other NCB establishments, and NCB Areas, where a great deal of trouble was taken to enable me to get a first-hand acquaintance with the working practices and the immense variety of conditions in the coal industry. I learned much, especially from those who guided me and talked to me at length, but also from many to whom I chatted as I passed, most of whose names I never knew.

There are some individuals to whom I have particularly large obligations. My most continuous help arose from the readiness of the NCB to second one member of staff to work with me as my research assistant. This role was first taken for a few months by Miss Ruth Wood (afterwards Mrs Macdonald), and then Dr Mark Pegg joined me for nearly three years. He located and put before me a vast amount of documentation and much of it was, with great care and clarity, annotated by him. His presence in NCB headquarters and familiarity with its ways continuously smoothed my path. He regularly discussed with me the interpretation of the problems we examined, and he read and commented in detail on the whole of the first draft of my book. His contribution was invaluable. Without it, the book would have been very different and would have missed some essentials, and would have taken much longer to prepare. I thank him for all his help and not least for the easy and friendly way in which it has always been given. I have also benefited from general discussion with the authors of the other volumes in the history and with Professor Peter Mathias, who read the first draft of my volume and made some useful suggestions.

Numerous people now or formerly in the service of the NCB assisted me in various ways. It was particularly helpful to be given first-hand accounts of the earlier years of nationalization from some of those deeply involved. Sir Joseph Latham and the late Sir Geoffrey Vickers talked freely to me, and Mr Michael Roberts familiarized me with his work as personal assistant to Lord Hyndley. Sir Andrew Bryan discussed his activities both as HM Chief Inspector of Mines and subsequently as a member of the national Board. Mr W. J. Charlton recounted for me his experiences as an Area General Manager and compared the conditions of colliery management immediately before and immediately after nationalization. Others assisted me to a better understanding of some later events and aspects of policy. I had helpful interviews with Mr R. T. Arguile, Mr H. E. Collins, and Mr R. V. Findlay. Mr John Barlow introduced me to the central records of the NCB. Mr Malcolm Edwards and Mr Michael Parker made valuable comments on the first draft of the book and saved me from several errors.

I was fortunate in the support of successive NCB chairmen. Lord Ezra has taken a constant interest in the progress of the history. He discussed a number of questions with me and made useful suggestions about my text. Sir Norman Siddall allowed me to draw on his knowledge and his time, and gave me a lot of encouragement. I benefited, too, from his close attention to all that I wrote. To the Secretary of the Board, Mr David Brandrick, my debt is especially large. He gave me the fruits of his long experience, which he discussed at length with me; ensured that I could range widely over the affairs of the NCB; kept constantly in touch with the progress of the book; and provided me with most helpful and undogmatic commentary on it. Other members of his department have also given great assistance, in particular Mr Martin Povey, the Assistant Secretary, and Dr John Kanefsky.

Help in a different form came from the Department of Energy which, in 1981, permitted me to use the files of the Ministry of (Fuel and) Power under a fifteen-year, instead of the normal thirty-year, rule, and which gave me temporary accommodation in the Department while I was studying the files which were closed to the public. Although many (but not all) of the files that I found most informative have since become publicly available, it greatly helped the continuity of my work to be able to consult them when I did.

To all the many, both named and unnamed, who have assisted me in such various ways, I offer my thanks.

Finally, one fundamental point should be emphasized. This book is the concluding volume of a history commissioned by the National Coal Board. Unlike all the other volumes, it is concerned with the affairs of the commissioning body. But, just like all the other volumes, it is an independent study. I have had the benefit of discussion with well-informed people who, by evidence and argument, sought (sometimes, but not always, successfully) to persuade me of the soundness of particular interpretations. But nobody tried to compel me about anything. The choice of matter for inclusion or omission, the manner of presentation, and the judgements, all are mine. They have no 'official' status and I take responsibility for them.

WILLIAM ASHWORTH

*June 1985*

# Contents

# Maps



# Diagrams

# Illustrations

# Tables



APPENDIX TABLES

# Weights, Measures, and Currency

During the period covered by this history two important but confusing changes took place: the decimalization of the pound sterling and the gradual, though incomplete, adoption of the metric system in the UK. The NCB formally adopted metric units from 1 April 1978. In this volume decimal currency and metric units of weight, length, and area have been treated as the norm. They have been used wherever continuous series or long-term comparisons are given, and spot figures have been converted from the earlier to the later units wherever this seemed helpful, though conversion has not been thought necessary in all cases. Direct quotations involving quantitative statements expressed in the earlier units have been left in the original form. Some quantitative materials that have been used were collected only in groups whose limits were round numbers in terms of the earlier measures. As conversion would require the ranges to be expressed in very odd fractions, such figures have been left in their original form. For some items where no change has been introduced in systems of measurement, figures have been given in the older units throughout. This applies mainly to the capacity of equipment used in opencast mining, much of it American in origin, where details are given in cubic yards and short tons (2,000 lb.), as in the suppliers' own ratings.

Any figures of currency, weight, length, or area which, for the reasons given, have been left in their original form, may be converted to the later standards by using these equivalents:

2.4*d.* = 1p = £0.01
1*s.* = £0.05
1 ton = 1.016 tonne
1 foot = 0.305 metre
1 yard = 0.914 metre
1 acre = 0.405 hectare

# Abbreviations in Text and Notes

| | |
|---|---|
| AB | Anderson Boyes Ltd. |
| AFC | Armoured Flexible Conveyor |
| AG | Aktiengesellschaft |
| AGR | Advanced Gas-cooled Reactor |
| *Ann. Rep.* | *Annual Report* |
| APEX | Association of Professional, Executive, Clerical and Computer Staff |
| ATM | Advanced Technology Mining |
| BACM | British Association of Colliery Management |
| BCORA | British Colliery Owners' Research Association |
| BNOC | British National Oil Corporation |
| BSC | British Steel Corporation |
| CapCoal | Capricorn Coal Developments Pty Ltd. |
| CBI | Confederation of British Industry |
| CEE | Central Engineering Establishment |
| CEGB | Central Electricity Generating Board |
| CEPCEO | Comité d'études des producteurs de charbon d'europe occidentale |
| CINA | Coal Industry Nationalisation Act 1946 |
| CISWO | Coal Industry Social Welfare Organisation |
| Conoco | Continental Oil Company |
| COSA | Colliery Officials Staff Association |
| CRE | Coal Research Establishment |
| DCF | Discounted Cash Flow |
| DSIR | Department of Scientific and Industrial Research |
| ECE | Economic Commission for Europe |
| ECSC | European Coal and Steel Community |
| EEC | European Economic Community |
| FIDO | Face Information Digested On-line |
| f.o.b. | free on board |
| HC | House of Commons |
| HC Deb. | House of Commons Debates (Hansard) |
| HL | House of Lords |
| ICI | Imperial Chemical Industries Ltd. |
| IEA | International Energy Agency |
| k.p.h. | kilometres per hour |
| MFGB | Miners' Federation of Great Britain |
| MINOS | Mine Operating Systems |
| MMC | Monopolies and Mergers Commission |

| | |
|---|---|
| MRDE | Mining Research and Development Establishment |
| MRE | Mining Research Establishment |
| m.t.c.e. | million tonnes of coal equivalent |
| NACODS | National Association of Colliery Overmen, Deputies and Shotfirers |
| N&C | Northumberland and Cumberland |
| NBPI | National Board for Prices and Incomes |
| NCB | National Coal Board |
| NPLA | National Power Loading Agreement |
| NSHEB | North of Scotland Hydro-Electric Board |
| NUM | National Union of Mineworkers |
| OCDQ | Overseas Coal Developments (Queensland) Ltd. |
| OEEC | Organization for European Economic Co-operation |
| OMS | output per manshift |
| OR | Operational Research |
| PEP | Political and Economic Planning |
| PIT | Performance Improvement Team |
| PMF | Progressive Massive Fibrosis |
| PRO | Public Record Office |
| PVC | Polyvinyl Chloride |
| PWR | Pressurized Water Reactor |
| RAG | Ruhrkohle Aktiengesellschaft |
| ROLF | Remotely Operated Longwall Face |
| ROM | Run-of-mine |
| SC | Select Committee |
| SMRE | Safety in Mines Research Establishment |
| SR&O | Statutory Rules and Orders |
| SSEB | South of Scotland Electricity Board |
| TS | Terms of Settlement |
| TUC | Trades Union Congress |

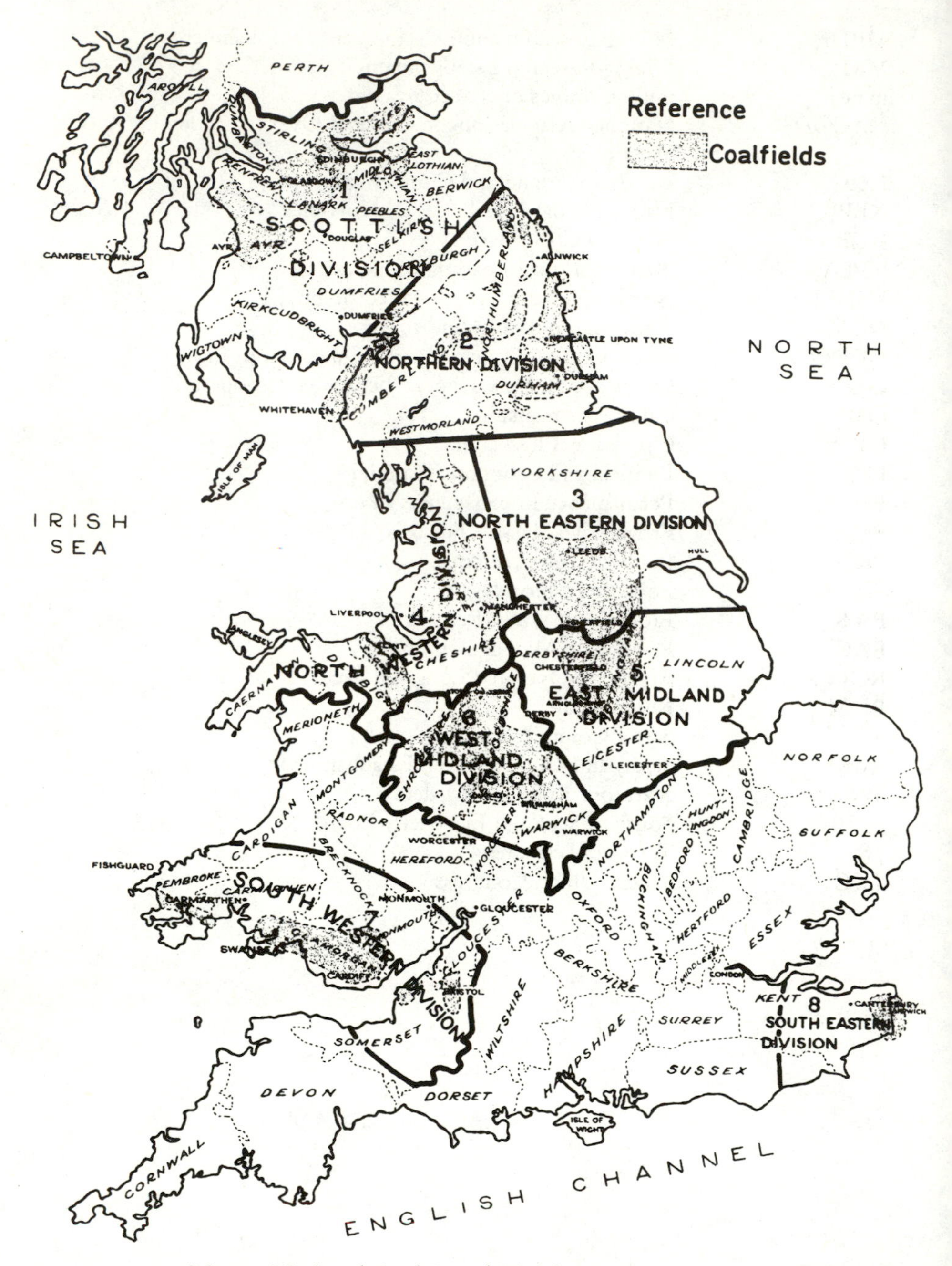


Map 1. National Coal Board Divisions, 1947

Map 2. National Coal Board Areas, 1982

PART A

# CONSTRAINTS AND OPPORTUNITIES

CHAPTER I

# The Inheritance

## i. The coal industry on the eve of nationalization

Anyone who looked at the coal industry in 1946 was likely to have two strong impressions. One was a feeling of anxiety caused by its generally deteriorating performance and the many signs of mistrust among those engaged in it. The other was an awareness that this was an extraordinarily varied industry in which most national generalizations had to be qualified because of the many differences in natural conditions, working practices, social traditions, and business organization.

The war years had shown a record of decline of almost every measurable kind. Total saleable output of deep mined coal had fallen every year from 231.3 million tons (235.0 million tonnes) in 1939 to 174.7 million tons (177.5 million tonnes) in 1945, and the gradual increase of opencast production, from its small beginning in 1942 to 8.1 million tons (8.2 million tonnes) in 1945, offset only a little of this drop. Even the small recovery of 1946, when the combined deep mined and opencast output reached 190.0 million tons (193.0 million tonnes), was quite inadequate for current needs, so that distributed stocks of coal were reduced for the fifth successive year, despite policies to restrict uses that were not absolutely essential. By the end of 1946 these stocks were down to the unsafe level of 8.5 million tons (8.6 million tonnes). Between 1939 and 1945 face output per manshift (OMS) had fallen from 3.00 tons (3.05 tonnes) to 2.70 tons (2.74 tonnes), overall OMS from 1.14 tons (1.16 tonnes) to 1.00 tons (1.02 tonnes), absenteeism (measured as a percentage of possible shifts) had risen continuously from 6.9 to 16.3, and total costs of production had more than doubled in terms of money. Although there had been wartime measures to keep workers in the industry and to direct men to it, and although miners' wages had been increased relatively to those in many other industries, there was obviously a declining willingness to look to coalmining for a livelihood. The number of workers on colliery books had been 766,000 in 1939 but was down to 692,000 in 1946, the lowest number for many, many years.[1]

[1] NCB *Ann. Rep.* 1947, 2; Kirby 1977, 172.

Anxiety aroused by this record was liable to be all the greater because in most respects the deficiencies appeared to be associated with long-term tendencies in the industry, rather than with the temporary difficulties of wartime. Though there had been some improvements in productivity in the inter-war years, they had been much less than those in other countries which raised large amounts of coal. The peak of output reached in 1913 had never been regained. The equality or superiority in productivity that the British industry then displayed in relation to that of nearly all other countries except the USA had long been lost. From the mid-1920s, when wartime declines had been overcome, to the cyclical maximum of 1936, OMS rose by 118 per cent in the Netherlands, 81 per cent in the Ruhr, 54 per cent in Poland, but only 14 per cent in Britain. Any general international competitive advantage of the British industry had disappeared. Natural conditions for mining were nowhere near as good in Britain as in the USA or Australia, but conditions in the coalfields of the Netherlands and the Ruhr were, on average, no easier than in Britain. Yet by 1936 OMS averaged 35.94 cwt. (1.83 tonnes) in the Netherlands and 33.66 cwt. (1.69 tonnes) in the Ruhr, compared with only 23.54 cwt. (1.20 tonnes) in Britain.[1]

Difficult labour relations—frequent antagonism of workers to their employer and their job—had long persisted, and, though these were to some extent characteristic of coalmining all over the world, they were in Britain a source of very serious difficulty in the maintenance of a smooth flow of output, without which efficiency was impossible. In the thirteen years from 1927 to 1939 inclusive there were only three in which the number of days lost by disputes in coalmining was less than 30 per cent of the total for all British industry, and eight in which it was more than 40 per cent.[2] The miners' unions remained whole-heartedly opposed to the existing structure and ownership of the industry. They regarded the period since the 1926 general strike as 'lean years' and continued to support the tradition embodied in Keir Hardie's nationalization bill of 1893. In its preamble this had declared that 'the present system of working mines as private concerns leads to great waste of our mineral supplies, and to strikes and lock-outs'.[3] Levels of pay and status sharpened these disaffections. In the hierarchy of industrial wages the miners ranked eighty-first in 1939. It was an indication of how much was

[1] Ministry of Fuel and Power, *Coal Mining: Report of the Technical Advisory Committee*, Cmd. 6610 (subsequently cited as Reid Report 1945), 29.

[2] Haynes 1953, 35 and 60.

[3] Arnot 1979, 90 and 105.

needed to restore their sense of fairness that, by 1944, successive public enquiries had had to recommend improvements that raised them to fourteenth position.[1]

Partly in response to these dissatisfactions, the miners agreed in 1944 to a fundamental change in their trade union structure, which operated from the beginning of 1945. Questions of industrial relations had long been settled on a district basis and miners had been organized in local or regional unions, which were loosely associated in the Miners' Federation of Great Britain. There was a strong belief that competition had made it possible for one union to be played off against another, to their general disadvantage; and wartime wage awards by the government, which applied nationally, had accustomed miners to see benefits in *general* wage settlements. So it was agreed to replace the MFGB by a National Union of Mineworkers, with responsibility for negotiating national agreements on wages, hours, and related questions; and the eventual adoption of a common national wages structure was proclaimed as an objective. Yet most of the day-to-day detail of union activity still depended on matters which varied from district to district, and much of the old federal character of mining unionism still remained. The NUM had more than twenty areas, each of which had its own funds and could register separately as a trade union.[2] So labour problems, difficult enough in any case, involved additional risks because the union organization was new and untried and could be subject to misunderstandings between the unfamiliar centre and the familiar areas.

Thus the coal industry showed many symptoms of dissatisfaction and relative or absolute decline. It was also still struggling with the effects of recent institutional changes and the search for improvements in them. The verdict of coal's wartime historian is neither surprising nor seriously to be questioned: 'No other major British industry carried so many unsolved problems into the war; none brought more out.'[3]

Yet it would be misleading to look only at the broad generalizations, however central they were. The coal industry was no monolith. No one will ever understand it if he fails to appreciate the strength of a sense of common interest among miners; but he will go just as far astray if he forgets that the industry is made up of individual people, communities, institutions—even individual coal seams with their own gift of perversity—and that these are almost bewilderingly diverse.

[1] Kirby 1977, 180 and 185.

[2] The changes in union structure are discussed more fully in Chap. 11, section ii below.

[3] Court 1951, 391.

A particularly clear aspect of the diversity was the dispersion of ownership and location of the mines. Immediately before nationalization there were in Great Britain about 1,470 collieries, of which 481 met the statutory definition of 'small mines';[1] that is, they had only a small underground labour force, normally not exceeding thirty. The ownership of this large number of mines was spread among more than 800 firms and individuals,[2] so that the 'average' undertaking in the coal industry owned rather less than two mines. Of course, the arithmetic average was not a representative figure. The small mines, which were about one-third of the total number, produced only just over 1 per cent of the output, not quite two million tons per year.[3]

By contrast, there were some large firms with great resources and several large mines. A few, with rough indications of their scale, may be cited as illustrations. A notable example was Doncaster Amalgamated Collieries Ltd., with six collieries and a labour force of well over 16,000; and on a slightly smaller scale, not very far away, were Amalgamated Denaby Collieries Ltd., with five, and Bolsover Colliery Co. Ltd., with six collieries, each firm employing some 9,000 men. A comparable instance in Northumberland was the Ashington Coal Co. Ltd., which owned five collieries and employed over 9,000 men. In Lancashire was one of the greatest firms in the industry, Manchester Collieries Ltd., with fifteen collieries, a capital of £7,000,000, and a labour force of over 14,000. There was also the Wigan Coal Corporation Ltd., with nearly 12,000 employees in fifteen collieries, but this firm illustrated an emerging trend for some large firms to spread their interests geographically. Although the rest of its undertakings were in central Lancashire, its largest colliery was at Manton, near Worksop. In much of Scotland and South Wales many of the mines were relatively small and the bigger firms operated an unusually large number. In Scotland, for instance, Bairds and Dalmellington Ltd. had twenty-three collieries with around 9,500 employees, and The Fife Coal Company, which was generally accounted one of the strongest firms in the industry, owned sixteen

[1] The figure is approximate. A few collieries which were not operating on vesting date never operated again and are not now reckoned in NCB totals. The first NCB annual reports were not ready until mid-1948, and the first three reports all give 980 as the number of collieries operated by the NCB, who did not operate the small mines. NCB *Ann. Rep.* 1948, 230, states that this figure applies to 1.1.1948. The total was later revised to 958 for 31.12.47. As closures exceeded openings by 12 in 1947, this implies a total of 970 for 31.12.46. Differences of classification could give a rather higher figure. The figure of 481 small mines is from NCB *Ann. Rep.* 1946, 11.

[2] *Ann. Rep.* 1946, 2.

[3] Ibid., 10.

collieries and employed over 8,000 men. In South Wales was Powell Duffryn Ltd., the largest firm in the coal industry. It operated 49 collieries with nearly 27,000 employees, and through subsidiaries (of which Cory Brothers & Co. Ltd. was the largest) it had another twelve collieries which brought its total colliery labour force to over 32,000. South Wales was also the scene of activity of other amalgamated firms. Amalgamated Anthracite Collieries Ltd., though burdened with half a dozen closed collieries, employed over 8,500 in nineteen others, and Partridge Jones & John Paton Ltd. also had nineteen collieries, with over 12,000 workers.

Mention of the latter is a reminder that a significant proportion of the coal industry belonged to firms by no means exclusively concerned with coalmining. Some of the largest coalmining firms had even larger interests in steel or engineering. Dorman, Long & Co. Ltd. owned twelve collieries with over 12,000 workers. The United Steel Companies and subsidiaries had fourteen collieries with over 8,000 workers, and Richard Thomas and Baldwins Ltd., directly and through subsidiaries, had eleven collieries with a labour force of about 9,000. Such firms were spreading their interests geographically as well as over a variety of products. Dorman, Long, though active mainly in Durham, owned the large Upton colliery, near Pontefract. The United Steel Companies were the chief coalmining firm in Cumberland but also had several collieries in South Yorkshire. Richard Thomas and Baldwins, with major activities in South Wales, had a controlling interest in The New Sharlston Collieries Co. Ltd., with its two collieries near Wakefield.

There was thus an appreciable element of large-scale organization in the coal industry. The fourteen firms (less than 2 per cent of all firms) mentioned in the two previous paragraphs had about 168,000 colliery workers, which is almost one-quarter of the entire labour force of the industry, and there were several other firms not greatly different in size. Moreover, even where ownership appeared widely diffused among smallish firms there was sometimes more consolidation or association than was immediately apparent. Quite apart from the various forms of association which arose, directly or indirectly, from the statutory measures of the 1930s, there were small firms that were effectively linked by overlapping directorates, and there were separate undertakings which formed joint companies to pursue some common purposes. For example, fifteen firms, producing 90 per cent of the output of the North Staffordshire district, jointly formed North Staffordshire Land and Minerals Ltd. And sometimes a firm of moderate size was in a position to co-ordinate much of the mining of a particular locality. Hargreaves Collieries Ltd., though its labour force was less than 4,000, owned thirteen collieries which produced most of the output of the small Burnley coalfield.[1]

It was important for the well-being of the coal industry that there should be firms that were large enough, and appropriately located, to be able to plan developments rationally, and financially strong enough to provide the resources that were urgently needed for the maintenance and restoration of the industry. Three or four such firms set a notable example of technical and managerial advance. But it was in no more than a substantial minority of the industry that so favourable a structure was to be found: in 1943 90 per cent of total output came from 46 per cent of the mines, but these mines were divided among 353 separate firms.[2] It was perhaps a mixed blessing that some of the large firms, though they had additional sources of revenue, had very large claims on their funds for activities outside the coal industry. Nearly one-third of coal output came from firms with interests in the steel and/or engineering industries.[3] In large parts of the industry, firms were neither financially nor locationally well placed to tackle the problems confronting them, and there was often a conflict between the short-term interests and capabilities of the individual firm and the long-term needs of efficient development in the industry. The Reid Committee expressed the difficulties succinctly:

> There are mines ... which should be closed down and their reserves worked from adjoining collieries.... There are undertakings which have a lease of coal that could be worked to better advantage by another undertaking; and undertakings whose mines are widely spread through a district, and even among several districts. There are new sinkings required where the reserves which should be worked from them are leased to two or more undertakings; and new sinkings where ... the cost of the shafts, plant and development is likely to be beyond the resources of the undertaking owning the leasehold.... In these circumstances, it is evident to us that it is not possible to provide for the soundest and most efficient development and working of an area unless the conflicting interests of the individual colliery companies working the area are merged together into one compact and unified command of manageable size.[4]

[1] The information in this and the preceding paragraphs is based on the directory of Colliery Owners of Great Britain and the note on colliery amalgamations in *Colliery Year Book* 1946. The figures can be only approximate, for colliery employment fluctuates and the returns from different owners are unlikely to have been on a completely uniform basis.

[2] Reid Report 1945, 122.

[3] Lee 1954, 11.

[4] Reid Report 1945, 137–8.

The significance of dispersed ownership was reinforced by the dispersed location of coalmining and the need to adapt to changes in it. The British coalfields were widely scattered, but may conveniently be looked at in four or five groups. In Scotland the main field extends through the central valley and north-eastward to Clackmannan and Fife and there are also large fields in Ayrshire and in Mid- and East Lothian. In northernmost England the coalfield of Northumberland and Durham was for long the most productive in the country, and there is a small field on the Cumberland coast. East of the Pennines is the very large coalfield of Yorkshire and the East Midlands. West of the Pennines coal is found less continuously but there is a succession of fields in Lancashire, North Staffordshire, South Staffordshire, and Worcestershire, and there are small detached fields in North Wales and Shropshire, as well as a Warwickshire field in the east of the county. In the south, where physical conditions generally tend to be most difficult, the large field is that of South Wales, and the others, the Forest of Dean, Bristol and Somerset, and Kent are all small. Since the First World War, and more strikingly since 1930, many of the older producing districts had seen their share of national output decline, while the increased share came from Yorkshire and the East Midlands. Table 1.1 shows the changing sources of national output.[1] Yorkshire has displaced South Wales as the largest British coal-producing district in 1931 and retained that position at the end of the Second World War.[2]

Table 1.1. *Regional shares of coal output, 1913-1947*
(percentages)

| | 1913 | 1930 | 1947 |
|---|---|---|---|
| Scotland and Northern England | 34.4 | 33.1 | 31.5 |
| East Pennine | 26.9 | 31.4 | 39.5 |
| West Pennine | 17.0 | 14.7 | 16.2 |
| South Wales and S. England | 21.7 | 20.8 | 12.8 |

*Source*: Figures from A. M. Wandless, 'The British Coal Resources', in Manchester Joint Research Council 1960, 53.

[1] A. M. Wandless, 'The British Coal Resources., in Manchester Joint Research Council 1960, 53, gives further information about the location of different categories of coal.

[2] Ministry of Fuel and Power, *North Eastern Coalfield Regional Survey Report* (1945), 21.

Within individual coalfields, whether they were relatively gaining or losing, there were also changes in the location of mining, usually with older areas of shallower workings contributing a diminished share of output. In Scotland, for example, the central coalfield (which was much the largest) contracted its output by more than 20 per cent between 1920 and 1939, and Lanarkshire (where in the late nineteenth century more than four-fifths of the output of this field had been raised) reduced output by a third, whereas the other main coalfields produced more in 1939 than in 1920.[1] In Yorkshire the location of the largest coal workings had long been moving eastward and southward, and more and more from the exposed to the concealed coalfield.[2] Although many of the statistical shifts in the regional distribution of output were affected more by the incidence of contraction than of new development, such greater detail indicates the importance of the industrial structure. The needs and opportunities of development were often in new locations and usually involved deeper mining. Thus even the arrest of the industry's decline required developments that were within the scope only of undertakings strong enough to break out of past constraints and to invest in schemes of high initial capital cost.

The variety of conditions in different coalfields could be illustrated in detail in many ways. At this point it is enough to note that this variety offers a warning against easy explanations, but provides some suggestions about what underlay the many indicators of decline and weakness that caused so much anxiety in the 1940s. With such diversity it is unlikely that all coalfields could have had enough of the sorts of institutions, resources, knowledge, and attitudes that could cope with the combined technical, economic, and social tasks that had emerged in the coal industry. In the mid-forties the condition of the industry was thoroughly investigated, to the enlightenment of public as well as professional opinion. It was made evident that there were some general characteristics of British collieries which held back their performance, though the detailed incidence of these characteristics, and the opportunities and will to change them, varied from place to place and firm to firm.

The basic features of British mines, which caused them to fall behind their foreign competitors in efficiency, were technical. It has already been noted that the location and layout of individual mines in relation

[1] Scottish Home Dept., *Scottish Coalfields: Report of the Scottish Coalfields Committee*, Cmd. 6575 (1944), 23 and 54–5.

[2] *NE Coalfield Regional Survey Report*, 10 and 18.

to the accessibility of coal reserves was often unsuitable for the most complete and economical working of the coal. This was to some extent a legacy of the former private ownership of minerals, because the pattern of ownership had often led to the creation of small and awkwardly shaped leasehold areas. The taking over of coal royalties by the state was too recent to have made much difference. The lessor had changed, but the boundaries of leasehold areas mostly remained as they had been. But there were also serious technical limitations in the design and equipment of most mines. In the period since the First World War two important innovations had been widely, but by no means universally, adopted. These were coal cutting machines, which undercut the base of the seam, after which the coal was blasted down; and mechanical conveyors along the coal face. The cut coal was usually loaded on to these by hand and was then delivered by them to some other means of haulage at the end of the face. The results varied curiously. By 1939 North Staffordshire was the district with the highest proportion of output cut mechanically (95 per cent) and face conveyed (93 per cent), and had achieved the greatest increase of any district in face OMS since 1927. Yet Scotland in this period increased the proportion mechanically cut from 56 per cent to 80 per cent and the proportion face conveyed from 25 per cent to 59 per cent, and experienced a small decline in OMS. Over the country as a whole, though these technical advances made a useful contribution, the results were disappointing, and the old exporting coalfields of Durham, Northumberland, Scotland, and South Wales had achieved hardly anything to improve face OMS, though Northumberland had the second highest proportion of coal cut mechanically.

Other changes were needed and were seldom adopted. The movement of men, stores, and coal between the face and the surface changed little. As few new shafts were sunk, the average distance between pit bottom and coal face was increasing and most miners had to walk it. In 1943, out of 615 collieries with at least 250 men underground, 407 were working faces at least 2,000 yards from the shaft bottom, with no provision for manriding; and it was not unusual for a man to be at the face for only five hours of a seven and a half hour day. In the subsidiary roads underground the use of pony haulage was still common, though the number of ponies had been halved between 1924 and 1938, when it was 32,500. Most of the movement of coal underground, in both main and subsidiary roads, was in tubs which were attached to an endless rope haulage, a system which by 1940 was regarded as obsolescent in most other large coal-producing countries. The tubs were not of sizes standardized for the industry. At the shaft bottom they were loaded into the pit cage and wound to the surface, usually by steam power, though the use of electricity was increasing. Few British mines loaded the coal into large skips for winding, although skip winding greatly reduced the proportion of dead weight to weight of coal that had to be wound, and obviated the need to limit the size of tubs to what would conveniently go into the cage. Locomotive haulage was little used underground, although it was well established in some other countries, and conveyor belts along the main roadways were rare, though not unknown. (They had the disadvantage at this time of being restricted to a maximum length of about 1,500 yards.) So haulage remained slow and wasteful of labour, and set limits of capacity to what it was worth while to equip the faces to produce. In Britain 25 per cent of all underground workers were engaged in haulage and it was calculated that underground OMS could have been raised by one-third in 1939 if every haulage worker had moved as much coal as in the Dutch mines.

Things went on like this because it was difficult to alter them without much costly reconstruction of mines, and especially the driving of new underground roadways. In British mines roadways were driven in the seam and followed its contours, so that a roadway was often a circuitous switchback with sharp curves and frequent changes of gradient. Locomotives could not cope with such conditions. They needed the fairly straight roads with only slight gradients, which in other countries it was common to drive through the strata. The reorganization of underground workings was severely limited because many mines were too old or too small to justify its cost; or because ownership was too scattered to enable one replanned mine to be made the efficient replacement for several of poor quality; or because owners believed that modernization was made unprofitable by workers who refused to change old practices; or because many firms lacked adequate financial resources; or because few mining engineers could both fully perceive the potentialities of improvement and devise persuasive schemes for their realization.[1]

To describe the coal industry of the mid-forties in terms such as these is to emphasize its shortcomings of quality by contrasting what was with what might have been, as demonstrated by experience elsewhere. This was the approach of the reformers, headed by the Technical Advisory Committee under Charles Reid, which reported to the Minister of Fuel and Power in March 1945. It might be contended that the

[1] Reid Report 1945, *passim* gives the best description and analysis of these and related problems; see esp. 37–8, 65–78, 93–5, and 120–1 for points raised in the foregoing paragraphs.

members of this committee had gained their experience with large firms of great resources and paid too little attention to the achievements, as well as the problems, of less well-endowed undertakings.[1] But they knew that most new output and most advances in efficiency could come only from large and, in many cases, reorganized units, and they made out their case brilliantly. The Reid report was the strongest influence in persuading public and governmental opinion, irrespective of political views about ownership, that the revived coal industry that was so urgently needed could not be achieved without drastic change in the existing organization.

Yet many in the industry thought that much more recognition was due to what was still being achieved, and believed that it could maintain an expanded output into an indefinitely extended future with no great departures from present arrangements. More or less simultaneously with the Reid enquiry, committees were set up for every coalfield region in England and Wales[2] with these terms of reference: 'To consider the present position and future prospects of coalfields in the [named] Region, and to report—(*a*) what measures, apart from questions of ownership, form of control or financial structure of the industry, should be taken to enable the fullest use to be made of existing and potential resources in the above coalfields and (*b*) in this connection what provision of housing and other services will be required for the welfare of the mining community.' Each committee drew a majority of its members from the management and the trade union officials of the coal industry in its region, so the reports should reflect the views of those actively involved.

On the whole, the reports suggest a belief that a successful future for the coal industry was attainable with only gradual adaptation, provided that more labour could be attracted. The terms of reference may have precluded some expressions of reforming ideas. Both in the report for Durham and that for Northumberland and Cumberland the trade union members protested against the terms of reference, and in the latter report said, 'they express themselves satisfied, on the evidence before

[1] Of the seven members, H. J. Crofts (formerly of Chatterley-Whitfield Collieries Ltd.) and H. Watson Smith (Managing Director of The Hardwick Colliery Co. Ltd.) were associated with medium-sized firms; the rest with some of the largest firms.

The government published a popularized summary of the committee's conclusions in a pamphlet entitled *The Future of the Coal Miner–How coming changes mean a new deal and new opportunities for all who work in the coal industry*.

[2] There were ten committees. Most reported in 1945, a few not until 1946. A separate committee had already reported on the Scottish Coalfields in 1944.

them, that under present ownership the pits have not been maintained in a proper state of organisation and efficiency, and are convinced that this will only be possible under full state ownership and control.'[1] In other respects, ideas were less radical. In many regions major technical change was regarded with more suspicion than welcome. In Northumberland and Cumberland it was remarked that experience over the years had been that the introduction of machinery had increased neither total output nor OMS, so some committee members thought no sweeping changes could come from further mechanization, though improved haulage was favoured; yet opinion at fifteen out of eighteen Northumberland collieries asked was opposed to skip winding. In Kent and in Bristol and Somerset the low proportion of coal cut mechanically had been getting even lower in the last few years, a change for which no satisfactory explanation was offered in the latter field. In Kent it was claimed that the coal was too friable for mechanical cutting and, even if this were not so, it would not be worth while to mechanize without a change in the wages system. By contrast, the Midland coalfields, already highly mechanized, were enthusiastic for more. They were anxious to try out more American machinery, to experiment with power loaders and with electric trolley locomotives, and they suggested that skip winding should be installed in all new shafts and all existing large shafts. In Durham there was support for driving new and enlarged roadways by modern methods and a general increase in mechanization, with a suggestion that these improvements could be adopted with the necessary urgency only if the government financed them.

For the most part, however, developments were expected to be steady and on lines already familiar. Collieries had identified the coal reserves within their own leaseholds, at least approximately, and expected to deplete them at about the same rate as before, perhaps a little faster if the market required it and more miners could be found. The normal life of a mine was expected to be long, and only in a small minority of cases were new mines thought likely to be the best way of working known reserves. Some assumptions have acquired a certain irony with the passage of time. Of the thirteen collieries in the Bristol and Somerset field, seven were forecast to remain in production for over 100 years. (In fact, the last one closed in 1973.) All four Kent collieries were given a forecast life of 100 to 300 years. South Wales was somewhat more realistic with the calculation that thirty-five collieries would be

[1] Ministry of Fuel and Power, *Northumberland and Cumberland Coalfields Regional Survey Report* (1945), 44.

exhausted in less than ten years and another eighty-one in less than twenty-five years; and four of the six largest mines in the Forest of Dean were expected to be exhausted within twenty-five years. Where prospects for expansion were greatest, a minority of firms had plans for major new developments, but they were usually thinking some years ahead. In Yorkshire there were 125 separate colliery undertakings, of which twenty-two had plans for the reconstruction of shafts or the sinking of new shafts or the construction of new drifts, but in 1945 the only schemes in progress were two for new drifts, and nearly all the schemes were scheduled to begin later than 1950. Only rarely was there the leap of imagination which recognized that a coalfield had possibilities free from the constraints of existing leaseholds and practices. But at least in the north-west there was an attempt to establish where there were possibilities of new areas to work and a suggestion for their further exploration by additional borings. And in Yorkshire and the North Midlands there was a clear recognition that nobody knew how far east the coalfield extended, nor what it was like to the east, that it needed to be explored, and that any large increase of output would have to come from new pits to the east of existing workings.[1]

It was, of course, entirely sensible to recognize that most collieries would have to go on as they were for the immediate future, with only gradual development along familiar lines. Their output was needed while they could maintain it, and to refrain from any drastic reconstruction was often the most economical way to operate a mine in the short and medium term. This was particularly so in fairly small mines in which the faces were never going to be at enormous distances from the pit bottom, and in which coal reserves were so distributed that new faces could be opened up by driving short roadways in a new direction from the pit bottom or by deepening shafts only a little. What was unrealistic was to assume that many mines could continue for decades without losing the benefit of such conditions, or that when they had become uneconomic they could be replaced by mines which would operate under similar conditions for more than a very few years.[2] There was thus some degree of conflict between the shorter and the longer

[1] These two paragraphs are based on the ten *Regional Survey Reports* as a whole (1945–6); particular refs. to *Northumberland and Cumberland*, 3 and 49; *Kent*, 28 and 31; *Bristol and Somerset*, 33–4 and 54; *Forest of Dean*, 38; *Midland*, 23; *Durham*, 21 and 26–7; *South Wales*, 115; *North Eastern*, 9, 12, 16–17, and 31–3 (the 'north eastern region' was Yorkshire plus four Nottinghamshire collieries); *North Western*, 20–1 (the 'north western region' covered the Lancashire and Cheshire and the N. Wales coalfields); *North Midland*, 10–11 and 18.

[2] Cf. Griffin 1977, 169–70.

term in the needs and prospects of the coal industry. The reformers knew that eventually, without a vast amount of reconstruction and re-equipment, productivity would decline and costs rise ever more seriously, and that this problem was not one that could be long ignored. Others believed that the output of the whole industry was indispensable and that in most of it there was little financial possibility of major reconstruction. For example, it was argued in 1945 that it was too expensive to attempt much in the way of new sinkings and reconstruction in the deeper parts of the Lancashire coalfield. To sink a 21-foot diameter shaft to 1,200 yards would cost £500,000 at current prices, and the total cost of development at that depth to produce 4,000 tons per day would be at least £1.5 million, which was not a commercially attractive proposition.[1] Others again, while recognizing the need for heavy investment, greatly underestimated what was involved in restoring the efficiency of the coal industry and ensuring its long-term future. In 1945 the Mining Association of Great Britain established in a special enquiry that the member firms had £100 million of liquid resources available for investment in the coal industry and this was presented as an indication of how well prepared they were to meet the future.[2] Some firms clearly were well placed to carry through large schemes of improvement, but the resources available were far short of the urgent needs of the whole industry. When these were surveyed in detail and brought together in 1950 in the first *Plan for Coal*, the capital cost of what was needed was put at £635 million at mid-1949 prices, and by 1955 a sum equivalent to £248 million at that price level (£20 million less than planned) had been spent.[3] The contrast, both in physical and financial resources and in attitudes of mind, between preparedness for the short term and that for the longer future, is striking and was bound to have a deep influence on the industry's subsequent history.

The one physical resource about which, at least in national terms, everyone felt equally confident was the supply of unmined coal. The Coal Act of 1938 had abolished the private ownership of coal deposits and of coal royalties. The ownership of coal royalties, and hence the power to grant leases for the working of coal, passed to a new statutory body, the Coal Commission, with effect from 1 July 1942.[4] One of the hopes was that, as mineral leases fell in and colliery owners had to seek

[1] *NW Coalfield Regional Survey Report*, 62.
[2] Lee 1954, 207–8.
[3] NCB, *Investing in Coal: Progress and Prospects under the Plan for Coal* (1956), 5 and 12.
[4] Kirby 1977, 159–61.

new leases, it would become possible for the Coal Commission to impose such conditions, and devise such boundaries for the areas of the leaseholds, as would ensure the most complete and efficient working of the available coal reserves. How large these were and how they were distributed obviously had a great influence on the future prospects of the industry. To try to establish the relevant facts was a somewhat speculative exercise but important enough to receive a lot of attention.

The answers produced in the 1940s were more cautious but probably more useful than those provided earlier, by the Royal Commission of 1905 for instance. This body had published some very large estimates which were regularly cited thereafter. Each coalfield's regional survey committee produced new figures of reserves as part of its consideration of potential resources. The committees were able to refer to very recent figures produced by the DSIR's Fuel Research Coal Survey from information lodged by collieries as a result of the 1938 Coal Act and from some physical sampling of coal seams. Most of the committees produced their own amended estimates with the assistance of further direct enquiries from collieries. Their methods of estimation differed in detail but roughly comparable figures can be derived from their efforts. These figures, produced in 1945 and 1946, relate to known workable seams in leaseholds already worked by collieries or, in some cases, likely to be worked within the next ten years. Everywhere it seemed clear that production was under no threat from lack of coal suitable and accessible for working, but it was also clear that marked changes in the location of the main centres of the industry could be expected to go on steadily.

The estimated reserves for the whole of Britain came to 35,370 million tons, and rather more than one-third of them were in the one continuous coalfield of Yorkshire, Nottinghamshire, and North Derbyshire. Roughly another third were in South Wales, Durham, and Scotland, and it was notable that most of the Durham reserves were near the coast and under the sea and most of the Scottish reserves were in Fife and the Lothians, not in the central coalfield further west, where most past production had been. What seemed to be indicated was that a successful coal industry would have to continue more strongly the recent switch of production eastwards and, in particular, to go on drawing a bigger proportion of its output from Yorkshire and the East Midlands. This impression is reinforced when it is recalled that the last-named coalfield had, to its east, the most promising unworked area for exploration, not taken into account in the current estimates of reserves, and when it is noted that the estimates for some of the other fields included coal more difficult to work. Though the regional surveys were allowed to include seams 18 inches thick or more in their estimates, most of them excluded the thinnest seams because they were never likely to be worked; but more than half the small reserves of the Bristol and Somerset field were in seams less than two feet thick, and even the moderately abundant reserves of Lancashire included just over 20 per cent in seams as thin as this.[1] All these estimates appear to have been carefully made and to give a sensible assessment of the long-term availability of supplies. The DSIR Coal Survey's total estimate in 1946 (of which the regional survey report figures were partly but not wholly independent) was 35,500 million tons, and thirty years later, after an unprecedented amount of exploration and some narrowing of the concept of 'recoverability' to allow for changes in mining technology, the NCB were using a figure of 45,000 million tons of recoverable reserves.[2] So there was reasonable consistency in the estimates.

Another aspect of the supply is its quality and character. For the basic purpose of providing heat from a fire any coal may be used, though some coals will do the job more efficiently than others, but coal is not a completely homogeneous commodity. For some purposes only certain types of coal will do at all, and for others it is greatly preferable to use some types rather than others. It is common to rank coals according to their content of volatile matter and their caking properties, though a more detailed description of their characteristics must also refer to such things as their content of ash, sulphur, and moisture and their calorific value. Low volatile coals, with less than about 13 per cent of volatile matter, will not cake sufficiently to make coke of any commercial value, but there are also some high volatile coals which are non-caking. Coals that will cake are not interchangeable for all uses. The iron and steel industry needs very hard coke, resistant to crushing, especially for foundry use, and only a narrow range of coals will produce the appropriate qualities. Many coals which are excellent for gas-making yield a coke that is quite satisfactory for domestic purposes but too soft and friable for metallurgical use. Coals of low volatility are given a high ranking, high volatile coals a lower ranking which gets lower as the caking properties diminish. The specialized qualities of high-ranking coals are

[1] Information from the sections on reserves in all ten *Regional Survey Reports* and *Report of the Scottish Coalfields Committee*. Scottish figures appeared in 1944. To make them comparable with the rest their component of 'probable additional reserves', which were largely unproved, has been excluded from the total.

[2] Ezra 1978, 81–2.

partly attributable to their subjection to more extreme geological processes during their formation. They are therefore often found in locations with severe faulting and folding, so that these coals may be particularly difficult and expensive to mine. Though, under the most efficient forms of combustion, the highest ranking coals will produce more heat than the lowest rank, high rank does not necessarily mean better quality in all respects. A low rank coal, with comparable attention to preparation (or lack of it), may be cleaner and come in more suitable sizes than one of higher rank.

Britain in the 1940s still had a demand for coal in all its principal uses. Steam raising in ships' boilers, for which many coals from South Wales had been particularly suitable, was the only use for which there had then been a permanent drastic decline. It was therefore helpful for future prospects if the reserves were varied in type. Information on this point must be even more speculative than estimates of total quantity, but current production certainly included many varieties of coal, and there was no reason to suspect that the variety in the reserves was much different, save in one or two respects. The main worry was about the prime coking coal used in the iron and steel industry. This coal had come almost entirely from South Wales and the western part of the Durham coalfield, and, though the reports of reserves in South Wales were optimistic, it was estimated that in west Durham, where the quality was quite outstanding, only 226 million tons remained and much was becoming increasingly difficult to work. The small amounts of coal with similar properties, in such other fields as the Kent, Bristol and Somerset, and Burnley, were mostly mined by undertakings without the resources to make the best of them and tended to be dissipated in other uses, though the Lancashire Foundry Coke Company was one outlet for coal from the Burnley field. Kent coal had more dirt than most and was sold in local markets, more than 40 per cent of it to industry, mainly cement and papermaking. Bristol and Somerset coal reduced its quality because more than half was sold just as it came out of the mine, with no preparation whatever, and more than 40 per cent went for gas making. There was, then, some prospect of difficulty about supplies of prime coking coal and a consequent need to develop the more abundant resources of other strongly caking coals and adapt them in metallurgical use.

In all other respects supply prospects were satisfactory. Britain almost totally lacked the brown coals and lignites which some European countries had abundantly available for low-grade uses, but within the apparently narrow range of bituminous and semi-bituminous coals and anthracites Britain had probably a better distribution of coal types than any other European country. It produced about 8 per cent of very high rank low volatile coals (anthracite and low volatile steam coals), 8 per cent of prime coking coal, and 37 per cent of other caking coals, the rest being low rank coals. Most of the high ranking coals, apart from the prime coking coal of Durham, were in the southern coalfields. Scotland, Northumberland, and most of the Midlands had mainly low rank coals. Elsewhere there was greater variety. It was valuable that the most abundant reserves, those of the Yorkshire and East Midlands field, provided for nearly all uses; Nottinghamshire and North Derbyshire had mainly low rank coals suitable for general purposes, while Yorkshire had, except for the high ranking, very specialized coals, plenty of nearly everything, including much strongly caking coal.[1]

### ii. The terms and conditions of nationalization

It was evident that at the end of the Second World War the coal industry was full of contrasts. It had great resources, assured supplies of its raw material, a certain demand for all it could produce; yet a present and a recent past of decline and friction. It had immense potential but was performing nowhere near that potential and was falling increasingly behind the international leaders of the industry. To realize its full potential it needed great changes that would be costly and would be painful to some of its members, many of whom showed a strong preference for the status quo; and such changes, if accomplished, might well be relatively adverse to some coalfields. It had a very complex structure which made it easy to frustrate would-be large-scale innovators, but which also left much of the industry in the hands of powerful and experienced firms; and it was deeply intertwined with the activities of other industries. An industry with all these characteristics could not escape severe problems and the problems were unlikely to be resolved without major alterations in organization, staffing, and finance. Yet alterations of these kinds were bound to complicate immediately the tasks of management and adaptation, whatever hopes of eventual benefit they might offer.

In fact, the first alterations were imposed by the government and took the comprehensive form of the nationalization of the entire coal

[1] Wandless, 'The British Coal Resources', in Manchester Joint Research Council 1960, 47–55; *Regional Survey Reports*, esp. *Durham*, 46, *South Wales*, 15–22 and 106–7, *Kent*, 7, *Bristol and Somerset*, 16–23. It is noteworthy that the *North Western Report* was exceptional in totally ignoring the uses and characteristics of coals, although Lancashire had a very varied range of coal types.

industry. It was an innovation often mooted before, long and fervently hoped for by the miners' unions, hitherto always rejected but now adopted rather quickly.[1] The Coal Industry Nationalisation Act (CINA) was passed on 12 July 1946, the second and in many ways the largest and most experimental of the nationalization measures adopted by the Labour Government which had come into office the previous summer. It provided for the establishment of a National Coal Board in which, from a date to be determined, all the assets of the coal industry would be vested and which would thereafter be responsible for running the whole industry. The desire to get down quickly to fundamental change affected everyone concerned and, despite the lack of time and preparation, the NCB in November expressed their willingness to take over the industry from 1 January 1947, and the government made this the primary vesting date.[2]

It is important to establish just what passed to the new regime, and it is perhaps simplest to begin by pointing to some things that might conceivably have been done and were not. The NCB did not take over the colliery companies. These were not nationalized and did not lose their financial resources, and many remained actively in existence, pursuing lines of business other than coalmining. What the NCB did acquire from the private owners were the physical assets of the coal industry. Behind this apparently simple statement lie various complexities that need exposition.

The intention was to ensure that all the physical assets directly involved in the business of coalmining passed to the NCB; that the assets used by composite firms for their iron and steel business were not transferred, so that there was no 'backdoor' nationalization of part of the iron and steel industry; and that there was an opportunity for other ancillary assets of colliery companies to pass to the NCB. At first it appeared not too difficult to define which assets directly related to coalmining. The wages agreements operating since 1921 had required the net proceeds of the industry to be shared between owners and miners in agreed proportions and it had therefore been necessary to define 'the industry'. An arbitrator's award had listed what classes of activity were to be included in the industry for purposes of wage ascertainment, and the resources used in these activities were known as 'TS (terms of

[1] The arguments and events leading up to nationalization are discussed by B. E. Supple in volume 4 of this history.

[2] NCB 30th meeting, min. 316, 24.10.46; 31st meeting, min. 324, 29.10.46; 32nd meeting, min. 340, 1.11.46; 35th meeting, min. 382, 12.11.46; 37th meeting, min. 408, 19.11.46.

settlement) assets'. When a global sum of compensation had to be fixed for the nationalization of the coal industry, what was being valued was, in effect, the use of all TS assets by the coal industry as a going concern. In fact, there had been some anomalies of detail in the practical applications of the concept of TS assets in wage ascertainment. Some types of asset (private railways are one example) were treated in some districts as within, and in other districts as outside, the definition. The CINA made its own categorization of assets without using the phrase 'terms of settlement', with the result that the list of assets treated as so integral to the coal industry that the NCB must automatically have them all was not quite identical with the assets taken into account in fixing a global sum of compensation, though it was very similar.

Assets were divided into three groups. Those central to the operation of the coal industry passed automatically to the NCB on vesting day. They included the collieries, colliery coke ovens and manufactured fuel plants (except those used wholly or mainly to supply iron and steel works operated by the concern), coal preparation plants, colliery offices and workshops, colliery electricity plants, colliery railways and canal wharves and their loading and storage facilities, colliery merchanting property, welfare buildings such as colliery institutes, baths, and canteens, and stocks of colliery products. A second category consisted of assets which must be transferred to the NCB if either the NCB or the owners requested it. Among the principal such items were stores additional to those essential for colliery purposes, waterworks, wharves other than canal wharves, colliery houses, farmland, and farm properties. In the third category were items which either the NCB or the owner could ask to be transferred to the NCB, but to the transfer of which the other party could object, with an arbitrator's decision as binding if voluntary agreement could not be reached. This category included nearly all other property associated in some way with colliery activities. The most important examples were probably brickworks, vehicles (chiefly railway wagons) used outside the collieries, and merchanting property of associated undertakings.[1]

The assets which were not compulsorily transferred to the NCB had to be registered by their owners in 'statements of interest' which would enable options of transfer to be exercised at some time later than the primary vesting date, and compensation calculated. The preparation of the statements of interest in full could not be done quickly and, to save time, many assets that could be clearly identified were transferred to the

[1] Chester 1975, 91–102; Townshend-Rose 1951, 46–7.

NCB by agreement before the relevant statements of interest were completed and registered. The general policy of the NCB was to seek or accept ownership of most assets of which the transfer was optional. A few of these, notably wharves (other than canal wharves) and private harbours, were covered by the global sum for compensation, and it was pointless to decline ownership of things for which compensation had to be paid in any case. The colliery companies had found most of the other things advantageous for the running of their colliery business and the NCB expected to find them similarly helpful. This applied even to some properties at a distance from the collieries. For instance, some colliery companies were large landowners not only because they expected to have to extend their surface buildings to new sites some day, but also because to own the surface land oneself was often the cheapest way to deal with claims for damage by subsidence. If this took colliery companies into farming, it was not necessarily against the balance of advantage. But there were some properties about which the NCB proceeded cautiously. Brickworks were not wanted for their own sake, but some brickworks drew on collieries for all or part of their raw material and were operated as part of the colliery business, and these the NCB decided to take over. The property used for retail coal business presented a problem because, while it came automatically to the NCB if a colliery company was directly engaged in retail trade, it did not do so if the retail business belonged to a subsidiary. The NCB opted to possess such property only if its use was closely intertwined with other colliery operations, and this could not be established in any individual case until the statement of interest had been completed.[1]

As a result of the automatic transfers under the provisions of the CINA and the policy of trying to exercise options as quickly as possible, the NCB by the end of 1947 had acquired a vast array of assets, including a few oddities that might have been dispensed with if, under the new arrangements, they had not ceased to have any rational place in the businesses of their former owners. The main assets were:

(a) Nearly 1,500 collieries with their stocks of products and stores, their plant and equipment, including waterworks and power stations, with 60,000 acres of land occupied for colliery and ancillary purposes.

(b) 30 manufactured fuel and briquetting plants, of which 20 were

[1] NCB *Ann. Rep.* 1946, 8–9.

transferred under CINA and 10 were bought from the Ministry of Fuel and Power.

(c) 55 coke ovens, which produced over two-fifths of the country's hard coke; tar distillation plants; benzol recovery, sulphuric acid, and pyrites recovery plants.

(d) 85 brickworks and pipeworks, which produced in a year 420 million bricks, of which 94.5 million were used in the collieries and the rest sold.

(e) 1,803 farms covering 233,000 acres (124 of the farms having been run directly by the collieries); also sports grounds and undeveloped housing land, and 55,000 acres of miscellaneous land.

(f) 140,000 houses and 27,000 farmhouses and agricultural cottages; some entire colliery villages were owned, and the property there included schools and village halls.

(g) Other buildings including offices, 275 shops and business premises, swimming baths, a cinema, a slaughter house.

(h) Miscellaneous assets including private railways, wharves, 60,000 railway wagons for internal use at collieries, coal-selling depots, retail milk rounds, a holiday camp, and a cycle track; 177,000 main line railway wagons, which were transferred to the Transport Commission on 1 January 1948 as a result of the nationalization of the railways.[1]

One other asset, the coal itself, was already publicly owned under the provisions of the Coal Act, 1938. This had been vested in the Coal Commission, though the latter had alienated a minute amount of it to private ownership. With this tiny exception, the CINA provided for all coal which was being mined, still unmined, or still undiscovered to be owned by the NCB from the primary vesting date, and, as a consequence, the Coal Commission was abolished soon afterwards. The NCB thus combined the responsibilities of working and owning the coal, though their monopoly of coal working was not quite complete. Anyone carrying on any other activity could dig and carry away coal if the other activity could not be pursued without removal of the coal, the NCB could license private operators of small mines, and they could also license the mining of coal with other minerals if the quantity of coal present was so small that it would not have been worked except as ancillary to the working of the other minerals. The assets of those free miners of the Forest of Dean who operated as individuals were also excluded from

[1] NCB *Ann. Rep.* 1947, 10, 74, 80–3, and 141; *Ann. Rep.* 1948, 81 and 139–40.

nationalization. These were very small exceptions and the longstanding separation of functions between the ownership and working of coal was effectively ended, save in one respect.[1] Opencast production of coal had been conducted since 1942 by a succession of government departments, under powers conferred by emergency regulations, and this arrangement continued. The Ministry of Fuel and Power was the responsible department from June 1942 to January 1943 and again from 1 April 1945 onwards. In October 1946 the Minister of Fuel and Power asked the NCB to consider taking over opencast production and suggested that, if they agreed to do so, responsibility should be transferred not later than the middle of 1947. But the Board were unenthusiastic because there was a loss of 16 shillings per ton on opencast production and this was expected to worsen when proper rehabilitation of sites had to be carried out. Moreover, most people thought opencast production would cease in a few years. So things were left as they had been, with a tentative suggestion that the NCB might take over opencast production at the beginning of 1949 and perhaps run it down from 1951 to 1954.[2]

For the NCB there was a price to be paid for all the assets acquired. The coal itself had been valued after the 1938 Act when the mineral owners had been compensated in Coal Commission stock for the loss of income from their mineral rights. These rights were then valued at £66.45 million, to which the government added £10 million to cover the costs of registration of title and other incidentals.[3] When the ownership of the coal passed to the NCB, the owners of Coal Commission stock were compensated in a new Treasury stock to the value of £78,457,089. A small part of this sum represented the value of the Coal Commission's buildings and office equipment, and the rest, amounting to £78,442,369, was treated as the value of the minerals. Some small additions had to be made to this. A few colliery undertakings had been mineral freeholders and, after mineral rights were nationalized, they had chosen not to receive capital compensation but instead to be allowed to continue working their former mineral holdings at a peppercorn rent. For the loss of this privilege they were entitled to compensation which counted as part of the capital valuation of the minerals. In this way the latter figure increased to £80.888 million.[4]

[1] Chester 1975, 92; Townshend-Rose 1951, 42 and 46.

[2] NCB 28th meeting, min. 283, 18.10.46; 29th meeting, min. 304(*a*), 22.10.46; 35th meeting, min. 378, 12.11.46; 108th meeting, min. 1,361, 15.7.47; 110th meeting, min. 1,396, 22.7.47.

[3] Kirby 1977, 257.

[4] NCB *Ann. Rep.* 1947, 130; Chester 1975, 257.

The valuation of the operating assets was a complicated matter, which took longer to complete than anyone had expected. The TS assets were fairly easy to deal with. A tribunal of three, with Lord Greene, the Master of the Rolls, as chairman, was asked to determine a global sum as compensation for these. All concerned agreed that the assets should not be valued as separate items but that the businesses using them should be valued as going concerns; i.e. the tribunal should estimate the future net annual revenue from the use of the assets and capitalize it at what it judged a fair number of years' purchase. The tribunal fixed the global sum at £164.66 million, but never explained how it reached this figure.[1]

The delays occurred in sharing out the global sum among the many claimants upon it, and in determining the value of subsidiary assets transferred to the NCB. There were also two much smaller items to be settled. In order to discourage any interruption or delay in the execution of urgent investment projects, there was provision for the refund of capital outlays made by colliery undertakings after 1 August 1945 on any assets subsequently transferred to the NCB. Composite undertakings could also claim compensation for any increase in the overheads borne by their remaining business as a result of the severance of their coal industry assets.

Questions of compensation were settled by a Central Valuation Board (established in September 1946) and twenty-one District Valuation Boards, each of which dealt with a district usually more or less the same as had been used for wage ascertainment. The District Boards were constituted in June 1947. The Central Valuation Board had to allocate the global sum among the districts and exercise general supervision over the methods of valuation used by the District Boards. Each of the latter had to allocate its district's share of the global sum to the individual firms (which involved it in preparing valuations of every colliery undertaking), and it had to value all the ancillary assets transferred to the NCB, for which purpose it needed both accounting and technical reports.[2] The amount of paperwork and the consumption of the time of highly skilled people were immense. Even the Central Board, in order to share out the global sum, needed information built up from details about every colliery undertaking in every district. A typical investigation, that into the Nottinghamshire District's claim on the global sum,

[1] Chester 1975, 250, cites two rather light-hearted guesses by an official about the way the tribunal might have reached its figure.

[2] Ibid., 251; PRO, POWE 36/184.

involved the presentation of documentary evidence under twenty-nine different headings. As the global sum was already fixed and each District tried to claim an entitlement to as much as possible of it, there had to be hearings at which other Districts could argue against a claim if they thought it would leave too small a share for others. For example, when the Nottinghamshire District claim was presented, the South Wales and Monmouthshire District appeared formally as respondent and was, in effect, questioning the size and details of the Nottinghamshire claim on behalf of all other Districts. Nottinghamshire claimed that it deserved about one-fifth of the global sum and put forward an engineer's valuation of £34.75 million and an accountant's of £30 million. South Wales sought to show that this was an exaggerated claim, based on forecasts of a fantastic increase in output, and suggested that a realistic valuation was £20 million or possibly even £17.5 million. In the end Nottinghamshire was awarded only £17.983 million from the global sum, though this was the third largest District allocation.[1]

It was a useful saving of effort that the preparation of material by individual undertakings to support the District's claim for a share of the global sum also provided readymade data for the sharing out of the District's allocation. But there still remained the laborious task of valuing the ancillary assets; and there was the further complication, not only for these but also for the coal industry assets paid for out of the global sum, that individual firms were not the units for valuation, because some firms had collieries and other assets in more than one district. So compensation was assessed in terms of 'compensation units', and the assets of each company were usually divided into several compensation units. For each compensation unit a firm received (i) out of the allocation from the global sum to the District in which the unit was included, a fraction equal to the ratio which the coal industry value of the compensation unit bore to the sum of the coal industry values of all compensation units in that district, and (ii) the amount found by the District Valuation Board to be the value of the compensation unit for 'subsidiary purposes' (i.e. ancillary to the use of TS assets).

Valuation in this way had some serious drawbacks. First, the preparation and submission of the necessary data were expensive. The Ministry of Fuel and Power paid £4,481,000 towards colliery owners' costs in this exercise and incurred costs of £2,086,300 on its own behalf. Second, it was very slow. Some delay had been expected and, to cover this, there was provision for revenue payments to be made for a two-year period in

[1] PRO, POWE 35/243, 35/244, 42/4, 42/21.

anticipation of eventual capital payments in satisfaction of compensation awards. But this provision had to be extended because many of the settlements had to wait far longer than two years. Eventually, during 1949–50, with the approval of District Valuation Boards, some firms got together to agree on the division of District shares of the global sum among themselves and, by the end of 1951, over £96 million of the global sum had been shared out in this way. But there still remained much to settle about the ancillary assets. In fact, a final statement about the compensation paid for the nationalization of the coal industry's assets was made to the House of Commons on 27 January 1966, though all but a few items had been dealt with by the mid-fifties. The eventual payments are set out in Table 1.2.

Table 1.2. *Compensation payments for nationalized assets*

| | £ | *s.* | *d.* |
|---|---|---|---|
| Coal industry (TS) assets (global sum) | 164,660,000 | 0 | 0 |
| Subsidiary assets (including stocks of products and stores) | 126,647,804 | 2 | 6 |
| Additional overheads caused by severance | 689,137 | 0 | 0 |
| Capital outlay refunds | 17,656,235 | 6 | 3 |
| Total operating assets | 309,653,176 | 8 | 9 |
| Minerals | 80,888,000 | 0 | 0 |
| Total operating and raw material assets | 390,541,176 | 8 | 9 |
| Additional compensation | 3,824,000 | 0 | 0 |
| Total compensation | 394,365,176 | 8 | 9 |

*Note*: The 'additional compensation' relates to liabilities of colliery companies which the NCB found it expedient to take over, although they never admitted legal responsibility for them. Chester, relying on figures published in 1954 and 1956 and ignoring later adjustments, gives slightly lower sums for subsidiary assets and capital outlay refunds. On the basis of official files he gives (240–58) a much fuller discussion of compensation principles and procedures than there is room for here. See also A. Mozoomdar, *Compensation for the nationalisation of industries* (unpublished Univ. of Oxford D.Phil. thesis, 1953), chap. 6.

*Source*: PRO, POWE 35/244; HC Deb. 27.1.66, col. 135; Chester 1975, 251–7.

Most of the compensation was paid in government stock but some £66 million was paid in cash, mainly for capital outlay refunds, stocks of products and stores, and main line railway wagons. Because the latter were vested in the NCB for such a short time it was agreed that compensation to the owners should be calculated in accordance with the terms of section 30 of the Transport Act, 1947 and charged to the NCB. In turn, the NCB subsequently received 3 per cent Transport Stock, which was sold.[1]

Whether the coal industry was taken over at a valuation which neither favoured nor unfairly burdened the NCB it is difficult to say. The valuation of the minerals was such as to give the holders of Coal Commission stock compensation for its current value. If there was any unfairness it had occurred earlier, when coal royalties were nationalized under the Coal Act, 1938. The owners had then claimed that the annual royalty income should be capitalized at 25 years' purchase, giving £112 million, but an arbitration tribunal awarded them only 15 years' purchase. This produced a lower sum than the government had already offered, which was probably why the government found supplementary reasons for adding £10 million to the arbitration award.[2] The advantage to the NCB of the transfer of mineral ownership was that they were freed of the cost of royalty payments in current operations, and the continuation of the modest capital valuation meant that the related interest and amortization charges were light. If the valuation of minerals had been treated in any way other than as the capitalization of a former revenue item, then any resulting figure would have been quite arbitrary. In one sense, £80 million (or the 1938 proposal of £112 million) as a value for all the vast reserves of ultimately recoverable coal was preposterously low. But, as much of the coal was still undiscovered and most of it not immediately accessible, there was no means of putting any realistic present valuation on it. The one unsatisfactory aspect of this state of affairs—potentially a serious problem—was that the assumption of a low financial value for very large reserves of coal could distort economic judgement about the optimum energy policy and the rate of depletion of different fuels.

Greater realism was practicable when dealing with the operating assets of the industry. Here the main problem was to use past experience (and the Treasury insisted that it ought to be the most recent peacetime experience, undistorted by the market abnormalities of war) to establish

[1] Chester 1975, 253 and 258; NCB *Ann. Rep.* 1948, 139–40.

[2] Chester 1975, 240–1.

the likely net maintainable revenue in the immediate future, and convert this into a capital sum. Undoubtedly, the Mining Association, representing the colliery owners, believed that the coal industry (TS) assets were worth over £200 million, and was disappointed by the £164,660,000 awarded. But its belief was not necessarily justified. There was fairly general agreement that pre-war experience suggested a net maintainable revenue of £11 million, though the Mining Association claimed that this had been adjusted to allow for the contribution of current assets which were not being transferred to the NCB, and the Treasury thought that it had not. The government believed that any sum calculated by capitalizing net maintainable revenue would have to be reduced by the amount of non-transferable current assets, which it put at £24 million (the 1944 value), whereas the Mining Association objected that in 1938 it had been only £12 million. The Mining Association also maintained that the coal industry assets were capable in the post-war period of producing a much larger output (240 million tons), which would increase the net maintainable revenue to £21 million,[1] but this was a highly optimistic speculation.

A plausible guess at what might have been done with these data could run something like this. Accept £11 million as the figure for total net maintainable revenue and capitalize it at $16\frac{2}{3}$ years' purchase, which was what the Mining Association had requested, and thus reach a figure of £$183\frac{1}{3}$ million; take the value of current assets as midway between the government and Mining Association estimates, and add £$\frac{2}{3}$ million to allow for changes since the end of 1944; subtract this sum from the previous figure. This exercise produces a capital valuation of £$164\frac{2}{3}$ million which, if rounded to the nearest £10,000 below, is the same as the global sum actually awarded.[2] There is nothing very severe in this calculation, which leans as much towards the Mining Association as the government. Hindsight (as well as reviews of the technical state of the mines at the time) suggests that many collieries, just to ensure the long-term maintenance of their productive capacity, needed to subtract much more than they had been taking from gross revenue for depreciation and development reserves. In other words, their accounting policies indicated a net maintainable revenue somewhat higher than was justifiable. It may well be that the wish to give something to everybody out of a

[1] Chester 1975, 245–9.

[2] This is pure guesswork but the assumptions are more straightforward than those underlying the guesses cited in Chester 1975, 250. The current yield on coalmining shares was just over 6 per cent, which gives a rough plausibility to a multiplier between 16 and 17 years' purchase.

limited global sum meant that the small number of highly efficient firms got less than they deserved, but, for the industry as a whole, the global sum seems to have been rather on the high side; and, indeed, the lowest figure put to the arbitration tribunal by the government had been only £132 million and, in earlier discussions, the Treasury had suggested amounts well below that.[1] The high proportion which the sums eventually paid for the subsidiary assets bore to the amount paid for the coal industry assets suggests that the subsidiary assets also were not ungenerously valued. There can be no justification for thinking that the NCB acquired the coal industry at a cut price.

They did, however, acquire with the industry some difficult problems which may not have been very costly in a direct sense, but which, if left unsolved, were bound to reduce the potential of the physical assets. Perhaps the largest of these problems was that of administrative structure and management. When the NCB took over the coal industry it was, in terms of numbers employed, the largest single industrial undertaking in the non-communist world. Not only did it have this rather frightening uniqueness, but it was in sharp contrast with the previous state of its own industry, where ownership had been widely dispersed. No one had made any plans for this transition, or even given it much thought. The Minister of Fuel and Power later recalled: 'For the whole of my political life I had listened to the party speakers advocating state ownership and control of the coal mines. . . . I had believed, as other members had, that in the Party archives a blue-print was ready. Now, as Minister of Fuel and Power, I found that nothing tangible existed. There were some pamphlets, some memoranda produced for private circulation, and nothing else.'[2] And in a speech in 1945 he had said: 'I have been talking about nationalisation for forty years but the implications of the transfer of property have never occurred to me.'[3] Exception might be taken to these statements if they related to the way political principle might shape legislation, but, as comments on the lack of foresight about administrative and managerial arrangements, they are entirely justified.

It is true that the whole coal industry had been subject to government direction since 1942 and there were hints to be got from the administrative experience. But most of them were negative hints. The day-to-day running of the industry had been left to the colliery companies and their staffs, and a clear, easily workable relation between them and the

[1] Chester 1975, 245, 246, and 249.
[2] Shinwell 1955, 172.
[3] Rogow 1955, 155.

government appointees responsible for general policy had not been established.[1] Some new system had to be devised, and many local adjustments of detail made within it. All this had to be done against a background of extravagant expectations and with some shortage of staff with suitable experience. The government took it for granted, as the Minister of Fuel and Power publicly indicated, that public ownership of the whole industry was bound to produce great savings. The miners had looked to nationalization as the promised land for so long that they now thought of it as something to come 'at a wave of a ministerial wand', something which would immediately give them better conditions, yet enable them to produce all the coal that was needed without departing from familiar practices.[2] But none of this could happen overnight, and it could not happen at all unless someone made it do so, by effort and thought. There were too many things that needed to be done quickly, and too few people to do them.

The NCB experienced appalling overwork in the half-year before the primary vesting date. The board held long meetings twice a week, the members worked very long hours, and summer holidays went by default. Yet most of their effort had to go on the creation of an administrative structure, because none existed, the appointment of staff to it, and the settlement of relevant conditions. Very little attention was given at board level to the immediate problems of coal production, though there were, of course, delegated responsibilities for this.[3]

Staff were taken over with the industry. The CINA provided for the NCB to inherit the contracts of service of all employees in the industry. This provided a continuity without which management would have been impossible, and it ensured that the day-to-day running of the collieries went on with no hiatus. But difficulties remained. The larger scale of organization required the creation of posts carrying greater responsibility. The filling of them meant that, out of a group whose members had had similar status, a minority moved to a higher position.[4] Among the majority it was tempting for the disappointed not to remain with the NCB, and this was often easy for those in a composite firm that could give them posts in its continuing business. Such firms, in any case, could

[1] Court 1951, 241–3.

[2] Chester 1975, 246; Shinwell 1955, 172 and 174.

[3] It is recalled by one of them that in the NCB's first year it was common for senior staff to work until about 10 p.m. Mondays to Fridays and until 2 p.m. to 3 p.m. on Saturdays. I have noted only two occasions in 1946 when the Board discussed the direct problems of improving coal production.

[4] NCB *Ann. Rep.* 1947, 27–31.

seek to retain their ablest people and usually succeeded. In 1955 the hindsight prompted by nearly a decade of experience produced these comments:

The number of [able and experienced officials] is quite inadequate. The main reason for this state of affairs is historical. Between the wars . . . the industry failed for many years to get the share it ought to have had of the nation's brain-power and executive ability. . . . When the industry was nationalised it lost many of its administrators. In particular, most of the Managing Directors of the larger undertakings did not choose to come into the service of the Board. Thus, the industry lost virtually a complete level of management at the moment when, because of the great size of the new undertaking, managerial and administrative talent was needed more than ever before.[1]

Such limitations in the inheritance of staff were bound to cause their greatest difficulties in the early months of nationalization because so many of the old institutional arrangements came to an end and had urgently to be replaced, while the normal operation of the industry had to be maintained and, as soon as possible, improved. Among the functions for which new institutions had to be devised were labour negotiations, the selling of coal, the finance of the industry at every level. Negotiations on wages and working conditions had for many years been conducted on the employers' side by twenty-three District Associations of colliery companies, which now ceased to exist. The commercial business of distributing coal to specific markets had been conducted by District Selling Schemes created under the Coal Mines Act of 1930. Their staff and duties passed to the NCB on the primary vesting date, but they had to be adapted to new conditions and new objectives. Perhaps fortunately in these circumstances, wartime needs had caused the Ministry of Fuel and Power to take responsibility for policy on the allocation of scarce coal supplies to different classes of user, and this arrangement continued unchanged for the time being. Financial arrangements had to be rebuilt since the NCB inherited no money at all from the colliery companies and the individual collieries would cease on vesting day to have even their old bank accounts. The NCB had to get their money, including their initial working capital, from the Treasury. New bank accounts, usually with the same bank as the previous owner had used, were opened for collieries, and banking arrangements were likewise made for the new institutions which were part of the administrative structure.

[1] NCB, *Report of the Advisory Committee on Organisation* (Fleck Report 1955), 23.

Some elements of previous financial arrangements had to be completely changed. To take one important example, there had been since 1942 a Coal Charges Account financed by a flat rate levy on each ton of coal produced and used to enable low cost districts to subsidize high cost districts by allowing the latter to draw out amounts which would enable them to meet such things as wage additions and still retain a 'standard credit balance'. 'Necessitous undertakings' could receive additional payments to keep them solvent and, as the levy proved insufficient to meet all the claims on the Account, the Exchequer had to supply extra funds, which by 1947 had amounted to a subsidy of £27½ million. All this was to come to an end. The Coal Charges Account was to be wound up and the NCB were instructed to run the industry without subsidy, though they would have to work out their own system of financial relations between high cost and low cost coalfields.[1]

Nor was this all. The NCB took over various formerly independent institutions which may have been small—in some cases much smaller than the importance of their functions required—but which called for all the time and effort that would enable them to operate successfully within a new context. Among them were the regrettably few research bodies, which included the British Colliery Owners' Research Association, the Coal Survey (which was transferred from the Department of Scientific and Industrial Research in August 1947), and the Sheffield Mines Mechanisation Centre. There were also links to be established with separate bodies integral to the industry, notably the Miners' Welfare Commission, a statutory body with members appointed by the Minister of Fuel and Power from the National Union of Mineworkers (NUM), the NCB (replacing the colliery and royalty owners), and independent persons.[2] These changes had all to take place in a setting where it still had to be worked out how the new and closer relationship of the industry and the government would operate.

Perhaps the pressures of a seemingly ubiquitous need to adapt and redesign were initially so concentrated on administration at the centre that they give a slightly misleading impression. In the coalfields continuity was much more obvious. The immediate task was unaltered: to go on raising as much coal as before and, if possible, more. It was performed for new owners but mostly in the same places, by the same people, by the same methods. Since the work was being done in a more

[1] NCB *Ann. Rep.* 1946, 2–3; NCB *Ann. Rep.* 1947, 94–6.

[2] Ibid. 1947, 34–5, 85, and 87.

hopeful atmosphere, it was reasonable to expect that this degree of continuity would at least end the retrogression, shown most obviously in the level of output, which had been so alarming in the war years.

But such a statement only highlights the difficulties in another way. Continuity, without innovation, was not enough; and the need to ensure continuity, simply to try to catch up with current demand, was an obstacle to the speed and amount of innovation. When the coal industry was nationalized, it had vast assets, both physical and human; but for many years the quantity and quality of those assets had not been sufficiently maintained, and many of them had not been used even with the degree of efficiency which general contemporary standards of equipment and training would have allowed. The best British practice in management, training, and technique, which was a worth-while example for the whole industry, was confined to a few firms responsible for a minority of production; and, in some features of mining practice and equipment, even they did not match the best foreign firms. In such conditions, especially in an industry whose product had inevitably to be sought and worked with gradually increasing physical difficulty, the persistence of former practices could only become, within a fairly short period, a step on the road to further decay. The acceptance of nationalization as a policy owed something to a recognition of this, but too much was expected to follow simply from a change of heart which it was prophesied that nationalization would induce. A change of heart was certainly needed but, by itself, it could make only a limited and short-term difference. For a fundamental and lasting change there had also to be drastic modernization through heavy investment in an industry that would be physically as well as administratively reorganized. The physical change and the heavy investment were not practicable until the new administrative structure had become established and proved workable. Even then they could be carried out only in ways that permitted the uninterrupted maintenance of current output, and they were unlikely to be fully adopted until far more people than in 1946 were convinced of the inescapable need for them. That conviction was most likely to come from the continuing pressure of market demand, a detailed examination (projected some way into the future) of the ways in which it might go on being met, and a demonstration that technical ingenuity was proceeding on practical lines that could make it economically worth while to invest in its application. The process was bound to be gradual.

Thus the NCB, when they took over the coal industry, acquired not only huge assets at a tolerable cost, but also huge problems the solution of which required expenditure on a scale which very few people then realized. The short-term and long-term problems were essentially complementary, but initially they could not avoid competing for attention and resources, and the short-term problems were bound to get priority. That left to be tackled, as soon as practicable, a fascinating task on which the future of the coal industry depended. The fruits of such a fundamental exercise took years to appear. Before that, developments could be directed to expansion and improvement along familiar lines, and it was reasonable to hope for clear signs of progress from this. But everything in the state of the coal industry in 1946 made it unwise to expect any more than this in the next few years, and those whose expectations were surrounded by a rosy glow of optimism were likely to experience some disappointment.

CHAPTER 2

# The Market for Coal

Two influences do more than any others to impose some sort of framework on the operations of an industry: the market for its products and the available techniques for turning out these products—the know-how. Neither is immutable and neither is beyond the power of businesses within the industry to alter. But both depend a good deal on conditions elsewhere and they can hardly ever be wholly controlled by producers. For periods which may extend over many years producers have to treat these influences as, in substantial measure, 'given', and seek to make the best of them. 'Making the best' involves adaptation to them by producers, but also, where practicable in the not too long run, attempts to modify them in ways which are advantageous to businesses within the industry. Before looking at the activities and organization, the successes and failures, of the coal industry it will therefore be helpful to consider these strong and partly external forces within which it found itself.

First, in this chapter, there will be an examination of the market for coal and the way it changed between the nineteen-forties and nineteen-eighties. This must necessarily involve a look at the market for energy in all its chief commercially available forms. The basic features of the course of demand for energy are fairly simple and so is the chronology of the emergence of competing sources of supply. Much more complexity appears when one seeks the determinants of the relative strength of the various components of the market, especially those which stimulated demand in one direction rather than another and those which bore upon the relation between cost and price.

Take first the simple fundamentals. The most obvious was the continuous growth of the demand for energy in the UK, as in the world in general, until the early seventies, after which demand fluctuated uncertainly around a level that initially represented near-stagnation and then, at least temporarily at the start of the eighties, some decline. There are various ways of expressing energy statistics on a common basis. It is convenient to convert the quantities of all primary fuels to 'coal equivalent', even though the results are less exact than they are made to appear, as different coals, different oils, gas supplies of different richness, each have an appreciable range of calorific values, and the yield of energy varies with the conditions of combustion as well as the innate properties of the fuel. Errors from such causes may roughly cancel each other and figures based on conventional formulae for the different fuels certainly give a good indication of the trend of change. In million tonnes of coal equivalent, UK energy consumption rose from 228 in 1950 to 269 in 1960, 337 in 1970, and 353 in 1973. In the next few years it was lower and recovered only slowly before turning down again: it was 340 in 1978 and reached a new peak of 356 in 1979, but it then fell to 329 in 1980, 317 in 1981, and 312 in 1982.[1]

At the beginning of this period coal was the primary fuel for nearly all the energy used, except in forms of transport other than railways and a rapidly diminishing proportion of shipping. Until the late nineteen-fifties only a small amount of substitution of other fuels for coal took place, but thereafter the state of competition changed dramatically. Until 1957 any decline in the share of coal depended on changes in the proportion of total energy devoted to transport and on the extent to which the supply of coal could keep up with the growth in the demand for energy. Coal had a near monopoly in the supply of energy for industrial and domestic use. It did not succeed in supplying quite all that was wanted and some switch to oil was looked on, even by the NCB, as a helpful second-best that enabled a gap to be filled. In the first ten years of the nationalized coal industry (1947–56 inclusive) other fuels met almost one-third of the total increase in inland consumption of energy.[2] Nevertheless, coal still retained a dominating position. The coal industry is sometimes regarded as having been in perpetual decline from its peak attained in 1913, but its greatest market loss had been in exports. The peak year for home consumption of coal was 1956, when it was 217.5 million tons (221.0 million tonnes).[3]

After that consumers turned increasingly to competing fuels, not because they needed a second-best to fill a gap, but because they found in them positive advantages of cost or convenience or both. The main competition to coal was from oil, though the planning of nuclear power stations had already begun. Nuclear power, however, experienced

[1] Department of Energy, *Digest of UK Energy Statistics*, 1984, table 1.

[2] NCB *Ann. Rep.* 1956, I, 34.

[3] Ibid. 1971–72, II, 84–5. The retrospective figures of the Dept. of Energy for 'inland consumption of primary fuels ... for energy use' are slightly lower, at 217.4 million tonnes, but still show the peak in 1956.

unexpectedly long delays in development and gained only a small market share. From the late sixties natural gas was available in rapidly increasing quantities at low cost and this greatly increased competition within the energy market. In the sixties and early seventies the abundance of alternative fuels was such that coal found its market rapidly shrinking not only relatively to that of its competitors but absolutely.

From 1973 conditions changed again because the price of oil was repeatedly and hugely increased, but the availability of choice among indigenous fuels became greater than ever as the UK developed its newly discovered oilfields under the North Sea. The demand for coal in the UK ended its precipitous slide but, as the total consumption of energy had stopped growing, there was no sustained growth in the demand for coal, though there was a roughly maintained market share. In terms of coal equivalent, coal supplied 91 per cent of UK energy consumption in 1948 and oil not quite 9 per cent; in 1958 coal supplied 79.9 per cent and oil 18.7 per cent. Then coal's share fell to 73.7 per cent in 1960, 58.3 per cent in 1966, 46.6 per cent in 1970 (still slightly larger than oil's), and 37.6 per cent in 1973, when it had been markedly surpassed by oil.[1] Subsequently the great change was the rapid increase in the share of natural gas, almost entirely at the expense of oil, while coal, by maintaining its market share, became once again the fuel making the largest contribution to UK inland energy consumption, but at a much lower level than in the fifties and sixties. In 1981–2, in million tonnes of coal equivalent, UK inland consumption was: coal 116.9, oil 113.1, natural gas 72.8, nuclear 13.7, hydro 2.1.[2] In percentage terms that is coal 36.7, oil 35.5, natural gas 22.8, nuclear 4.3, hydro 0.7.

Drastic changes in the energy market were not peculiar to the UK but were worldwide. The world's energy consumption is estimated to have trebled in the twenty-two years from 1950. In 1950 the consumption was shared thus: solid fuel (i.e. mainly coal and lignite) 60.3 per cent, oil 28.6 per cent, natural gas 9.5 per cent, and hydro power 1.6 per cent. In 1972 the percentages had changed to: solid fuel 31.2, oil 46.3, natural gas 20.2, hydro power 2.1, nuclear 0.2.[3] The UK market had some distinctive features. It had been dominated by coal to an exceptional degree and coal retained a slightly larger share in it. Coal's share, indeed, fell less rapidly in the UK than in continental western Europe, though more rapidly than in eastern Europe. But everywhere the trends were in the

[1] *Digest of UK Energy Statistics*, 1984, table 1. The long-term trends are illustrated in fig. 2.1.

[2] NCB *Ann. Rep.* 1981–2, 15.

[3] Calculated from Ezra 1978, 9.

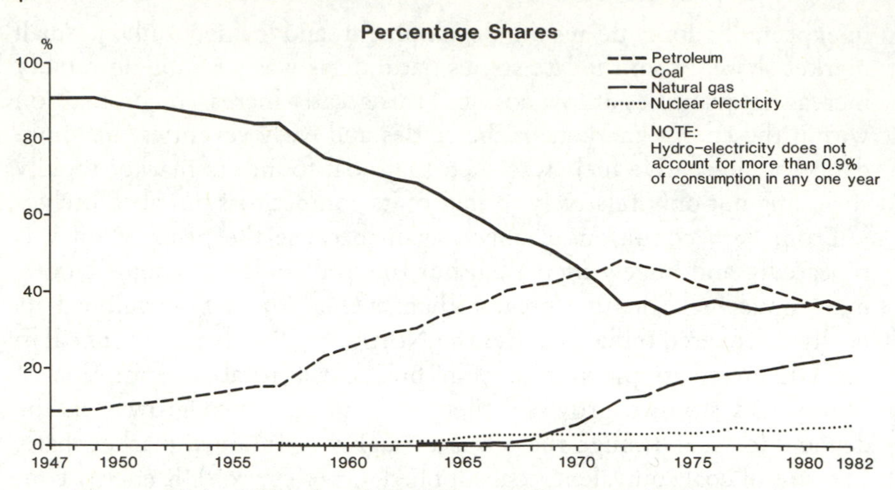


Fig. 2.1. United Kingdom Inland Energy Consumption, 1947–1982
*Source*: NCB and *Digest of United Kingdom Energy Statistics* 1984

same direction. This was important for the British coal industry, which was involved with the international market. In the late forties and early fifties continental Europe, in particular, had difficulty in supplying its energy needs and there were opportunities for other producers to export coal there at high prices. The increasing competition from other fuels caused these opportunities to dwindle. On the other hand, that competition stimulated some coal producers to seek more sales abroad.[1] The risk was thus developing that British coal might face some competition in its home market from foreign coal, if price and quality were acceptable, as well as from other fuels.

Changes in the level of demand for coal in the UK were associated with changes in the structure of demand, as Table 2.1 illustrates. A few additional details add to the significance of the figures given in Table 2.1. 'Gasworks and Coke Ovens' is not an altogether homogeneous category. Coke ovens have mainly produced hard coke for the iron and steel industry, with gas as one by-product. Gasworks have produced coke for domestic and miscellaneous commercial uses as an important by-product of their principal activity. The coals required by the two differ in type and quality, though with some overlap. The demand from coke ovens continued throughout the period, but the great drop in the final figure resulted from the rapid contraction of the steel industry which set in at the end of the seventies. The earlier

[1] Cf. Jensen 1967, 33–4 and 65.

Table 2.1. *Home consumption of coal, 1947 to 1980-1*
(million tonnes, percentages in brackets)

| Year | Power Stations | Gasworks and Coke Ovens | Domestic | Other Inland | Total |
|---|---|---|---|---|---|
| 1947 | 27.5 (14.7) | 43.2 (23.0) | 37.2 (19.8) | 79.6 (42.5) | 187.5 (100.0) |
| 1950 | 33.5 (16.3) | 49.6 (24.1) | 38.4 (18.6) | 84.3 (41.0) | 205.9 (100.0) |
| 1960 | 52.7 (26.4) | 52.2 (26.1) | 36.1 (18.1) | 58.8 (29.4) | 199.9 (100.0) |
| 1970–1 | 74.7 (49.6) | 28.6 (19.0) | 18.7 (12.4) | 28.7 (19.0) | 150.7 (100.0) |
| 1980–1 | 87.7 (72.6) | 11.3 (9.4) | 8.5 (7.1) | 12.7 (10.9) | 120.3 (100.0) |

*Source*: NCB *Ann. Rep.* 1981–2, 32–3.

changes originated mainly in the gas industry. The share of the gas industry in the coal consumption of the combined category was 53 per cent in 1947, 54 per cent in 1950, 44 per cent in 1960, and only 12 per cent in 1970–1.[1] In the following year its coal consumption fell by two-thirds to only 1.1 million tonnes,[2] and after that the gas-making market for coal virtually disappeared. The coal industry lost this market because of technical changes in the gas industry. Feedstocks other than coal were in use to a small extent in the fifties and became commercially attractive from the beginning of the sixties. Between 1961 and 1963 processes were developed which made town gas more cheaply from light petroleum distillates than from coal or from the methane which the Gas Council had already (by 1959) arranged to import in liquefied form. The result was that oil-gasification plant exceeded carbonizing plant in gas-making capacity by 1964–5, and oil-gasification reached a peak of output in 1968–9, when only 22.6 per cent of manufactured gas was made by carbonizing plants.[3] But in 1965 natural gas was discovered under the North Sea and the whole future of the gas industry was soon transferred to this basis.[4] By the end of the financial year 1967–8, North Sea and imported natural gas already totalled 13 per cent of gas supply. Thereafter the build-up of supply was very rapid and in 1970–1 the proportion was 70 per cent.[5] By 1972 only one coal-based gasworks was still left in

[1] Calculated from NCB *Ann. Reps.* for years cited, tables of inland consumption.
[2] NCB *Ann. Rep.* 1971–2, II, 104.
[3] Reid, Allen, and Harris 1973, 103–9.
[4] Ibid., 118.
[5] NCB *Ann. Rep.* 1968–9, I, 17; *Ann. Rep.* 1970–1, I, 15.

England and there were fifteen small works in Scotland, apart from the one Lurgi plant at Westfield in Fife,[1] using a process which makes a lean gas from low-grade coal without producing coke.

Other points worth noting concern the miscellaneous category of consumption headed 'Other Inland'. One important user was the railway, which in 1947 consumed 14.7 million tonnes, and the same in 1950. By the mid-fifties there was already an appreciable decline, though the total was still 13.0 million tonnes in 1956. Some improvements in fuel efficiency, reductions in locomotive mileage, but above all the replacement of steam by diesel and electric traction, which was very rapid at the end of the fifties and in the early sixties, thereafter progressively cut the railways' demand for coal. It was 9.0 million tonnes in 1960 but only 2.4 million tonnes in 1965–6, and was down to 600,000 tonnes in 1967–8. The railway had become negligible as a market for coal.[2]

The biggest item in the 'Other Inland' category was coal for industry. Of the decline in consumption which began in the fifties, some was the result of increasing electrification, which involved the indirect rather than the direct use of coal; but much was a result of competition from other fuels, principally oil, which was aggressively marketed. Except that the attraction of gas was stronger, much the same may be said of the domestic demand, where the heavy fall came a little later, partly because of the gradual extension of the areas regulated under smoke control legislation. For the same reason, in the domestic demand of the seventies and early eighties the share of naturally smokeless and manufactured smokeless fuels was higher than it had been earlier, and this should be borne in mind in interpreting the figures. For 1980–1 the 'domestic' total includes coal used for plants manufacturing smokeless fuel, whereas this item (relatively much less important before the sixties) previously appeared under 'Other Inland'. The interpretation of the figures is also affected because domestic consumption always includes miners' concessionary coal, an item outside the normal operation of the market.

One other example of a demand disappearing because of technical change should be cited. Most of the movement away from coal burning for ships had happened before the Second World War, but a useful small market survived into the nineteen-fifties. Coal exports were artificially restricted in 1947, but in 1948 the bunkering of foreign-going vessels took 5.5 million tonnes of coal, which was just over one-third of the combined total of exports and bunkers. In 1950 the total for bunkers

[1] NCB *Ann. Rep.* 1971–2, I, 26.

[2] NCB *Ann. Reps.* for years cited, tables of inland consumption.

was down to 4.1 million tonnes and was less than a quarter of the total of exports and bunkers. Ten years later the bunkering total was only 0.4 million tonnes, and it was recorded as a separate item for the last time in 1964–5, when it was a mere 3,000 tonnes.[1]

The only continuously growing demand for coal came from the power stations. By the seventies these were taking more coal than all other users combined. There was thus erected a very special market relationship between the coal industry and the electricity supply industry. In some respects this was very helpful for the former, especially because the power stations had one of the easiest types of demand to supply. Many of the larger users of the fifties presented problems for coal suppliers. Domestic consumers, railways, ships, and a few industrial firms with older designs of boilers all wanted mainly large coal, but, for a variety of reasons, large coal was a declining proportion of total output. Increasing competition from other fuels meant increasing pressure to keep down the costs of coalmining. Greater use of machinery, which was one response, had as one of its effects a further increase in the proportion of small to large coal. A customer who could burn increasing quantities of small coal, as the power stations could, was a blessing to the industry. Moreover, the domestic demand for coal tended increasingly towards smokeless varieties, which were scarce and were costly to mine; and the continuing demand from coke ovens was also for specialized and increasingly scarce coals. But, within fairly wide limits of tolerance, the power stations could burn the most ordinary general purpose coals which presented far fewer problems of supply.

In other ways the emergent new relationship with electricity supply was much less favourable for the coal industry, which was put into a weaker marketing position. Coal became more and more dependent on sales to only two customers (CEGB and SSEB) who together made up one large industry, whose demand was directed to the cheaper and less profitable end of the coal market, just at the time when those customers were ceasing to be exclusively dependent on coal supplies. The energy market had become characterized by shifting degrees of monopoly and monopsony, and the shifts were of a kind which suggested that the coal industry was experiencing a greater increase in vulnerability to pressure than the other members of the market.

The changes that have been described were mainly attributable to, or (more strictly) initiated by, basic commercial and technological

[1] Ibid., for years cited, tables of exports by destination.

influences. That does not mean that the energy market was a beautiful example of the hidden hand of perfect competition optimally adjusting the distribution of productive resources to the pressure of consumers' demands. There were *shifts* of market power and, on the whole, they favoured consumers, among whom competition decreased, more than producers, among whom it increased; but they were shifts within a market that was always highly imperfect. Its institutional composition necessarily made it so, and there was so much government regulation that the operation of the market was bound to be still farther from the conditions of perfect competition. It is desirable to try to get closer to the realities controlling the working of the market. The most direct route is to look at the activities and policies of government, for these not only acted directly on the market but also shaped the major institutions which made up the market. It is unnecessary at this stage to give a detailed narrative of the actions taken by governments and the statutory bodies which they set up. It is sufficient to indicate the forms of regulation that maintained a powerful influence over the market for extended periods.

First, there were the government controls of the coal market which existed before any of the energy industries was nationalized, and which continued long afterwards. The government had given itelf statutory control of the allocation of coal supplies to consumers at the outbreak of the Second World War, and coal remained under this system of control until 1958, well after comparable wartime controls had been removed from all other commodities.[1] For domestic consumers whose purchases were restricted to specific maximum quantities, both for the whole of each year and for particular periods within it, and who must get their coal from a particular merchant, this arrangement probably distorted the market appreciably. For industrial consumers, for whom there was an elaborate system of individual allocations which tried to share out coal proportionately to relative requirements of fuel at the expected level of operations, the departures from what would have been expected in an unregulated market were less severe, except in periods of acute

[1] NCB *Ann. Rep.* 1947, 68–70; *Ann. Rep.* 1958, I, 31–2; *Report of SC on Nationalised Industries*, 1958 (HC 187–I of 1958), esp. Minister of Power's memo on 'The Statutory Powers and Responsibilities of the Minister of Power in relation to the NCB', 1–2; PRO, POWE 26/434, 'Emergency Legislation: The Statutory Powers of the Minister of Fuel and Power'. Ministerial control was exercised under the Defence Regulations. The principal orders which remained in force were the Coal Supply Order 1942 and the Coal Supply (National Coal Board) Order 1947. Supplies to non-industrial (including domestic) premises were controlled under the Coal Distribution Order 1943.

shortage, as in the early months of 1947. Nevertheless, there were such departures, which showed themselves particularly in the availability of different qualities of coal to different consumers.

Even longer lasting were governmental influences on coal prices. Only over the retail price of domestic coal did the government have statutory power, but its general influence was continually exerted. At the outbreak of war the government had made a 'Gentlemen's Agreement' with the coalowners that prices would not be raised without prior approval by the government, and this agreement was extended to cover the individual prices of different qualities of coal. The National Coal Board undertook to continue this extended agreement and found themselves subject to irksome constraints from which they could not escape. The Minister of Fuel and Power in 1954 declined to give up the arrangement, though he agreed that the board might publish the correspondence in any future case where proposed price changes were not implemented because of his objections.[1] The NCB in 1957 informed the Select Committee on Nationalised Industries of the outcome of their ten proposals for price rises down to that date. One was refused outright, four were granted only for lower amounts than requested, and only four could be introduced on the date originally proposed.[2] The government rejected a recommendation of the Select Committee that any ministerial amendment to prices should be in statutory form. In the next ten years, however, the Ministry of Power did little to alter the board's proposals about prices, but the timing of price changes was still delayed, especially when statutory prices and incomes policies were in operation. The board found themselves having to submit price proposals to the Minister of Power, who consulted two Consumers' Councils, and sometimes also having to get the endorsement, after enquiry, of the National Board for Prices and Incomes. They complained that they needed greater commercial flexibility, the authority to react quickly to changing conditions, and suggested that the type of ministerial control exercised under the Gentlemen's Agreement was no longer appropriate. Subject to compliance with the early warning procedure required under the government's prices and incomes policy, they wanted a general

[1] *Report of SC on Nationalised Industries*, 1958, 37–8. Although the Minister there stated that no question had arisen about the Gentlemen's Agreement until the end of 1953, there had been several previous discussions at board level in the NCB about the possibility of trying to have it ended.

[2] Ibid., xviii and 134–5.

understanding with the Ministry of Power that (in limited defined fields) they could change prices without detailed consultation.[1]

In fact, the system continued as it was for the time being. But in the late sixties and early seventies, it was the attention given to coal prices within the general statutory prices and incomes policy that made more difference to market conditions than the long-standing non-statutory arrangements between the government and the NCB. The National Board for Prices and Incomes reported on coal prices in 1966 and again in two reports in 1970, with a supplement in 1971.[2] The Heath government, however, then abolished that board.

The role of the government in the determination of coal prices changed partly because of this, but more fundamentally because of the entry of the UK into the European Communities on 1 January 1973. The Treaty of Paris, which had established the European Coal and Steel Community, embodied principles of pricing, which included limitations on discrimination and requirements about the openness of price lists, and it placed the obligation of conformity on 'enterprises'. The rules made under the treaty clearly distinguished between the publication of coal and steel prices, which was an obligation of producers, and the publication of transport charges for coal and steel, which was an obligation of governments. In certain circumstances some residual powers of price fixing were conferred on the High Authority of the ECSC (which, by the time of UK entry, had been absorbed into the Commission of the European Communities), but, apart from that, coal and steel prices were a matter for the *producers*, provided that they kept within the rules of the Community, which the Commission interpreted fairly flexibly.[3] This meant that, from 1973, though the NCB still kept the UK government informed about their price proposals and weighed the government's comments, they were formally free to make their own decisions and implement them when they thought best. To that extent the market had become less regulated.

It had also become less regulated in one other way. The government's

[1] *Report of SC on Nationalised Industries*, 1967–8, III, *Appendices and Index* (HC 371–III of 1967–8), 226–7; an example of the timetable involved is on 229.

[2] NBPI Report no. 12, *Coal Prices* (1966), Cmnd. 2919; NBPI Report no. 138, *Coal Prices* (1970), Cmnd. 4255; NBPI Report no. 153, *Second Report on Coal Prices* (1970), Cmnd. 4455, and *Supplements nos. 1 and 2* thereto (1971), Cmnd. 4455–I and 4455–II. The history of coal prices is examined much more fully below, especially in Chap. 5, section ii, Chap. 6, section ii, and Chap. 7, section vi.

[3] *Treaty establishing the ECSC* (Treaty of Paris, 1951), Articles 61 and 62 for coal and steel prices, Article 70 for transport charges. For a clear statement of the effect of the relevant rules made under the Treaty, see Howard 1971.

wartime power to control imports and exports of coal, coke, and manufactured fuel was continued into peacetime. Control of exports was applied in times of scarcity of supply at home, but ceased to have any economic importance when that scarcity disappeared. The regulation of imports had more potential influence on competition within the home market. In practice until 1970 the government gave only the NCB a licence to import, which they were more or less bound to use at times when home production fell short of home needs. But that did not apply from the late fifties, the NCB from 1959 did not use their import licence, and the home market was thus protected from the possible competition of foreign coal.[1] In 1970 the government abandoned this policy and granted an open general licence to anyone to import coal.[2] From 1973 it was also bound to permit the entry of coal produced in the European Community, but production there was too low for that to make much difference. Indeed, the change of policy did not introduce unfettered competition. A wish to use the licence to import was confined almost entirely to two nationalized industries, electricity generation and steel, so the level of coal imports became inevitably a matter involving government policy.

The large role of nationalized undertakings in energy supply, the statutory duties placed on them, and their political and financial relations with the government had a permanent and pervasive effect on the working of the market. The CINA included provisions that had a strong influence on the market behaviour of the NCB. The Act included in its prescription of the duties of the NCB, 'making supplies of coal available, of such qualities and sizes, in such quantities and at such prices, as may seem to them best calculated to further the public interest in all respects, including the avoidance of any undue or unreasonable preference or advantage'.[3] It laid down also that the policy of the Board should (among other things) be directed to securing, 'consistently with the proper discharge of their duties' under the Act, 'that the revenues of the Board shall not be less than sufficient for meeting all their outgoings properly chargeable to revenue account . . . on an average of good and bad years'.[4] The wording of the Act thus indicated that the NCB had an

[1] Control was instituted under the authority of the Import, Export, and Customs Powers (Defence) Act 1939 and administered under the Imports of Goods (Control) Order 1940 and the Export of Goods (Control) Order 1951. An Open Licence to import, which the NCB received in November 1950, was amended to allow it to run on without the need for periodic renewal, which had hitherto appied. (PRO, POWE 26/434.)

[2] NCB *Ann. Rep.* 1970–1, I, 1 and 18.

[3] CINA (9 & 10 Geo. VI, c. 59), s. 1 (1)(*c*).

[4] Ibid. s. 1 (4)(*c*.).

overriding duty to produce the quantity and quality of coal that was in the public interest and that, in so doing, they had no obligation to make a profit, but only to break even over a period; and even this limited obligation was to be pursued only consistently with the duty to produce what was in the public interest. In addition, the reference to the avoidance of undue preference or advantage was, for many years, believed to restrict the board's opportunities to vary their prices in accordance with the circumstances in which a potential customer was situated. Lord Robens, when chairman of the board, complained in December 1965:

If I were a free agent and the CEGB were about to build a power station, say within 100 miles of a coalfield, and I wanted that business, and I was free from any statutory obligations, then commercially, if it was my own business, I would pick my best pit where I make 30s. a ton profit and I would take a nice order for a million tons a year from that pit and make myself a millionaire before the contract was over. Then I could give them a long-term contract. This is denied us.[1]

The word 'undue' in the Act offered some latitude, however, and, after coal began to lose sales to other fuels, some coal was sold at specially contracted prices to a number of industrial customers.[2] But until July 1979, 80 per cent of all coal was sold at list prices, the same price for similar coal to all buyers. After that the proportion greatly decreased as separate contract prices were negotiated for the electricity and steel industries, the largest coal users.[3]

The duties and policies prescribed by the CINA had a further influence. During the Second World War and until nationalization the government had accepted the obligation to try to ensure that the coal needed in the public interest was produced. To do this it found it necessary to operate the Coal Charges Account, through which the more profitable coalfields provided funds to cover the losses of others. When the NCB took over the industry the government terminated the Coal Charges Account, but the CINA left the NCB to produce all the coal the public needed, and also to break even financially, which the Coal Charges Account had failed to do.[4] The NCB were, however, not

[1] *Second Report of SC on Nationalised Industries: Gas, Electricity and Coal Industries*, 1966 (HC 77 of 1965–6), 41–2.

[2] *Report of SC on Nationalised Industries*, 1969 (HC 471 of 1969–70), II, 65 has the NCB's indication of this in evidence.

[3] NCB *Ann. Rep.* 1980–1, 10 points out that most of the Board's larger industrial and coking customers had been brought under special long-term pricing arrangements.

[4] Ibid. 1946, 3.

expected to make a profit, which they would have done if prices had been set high enough to cover the costs of the less efficient coalfields, which were still found indispensable while the aim was to maximize total national output. To steer a course between profit and loss, in such circumstances, the NCB were bound to continue and adapt the practice of cross-subsidy, which they had inherited. Their pricing policy thus had to try to recoup total costs on revenue account, by some system of national averaging. Outside commentators (including politicians and civil servants) wrote innumerable words in discussion of the advantages of marginal cost pricing of coal, though by no means without challenge;[1] but something different was made inescapable from the outset.

Moreover, though the pricing and cross-subsidy arrangements were associated with the combined requirements to maximize output and to break even financially, the effect was to build cross-subsidy (which had firm roots from the recent past) into the system. Even when maximum output was no longer the objective, the assumption was still widespread within and outside the industry that break-even through national averaging remained the appropriate maximum financial objective, and that total output should continue to come from areas and collieries with widely differing costs. The low cost producers were thus expected to continue to subsidize the high cost, as they had done in the forties and fifties and even, to some extent, under the selling schemes of the thirties.

The financial arrangements immediately before nationalization are a reminder that all governments at all times have the power to tax and subsidize and that this can be a strong influence on the operation of a market. On the whole the energy market was not much affected in the forties and fifties. The imposition of an excise duty of 2*d.* per gallon (0.18p per litre) on fuel oil in 1961 is often regarded as the beginning of a change. It was announced simply as a revenue duty, but it had a protective value (which contemporaries put at about 25*s.* per ton (£1.23 per tonne)) for industrial coal.[2] In the later sixties, and still more in the seventies, the variety and amount of government imposts and payments greatly increased and added to the complexity of the problem of

[1] Among instances illustrating the prolonged fluctuations in this argument are the White Paper, *Nationalised Industries: A Review of Economic and Financial Objectives* (1967), Cmnd. 3437, 8–9, which stated that cross-subsidies should be minimized and prices usually related to long-run marginal costs, and SC on Nationalised Industries, *National Coal Board: Observations by the Minister of Technology* (1970), Cmnd. 4323, 5, which rejected the SC's endorsement of that view and stated that 'the Government do not accept that long-run marginal costs are necessarily the only right basis for setting pricing policy in the current circumstances by the coal industry'.

[2] PEP [1966], 30; Robens 1972, 83 and 151.

determining the relation between real costs and prices in the energy market. But by then the financial transfers were only one element in an array of regulatory measures.

It was probably inevitable, as the use of energy is so basic to society, that the government would try to work out some kind of energy policy as soon as a continuing choice of fuels became apparent. The incentive to do this was all the greater, and the political feasibility all the easier, because so many of the energy industries were nationalized. Coal, gas, and electricity were followed by atomic energy, and then the government acquired new assets as a result of the discovery of undersea reserves of gas and oil. The coal industry found a bigger and bigger proportion of its sales going to nationalized undertakings, and much of its competition coming from others. The government, as the sole investor in these capital-hungry industries and as their ultimate guarantor and paymaster if they were not fully self-supporting, saw a necessity to try to regulate the relations among them. This involved attempts to prescribe their optimal market shares and to adjust their productive capacity to that prescription and to forecasts of the absolute size of the market. In practice, a particularly strong influence could be exerted by decisions on the power station programme. To the coal industry the maintenance of a large coalburning electricity generating industry became the main determinant of its own survival.

Once policy decisions of this kind were put into practice they could have a long lasting effect on the market shares of different fuels, so that these did not necessarily conform to contemporary relative costs. If nuclear power stations were built to meet expected increases in electricity consumption they were more or less bound to be used even if their costs turned out to be unexpectedly high; and their use precluded the introduction of that amount of new oil-fired or coal-fired capacity. If a number of oil-fired stations were built in preference to coal-fired, or vice versa, the new would have an advantage over the old, irrespective of relative fuel supply costs to some extent, because the plant was more up to date and more efficient; and that advantage could persist for years. In electricity supply, an industry with large daily and seasonal variations in load, it could mean that a particular fuel came to be less used not because it was dearer but because some of it could be used only by the oldest and least efficient plant. Comparable effects could be seen, on a smaller scale, in the domestic and general industrial markets. Heating systems are specific. The householder who installed gas-fired central heating was committing himself for years not to use sold fuel or oil instead. The industrial firm which replaced coal-fired by oil-fired boilers was not in a position to switch fuels again as soon as their relative costs changed. Thus a rising or falling trend in the use of a particular fuel could be started by a comparison of costs or an arbitrary policy decision within a quite short period, and could be prolonged because of the commitment to heavy investment in specific equipment.

These are important considerations because they are illustrated by what actually happened. It was well known that decisions made today have to be lived with ten and twenty years ahead, and the large decisions were invariably presented in terms of what purported to be rational expectations about future market conditions. But some expectations which were built into policy were never rational, and some economic forecasts went badly astray despite the great care that went into their calculation. The power station programmes provide confirmation. Early in 1955 the government announced a commitment to a civil programme of nuclear power and an initial four Magnox stations were ordered in 1956–7, with others to follow quickly. The government's proclaimed policy was that by the early sixties two-thirds of all new power station capacity should be nuclear. All this was attempted despite the absence of experience of even one nuclear station on the scale of all those then being ordered, and despite the misgivings of the Central Electricity Authority and the CEGB, which were well aware of the rapid fall in generating costs that was being achieved in coal-fired and oil-fired stations. It was a policy pushed by politicians, who seemed to regard nuclear power as a bright new toy, and by the Atomic Energy Authority, whose interests and experience were technological rather than commercial.[1] Construction costs rose, the programme was delayed and reduced, and in the end only about a fifth of the new power station capacity commissioned between 1960 and 1965 was nuclear, instead of the planned two-thirds.[2] Despite some serious corrosion problems, the Magnox stations proved safe and fairly reliable and achieved appreciable improvements in fuel efficiency. But none of them ever generated electricity as cheaply as coal-fired stations.[3]

Yet in the middle sixties another programme of nuclear power stations was begun and was used as an element in planning a major redistribution of the shares of the various fuels in the energy market. The

[1] Hannah 1982, 172–83, 229–30, 233–40.

[2] Ibid., 244.

[3] Ibid., 284. On generating costs see the admissions in evidence at the Public Enquiry into the proposed Sizewell PWR nuclear power station (*The Times*, 11 Mar. 1983).

new stations used the advanced gas-cooled reactor (AGR), and again several stations were ordered and begun when there was no commercial experience, anywhere in the world, of even one example of this type of power station. The only concession to caution was to order only half the number proposed by the Atomic Energy Authority. Again the technologists promised cheapness and were believed. In 1967 the government claimed that generating costs at the AGR stations already ordered (although estimates had risen over 5 per cent in two years) would be lower than those of the newest coal-fired stations, and that those still to be ordered would be cheaper, so bringing costs by the mid-seventies near those of an oil-fired station freed of fuel oil tax.[1] Such expectations were disappointed. The first AGR station to be ordered (in 1965) was Dungeness B, which took eighteen years to become operational. The completion of others took longer than planned, though the delays were not as bad as that, and they had many teething troubles. Above all, the reduction in costs was not achieved.[2] One result of this costly disappointment was that in the seventies the UK, unlike other industrial countries which had got better results by proceeding more gradually, ordered no new nuclear power stations. Thus in the sixties and early seventies nuclear power was taking business from other fuels (in those years mainly from coal) without any economic justification; and in the mid and later seventies, when the costs of energy from all other sources were rising, the UK had none of the cheaper types of nuclear power station in use elsewhere, which were more likely to have made an economic contribution in the new conditions.[3]

Fortunately, the installation of oil-fired generating capacity was better managed, apart from a temporary panic in 1970. Although it was originally more expensive to use oil than coal, there was a wish to ensure adequate supplies of fuel for the future, and in 1954 came a first response to pressure by the Ministry of Fuel and Power to have some additional oil-fired capacity. In the late fifties oil prices were falling rapidly and, though the temporary closure of the Suez Canal caused some hesitation, there was an increase in oil-fired and dual-fired capacity, achieved partly by building new power stations and partly by conversion of existing coal-fired stations.[4] But the switch to oil could be only gradual and, after the emergence of a potential surplus of coal, the desire to preserve the

[1] Ministry of Power, *Fuel Policy* (1967), Cmnd. 3438, 78.

[2] Hannah 1982, 244 and 284–5; *The Times*, 23 Mar. 1983.

[3] Hannah 1982, 285–6.

[4] Ibid., 169–71.

stability of what must for many years be the chief supplier of fuel slowed it down somewhat.[1] At the end of 1966 oil-fired and dual-fired made up 12.3 per cent of generating capacity, and the oil-fired stations then under construction were going to increase the proportion only slightly.[2] So the increases in oil prices from 1970, including the huge leaps of 1973 and 1979, which made it much more expensive to use oil than coal, could be countered by using more of the cheaper fuel, especially as there was a large surplus of generating capacity.

Electricity generation provides the biggest example of the way in which planning decisions taken at one time could alter the fuel market for years ahead. A government concerned to adapt supply from the nationalized energy industries to changing conditions of cost and demand in such a way as to permit the nation to secure the maximum economic return was impelled to go further, for it is not the way of governments to admit that ignorance is too great for the accomplishment of such a feat. The most comprehensive reassessment came in a White Paper on Fuel Policy in 1967, and was said to be due to the discovery of North Sea gas and 'the coming of age of nuclear power as a potential major source of energy'.[3] (The lowering of the age of majority had become a fashionable subject by this time.) No doubt the exercise was conducted with great care to co-ordinate all the complex interrelations involved and some of the forecasts may look impressive. It was estimated that between 1966 and 1975 coal consumption would fall by over 55 million tonnes, and coal's share of primary energy supply from nearly 59 per cent to 34 per cent.[4] In fact in 1975 the share was 37 per cent and the amount consumed was almost identical with the forecast.[5] But as the forecast estimates were openly proclaimed as the basis for policy planning and the policy was to equate actual productive capacity with forecast demand, there was a large element of self-fulfilling prophecy in all this. The forecasts of expansion proved very accurate for natural gas but much exaggerated for oil and nuclear power. Some of the premisses on which the rationale of the policy objectives was based looked very dubious. The cost estimates provided by the nuclear engineers were taken at face value, with the comment, 'nuclear power has emerged ... into a proven and increasingly competitive source of

[1] Ibid., 232–3 and 237–8.
[2] Cmnd. 3438, 15.
[3] Ibid., 1.
[4] Ibid., 36.
[5] NCB *Ann. Rep.* 1975–6, 6.

energy', which was untrue. On oil it was said that ultimate exhaustion of reserves was too far away to be treated as a constraint and that 'it seems likely that . . . pressure to force up crude oil prices will be held in check by the danger of loss of markets', a somewhat reckless assumption which was soon proved to be mistaken. The over-optimistic NCB forecasts of the rise of productivity in coalmining appear to have been rejected, but lower figures from the Ministry of Power were also much too high.

Further light on the making of fuel policy is thrown by the fuller subsequent treatment of the subject by Mr Posner, who was Director of Economics at the Ministry of Power when the 1967 White Paper was being prepared. He makes clear how well understood were the subtleties of the exercise and the nature of the risks of error which they created, and he makes out a strong case in favour of attempting such an exercise despite the risks. He also leaves the impression that far too little attention was paid to the crudities which could make nonsense of large parts of the resultant policy. In a book published in 1973, on the eve of the fourfold rise in oil prices, it is startling to read, 'One fuel, and one fuel only, is assumed to be available in very large quantities at a fairly fixed price, given sufficient notice—fuel oil.'[1] That needs to be associated with the decision to pay no attention to the future prospect of physical shortage of fuel oil because it was possible that any fall in the ratio of proved reserves to current consumption might have come about through a variety of causes that would make a larger amount of exploration financially unattractive for the time being.[2] The assumption appears to be that when exploration becomes financially attractive it always restores the level of supplies. But does it? The question is relevant to the policy of rapid depletion of North Sea gas supplies,[3] which was adopted. The advantages of rapid depletion include some economies from the scale of operation and, it is hoped, a more rapid growth of national product from cheaper inputs of energy. The underlying assumption is that when earlier exhaustion of the gas reserves arrives, compensation can be found by using a little of the increase in national product to buy energy which will be available from other sources. But if this assumption is falsified by the arrival of a greater scarcity or higher real cost of alternative energy supplies, then the future position will have been in-

[1] Posner 1973, 335.

[2] Ibid., 56–7.

[3] Ibid., 136 and 217–31 discusses in detail the arguments about the optimum rate of depletion of North Sea gas supplies.

accurately discounted and the contemporary energy market distorted. Mr Posner pointed out that the kind of analysis required for the formulation of a fuel policy is necessarily 'tendentious' because it assumes that optimum conditions are fulfilled in all other sectors of the economy, and he added that, once this assumption is abandoned, 'the morass of "second-best" arguments opens before us'.[1] Unfortunately, this morass is characteristic of the real world. To avoid being put too deeply or too often into hazard by the morass requires luck with the unpredictable, as well as intellectual skill.

The essential point is that, especially from the later sixties onwards, all the energy industries were operating in a market of which the characteristics were, in part, artificially determined not just by the permanent elements of imperfect competition, but by government trying to play God in the system without having divine knowledge or divine power. It may well be that coal was more handicapped by this than its competitors were, even though some of the assumptions about the future trend of its costs were too favourable, for the artificialities of policy were directed towards pushing coal down and its competitors up, as a reinforcement of what were believed to be virtuous emerging economic trends. As Mr Posner put it, the estimates of the marginal resource cost of coal were on the assumption 'that output will decline, that no new collieries will be sunk, that major reconstruction schemes will have come to an end: the industry will resemble an engineering production line, operating with free raw materials, in a particularly beastly environment, subject to a malign roulette wheel of geological chance.'[2] If the coal industry alone, whatever its future size, was to include no new capacity in replacement of old, its costs were likely to increase relatively to those of its competitors. But it seemed odd to establish such conditions as part of a policy which assumed that the cheapest 70 million tons of coal would remain among the cheapest fuel available.[3]

In the execution of any policy which calls for changes in the share of each fuel industry in total supply, the contractions can be achieved much more quickly than the expansions, because of the long lead times for the creation of additional capacity. So the policy adopted in 1967 was likely to have its greatest and earliest effect on coal's market share.

Coal was also the most vulnerable of the industries when policy was

[1] Ibid., 2.

[2] Ibid., 331.

[3] This appears to be the most likely implication of the estimates in Posner 1973, 327, although the warning is given that marginal resource costs are defined differently for different fuels.

determined by an assumption that current short-term conditions indicated approximately the long-term future for competition within the market. In the short term it is essential only to cover short-term costs. For deep mined coal these are a much larger proportion of total costs than they are for other fossil fuels. For oil and natural gas the discovery of reserves and the installation of plant for their extraction are very expensive, but, once these two stages have been accomplished, the actual operation is fairly straightforward and requires only a small amount of labour. But even after coal has been discovered and a colliery established, the working of the coal is an elaborate and labour-intensive activity. In a fluctuation of the market which tips the balance towards consumers, it is easier for its competitors to squeeze the coal industry by price competition than for coal producers to squeeze them, even if long-term total costs are similar. By the seventies, however, coal was probably not adversely affected by this feature of cost structure, at least in relation to oil. This was because the oil producers, far from seeking to take advantage of their lower ratio of short-term to long-term costs, preferred to use the strength of a cartel to enable them to keep prices far above the level of their short-term costs.

The change in relative prices made current reality very different from the assumptions on which British fuel policy had been based since 1967. No doubt it would be contended that this was because the political factors, leading to monopolistic association, had caused oil prices to rise well above total costs of production. But there is so much uncertainty about the true and relevant level of long-term costs that any such contention is dubious. It is at least as plausible to maintain that the oil prices of the sixties, which the fuel policy assumed to be permanent, were too low to cover all long-term costs. This may do no more than illustrate the extreme difficulty of forward planning on a national or global scale. It may, however, suggest some errors or omissions in the formulation of policy which contributed to the excessive contraction of coal's market share.

The case for the fuel policy is that policy-induced shifts in the market were based on recognition of wide differences in costs that would persist far into the future. When pricing was discussed there was fairly general agreement that prices ought to be based on long-term costs, even when there were disagreements on whether *marginal* cost pricing was preferable or practicable. The errors in forecasting major items in the policy suggest, however, that the data on long-term costs may have been missing or overlooked or inaccurate. Estimation of long-term costs was, indeed, very difficult for a variety of reasons. There were uncertainties about the most appropriate allocation of particular items of cost, not least where different types or qualities of output were jointly produced. There were conceptual problems. If marginal costs were used in the case of coal, it made a huge difference if they were defined as the avoidable costs of the highest-cost colliery in operation rather than the avoidable costs (including return on capital) of the next bit of new capacity; and there were comparable problems in the costs of other fuels. Average or total costs did not involve all the same problems, but they still presented difficulties. There was uncertainty about both the length of the long term and the rate of discount to be applied in calculating the present value of future outputs. The choice of figures was, of necessity, arbitrary to a high degree.

E. F. Schumacher, Economic Adviser to the NCB for twenty years from 1950, regularly maintained that, if long-term costs cover all that is needed to maintain the business intact, it is necessary to take account of a much longer period for extractive industries which are concerned with non-renewable resources. This is particularly because of the need to maintain an adequate supply of resources to work on, despite their continuous depletion, and to make allowance both for the rise in product value resulting from that depletion and for the costs incurred in adjusting to the physical aspects of depletion.[1] Not everyone would agree with the conclusions he drew, and probably most informed commentators would think that he exaggerated the difference between the fossil fuel and other industries in the choice of criteria for rational forward planning.[2] But at least he drew attention to areas of doubt which might have tempered over-confidence in the reliability of the 1967 fuel policy as a rational basis for immediate rapid action to hasten the decline in the use of coal.

Reference to past experience might have added to the case for caution. Over a very long term the price of fossil fuels had risen, and had risen faster than the real interest rate. If the extended long period that Schumacher favoured were taken into account, it might be expected that similar trends would be experienced. In this case it would be unsound to use low discount rates to estimate the present value of potential future output, and projections made by applying conventional economic criteria to current data would be similarly unsound. There

[1] Kirk 1982, chap. 1 and p. 85.

[2] This comment would not have surprised Schumacher, who remarked that conventional economics ignored the points he was putting forward.

would be further complications because changes in the relation of real prices to real interest rates have not been identical for all fuels and would probably not be identical over an extended long-term future, especially as there are large differences in the ratio of production to known accessible reserves. On the other hand, if a less extended long term were used (say, perhaps, the next ten years), there was a temptation to ignore the longer history and base policy on the experience of the sixties, when real prices of fossil fuels were not rising, and, indeed, those of oil and gas were falling appreciably. Current conditions might then appear to be a reliable base from which to project future policy.

Such considerations, doubtless, are relevant mainly to the question whether fuel policy was inadequately formulated or merely ran into bad luck with the unforeseeable. They do not alter the fact that the market was changed mainly by commercial forces and that fuel policy was merely added to them. They also do not alter the conclusion that, whatever the reasons, the projections of the fuel policy soon proved seriously wrong in some important respects and added to market irrationality rather than corrected it.

National fuel policy is presumably meant to overcome the deficiencies of an imperfect market by imposing on its dominant institutions decisions that serve the public interest better than those they would otherwise make. The record makes it appear doubtful whether that purpose was either achieved or achievable.[1] The signs are that only a selection of the inherent complexities was tackled and that, for what may have seemed good reasons at first, there was a strong tendency to project what were really only short-term features of a particular time. Once that practice had begun, it was tempting to change the 'long-term' policy with every large short-term fluctuation. The fuel policy of 1967 was largely abandoned by the mid-seventies. Then for a few years it was policy to restore to coal a larger share in an energy market assumed to be in continuous growth. But this policy also was not long pursued in a consistent way.[2] In practice, one effect of fuel policies was to add to the uncertainties of the energy market, which were already serious. And the uncertainties caused by policy were often the hardest to adjust to, because they appeared arbitrary and less predictable than market forces.

*

[1] Robens 1972, 206–33 for a hostile account of the errors and inadequacies of fuel policy in the sixties and early seventies.

[2] See, for example, among the documents issued by the Department of Energy, *Interim* and *Final Reports of Coal Industry Examination* (1974), *Energy Policy Review* (1977), and *Energy Policy: A Consultative Document* (1978), Cmnd. 7101.

In looking at market conditions as a major factor in the fortunes of the coal industry one needs to return to the basic simplicities. No other influence did so much to determine the contrasting phases of the industry's history. For ten years after nationalization the NCB had something approaching a monopoly, under pressure to maximize production and confident of a sale for all they could produce. For fifteen years after that coal was faced by sharp competition from other fuels in a period of generally falling world energy prices, and the effect of competition was enhanced because some of its best customers were diverted to other fuels by technological changes in their own activity. After 1973 energy prices generally were rising in real terms, and though this was true of British coal also, its competitive position in terms of relative price improved. But there still were strong competitors to be faced, and the energy market in which they all struggled had ceased to grow as it had been doing.

The three periods thus had very different conditions and the response of the industry to them is the major part of its history. Some elements of that response, for example the coal pricing policies and the various approaches to the control of production costs, modified the detail of the market conditions themselves; but these remained more strongly moulded by influences outside the industry than within it. The purpose of examining the assorted complications of the market (arising mainly from law, politics, natural conditions, and the exceptionally long time horizon that may be necessary for rational judgement of the relation of means to objectives) is to indicate both the definite limits to the choice of response and the influences which might predispose the coal industry to follow one type of policy even when others were open to it. The NCB did not constitute a market institution quite like those assumed in expositions of economic theory, for they were not required to maximize profits (or even to make profits), and they were required to serve the public interest, which is a difficult and controversial task. When they had a near-monopoly they were not in a position to exploit it for the maximum or optimum business advantage. When they were exposed to fierce competition they were not allowed to meet it on equal terms. At critical times they were shut out of business which competitively they would have secured, and were compelled to contract in favour of their competitors to an extent which went beyond the indications given by the relativities of the market. At other times they were given financial help to retain some business which short-term conditions would have caused them to lose, yet without any indication of how long or to what extent that help would continue. Competition makes prediction harder than monopoly and is therefore more testing for business judgement, but these differences were exaggerated because government policy towards the energy market became more erratic after competition emerged than it had been before. So, whether the business judgement was generally good or bad, it was likely to be more often at variance with the assumptions of government policy. And in such circumstances it was an irritating defence if the business results were presented as being in conformity with the long-term public needs that the NCB was required to meet. It must always seem unnatural to a government that its predecessors have delegated to a financially subordinate body the task of defining 'the public interest in all respects' and operating as that interest requires.

It is in such a context, so much less simple than the presentation of the operation of markets given in political speeches or elementary textbooks of economics, that the activities of the coal industry have to be seen, understood, and judged.

CHAPTER 3

# Technology and Exploration

## i. Introductory

Whatever the nature of the market, the key role in meeting its needs was taken by technology. The more drastically the market changed, the more the adaptation of the whole industry depended on innovation in its technology. These may seem to be large claims when it is borne in mind how influential are investment, management, and the skills and application of labour; but the centrality of technology to the progress of the coal industry had already been demonstrated by the Reid Committee of 1944–5.

The seemingly restricted terms of reference of this committee were: 'To examine the present technique of coal production from coal face to wagon, and to advise what technical changes are necessary in order to bring the industry to a state of full technical efficiency.'[1] Yet its report showed in some detail not only how particular shortcomings in the application of technology held back the industry, but the imperatives which technological requirements established for business organization and structure and for the training and deployment of labour.[2] No other document so strongly moved public opinion at the time towards reform and reconstruction of the coal industry, mainly because the report was concentrated on technical efficiency, irrespective of political ideology or administrative theory. Yet it was equally important in earning the respect of professionals and calling for their practical attention in many specific ways. Without great elaboration it looked at the principal features of mining operations and described prevailing British practices and the opportunities for improvement by applying what was already known. It also set out some of the problems which still remained to be solved even if current best practices were adopted. So at the time when the coal industry was nationalized, there was widespread awareness of

[1] Reid Report 1945, 1.

[2] Ibid., 137–40 for 'the conditions of success'; 113–18 touch on some implications for labour relations and deployment.

the gaps in technology both between what was achieved and what was attainable, and between what was currently attainable and what was wanted.

One of the important things about this awareness was that it was given a basis in highly specific detail. Though there are useful insights to be got from general and comparative studies, it is not realistic simply to take a set of generalizations about industrial technology and expect that, with a little local adaptation, they will indicate the lines of progress towards efficiency in coalmining. There are special features in this industry which greatly affect the kinds of technological achievement that are needed and the economics of technological change. It is particularly easy for the casual and uninformed observer, who sees the same pithead and its surrounding buildings on the same site decade after decade, to be misled into thinking of a colliery as a fixed workplace like a factory. The essential peculiarity of a colliery is that its underground workplaces—the locations where its product is worked and moved—have to be shifted not merely from time to time but every working day; and they may be anywhere within a radius of some eight kilometres from the pithead. New means of access to and disposal from these shifting places of production have to be continually created; otherwise the whole enterprise will stop. The peripatetic nature of the workplace has a further significance. Not only the location but the detailed physical characteristics of the workplace change frequently, and the nature of the changes, which may make work much more or much less productive, can often not be predicted very far ahead.

The need for production to be accompanied by a permanent task of creating access to new places of production, and the need to deal with the contingencies arising from the lack of uniformity and the unpredictability of working conditions, all tended to reinforce one of the technical features of coalmining: the use of a large number of men. Circumstances varied from district to district, but in 1947 it was usual for wages and other labour costs to be of the order of 70 per cent of the total cost of coalmining.[1] It followed that most technological change to increase the efficiency of the industry must aim to raise the productivity of labour. This was almost certain to depend on a large increase in capital equipment per man and improvement in the quality of the equipment. This, of course, was familiar over a wide variety of industries, but the coal industry was unusual in the degree to which it seemed capable of benefiting if appropriate methods of capital-intensification could be devised.

[1] NCB *Ann. Rep.* 1947, 164–85.

Familiar, too, was the problem of interrelatedness in successive processes in many kinds of industrial production, but, in this respect also, perhaps coalmining showed the problem with special acuteness. The link between working coal and providing physical outlets for its disposal clearly suggests this, but the need to relate every aspect of work in a mine to continually changing locations added to the complexity. Faceworking, roadway development, haulage, winding, ventilation, and lighting all needed to improve together, if the technological improvement of one of them was not to have its benefits reduced by the stagnant techniques of others. But the task of mutual accommodation was peculiarly difficult and made it likely that technological change would be piecemeal, with its contributions to greater efficiency cumulating only gradually.

An important implication of these characteristics was that improvements in technology were likely to promote increases in the size of collieries. Indeed, improved methods across the whole range of mining operations were probably dependent on increased colliery size for their general adoption. This was because the capital expenditure on integrated new technology was necessarily high enough to create additional overhead costs that could be recovered only from outputs larger than the average for existing collieries. The great variety of colliery sizes and the numerical preponderance of fairly small collieries immediately before nationalization have already been noted.[1] So even to implement the range of technological advances that the Reid Report saw as available presented difficult choices. The less the upheaval and cost of drastic changes in physical and financial structure were accepted, the slower and less comprehensive was technological improvement likely to be, and vice versa. It was not that fairly small mines, with no great innovations in most of their activities, could no longer be remunerative, but that the conditions which favoured them did not exist in more than a small part of the industry. There was, however, an obvious temptation to keep such mines going for as long as possible. There was also an incentive to look for specific new techniques and equipment which were so superior to those existing that it was rewarding to introduce them on their own, even though the potentially greater benefits of an integrated range of improvements were forgone because of a shortage of suitable collieries. In practice, it proved very important (for both better and worse) that ways were found of changing some particular operations without the immediate compulsion to change nearly everything else.

[1] Above, Chap. 1, section i.

## ii. The physical features of coalmining

Coalmining is carried on in peculiar conditions which it is not easy for the outsider to visualize or appreciate. To be able to see the significance of its technology, and to consider whether its output is being achieved as efficiently as can reasonably be expected, it is necessary to try to understand its physical characteristics and the nature of the operations they entail in order to supply the consumer with an acceptable product.[1]

With only tiny exceptions (such as the Brora coalfield, which is much younger), the coal worked in Britain occurs in geological formations of the carboniferous period. The relevant formations, in order of increasing age, are the coal measures, the millstone grits, and the carboniferous limestone. Where all of these occur the younger rocks are naturally found above the older, but in any particular locality the older rocks may be exposed because the younger have been removed by erosion, or displaced by faulting, or have never been present. Where all are present it is the coal measures which are most accessible because they are nearest the surface. The coal measures are commonly divided into upper, middle, and lower, not all of which are necessarily found in succession in one location: there may be unconformities at the base of any of them and hence there are gaps and displacements in the geological sequence. Coal does not occur significantly in the upper coal measures which, for this reason, are often called 'the barren coal measures'. Where coal is found in the various carboniferous formations it occurs as seams which, in relative terms, are like extremely thin slices of meat in a thick multidecker sandwich that has been cut with appalling crudity. Most of the coal worked in Britain is in the middle coal measures, with some in the lower coal measures, which contain fewer coal seams because the total thickness of the formation is usually much less than that of the middle coal measures. Coal seams occur in the millstone grit, but they are few, and it is normally not worth while to seek coal within or below this formation because of the excessive proportion of hard rock which is costly to penetrate.[2] Carboniferous limestone is, in many places, richer in coal seams, but they are attainable only if there is not much hard rock between this formation and the surface. Much Scottish coalmining is and has been in the carboniferous limestone, but only a very little has come from this formation in England and Wales.

[1] This chapter will not deal with the physical and technical questions which are peculiar to opencast mining. These are discussed in Chap. 8 below.

[2] Some coal in the millstone grit has been worked at the Westfield opencast site in Fife.

The accessibility of the coal depends also on the nature of the geological cover above the carboniferous formations. In some localities there is virtually no cover and coal seams crop out on or close to the surface. That is where mining is obviously easiest, but by the mid-twentieth century that was where coal was most likely to have been worked out, though, in fact, not all of it had gone. Where the carboniferous formations are not exposed the overlying material, in addition to various glacial deposits, is most commonly from the Permian and Triassic periods. This is usually not hard enough to offer a serious obstacle to mining, but in some places it is much permeated by water, and in others its thickness may cause insuperable difficulties. Mainly because of high temperatures and the pressure of overlying strata, 1,200 metres and 4,000 feet have, for most of the last hundred years, been taken as round figures indicating roughly the maximum depth at which coalmining can normally be undertaken. Refrigeration and heavy duty equipment have in recent years made it technically feasible to operate at greater depths, but the costs may be high, and there still are economic constraints on very deep mining. In special circumstances a few British mines have gone deeper (though at present the deepest British workings are about 1,000 metres), and in some other parts of the world appreciably greater depths have occasionally been reached. But, in general, coal at greater depths than this has to be regarded as uneconomic to work at present. The thickness of particular geological formations varies greatly from place to place, but those relevant to coalmining can often be much thicker than 1,200 metres. In Lancashire, for instance, the barren coal measures have a maximum thickness of about 1,400 metres and the middle and lower coal measures a maximum of about 2,100 metres, whereas in the main Northumberland coalfield the cover of barren coal measures never exceeds 230 metres, and in Yorkshire 200 metres. In Yorkshire, the Permian and Triassic cover can be up to 820 metres, and the productive middle and lower coal measures reach a maximum thickness of over 1,550 metres in South Yorkshire, then narrow to the south and east.[1] The essential point is that coal seams attainable in one part of a coalfield lie too deep for access in another part.

Various features of the coal seams themselves should also be noted. Not only do they form a very small proportion of the formations in which they are situated—in Cumberland they were estimated to form

[1] Ministry of Fuel and Power, *Regional Survey Reports*: *North Western*, 8; *Northumberland and Cumberland*, 15; *North Eastern*, 10–12. The Regional Survey Reports as a whole provide a useful short summary of the basic features of the geology of the English and Welsh coalfields.

5 per cent of the total thickness of the middle and lower coal measures, and in Yorkshire slightly less than that[1]—but they also display a lack of uniformity in almost every respect. Although they are such a small proportion of the total, the number of coal seams in any coal-bearing formation is often large, so that their average thickness is not great. In Britain only a small proportion of the coal is in seams 2 metres or more thick and, though a few thicker seams have been worked in the past, $3\frac{1}{2}$ metres is near to the maximum now. Seams less than three-quarters of a metre thick, which are numerous, tend to be awkward and very uncomfortable to work, though many seams as thin as 46 cm., or occasionally even less, have been worked for long periods. Coal seams are not of an even thickness over their whole extent, nor do they provide a continuous area of nothing but coal; parts of many seams have been washed out by the action of water or the intrusion of other geological formations, and it is common for bands of coal in a seam to be separated by narrow bands of dirt which it is impossible to avoid cutting along with the coal. Coal seams are easier to work if they are fairly flat, and coalfields with an above-average proportion of flattish seams, such as the East Midlands, have a physical advantage. But most coal seams do not lie nearly horizontally in the strata. They are on gradients which may in places be as steep as one in one and which vary greatly from one part of the seam to another. Some seams extend over very large areas, others may be traced for quite short distances and then disappear. Perhaps the one helpful characteristic is that there is often some concentration in the distribution of the seams. In a vertical section there may be a long distance with no significant coal at all and then a succession of seams with fairly short vertical distances between them. In such conditions it is practicable to work several seams at different levels in the same pit, and it becomes worth while to extract coal from seams which contain relatively small reserves when the same pit also has access to other and much more extensive seams.

Conditions for mining are greatly and adversely influenced by geological faulting, which causes large irregularities. Where there is a fault the strata on one side of it are thrust up or down. Work on a coal seam may have come to an end because the seam has reached a fault and reappears beyond it much higher or lower. Major faults are easy to observe and study, but their consequences for mining may be difficult to deal with because they are so irregular and so large. For example, the Lancashire coalfield has twelve faults which each have a downthrow of

[1] Ministry of Fuel and Power, *Regional Survey Reports*: *Northumberland and Cumberland*, 26; *North Eastern*, 12.

more than 1,000 feet (305 metres), yet one of them, the Upholland Fault, varies from a maximum downthrow of 1,280 metres to less than 60 metres only five and a half kilometres away.[1] But the varied directions of underground stresses also produce large numbers of minor faults, the details of which remain unknown until they are directly encountered. Such faults may displace a coal seam by only a few metres, but that is quite sufficient to cause a serious interruption to its working and sometimes a permanent halt. It may also be unsafe to mine across some faults because they upset the stability of the strata and alter underground drainage.

All these simple but various geological characteristics indicate very clearly some fundamental technical features of the coal industry. First of all, the physical media in which it must operate consist of water and of rock and other strata apart from coal to a greater extent than they do of coal, and the industry must be equipped to cope with all these media. Secondly, the coal itself is non-uniform and located in conditions which continually vary in an irregular way. The techniques employed have to be flexible enough to deal with an appreciable range of variations. Thirdly, the details of the irregularities are not usually made apparent in advance as a result of conventional operations. Mining therefore has to be planned to allow for contingencies, and there are advantages to be obtained from devising techniques which will provide earlier and fuller advance knowledge of changes in the physical conditions of operation. Fourthly, because geological irregularity and uncertainty make interruptions of production statistically probable, a colliery benefits from working coal in several locations simultaneously. Mine planning needs to allow for this wherever practicable and to take advantage of the frequent proximity of several coal seams.

When coalmining is to be undertaken the obvious start of the process, if the industry has already used up the outcrop, is to establish a suitable site for a colliery by exploration and drilling. Where mining has been long in progress it might be assumed that the location of coal deposits is well known and there is no longer a need for major exploration. This is partly true. In an old-established industry most of the additional coal for working is located in the course of development within existing mines; and at the end of the Second World War there was no general feeling of urgency about the need to search for additional coal reserves. But when the extension of mining is restricted to areas which have been worked

[1] Ibid., *North Western*, 9–10.

for a long time, there is an appreciable risk that most of the best opportunities will have been used up. So there may be advantages in looking for a fresh start in a new area. In fact, since 1950 there have been two phases of intensive exploration and drilling, and the coal industry has been able to make use of technological advances that were not all developed for its own peculiar conditions or in response to its own demands.

The next stage is to start gaining access to the coal by sinking shafts to a pit bottom or, if the depth is not very great, by driving drifts (sloping tunnels) from the surface. This also is unlikely to be the first subject to attract technological attention for, in a well-established industry, most additional working underground will make use of the very numerous shafts and drifts already in existence. But gradually, as more coal has to come from new or reconstructed mines, methods of shaft sinking and the use of drifts are bound to receive more attention.

The main focus of technological attention, however, has been the operations underground. Before the specific problems are considered in more detail, it is helpful to look at the general layout that makes possible these operations. From the pit bottom there must be at least two main roadways to the areas where coal is to be worked. Pairs of district roads branch off the main roads, each pair serving a district in which several coal faces will be made, and a pair of gate roads leads from district roads to each end of every coal face. The roadways must provide routes for men, materials, and equipment to be taken in and out, and for coal and stone to be taken out. From the pit bottom, roadways may be built in more than one direction to reach the coal so that faces in several districts may be operated simultaneously in the same mine. If additional roads in new directions are not made immediately, everything must be planned so that this can be done later without upsetting the activity of the mine. In a large colliery the same layout may be made at several levels, so that more than one seam can be worked, and additional tunnelling must then be undertaken to link the different levels.

The roadways accommodate the mechanized equipment of transport underground. That includes ropeways, railways with locomotive haulage, and conveyors. The transport system is also extended at both ends to provide for movement along the coal face and between the pit bottom and the surface by way of shafts or drifts. The ideal, approached with varying degrees of completeness, is to provide continuity of movement from coal face to wagon.

The shafts, drifts, and roadways serve several purposes simultaneously. They are the main routes of a mine's ventilation system as well

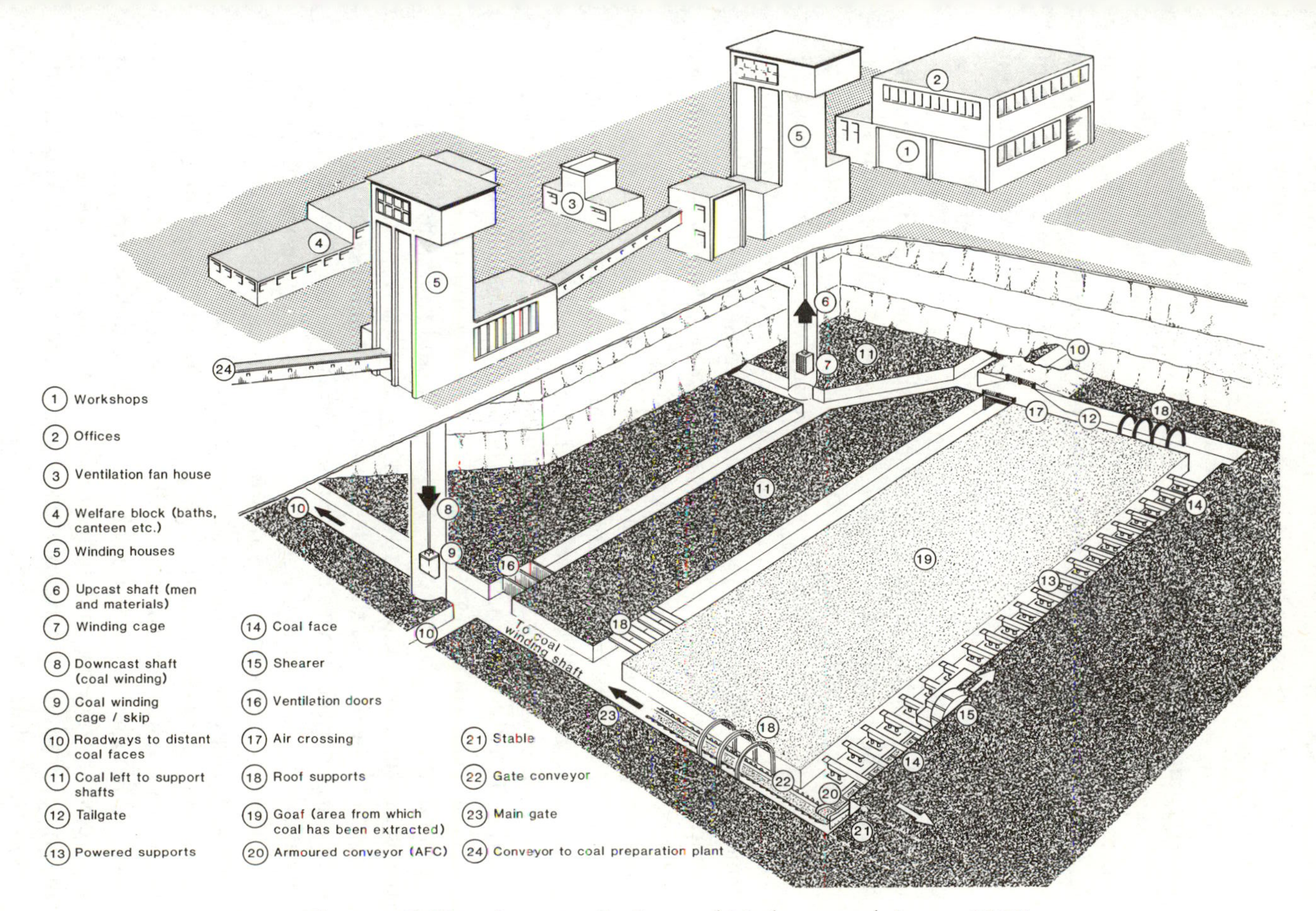


Fig. 3.1. Colliery Layout—Surface and Underground *Source*: NCB

as its transport, and have to be designed accordingly. Roadway layout must incorporate overcasts (air crossings) and airlocks so that the air which is drawn into the mine by surface fans has an unimpeded course from the downcast to the upcast shaft or drift. In every respect the underground layout must take account of the special needs of safety. For example, as much as possible of the electrical equipment is located where the air is purest, i.e. in the intake airway before it reaches the coal face, and sufficient space has to be provided accordingly; and roadways need to be located and constructed so that they are not at risk from sudden changes in underground drainage. Along the roadways also go the power supplies, telephone lines, and water mains, and, in the many cases where these are required, the pipes that drain away excessive methane. And as the precise position of the coal faces is continually changing, the details of the underground layout change with it, new roadways being made and parts of old ones abandoned, but always in such a way as to maintain the transport and ventilation systems and their relation to the fixed points of the shafts or drifts and the surface buildings. It is all rather elaborate and time-consuming to establish and maintain, especially as all roadways and faces, as long as they are in use, must constantly be supported to resist the pressures of the overlying and surrounding strata. As far as practicable, therefore, the size of the excavation is kept to the minimum that leaves economic operation possible. At and near the face that minimum is set by the thickness of the seam. Elsewhere, the variety of functions that have to be performed together usually calls for larger dimensions. Congestion in roadways that are much used for transport is false economy and the more mechanized and uniform the systems of transport, and the higher the speeds, the larger the dimensions of roadways need to be.

Much of the detail of the underground layout depends on the systems of mining adopted. It was common for British mines down to the mid-twentieth century to make most of their roadways in the coal rather than through the adjoining strata of harder rocks. As the coal usually followed irregular varying gradients the roadways often went up and down and round sharp bends. This was a physically easier and a cheaper layout to establish in the first place, but it often caused many operational difficulties and was identified as a source of low productivity and high cost in underground transport.[1] The alternative was to drive all mining operations towards one constant horizon at each level, regardless of the inclination of the strata. Some compromises might be made

[1] Reid Report 1945, 37 and 70–1.

but the outcome was a layout with roads which ran more or less straight for long distances and had only slight gradients. Such layouts became increasingly common in the second half of the twentieth century.

The other main variant was the method by which the coal itself was extracted. There were three principal systems: bord and pillar (with various other regional names), longwall advancing, and longwall retreating. In bord and pillar mining the method of extracting the coal is to drive forward into the seam a series of parallel bords or headways, and then a second series of headways at an angle (preferably a right angle) to the first. Thus at every intersection there is left a large square, or nearly square, pillar of coal to support the overlying strata, though some artificial support is also needed in the bords. When the far end of the area to be worked has been reached it is common, though not invariable, to start working back towards the pit bottom, extracting coal from the pillars that have been left. This may be by taking some coal from nearly every pillar, or by removing some pillars completely, perhaps along a diagonal line, while leaving others intact. This further extraction may take place in more than one stage, but at some point, as mining returns towards the pit bottom, enough coal is taken from the pillars for them no longer to give adequate support to the roof, which may be artificially supported for a time but is eventually allowed to collapse, except where it needs to be maintained in order to keep access to another working area. This inadequacy of support will occur before every pillar has been fully extracted, so one disadvantage of the system is that a significant proportion of the coal is never recovered. This is one, though by no means the only, reason why bord and pillar working was found in only a minority of British collieries by the twentieth century, though it remained the predominant system in the USA. Nevertheless the Reid Committee, while recognizing that only a limited proportion of British output was suitable for bord and pillar working, recommended this system in preference to all others wherever it was practicable and did not involve a much lower rate of extraction than longwall. The Committee found a higher OMS in bord and pillar and traced this back to a series of advantages that included a higher proportion of directly productive labour, easier supervision of work, and the fact that much of the proving of the working area was done in advance of coal getting.[1]

In fact, after nationalization, bord and pillar gradually became an even less usual system. This was mainly because, unless the rate of extraction was very low, bord and pillar gave less satisfactory roof

[1] Ibid., 39–41 and 45.

support and, as corollaries, made surface subsidence more difficult to predict and control, and became too unreliable for safety at great depths. It would usually be thought unwise to use bord and pillar at a depth of more than about 300 metres, and a high proportion of British coal came from greater depths than that. There was another influence from a series of technical innovations suited only to longwall working. So by the 1970s the use of bord and pillar was almost confined to those locations where a low rate of extraction had to be accepted. This was mainly in undersea mining (most of which was off the coasts of Durham and Northumberland), where enormous safety margins were applied to avoid the risk of subsidence and fissuring in the seabed. By this time NCB rules forbade longwall mining where there were less than 60 metres of carboniferous strata above the seam or 105 metres of all strata between the seam and the seabed. Partial extraction by bord and pillar was allowed if there were between 45 and 60 metres of carboniferous strata above the seam and the cover to the seabed was between 60 and 105 metres. Usually only 35 per cent of the coal was extracted, but in some conditions this was increased to 40 per cent.[1]

Although the longwall advancing system was the one least favoured by the Reid Committee it was the principal mining system in 74 per cent of British collieries at the time the Committee reported.[2] This would not have happened if the system had not had some strong positive advantages and the Reid Committee recognized that in many circumstances it was the only one which could be used. Once a development heading had been driven into the coal and had established a suitable direction in which a panel of coal could be worked, it was practicable to drive two parallel roadways out of that heading and establish a coal face extending from one roadway to the other. Coal could then be cut along the entire length of the face. That length varied a good deal from mine to mine and also from face to face in the same mine, but a representative norm in later twentieth-century conditions would be about 160 to 200 metres, much more than sufficient to explain the aptness of the term 'longwall'. An obvious advantage was that a high proportion of the coal could be extracted, though it could hardly ever be 100 per cent: for instance, some coal had often to be left at the top of the seam to provide safer roof conditions, and various irregularities in the seam caused some coal to be left; but the rate of extraction could normally be higher than with bord and pillar. Another characteristic of the longwall system is that there has

[1] NCB North East Area, *Ellington Colliery*, 4.

[2] Reid Report 1945, 42.

to be artificial support for the entire working area, although, as the face advances, support is withdrawn from the worked-out area behind it (known as the 'goaf'), except where access roadways have to be kept available. This need for support has both advantages and disadvantages and the longwall system presents some serious problems of efficient working which will be considered when face technology is looked at more closely. One reason why longwall advancing was subjected to criticism is its liability to serious interruptions in the getting of coal. This is partly because the face is always advancing into the unknown, and changes in physical conditions may be encountered without prior warning, and partly because there is so much additional work that has to keep up with the advance of the face. In particular, the roadways at the ends of the face have to be extended as the face advances and the driving of roadways may experience its own delays, which hold up everything else.

It was to deal with these latter problems that the longwall retreating system was advocated. The basic difference is that, in this system, the roadways between which a panel of coal is to be worked are driven to their full length before coal-getting begins. The face is established at the far end of the roadways and the panel worked backwards towards their origin. The great advantage is that, in general, the physical characteristics of the area to be worked are established before the start of coal-getting and the necessary roadways are already in existence at the ends of the face, however fast it is moving. Thus the risks of delay in coal getting operations are much reduced. The disadvantages are that there is much more capital expenditure on roadway drivage and support before there is any return from the production and sale of coal. The extra capital cost is all the more of a financial problem if roadway drivage is slow and there is a long delay before the new face can be brought into production. It was probably because of these mainly financial problems that the longwall retreating system was little used in Britain before the mid-twentieth century.[1]

In more recent years much more use has been made of retreat mining. In March 1971, out of the NCB's 839 major longwall faces only 46 were on the retreat system, but their average daily output per face was about one-third higher than that of advancing faces, and this difference increased in the next year or two. Ten years later retreat faces were 132 out of 645, and in March 1982 they were 122 out of 601.[2] By the early

[1] Ibid., 43.
[2] NCB *Ann. Rep.* 1981–2, 11.

eighties, however, improvements in the design of advancing faces had reduced the advantage of retreat faces in average daily output. The much greater use of retreat mining in the seventies was encouraged because the heavy investment in coal face machinery made interruptions in face operations more costly, and because improved techniques in roadway drivage made it practicable to avoid some of the earlier long delays in starting up retreat faces. There were also economies from some variations on the conventional retreat system. For example, a panel was sometimes worked in retreat between one roadway specially driven for the purpose and one already in existence, having been driven to serve a previously worked panel on its other side. And there were other variants making use of existing roadways or roadways in the course of creation for some additional purpose.[1]

### iii. Coal face operations

It can readily be seen that the underground transport system and the coal face operations occupied the central position in determining the efficiency of a coalmine. The Reid Committee argued that the key factor in attempts at improvement was underground transport. It went so far as to claim that Britain had introduced mechanization the wrong way round, by starting at the face when the start should have been at the shaft bottom.[2] Nevertheless, not only had mechanization begun like that but, for a good many years after the mid-forties, it was the coal face to which most innovatory attention was paid. That is a sound reason for beginning a closer look at technological change with the coal face, and it is simplest to concentrate on the commonest type, the longwall face.

One of the great problems at the coal face was the small proportion of time that could be given to cutting coal and, although by 1945 most faces were partly mechanized, there had been little progress in reducing this problem. Coal, whether it was got by hand or by machine, was cut on only one shift out of three. The seam was undercut and the coal above was loosened and brought down by pick or by drilling holes at intervals and inserting shots which were then fired. Commonly shotfiring was a task for the beginning of the second shift, whose main function was to load the coal that had been brought down and get it away from the face to the mine's haulage system. The third shift had to get everything ready

[1] Brief discussions of the main systems of mining can be found in Reid Report 1945, 39–45 and Griffin 1971, 6–10 and 47–53.

[2] Reid Report 1945, 37.

for the next coal cutting shift. A new area of roof had been exposed and this had to be supported by building stone packs alongside the roads leading to the face and by setting new lines of pit props close to the face. All equipment had to be moved up to the new line of the face and the roadways at the face ends had to be further advanced or enlarged. Mechanization had helped to increase output somewhat but had not changed the balance of operations. By 1945 72 per cent of the coal was cut mechanically, mainly by chain coal cutters. These had a horizontal jib, round the edge of which was a continuous chain on which picks were mounted. As the jib was hauled along the face near its base the chain was in circulating motion and its picks made the required undercut. In some collieries hand-cutting was assisted by the use of pneumatic picks, though these were never as common in Britain as in continental Europe. The increased amounts of coal cut could be handled on the second shift because a rubber-covered conveyor belt was installed along most faces, and in 1945 71 per cent of coal was mechanically conveyed along the face. It was also common to have a conveyor from the end of the face to the mine's main haulage system. But all the coal had to be loaded on to the face conveyor by hand.

The equipment and working arrangements just described remained common until about 1960, with a further diminution in the proportion of faces worked entirely by hand. But there were serious weaknesses, especially because the cyclical nature of the operations meant that if one shift fell short on its task, for instance as a result of absences, all the next day's work was delayed. In some ways the partial mechanization made the arrangements even more vulnerable because of the increased amount of equipment that had to be moved by the third shift. Among other things, it was a very time-consuming task to dismantle the face conveyor, thread it through a line of pit props, and reassemble it alongside the face, and this had to be done every day. What were needed were machines for some of the remaining hand operations, especially the loading of the cut coal, and the performance more or less concurrently of operations which hitherto had been done in cyclical sequence. The meeting of these needs is the essence of the technological revolution which has occurred since 1940 and which had its main phase from the mid-fifties to the early seventies.

An early contribution came from the Meco-Moore cutter-loader produced in the 1930s by the Mining Engineering Company, of Worcester, from a design by M. S. Moore. But the benefits were few because, in its original form, this machine could only undercut the coal and then, after the coal had been blasted down, make another run to gather it up. During the Second World War, with government encouragement, the firm of Anderson Boyes Ltd. (assisted by the original manufacturers and by the Bolsover Colliery Company in whose Rufford Colliery all trials were carried out) developed the Meco-Moore into what was virtually a new machine. It had two horizontal jibs, one cutting at the base and one about the middle of the seam and, at the back, a vertical cutter which sheared the loosened coal from roof to floor. As the coal was cut, some fell into a trough-shaped belt at the back of the machine and was thence delivered sideways on to the face conveyor. There was also a bar with picks arranged in spiral formation. As this moved along in the bottom coal (which it helped to break up) the picks set up a spiral movement of coal on to the face conveyor. The AB Meco-Moore was thus the first machine to cut and load simultaneously, and it started to lessen the cyclical character of face operations. But it worked satisfactorily only in seams that were at least 1 metre thick, level or only slightly inclined, and located in strata free from faulting. These were not common conditions outside the East Midlands, so progress in combining cutting and loading was slow. From 1943, when the AB version first came into use, until 1948, 70 machines were produced.[1] The maximum use was not achieved until 1957 when 165 machines were in operation at the end of the year, and in the sixties the Meco-Moore was in decline and disappeared in 1966.[2]

Alternatives to the Meco-Moore were slow to appear. The firm of Mavor and Coulson had, since the beginning of the Second World War, been developing a machine known as the Samson Stripper, on which great hopes were placed. It cut a strip of coal as it moved along the face and delivered it on to the conveyor. But it met with one snag after another and discussions of it in the NCB's early years seem to fluctuate between impatience and despair. A couple of experimental models were in use from 1947, but ideal conditions were always wanted and seldom found and the maximum number of these machines ever used at one time was only 7, in 1952. Much attention was given to immediately available American machines, but almost all these were designed for bord

[1] Carvel 1949, 61–9.

[2] 'Disappeared' can be taken literally. Not one complete Meco-Moore machine has survived, even for museum purposes, which is one of the problems of making familiar the details of this pioneering innovation. Some accounts of the machine and its history can be found in Carvel 1949, 61–9; R. F. Lansdown and F. W. Wood, 'Coalface Machinery Developments', in *Colliery Guardian*, *National Coal Board*, 1957, 42; Shepherd and Withers 1960, 19 f.; Various Authors (Vielvoye and others) for NCB, 1976, 47.

and pillar working and were not usually adaptable to longwall. They were used where practicable, but could serve only a small proportion of British coal production. In fact, the most successful short-term expedient was to adapt the older conventional coal cutters by attaching to them a loading device in which a series of scraper-bars, known as 'flights', moved along in the cut coal, pushing it on to the face conveyor. The adapted cutters were named 'flight loaders' and the number in use rose rapidly to 220 in 1957, after which it gradually declined, though flight loaders have remained in use on some of the few faces which it is not practicable to mechanize fully.[1]

In any case, none of these machines helped with the problem of the third shift in the cycle. If anything, they made it more difficult, by increasing the quantity and size of equipment that had to be repositioned before each cutting shift could begin. The requirement was to get more clear space at the face so that equipment could be moved freely without its path being blocked by long lines of props set close to the face, and within that space the repositioning of equipment needed to be done quickly and easily, which meant mechanically. All this had to be accomplished without in any way weakening support of the roof and surrounding strata.

The first contributions came from Germany. During the Second World War the *Gewerkschaft Eisenhütte Westfalia*, at Lünen near Dortmund, designed and introduced the *Panzerförderer*,[2] a new kind of face conveyor, which in Britain (where a few specimens were tried soon after the war) became known as the AFC (armoured flexible conveyor). This is one of the key inventions of modern mining and by the sixties it was being installed on all major longwall faces in Britain. Since its first appearance it has been much improved in detail and increased in power, by developments in both Germany and Britain, but the principles of its design have remained unaltered. The general layout is shown in Figure 3.2. An AFC is the same length as the face which it serves. It is made up of a series of heavy duty steel trays, known as 'intermediate line pans', which carry the load of cut coal and are loosely bolted together. The line pans are connected at one end to a drive unit and at the other end to a return unit. Across the full width of the conveyor, at intervals of about a metre, are scraper bars ('flights') which are attached to a chain assembly. This usually consists of two chains running in channels at each side of the conveyor, and returning on the underside of the line

[1] Flight loaders are described in Shepherd and Withers 1960, 99–106.

[2] *Colliery Guardian*, vol. 231, no. 4 (Apr. 1983), 197.

pans so that each chain forms a continuous loop. The chains are driven by sprockets mounted in the drive unit and these are powered by electric motors, usually at least one in the drive unit and one in the return unit. As the chain assembly is driven, the scraper bars attached to it push the cut coal along to the end where it falls on to a loader in the main gate (i.e. the principal of the two roads leading to the face; the other is the tail gate). The conveyor is operated in one direction only, towards the drive unit.

The AFC brought great advantages over the belt conveyors previously used on faces. Its armoured steel structure made it much more robust, it could carry the great weight of coal which fell when a long face was cut quickly, and it was strong enough for the cutter-loader machine to be mounted on top of it and run along it. This last feature brought about an economy of lateral space in front of the face and also simplified some of the problems of directing the cut coal on to the conveyor. The segmented design of the AFC enabled it to be moved gradually to a new position by 'snaking'. As a cutter-loader moved along the face, part of the AFC could immediately be pushed up to the new line of the cut length of the face, while the remainder stayed in its old position opposite that part of the face which had not yet been cut. Thus, during this operation, the AFC had in it a curve which continually moved along until the cut was complete and the entire conveyor had been moved up straight against the new line of the face. It is a movement very similar to that of a flexible watch bracelet which twists at various points as it is moved on or off the wrist.

The robustness of the AFC also enabled a short version of it to act as a stage loader at the end of the face. The coal must be carried away along the main gate which is at right angles to the face, and therefore the face conveyor must deliver it at right angles on to the conveyor in the main gate. This would cause very severe wear and tear if this conveyor were a belt. So the right-angled delivery was made on to the armoured stage-loader which carried the coal a short distance and delivered it end-on to a belt conveyor, a much smoother movement causing less wear and tear.

One of the wartime supplementary innovations from Germany was a cutter-loader designed to run on top of the AFC. This was the coal plough. Its basic principle was that of a knife which was hauled along the face, cutting or planing off a wedge of coal which diversionary plates threw on to the AFC. One such machine was obtained in 1947 and installed in the Morrison Busty colliery in County Durham, where it was found generally satisfactory. In the fifties there was a steady

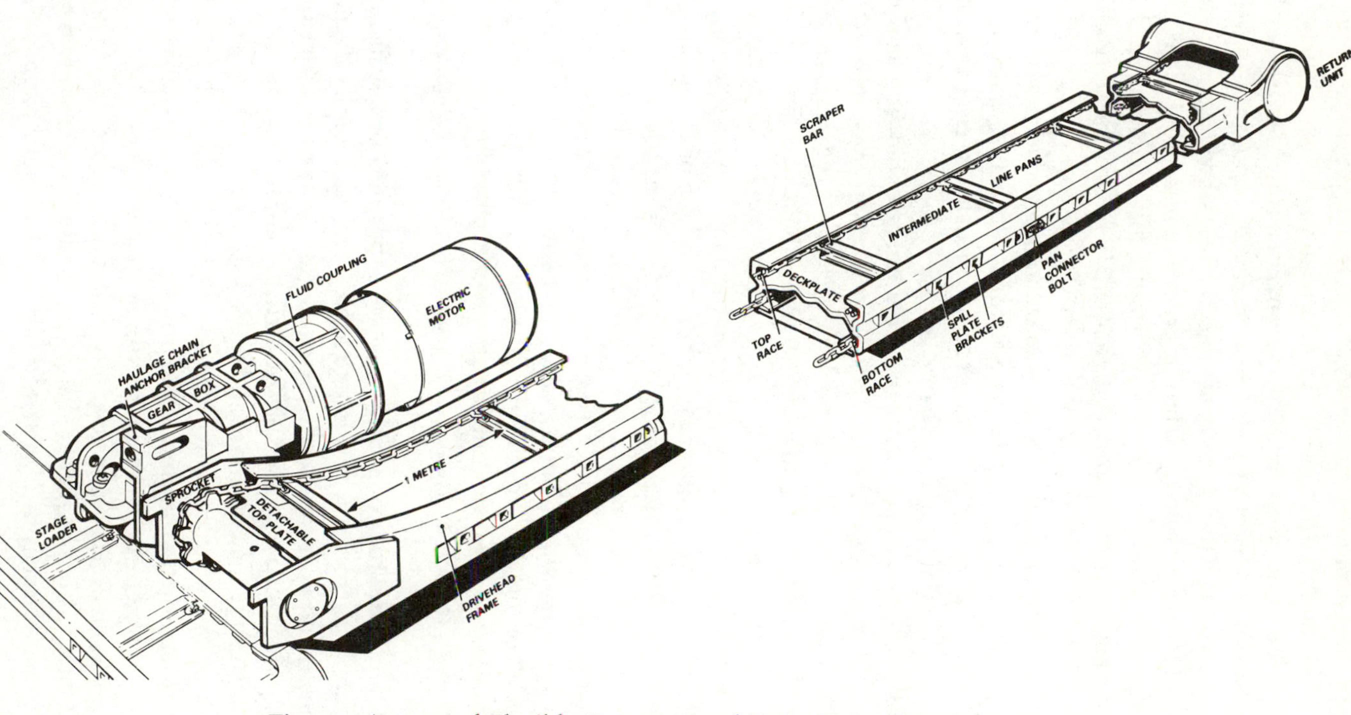


Fig. 3.2. Armoured Flexible Conveyor and Drive Unit—General Layout
NCB Manual, *Armoured Flexible Conveyors*, 1982 edn., pp. 2–3

increase in the use of ploughs, which underwent appreciable changes in design, and they reached their maximum use in 1963 when there were 190. But, although some use has always remained for ploughs, they have been found suitable for only a restricted range of conditions. The early ploughs mostly had stepped knives and moved slowly to cut a fairly thick wedge. Later versions cut thinner wedges and kept up their output by cutting more rapidly. They were found more adaptable. But in general, ploughs worked less well than in Germany because a high proportion of British coal is harder than German and they were best suited to the softer seams. Nevertheless, ploughs had an attraction wherever they could operate satisfactorily, because their cutting method produced a much bigger proportion of large coal than most other types of machine could provide.

But all this new German-designed equipment could be used effectively only if and where the face could be left with a prop-free front. Some progress towards this was made at the end of the war by the firm of *Gutehoffnungshütte AG*, of Oberhausen, which developed yielding props to which were attached link bars at right angles. Thus the props could be set well back from the face, and the roof supported on the link bars. But the design was not very satisfactory and the props and bars had to be reset by hand after every coal getting shift. A solution was approached by various British developments. In 1946 Dowty Mining Equipment produced the first hydraulic props.[1] A hydraulic prop consists of two tubes which can be raised or lowered telescopically by a pumping action on the hydraulic fluid inside it. Thus it can be released and reset in a new position much more quickly than any other kind of prop. By 1956, 400,000 hydraulic props and 200,000 other yielding props were in use.

Later steps were to group the hydraulic props into preformed units and to supply them with power to enable them to be self-advancing. Thus the hydraulic prop was developed into a system of powered supports, which became available in a fairly simple form in the early fifties and steadily became more and more sophisticated. While the use of individual hydraulic props with bars grew rapidly through the fifties, the main transition to fully developed powered support systems came in the sixties, especially between 1965 and 1967, when the number of complete face installations rose from 326 to 725, not far from the maximum of 803 reached in 1972. A powered support has anything from two to six (or occasionally seven) legs, each of which is a hydraulic prop, set on a

[1] *Colliery Guardian*, vol. 231, no. 4 (Apr. 1983), 181.

base and covered by a roof beam or canopy, which is cantilevered well ahead of the foremost legs and in this way carries the roof load while leaving in front of the face an adequate space for the AFC and cutter-loader. The base of the powered support has set into its front a ram, which can engage on the AFC and then push forward the AFC and pull forward the powered support itself. The power for this purpose comes from equipment at the end of the face which supplies an oil and water emulsion through the hydraulic flow lines with which each powered support is equipped.

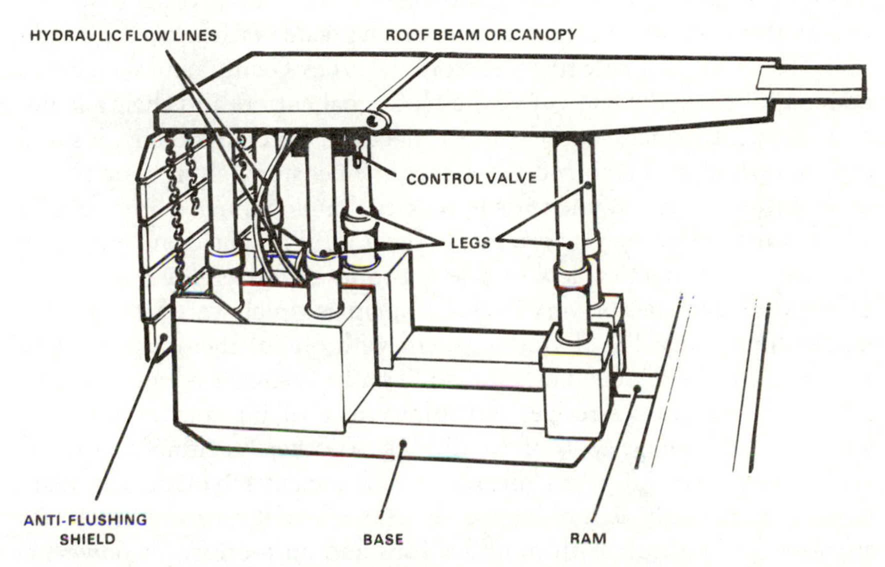


Fig. 3.3. Powered Roof Support—Basic Components
NCB Manual, *Underground Support Systems*, 1979 edn., p. 28.

In the seventies it became increasingly common to add an anti-flushing shield at the rear of a powered support. ('Flushing' is the fall of debris from the roof or from the excavated waste area behind the supports). This gave protection to equipment and made it easier for face workers as they moved about. It also led to the design of shield-type supports in which the shield became an integral part of the support structure, with linkages to the base adjustable so that the forward tip of the roof canopy remained at a constant distance from the coal face as the support was raised or lowered. This was a design that provided extra resistance to lateral movement of the roof strata.

The great achievement of powered supports was that the provision of support for the strata newly exposed by coal cutting, and the removal of the conveyor to the new line of the face, which used to take so much time and physical effort, could be accomplished with the use of hydraulically transmitted power while coal cutting was still proceeding. So the way was clear for coal to be won on two, or sometimes even three, shifts per day instead of only one. A simplified indication of how the face was laid out and operated in these new conditions is given in Figure 3.4.

To take complete advantage of the AFC and the prop-free front required great changes in cutting and loading machines in order to get the potential continuity of operation in a much greater variety of face conditions. This was not likely to come from the type of equipment which was dominant before the late 1950s: the chain coal cutters and their adaptations as flight loaders, and the more elaborate machines using jibs and chains, such as the Meco-Moore or even the Gloster Getter, which, with more cutters, was intended for harder coal and thinner seams, and of which a maximum of 31 were in use by 1956.[1] Ploughs came gradually into use where practicable, but that was only in favourable conditions. There were machines on very similar principles which were intended for harder conditions. The disappointment with one of them, the Samson Stripper, has already been mentioned. It had a vertical hydraulic jack to give it a holding for a stronger thrust into the coal, but this often caused unacceptable damage to roof conditions. Another machine on similar principles was the Huwood Slicer, of which the cutting blade was made to oscillate strongly as it ploughed along the face. It received much less attention at first but, with modifications and an increase in power, it proved to be a useful machine which was in use from 1951 to 1971, with a maximum of 32 machines in operation in 1960.[2]

The great innovation was the Anderton shearer loader.[3] James Anderton was the Area General Manager of the St Helens Area of the NCB. In 1952 he adapted a coal cutter by equipping it with a set of disc cutters,

[1] The Gloster Getter was invented by W. V. Sheppard, who became Area General Manager of the NCB Bolsover Area and in the later years of his career was Deputy Chairman of the NCB. It was developed by the American firm of Joy Sullivan Ltd. and was in use from 1953 to 1961 inclusive. See Lansdown and Wood in *Colliery Guardian*, *National Coal Board*, 1957, 42 and NCB, *The Bolsover Story* (1951), which includes photographs of the prototype.

[2] Lansdown and Wood in *Colliery Guardian*, *National Coal Board*, 1957, 41–2; Shepherd and Withers 1960, 126–8 and 245–8.

[3] A very detailed study of the invention and introduction of the Anderton shearer loader was made by the Science Policy Research Unit of the University of Sussex. The results of the study were set out in Townsend 1976. A much shorter version appeared in Pavitt (ed.) 1980, 142–58.

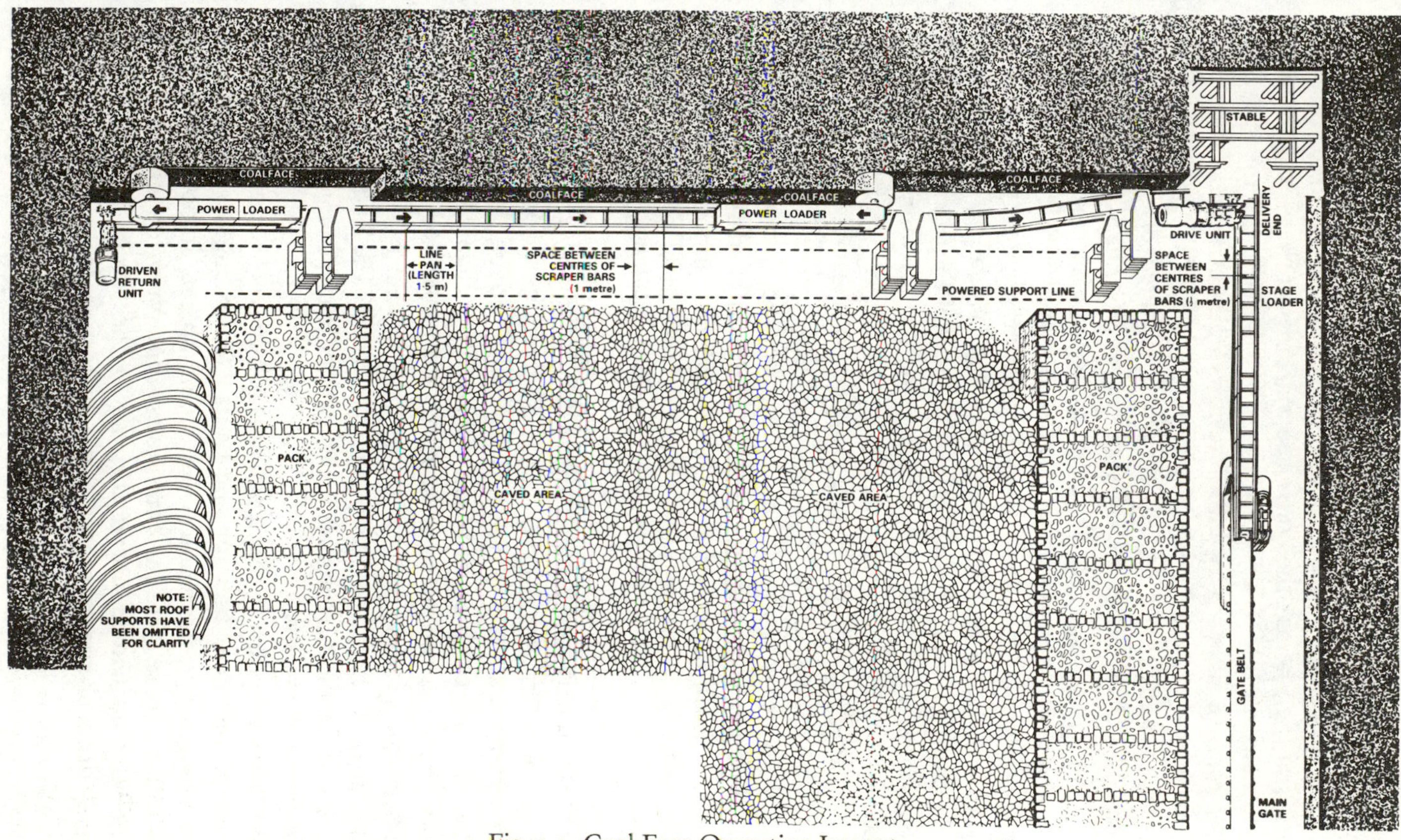


Fig. 3.4. Coal Face Operating Layout
NCB Manual, *Armoured Flexible Conveyors*, 1982 edn. pp. 4–5

standing vertically, to replace its chain jib, and behind the discs he placed a set of plough vanes which diverted the cut coal on to the conveyor, and then he put the converted machine on the AFC and had it hauled along the face by a power-driven rope. Disc cutters, much used in late nineteenth and early twentieth century coal cutters, had long been unpopular and regarded with suspicion, but Anderton had turned the discs at right angles and his experimental machine worked better than anyone dared hope when it was put on trial at Ravenhead Colliery. Very soon the discs were replaced by a rotary drum with picks fitted on its circumferential surface, so the machine operated rather like a bacon slicer running along the coal face with the drum slicing off coal as it was pulled along the AFC. The potentialities of the Anderton shearer loader were perceived very quickly. The Production Department of NCB headquarters took over responsibility for its development, and the principal manufacturers of coalcutting machinery hastened to get involved. Anderson Boyes and British Jeffrey Diamond both produced their own versions for the NCB, and both they and the German firm of Eickhoff, which was mainly interested in non-British markets, negotiated production licences from the NCB, who applied in 1953 to patent the machine. All three firms contributed detailed innovations and the Anderton shearer loader soon emerged from its start as a half improvised conversion to become an established and independent machine in several versions.

It came quickly into productive use. In 1956 there were 215 in operation and in 1957, when there were 347, it produced for the first time more coal than any other type of machine. Although there was then a hiatus in the growth of its relative share of mechanized output, that growth was resumed in 1962. Its share of mechanized output passed 50 per cent in 1966, 60 per cent in 1968, 70 per cent in 1971, and 80 per cent in 1977.[1] In fact, from some time in the sixties the Anderton shearer loader was the most important type of cutter loader in longwall mining throughout the world.

It underwent some fundamental changes. In its earlier forms it could operate in only one direction and had a fixed drum. In 1963 machines were brought into operation which would operate in both directions, and by 1971 these were the commonest type. Very few unidirectional machines remained in use at the end of the seventies. In 1969 the ranging-drum shearer appeared. Now the cutting drum could be raised

[1] Slightly different figures can be found in different NCB sources, the discrepancies arising from differences of definition.

or lowered on an arm as it proceeded along the face. Thus it could be steered to follow the variations in the line between the coal and the stone above or below it, and so cut the maximum amount of coal and the minimum amount of stone practicable. By 1976 this had become the most numerous type of shearer and by that time many of the machines had two ranging drums, one at each end. Since it was also becoming increasingly usual to have two, or even three, shearers on one face, the opportunities for more flexible working and for periods of very intensive output were growing. From 1977 there was also some use of floor-based, instead of conveyor-mounted shearers. These machines were smaller in size and often ran on a specially designed track. They were particularly useful in the efficient cutting of thinner seams, with thorough and accurate cutting of coal down to the base of the seam. For the conveyor-mounted models, which (except for the early trials with a rope) had been hauled by chain, chainless haulage, often by a rack and pinion mechanism between the AFC and the shearer, began to be adopted from 1976, but this was a contribution not so much to improved working as to safety, since a chain breaking or whipping under tension was very dangerous for face workers.

The Anderton shearer loader had perhaps two main disadvantages. One characteristic which it shared with other recent cutting machines was that the thickness of the slice taken off the face at one cut (what is called 'the width of the web') was much less than was achieved by earlier methods. Chain coal cutters often undercut to a depth of as much as 2 metres, but the Anderton shearer loader, though it cut much more deeply than a plough, commonly cut a web of only 56 cm. (22 in.). Of course, it more than compensated by cutting much more quickly. Nevertheless, in the late seventies, when stronger supports were available (sometimes with devices to hold part of the face itself as well as the roof), some shearer loaders were given additional power (up to 600 kW) and redesigned to cut a web of one metre or more. This was still very much the exception, but a web of about 70 cm. became common.[1]

The other disadvantage was that shearing reduced most of the coal to very small size and created a lot of dust. It was necessary to give a lot of attention to dust suppression at the face. On every shearer loader the power cable was accompanied by a water carrying hose and water was sprayed at the tips of the picks as they bit into the coal. The prevalence of small coal was regarded as a marketing limitation in the earlier years of the shearer loader and was doubtless one reason why other types of

[1] NCB, *High Technology in Coal* (n.d. [1982]), 15.

machine continued to be used in large numbers. But in the early sixties it was becoming evident that it was among the users of large coal that demand was shrinking fastest, while the use of small coal was increasing,[1] so the characteristics of the shearer loader were henceforward more in line with market needs.

One other type of cutting and loading machine was used a great deal. At the same time as Anderton was trying to design a workable shearer loader, Anderson Boyes Ltd. were busy with a trepanner, of which the first prototype was produced in 1952. More regular operations began in 1953 and the number in use built up rapidly from 33 in 1957 to 175 in 1960 and 303 in 1964. Thereafter its popularity gradually declined, but the number in operation did not fall below 100 until 1978. More than 20 per cent of mechanized output came from trepanners in every year from 1961 to 1970 inclusive. The main cutting head of the trepanner is an open cylinder, usually 0.85 metres (34 in.) in diameter, with picks around its lip. As the machine moves along the face, the cutting head rotates and takes a cylindrical core of coal in the same way as a surgeon takes bone from a skull in a trepanning operation. Because of the open structure of the cutting head, the coal breaks into much larger pieces than when it is sheared. A rotary head of this kind leaves some coal uncut and also makes it necessary for the face to be squared off after the cut, so the machine has several subsidiary cutters including jibs at floor and roof level and one (like the Meco-Moore) for cutting vertically at the back of the coal. The earlier trepanners had only one rotary head, but were later redesigned with one at each end, so that they could cut in both directions. All trepanners were originally floor-based and so were suitable for fairly thin seams, but by the seventies many had been mounted on the AFC.[2]

In the sixties an attempt was made to combine two types of cutter by building a trepanner which had a shearing drum as one of its supplementary cutters. The first trepan-shearer operated in 1960 and its use rapidly increased, with 130 operating in 1966. In each year from 1964 to 1967 inclusive trepanners and trepan-shearers together produced almost one-third of the mechanized output of coal. But the trepan-shearer never fully matched the performance of either of its parents and after 1967 its use rapidly declined. The last two machines of this type were withdrawn during 1976.

[1] Cf. Chap. 2 above, p. 43.
[2] Various Authors 1976, 83.

During the sixties the mechanization of coal face operations was carried almost to completion. The proportion of total coal output cut and loaded mechanically rose from 25 per cent in 1957 to 50 per cent in 1962 and 90 per cent in 1968. Such a change inevitably had large effects on closely related activities. This was most obvious at the face ends, especially in two respects. One was the amount of space that had to be provided for the additional equipment, notably the drive unit of the AFC. The other was that a gap or rebate had to be made at the end of the face, so that the face presented a profile to the cutter loader, for only in such a gap could the machine commence its cutting run. The space made for this purpose was called a stable, and much of the work of cutting stables had to be done by hand, which reduced the benefits of face mechanization. In the seventies, therefore, much attention was given to the elimination of stables. When more powerful ranging-drum shearers became available the cutting drums might be directed so that the machine could cut a profile for itself. In other cases the roadway at the end of the face could be extended beyond the face to make a T-junction with it, or, more simply and cheaply for the time being, a half-heading could be made in advance of the face, only half the width of the roadway and only at the height of the seam; or, if mining was on retreat, the roadway was already in existence. In any of these cases the roadway supports could be adjusted to leave room for the AFC drive unit and the cutter-loader to be brought right back into the roadway. The machine could then start its cutting run in the coal at the junction of the face and the roadway.[1]

Such changes involved additional challenges which were not all solved at once. New layouts at face ends were not practicable without changes in methods of support. In general there was more that had to be supported in a greater variety of ways, and the location of supports had to be changed more quickly. This still involved a good deal of hand-setting, but by 1982 some 100 face ends were operated as 'propless'. The provision of more space and more rapidly advancing headings at face ends depended on improved techniques in making and maintaining roadways. As long as most face-end work was done by hand there was an incentive to leave the roadheads there as small as could be tolerated, and the most efficient roadmaking equipment was unlikely to be made available there in quantity until the colliery's major roadway programmes were fully equipped. It was not until the late seventies that change in this

[1] NCB *Ann. Rep.* 1976–7, 9 briefly discusses problems and recent progress in face-end operations.

respect was substantial. Then it became more usual for the gate roadway to the face end to be advanced to its full height at the same time as the face advanced, i.e. the roadway roof was 'ripped in-line' with the face. Ranging-drum shearers were able to assist in this where part of the roof was in the coal, and various types of ripping machine became available in increasing numbers. By 1982 1 in 5 advancing face ends was proceeding with in-line ripping.[1] This is an indication both of the extent of the obstacles which still remained to reduce the full benefits of mechanized mining and of the pervasive influence of tunnelling methods on the technical efficiency of colliery operation.

### iv. Tunnelling and shaft sinking

Tunnelling is probably the most ubiquitous feature of coalmining. Not only does the physical construction or reconstruction of a mine consist mainly of tunnelling, but tunnelling is constantly needed in order to keep a mine in operation. In the early nineteen-eighties it was reckoned that for every 1,000 tonnes of coal extracted it was necessary to drive 5 metres of new tunnel in order to gain access to replacement capacity. And repeated work has to be done to keep many tunnels usable for as long as they are needed. The extreme pressures of the overlying and surrounding strata necessitate extra provision for support and maintenance throughout a mine, but especially in the gate roads leading to the coal faces. Because these roads are in ground disturbed by the adjoining face operations, the pressures cause the roof and the floor to start to converge and from time to time it is necessary to restore their height both by ripping (removal of stone from the roof) and dinting (removal of stone from the floor). Both new drivages and roadway maintenance are expensive and time consuming operations in which labour-saving and faster techniques have been much sought. Civil engineers had long had a good deal of machinery at their disposal but it was not all readily transferable to use in coalmines without loss. Everything electrical, for instance, had to be flame-proofed, which made it much heavier, so tunnelling equipment in mines tended to have a poorer power-to-weight ratio.

In fact, for many years most of the driving of tunnels was done by drilling holes and blasting out the material. In the first ten years of the NCB there were two important changes. One was some improvement in drilling shot holes and the technique of detonation. The other was

[1] NCB *Ann. Rep.* 1981–2, 11.

that nearly all the material blasted down was removed mechanically instead of by hand. The physical arrangements for tunnelling had a good deal in common with those of bord and pillar mining, and the sort of power-loading equipment available for the latter in the forties could be used, with or without adaptation, in tunnelling. By 1957 mechanical loading was used on nearly 90 per cent of all drivages.[1] This had indirect as well as direct importance because most tunnelling work is done in very confined spaces and anything that keeps down the amount of cluttering material makes the efficient deployment of labour somewhat less difficult. From 1953 to 1960 the average volume of material extracted per manshift increased by 76 per cent and the average weekly rate of advance of tunnels by 66 per cent, even though the average cross-sectional area was 30 per cent larger.[2]

But without new types of machinery for cutting in stone, rates of advance were still slow (about 12 metres a week) and were a serious retarding factor in the large-scale reconstruction of mines that the Reid Committee had said was desirable for efficient operation. Only in the sixties did new machines appear in any numbers. Initially then it was chiefly ripping that began to be mechanized, and over 100 machines were in use by 1967, after which there was also a rapid increase in the supply of roadheaders. Even so, at the end of the sixties there was worry about the slow progress in speeding up face-end work and tunnel drivage. In 1969 the NCB launched a 'pacemaker drivage' campaign (though with only modest results) and intensified the search for a not too costly heading machine, capable of dealing with all kinds of rock, for use at face ends. In the same year the first few dintheaders came into operation, and experiments were made in fitting ripping machines with the cutting elements of roadheaders.[3]

The most rapid improvements both in design and use came in the mid-seventies. In September 1975, 954 development roadway machines were in use (222 ripping machines, 240 dintheaders, and 492 roadheaders) and this was an increase of 205 in 12 months. The greater proportion of roadheaders was striking as these could generally work in harder rock than ripping machines. Improvements were also made in the methods of erecting supports, which were usually steel arches. Machines were adapted to move and hold these in place so that they could be rapidly fixed by men working in safe positions. The NCB's

[1] Ibid. 1957, I, 24.
[2] These figures relate only to tunnels exceeding 200 yards (183 metres) in length.
[3] NCB *Ann. Rep.* 1969–70, I, 8.

own Mining Research and Development Establishment (MRDE) took the lead in efforts to design still more effective drivage machines, and worked in association with several manufacturing firms. One type of machine resulting from these efforts was a full-face tunnelling machine with a rotating head containing a large number of wheels of the glass-cutter type and big enough to cut the roadway to its full size as it moved forward. Though the early prototypes had some successes, they also encountered serious problems which restricted their early use. But very powerful machines on somewhat similar principles were also developed by American firms, and one of these was successfully used to drive one of the main underground roadways linking the five mines of the new Selby complex in the early nineteen-eighties. It achieved some weekly advances of more than 100 metres.[1] Other roadheaders were enabled to work more continuously by being given complete roof cover by a canopy between the face and the permanent supports. Under the canopy men could set steel arches and line the roadway at the rear of the machine while its head was still cutting and the rock was being loaded away. There were also machines with a more powerful cutting head on a boom (in one case with twin booms whose two heads rotated in opposite directions), for use where very hard rock had to be drilled.

Roads that had to be driven wholly or mainly in coal were generally less of a problem, because less powerful and elaborate machines could cope with the conditions, and because face machines designed for bord and pillar mining, of which more powerful and sophisticated types were brought into use in the late seventies, could also be used for drivages. But roads to be driven in thin seams gave some difficulties, and it was mainly to overcome these that the MRDE and Dosco Overseas Engineering Ltd. designed a new in-seam miner, which proved to be extremely versatile. Variants were produced for different conditions and the machine was much used for driving new face lines, stable elimination, and face-end working.[2]

There still remained a good deal of tunnelling by conventional methods of drilling and shotfiring. At the beginning of the eighties 45 per cent of all the tunnels being driven in stone used conventional methods (and tunnels in stone were about half of all tunnels), but these tended to be the shorter and smaller tunnels. At any one time about

[1] NCB *MRDE Ann. Rep.* 1974–5, 10, and *MRDE Projects* 1975–6, 19; NCB, *High Technology in Coal*, 15.

[2] NCB *MRDE Ann. Rep.* 1977–8, 11, and *MRDE Projects* 1977–8, 9; NCB *Ann. Rep.* 1976–7, front cover and 9.

1,000 roadheading machines were at work, and in 1981–2 they accounted for 77 per cent of the total distance driven, including 70 per cent of the distance wholly or mainly in stone. Without these fairly recent changes, much of the advantage of technological progress elsewhere could have been lost, for the need was for more tunnels, driven more quickly. A highly mechanized coal face was worked out, on average, in rather less than eighteen months, which meant that about 450 new faces had to be opened up in replacement every year. Changes in transport created a demand for many more long surface drifts and for more underground bunkers. Changed technology meant that the contraction in the size of the coal industry since 1960 was not matched by any comparable contraction in the need for tunnelling. Indeed, in the late seventies and early eighties, when efforts were in progress to compensate for an earlier neglect of development work, the amount of excavation was increasing. The length of tunnels driven was 482 km. in 1979–80, 527 km. in 1980–1, and 548 km. in 1981–2. In other words, the length of tunnels driven *each year* in British coalmines exceeded the entire existing length of tunnels in the London underground railway system.[1]

Shaft sinking involved some comparable engineering work, but had additional problems of its own and did not have such a continuous history of innovation as tunnelling. New shafts were needed mainly for new collieries, but were also necessary for some major reconstructions. The best modern collieries of the nineteen-forties often had adequate shafts of 21 feet (6.4 metres) diameter, but many others were much less: 16 feet (4.9 metres) was not uncommon and was by no means the smallest size. Shafts of this size did not permit coal to be brought to the surface quickly enough if steps were taken to achieve much greater output below ground, so the sinking of an additional shaft was sometimes an essential supplement to the increase of productive capacity below ground, though it was so expensive that it was usually not undertaken unless a large increase of output was expected to be maintained for many years. Even in the early eighties the *average* diameter of coal-winding shafts was still no more than 5.7 metres.

Shafts were sunk mainly by drilling the area to be excavated, inserting and firing explosives, removing the debris, and lining the sides of the shaft with concrete as sinking proceeded. It is a procedure very susceptible to interference by water and this contributed to the extreme

[1] NCB *Ann. Rep.* 1980–1, 18 and 1981–2, 11; NCB, *High Technology in Coal*, 11.

slowness of shaft sinking, which was causing serious concern by the later fifties. Efforts were made to use foreign experience. In 1954 the German engineering group of Thyssen came to work on shafts at the new Cynheidre anthracite mine and obtained enough additional contracts to cause it to establish a British subsidiary company.[1] In 1958 specialists from South Africa, where shaft sinking was commonly faster, were brought to work with the main shaft contractors at the new Parkside Colliery in Lancashire and achieved much faster progress.[2] The NCB and the Dutch State Mines pooled experience in 1959 and 1960 to try to establish the nature of the forces exerted on shaft linings.[3] Some of the reliance on machinery which had paid off in other operations was adopted in shaft sinking. By 1958, for instance, hand loading of spoil in new shafts had been entirely replaced by mechanical methods which doubled the speed of clearance, and dirt disposal on the surface was speeded up. The methods of dealing with water-bearing strata began to be applied rather more effectively. The two methods were either to block off the intrusion of water by cementation or to freeze all the ground in the area of operations and thus leave the water to be dealt with whenever was most convenient at a later stage in the work. The freezing method had been practised for many years in continental Europe but was little used in Britain. Between 1947 and 1958 it was used for only 1 in 5 of the new shaft sinkings.[4]

Shaft sinking was not a great success story, but in one way and another the work was improved in speed and quality. In ten years from 1947 to 1956 only 19 new shafts were completed, but 33 more in the next five years, and the number completed between 1958 and 1961 was greater than the number already started but still incomplete in 1957. The improvements were maintained and there was henceforward a greater commitment to the freezing method. But after 1961 the contraction of the coal industry reduced greatly the demand for new shaft sinkings. By the later seventies when there was again a commitment to colliery construction and reconstruction on a large scale, it had become commoner to use drifts for much of the access to mines from the surface. In 1965 ground freezing was used for the first time in the construction of a surface drift, and freezing techniques were improved in the same year by the use of liquid nitrogen. Experience was gained in freezing larger

[1] *Colliery Guardian*, vol. 231, no. 4 (Apr. 1983), 139.
[2] NCB *Ann. Rep.* 1958, I, 18.
[3] Ibid. 1960, I, 19.
[4] Ibid. 1958, I, 18.

excavation areas than before.[1] When major activity was resumed, in 1977, for the construction of the Selby mine complex, knowledge and equipment had advanced a good deal. There ten shafts were sunk and two drifts were driven simultaneously in the four-year period to 1981. The freezing method was used in all twelve cases, improved detonators (not susceptible to water) were used, and much bigger equipment was available for removing the debris to the surface. Physical movement and control were aided by the use of a four-deck movable platform, suspended in each shaft and equipped with winches and electrical switchgear. In those shafts which needed very thick lining, because they passed through heavy water-bearing ground, shaft sinking proceeded at an average of about 10 metres a week, and in easier conditions a best performance of 129 metres in 31 days was achieved.[2] Down to 1957 the best achieved in any conditions was about 46 metres a month, and rates were usually much lower.[3] With improvements of that order it was practicable to plan shafts of wider diameter than before in order to reduce the risk of delays in coal clearance as the daily output per face increased. In the early eighties the construction of a new 8-metre diameter shaft was in progress at Maltby colliery.[4]

## v. Transport

When the Reid Committee complained of the waste of manpower through inefficient underground transport and called for the use of locomotives and trunk conveyors wherever possible instead of rope haulage, it seemed to be asking for reforms that could come only slowly because so many mines lacked a general layout and quality of roadways that would be suited to better methods. But, in fact, there was a rather quicker response than there was in the methods of face working, though improvements in transport were very unevenly distributed among coalfields. Most of the East Midlands, much of South Yorkshire, and some parts of Northumberland had fairly level seams and consequently fairly level and straight roads. Locomotive haulage along the whole or part of the main roads in such conditions was introduced within a few years. There were only 80 locomotives in British coalmines in 1947, but nearly 700 by 1955, and 906 in 1957, most powered by diesel engines and some

[1] *Colliery Guardian*, vol. 231, no. 4 (Apr. 1983), 208.
[2] NCB, *High Technology in Coal*, 7.
[3] NCB *Ann. Rep.* 1957, I, 23.
[4] NCB, *High Technology in Coal*, 7.

by electric batteries.[1] An experiment at Sandhole Colliery with the electric trolley locomotive favoured by the Reid Committee was followed up on only a very small scale.[2] Such locomotives could be used only with a special dispensation from the Mines Inspectorate and only around 1980 was there a move to reintroduce them in a few collieries with specially suitable conditions. In most circumstances there was thought to be too much risk from sparking between overhead contacts. Where locomotives were introduced, working time could be increased by the provision of manriding trains, though the greatest gain was in the movement of coal all or part of the way to the pit bottom in much larger quantities at a time: by 1955 15,000 new mine cars of 30 cwt. (1.52 tonne) capacity or more had been introduced.

Rope haulages continued to be greatly used, and an appreciable improvement in their efficiency was not too difficult to achieve in a great many collieries. There was much more standardization in the sizes of rope-hauled tubs, though there could not be an approach to uniformity because the tubs had to be wound to the surface in cages of such varied dimensions. Circuits for the tubs at pit bottoms were redesigned and the number of loading points reduced to save manpower. Between 1948 and 1955 there was a reduction of not quite 5 per cent in the number of manshifts in underground transport per 1,000 tons. Redesigned pit-top circuits matched those at the pit bottom and contributed to greater savings in surface labour. Even so, the minor reforms had nothing like the effect of thorough reconstruction. In reconstructed pits in the same period the reduction in manshifts devoted to underground transport was not 5 per cent but more than 50 per cent.[3]

Rope haulages in later years underwent some striking improvements, although fewer and fewer collieries used them for hauling coal in tubs. They remained in many circumstances indispensable for taking supplies along gate roads to the faces and development headings, and very efficient systems were installed where there was a haul from point to point on a constant gradient. In the seventies new methods of electronically operated remote control for rope-hauled vehicles were developed. Although the power still came from a stationary electric motor at one

[1] F. Marsh, 'Haulage and Handling', in *Colliery Guardian*, *National Coal Board*, 1957, 18; NCB *Ann. Rep.* 1957, I, 19.

[2] NCB *Ann. Rep.* 1956, I, 52.

[3] Numerous examples of improvements are described by Marsh, in *Colliery Guardian*, *National Coal Board*, 1957, 17–23.

end of the transport run, the operation of the motors could now be completely controlled by a driver in a cab on the train. New types of train were introduced both for manriding and for the transport of stores and equipment. Some of these used the two-rail track long customary for rope haulages, but some were on a monorail system with trapped wheels as a preventive of derailment. Some started at the surface instead of the pit bottom and carried men and stores along a surface drift. Such rope-hauled trains originally operated at speeds of about 15 k.p.h., but experience indicated that they could be safely used at their maximum speed of about 24 k.p.h. These speeds were lower than those of a cage in a shaft or of a diesel or electric locomotive on an extended straight run; but a rope-hauled vehicle had greater versatility. It could be operated in roadways which could never be used by locomotives, and if it operated from the surface it provided some additional time-saving continuity. Older types of rope-hauled manrider, without remote control, were more numerous and slower, but these also were freed from some delays by the widespread introduction of radio signalling between the train guard and the driver in the engine-house, which might be several kilometres away.[1]

The capacity to move more in a given time and to move it in a more uniform flow with the minimum of interruption was always the most desired objective. To its achievement the greatest contributions came from improved winding methods and from belt conveyors. In 1947 nearly all coal was wound to the surface in the tubs in which it came to the pit bottom. These were pushed into the pit cage side by side, very often on two decks. Four tubs at a time was probably a fairly representative figure of the capacity of a cage. A four-deck cage accommodating eight 15 cwt. (0.76 tonne) tubs was unusually good. Many of the winding engines were steam-driven. Larger cages were installed in some cases where the shaft permitted this, and in others cages were redesigned to take similar or slightly larger quantities of coal in fewer and larger tubs. Steam winding engines were replaced as a matter of general policy and 320 electric winding engines, which were much more efficient, were installed from 1947 to 1956. By the seventies only a few steam winding engines remained in operation. Skip winding began to be used rather more in the late forties, though its adoption was slow at first. It had great advantages. The capacity of a skip is less restricted than that of a cage by the size of the shaft, the ratio of payload to dead weight is much higher, and the loading and unloading are quicker. By 1955 33 skip

[1] NCB *High Technology in Coal*, 21–5.

winding plants had been installed, the largest skips then having a capacity of 12 tons (12.2 tonnes), but in that year only 5 per cent of national output was wound in skips.[1] Thereafter, skip winding was adopted in many more collieries, especially where reconstruction involved the widening of a shaft or the sinking of a new one; but it did not become universal.

In the long run it was conveyors, especially belt conveyors, that made the most difference to mine haulage. Though they had been used underground for many years, belt conveyors in the middle of the twentieth century still had some serious limitations. Because of the restrictions on their length and their inability to operate round bends, any extensive system of conveyor haulage had to have numerous transfer points where coal passed from one belt to another, with a liability not only to spillage but to degradation of the coal; and belts were unsuited to steep inclines, though there seems to have been a tendency to underestimate the degree of inclination at which a belt conveyor could operate safely and efficiently. But belt conveyors were well suited to carry coal along the gate roads from the face to the main roadways, though for many years the coal usually was there transferred to some other means of haulage. In the gate roads belt conveyors rapidly replaced ponies and rope haulages. The number of horses underground declined to 17,000 in 1950 and 7,750 in 1960, and by 1980 ponies were still used by only a very few mines, mostly in Northumberland and Durham, where the complex choice of routes, created near to the faces by bord and pillar mining, retained some advantage for their adaptability in hauling supplies to the face. There were also a few ponies in South Wales. The total length of conveyors installed in all underground roadways more than doubled in the 8 years to 1956, when there were over 3,200 km. There was, however, a serious setback. Belting was made of various woven textiles covered with rubber, and the dangers of this were revealed by a major disaster in September 1950 when 80 miners were killed at Creswell Colliery in Derbyshire by a fire and associated fumes resulting from the frictional heating of torn belting.[2] A replacement for rubber-covered belting had to be found and various types of belting material using PVC were developed and gradually adopted.

For the more general use of belt conveyors underground ways had to be found of increasing their length and capacity. By 1953 a conveyor of a type originally designed for a maximum length of 1,500 yards

[1] Marsh, in *Colliery Guardian*, *National Coal Board*, 1957, 20–3; NCB *Ann. Rep.* 1956, I, 52–3.

[2] The official report on this accident was published as Cmd. 8574.

(1,372 metres) had been made one-third longer,[1] but much greater lengths were desirable if belts were to be used much for trunk haulage. A basic difficulty was that the belt which carried the load also had to transmit the power, and searches for a design to overcome this were in progress in the late fifties.[2] Long belts began to be constructed to designs which left the belt to bear only the load of the coal or other material it was carrying, while all the tensions from the drive were transmitted in the supporting cables. Some later designs had steel strands woven into the belt to transmit the tensions and these were particularly used to cope with the very strong tensions of long uphill runs. Belts were equipped with much more powerful electric drive units, and the strength of belt fasteners was greatly increased.

The application of improved tunnelling techniques meant that by the seventies many more collieries had long lengths of straight main roadway underground, and an increasing minority had drifts between the surface and the pit bottom. These were physical arrangements ideally suited to trunk belt conveyors, once it had become possible to manufacture these in very long lengths. In 1970, when four Scottish collieries were linked in the Longannet complex, feeding coal direct to the Longannet power station, a cable belt conveyor of 8.85 km. was installed, at that time the longest in the world. It had a 1,500 kw. drive unit and was designed to carry 20,000 tonnes per day. Thereafter it became usual for coal to be carried underground all the way to the pit bottom by belt conveyor, with bunkers at strategic points to enable the flow to be evened out or to store coal during any temporary interruption of transport. By 1980 more than 95 per cent of coal output was carried in this way. At the pit bottom the coal usually passed directly or by way of a bunker either to a skip for winding to the surface or to a conveyor belt along a drift to the surface, very often straight into the coal preparation plant, occasionally even direct to a delivery point on the railway. From the mid-seventies the construction of a new surface drift to enable the entire output to be carried in this way was becoming a regular feature of new investment schemes in large collieries. The new Selby mine complex was provided with two conveyors each 14.7 km. long to carry the entire output of five collieries at a speed of 7.6 metres per second.[3]

[1] Sutcliffe 1955, 129; other examples of belt improvements around the same time are described in 129–33 with illustrations.

[2] NCB *Ann. Rep.* 1956, I, 52 and 1957, I, 19.

[3] NCB, *High Technology in Coal*, 25.

Belt conveyors by this time were making two other contributions. One was to the movement of men underground. Provision for manriding on conveyor belts was first made in 1970, and since then (subject to detailed rules about roadway clearance, gradients, speed of operation, and other aspects of safety, control, and maintenance), belt conveyors have been found increasingly useful for carrying men short distances on the last stages of the journey to the face. The other contribution, more familiar, was in the final stages of movement on the surface, where coal was carried not only into but also out of preparation plants to overhead bunkers from which, in more and more cases, it was rapidly loaded by automatic equipment into trains moving slowly underneath.

Thus belt conveyors began to realize their potential for moving coal by a single method in a continuous flow from the face to the surface and out of the colliery. When Richard Sutcliffe in 1905 first proposed the use of belt conveyors underground, which was then quite unknown, he ended his reply to the discussion by saying 'he also thought that conveyors would raise coal from mines 1,000 or more yards deep, through drifts 4,000 or more yards long and pour it into waggons at the mine mouth, but the expense of such conveyors would be large'. It took nearly forty years to achieve that objective on even a small scale, when in 1944 coal was first carried continuously on belts from face to sizing screens on the surface at Bullcliffe Wood, a smallish drift mine near Wakefield, producing about 65,000 tonnes a year.[1] It took more than another quarter-century to make the same sort of scheme feasible and commercially sensible for large collieries. But by then the technical efficiency of coal clearance had been transformed.

The other main aspects of underground transport did not show quite such an improvement, especially the movement of men. There was much more provision for manriding and the various types of mechanized transport were quicker than they had been. But no one was carried the whole way from surface to face without some interruption and change from one means of movement to another; and the average distance from pit bottom to faces had much increased. So the sort of loss of time in getting to the workplace which the Reid Committee had deplored was far from unknown thirty-five years later. The effect of technological advances had been to make it possible to exploit more distant reserves without any additional loss of time, rather than greatly to increase the proportion of the working day that could actually be spent at work.

[1] Sutcliffe 1955, 120 and 122.

### vi. General electrification

Practically every aspect of technology in the coalmining industry came to take it for granted that electricity was available wherever it was wanted. But this was a recent state of affairs. Before the Second World War most, though by no means all, collieries which used machinery at and near the coal face were driving it by electricity rather than the compressed air which had been common a little earlier. But for other mechanized functions, including winding, pumping, and various activities on the surface, there was a good deal of reliance on steam power. In 1949 only a little over half the power used at collieries was electrical and most of the rest came from the direct use of steam, much of it raised in very inefficient boilers. In that year the NCB worked on a national policy for the supply of electricity to collieries. Total electrification was concentrated on collieries undergoing reconstruction and others with an expected long life. The intention was ultimately to get rid of all uneconomic steam raising plant and obsolete colliery generators and to supply electricity from colliery power stations which would burn low grade coal on site.[1] In the event it proved more economic to take advantage of large-scale operations and use power from the national grid, so only at Barony and Grimethorpe collieries were new power stations built and the first of these was owned and operated by the SSEB. But the electrification programme went ahead. By 1956, 462 collieries were completely electrified,[2] and between 1950 and 1960 colliery consumption of coal was more than halved. By the sixties progress had gone most of the way towards the ubiquitous availability of electricity in collieries.

As electricity became more generally available there was a gradual tendency to take it more for granted that it should be used in a greater variety of ways. Two of the functions most affected in the not very long run were lighting and communications underground. On roadways, mains lighting increased by a third from 1947 to 1951. At the face most improvement came first from the adoption of much higher lighting standards for battery-powered hand and cap lamps and the rapid substitution of cap lamps for hand lamps as far as possible. But experiments were made with mains lighting on the face. At first they were not very successful because fittings were liable to be shattered by shotfiring and there were difficulties in designing flame-proof fittings. But shotfiring, which increased in the post-war years, became less necessary as new

[1] NCB *Ann. Rep.* 1949, 61–2.
[2] Ibid. 1956, I, 37–8.

types of coal-cutting machines were introduced, and the problems of light fitting designs were solved. In 1952 colliery managements were told they could order certain types of fluorescent lighting equipment where conditions were suitable, and experiments with tungsten filament lamps on the face continued.[1] Although full face lighting remained unusual, since 1947 lighting at and near most workplaces and in the roadways in their vicinity has improved enormously. No change has done more to transform the physical impression given by a coalmine and it has been a major factor in easing the conditions in which all mine work has to be done.

In 1952 it was agreed that underground communication had been neglected and needed to be greatly improved. Thoughts turned to the great advantages of a fully automatic telephone system, with enough capacity to serve a large mine, and to the design of such a system that would be flameproof and intrinsically safe. Electronic devices were examined as aids to the solution of the problems; and, though normal radio communication underground was then believed to be impracticable, experiments were made in using for radio purposes such conducting media as winding ropes in shafts or pipes running through underground workings.[2] From such beginnings elaborate and detailed telephone and other communication systems were worked out and installed in one colliery after another. By 1970 such installations were regarded as basic items for a modern, efficient colliery. There was a gradual but fundamental change of outlook. When electricity was routinely available those concerned with the design and operation of mines began almost routinely to think first in electrical and then in electronic terms, with a consequent great and cumulative influence on the way the daily work of a colliery is conducted.

### vii. Automatic monitoring and control of operations

It was in the early sixties that a practical interest began to be taken in the possibility of electronic control of mining machinery from equipment located in the main gate. In this way it was believed there could be a saving in labour, a large increase in output per man, and an improvement in the environment for face workers. The concept of the Remotely Operated Longwall Face (ROLF) was first tried at the end of 1962 at Newstead and Ormonde collieries in the East Midlands. Two years later it

[1] NCB *Ann. Rep.* 1951, 24 and 1952, 28.
[2] Ibid. 1953, 42–3.

was decided that a new colliery at Bevercotes in Nottinghamshire, originally intended for conventional operation, should instead use electronic techniques for a complete integrated system of remote control, which would eventually be applied to all operations from face to surface. This colliery began operations in February 1967 on the basis of 18 production shifts per week. In the meantime ROLF was tried in various ways at several other collieries, but the number of such experiments rapidly declined in the late sixties. It was clear that the problems involved were too complex for the existing state of knowledge. Among other things it proved impossible then to steer a coal cutting machine accurately from a remote point. By 1969 the NCB concluded that it was better for the present to concentrate on exploiting existing machinery to the full rather than instal complex and costly automatic systems. In 1971 Bevercotes colliery was converted to conventional operation.[1]

These experiments had been disappointing, but not wholly fruitless. They had shown what needed to be done in power-loader steering and in the automatic movement of powered supports.[2] Indeed, with the development of all-hydraulic support control systems more progress was possible. After the adoption in 1974 of a new concept of Advanced Technology Mining (ATM) and its gradual application in a few collieries of great productive potential, one of the main emphases was to improve on the hydraulic batch control systems which then enabled the movement of up to six supports in sequence from one control point. A specific achieved objective was, with a return to electro-hydraulic techniques, to increase the number of supports which could be controlled, singly or in batches, by one operator using a push-button system. The idea of a nucleonic vertical steering system for power loaders, which had emerged from the earlier experience, took longer to apply satisfactorily. By the early eighties, however, a number of machines were equipped with guidance systems recording the different gamma radiation of stone and coal in order to keep the cutting drum close to the edge of the coal without going into the stone above or below it.[3] As they relied on the detection of differences in low-level natural radiation, such machines were more acceptable on health grounds than some of the earlier unsuccessful devices which had included their own radioactive emitter.

By 1973 the NCB were swinging back to the belief that new methods

[1] Ibid. 1962, I, 10; 1963–4, I, 17; 1964–5, I, 9; 1965–6, I, 11; 1966–7, I, 10; 1967–8, I, 13; 1968–9, I, 8; 1969–70, I, 7.

[2] Ibid. 1971–2, I, 49.

[3] NCB, *High Technology in Coal*, 15 and 19.

in mining, including automation and remote control, would have to be developed and would be the source of much increased output per man in the medium term. But the approach was now much more cautious, step by step with each step carefully tested in operation, with much attention to fuller automatic monitoring before automatic control was attempted. This approach was much assisted by the new availability of small computers and the development of electronic sensors which could be built into various items of equipment and transmit information to the computers. The earliest comprehensive successes of this approach began to appear in 1975 with automatic monitoring of coal clearance systems, and the application of the computerized information to the control of the conveyors and the automatic operation of underground bunkers from a surface control room; and with the remote monitoring of some features of the mine environment, notably the rate of airflow and the proportion of methane. All these new methods were developed in detail by the MRDE, and the sequel in 1977 was the adoption of the set of standard computer programmes known as MINOS (mine operating systems), which by this time included an automatic control system for underground pumping as well as the programmes for coal clearance and the mine environment. Each colliery where MINOS was installed had a control room with display units on the surface and could use one programme with or without the others. Within the next two or three years the programmes were extended, with one for coal preparation and a particularly influential one, known as FIDO (face information digested on-line), which recorded and analysed production information at the face. The earlier programmes were enlarged to deal with more types of equipment, and a start was made with the storage of information so as to permit the computer analysis of trends over an extended period and help to identify the precise nature of persistent operating problems or successes.

By 1982 FIDO was operating on 82 faces at 28 collieries, the MINOS conveyor control programme at 27 collieries, ventilation monitoring at 5, and provision for the long-term storage and analysis of MINOS data had been made at 6 collieries and was in process of extension to 15 more.[1] It was still very much a minority of the coal industry to which these new techniques had been applied, but it was a minority from which most was hoped for the future. An industry which had come to rely on intensive mechanization for its most efficient production needed

[1] NCB, *High Technology in Coal*, 17; NCB *Ann. Rep.* 1973–4, 16; 1974–5, 20; 1975–6, 24; 1976–7, 20; 1977–8, 12–13 and 26; 1978–9, 25; 1979–80, 26; 1980–1, 18 and 27; 1981–2, 25–6.

to seek ways of making the most of its expensive equipment. The object of the new methods was to improve the management of subsystems in the industry by monitoring their operation and the condition of their equipment, identifying faults, and recording delays. It was by no means the only thing that needed to be done in order to optimize plant utilization, but it was a significant contribution which seemed to have much potential for further development.

### viii. The quality of the product[1]

Coal as it came out of the mine was not always in a state which made it readily saleable. It was mixed with dirt and stone in widely varying proportions, different sizes were mixed together indiscriminately, and the physical and chemical properties of particular batches of coal were not necessarily suited to the requirements of particular customers.Changes in the techniques of mining altered these conditions in ways which needed compensating action. For instance, on a national basis, in run-of-mine (ROM) coal the proportion of dirt rose from 10 per cent in 1947 to 30 per cent in 1962 and 40 per cent in 1978. The proportion of coal over 100 mm. in size declined from 31 per cent in 1947 to 13 per cent in 1962 and 8 per cent in 1978, while the proportions not exceeding 25 mm. in size were 40 per cent, 67 per cent, and 80 per cent respectively. These changes were almost wholly attributable to the growth of mechanized mining.

There was never an easy acquiescence in the sort of deterioration revealed by these figures. Indeed, there were periods of very intensive effort devoted to study of the causes of the deterioration and their possible removal. Quite apart from the policy, in the late fifties and early sixties, of increasing the number of ploughs and trepanners in order to produce more large coal, there were very many attempts to adjust details of operational practices at the face and of the design of coal-cutting machines and their components. Experimental techniques for steering coal cutters belong to the larger approaches to the problems, but more widespread were some of the simpler modifications, such as changes in the detailed design of picks and their arrangement on the shearing drum, and variations in the speed at which coalcutting machines were run. Whatever success was achieved by some of these

[1] Some of the problems of product quality, which involved prolonged research and development, concerned products made from coal, notably coke and manufactured smokeless fuels. The problems are not dealt with here but in Chap. 9.

changes, they could not do more than reduce or delay some of the adverse incidental effects of mechanized mining. They could not reverse the rising trends in the proportion of small coal and dirt.

There were various ways in which ROM coal could be made more suitable for sale and use. One was by coal preparation which involved cleaning and screening by size. Another was by adaptation of the appliances in which the coal was to be burnt. Both deserve attention, but it was the former that had the most immediate and general influence.

The NCB inherited a very neglected and non-standard stock of coal preparation plants, of which ninety-two were more than 30 years old (including one that was 66 years old) and nearly all had been allowed to run down in wartime.[1] It is not surprising that, although the proportion of dirt mined with the coal was much lower than it became later, users maintained for several years continual and justified complaints about dirty coal. In 1947 nearly 35 per cent of all deep mined coal was cleaned by hand, although it was soon established that this was neither an economic nor an effective way of dealing with coal below 100 mm. in size, and nearly 19 per cent was sold untreated. The most urgent task was to reduce hand cleaning as quickly as possible, partly by achieving a better match of coal and user so that more coal could be sold untreated, and partly by building a lot of new coal preparation plants. The proportion of untreated coal grew steadily until the mid-sixties when it reached 31 per cent. This was possible mainly because of the relative and absolute growth of the power station market. But the increasing proportion of dirt meant that this trend could not continue indefinitely, though new methods of blending and regular sampling for quality avoided the need to treat all coal in a preparation plant, especially as power stations could tolerate more ash than other users. The large decline in coal output also made it easy to maintain adequate capacity for coal preparation. By 1978 untreated coal was down to 15 per cent of the total.

The new plants in the late forties and fifties needed to apply techniques which were already known, but in some cases little used previously in Britain; to have a larger capacity than their predecessors; to be operated by trained people; and to be properly maintained. From 1947 to 1956, 200 new plants were brought into operation, and this total reached 377 by 1964. Though some small plants were built at small collieries, the average throughput per plant increased from 82 tonnes per hour in 1947 to 126 tonnes in 1956, but a few had a capacity of over

[1] A. Grounds and A. A. Hirst, 'Coal Preparation', in *Colliery Guardian*, *National Coal Board*, 1957, 64.

600 tonnes per hour and the biggest, built at Manvers Main to handle the output of four large collieries, could treat 1,340 tonnes per hour. As the proportion of very fine coal increased, changes in processes were found necessary not merely to clean it but to dry it and to ensure that it was recovered and did not go to waste on the colliery tip. In particular, by 1955 37 froth flotation plants had been included in complete new plants for this purpose and 52 more had been added to existing plants. The froth flotation process used various chemical additives to separate the fine coal from the tailings. The latter were left as 70 to 82 per cent ash, which could safely go on the tip, while the fines, having now a much reduced ash content, could be mixed back with other coal which had been subjected to cheaper preparation processes. Froth flotation was originally used only for high quality coals, but was more widely applied after high proportions of dirt and fine coal became general, particularly as untreated fines often cause problems for the handling equipment of large users. For some of the very small coal slightly larger than fines, new types of dense medium were introduced in the cleaning process to attract the shale more completely away from the coal.[1]

Hand cleaning was reduced to small proportions in the course of the sixties and ceased altogether after 1971. With the simultaneous fall in the proportion of untreated coal, the role of preparation plants became larger than ever before. It became usual to have fewer and larger plants, some of them serving several collieries. The commonest technique was to clean most of the coal by treatment in a suspension of magnetite in water, so that the coal floated and the shale sank, and to recover the coal below 0.5 mm. by froth flotation. A throughput of 500 tonnes per hour was common, and the efficiency had become high enough for the discard to contain only 2 per cent of coal on average. Coal preparation plants provided for regular sampling of their output and the analysis of the samples for ash and moisture. By 1982 fourteen plants had been equipped with the standard programme of computer control, which extended to blending to specified quality standards, as well as cleaning.[2]

Preoccupation with the character of combustion equipment was not originally a characteristic of the coal industry. Customers chose their appliances to suit themselves. It was known that certain types of user preferred certain types and sizes of coal, and the coal industry tried to supply what they would buy. The situation changed to some extent

[1] Ibid. 64–8; NCB *Ann. Rep.* 1956, I, 14–16; Dell 1976, 43–5.

[2] NCB, *High Technology in Coal*, 29.

when the industry could not supply all that was wanted of some grades of coal. It changed more drastically when customers could choose a fuel other than coal and when new methods of production caused the industry to supply fewer grades of coal and, in particular, a much greater proportion of very small coal. Then the coal industry had a strong incentive to establish and demonstrate that what it could supply was well suited to the customers' needs. A good deal of the work of the NCB's Coal Research Establishment (CRE) was devoted to the tasks arising from these conditions. Much effort went into the adaptation of the fuel itself: new types of smokeless fuel for the domestic market, new blends of coal to make metallurgical coke for the steel industry. But much also went into the design problems of domestic stoves and fires, industrial boilers, and mechanical stokers for industry. Earlier research concentrated on appliance testing for manufacturers and the development of solid fuel central heating systems. In the later sixties much research went into a project to burn bituminous coal in domestic grates with very low smoke emission.[1]

More elaborate attention was given to industrial equipment, partly because of the need to modernize colliery coalburning equipment. New types of boilers were developed to burn low grade coal and smalls. But most influential was the necessity to improve mechanical stoking and ash disposal if coal was to retain much attraction for the industrial market.[2]

The greatest concentration of effort and most original work went on fluidized bed combustion. Research began in the fifties but pilot plants could not be installed until the seventies. The method is to inject coal into a hot, churning bed of particles, such as silica sand or coal ash, which is pre-heated and made to circulate turbulently, as though it were a fluid, by the upward passage of a powerful air current. In this mixture the coal burns very efficiently and sustains the bed temperature as well as giving off usable heat. Such a boiler is less selective than most in the type of coal it can burn, it operates at a lower temperature than conventional boilers and therefore causes less internal corrosion and abrasion, it prevents the sintering of ash and therefore facilitates ash removal, and if some limestone or dolomite is included in the bed materials it can reduce the emission of sulphur dioxide into the atmosphere. The absence of moving parts in the furnace reduces maintenance tasks and costs, and it also has smaller physical dimensions than a conventional

[1] NCB, *CRE Ann. Rep.* 1971–2, 14–17 and 1973–4, 16–19.
[2] Ibid. 1976–7, 18–22 and 1977–8, 18–19.

furnace supplying the same output of heat or steam. The first commercial applications were for industrial and agricultural drying where all the heat was drawn off by the combustion gases and excess air. Other designs of boiler, in which the CRE collaborated with manufacturers, were at the prototype stage at the end of the seventies, including one installed for a district heating scheme for over 2,500 premises in Edmonton. Experiments were also in progress with fluidized combustion under pressure, which offered hopes of high efficiency in electricity generation by having combined cycles of steam and gas turbines both powered by the same combustor. The main trial was in an installation at Grimethorpe operated and managed by the NCB (through a subsidiary company) on behalf of the International Energy Agency, which was jointly funded by the governments of the UK, USA, and Federal Germany. It was commissioned in 1981.[1]

Fluidized bed combustion was a technique pioneered to improve the efficiency and convenience of coal as an industrial fuel. It had some promising successes, but it was slow in being taken up, and in the early eighties it remained uncertain how strong its influence would be.

### ix. The search for more coal

Although in 1947 everyone was satisfied that the coal industry had ample supplies of its raw material available for many decades ahead, coal exploration was an immediate necessity. Much of this was for the purpose of proving in detail the reserves of existing collieries, as this work had recently been neglected. But the industry also needed long-term reconstruction, which involved the establishment of new collieries and the sinking of additional shafts at existing collieries which might work reserves outside their previous takes. To plan this reconstruction efficiently needed much fuller and more detailed knowledge of the location and nature of additional coal reserves and of the geological and hydrological conditions in their vicinity. A large programme of boring was the way to acquire such knowledge. No comprehensive project of this kind had ever been attempted before and the lack of it was one reason why many existing collieries were awkwardly sited in relation to their coal reserves. The work of the Coal Survey and the Geological Survey had both contributed invaluably to the stock of relevant information, but it had necessarily been piecemeal. The absorption of the

[1] NCB *Ann. Rep.* 1979–80, 27; and 1981–2, 27; NCB, *Fluidised Bed Combustion* (n.d. [1979]), *passim*; NCB, *High Technology in Coal*, 33–5.

Coal Survey by the NCB gave an opportunity for some closer integration of scientific and immediate commercial interests in coal surveying.

On the other hand, there were serious limitations to the amount of relevant detail that could be accurately established. There was far less experience of exploratory drilling for coal than for oil or metalliferous minerals and it was not practicable to transfer all the methods of the other industries to coal drilling sites. Coal was usually sought at greater depths than metals but much less deep than oil. For oil the first object was to establish the existence and general features of large basins in which oil was likely to be trapped, whereas for coal the immediate practicalities called for the establishment of much more detailed geological evidence. Most boring for coal was done with a diamond bit attached to rods containing a corer. This method was rather slow and was so expensive that boreholes could be made only at fairly wide intervals, but it was considered essential in order to get the maximum amount of detail from the boring, and over a period of a few years much improvement was achieved in the recovery of complete cores intact. These were subjected to very detailed analysis. Moreover, there was progress in obtaining additional information. For instance, it became common practice to log additional information obtainable during drilling, such as the rate of penetration, because (with the bit subject to constant pressure and rotating at a constant speed) this varies with the type of rock being traversed. Thus it is possible to check more accurately the position and thickness of particular rocks, including coal seams. It was also possible to get usable information from sondes placed in open holes for such purposes as taking closely spaced samples from the sidewalls of the hole or ascertaining some of the physical characteristics of the strata. In the latter case some open boreholes, without coring, could be made, as in oil exploration, and this was cheaper and quicker.

Nevertheless, there remained difficulties from the lack of much-needed information about the precise details of geology (especially of faults) in the area between boreholes. Attempts were made to remedy this by the use of geophysical methods, such as seismic or, occasionally, gravity surveys, but the results were generally disappointing. At this time no seismic survey could detect a coalfield fault with a throw of less than about 150 feet (46 metres), and much smaller faults than that could greatly reduce the value of a coal seam.[1]

[1] For the techniques used at this time and their limitations, see G. Armstrong and A. M. Clarke, 'Exploration and exploitation in British coalfields', paper read at inter-university

One notable and fairly successful innovation was made. In 1953 contractors were engaged to start off-shore boring from the deck of a tower floated out and anchored. This was in order to establish fuller information about coal reserves under the seabed off Durham and Northumberland and in the Firth of Forth. No project of this kind for any purpose had previously been attempted in British waters, though there had been undersea exploration drilling for oil in some parts of the world.[1] The NCB were anxious to make this experiment because the reserves under the sea were needed as replacements for worked-out inland pits, especially in Durham, and because previous exploration had had to be done from established collieries near the coast and had proved both inadequate and misleading. It took some time to get this project into operation but drilling began in 1955 and continued for ten years. During that time 15,500 metres of drilling were completed from 22 offshore locations.[2]

Despite all the problems and limitations, exploratory drilling increased rapidly from 1947 to a maximum of 83,000 metres in 1955. It remained high in 1956 and 1957, but thereafter was greatly reduced as the coal industry moved from expansion to contraction. Much of the information acquired from drilling was used to guide the way in which existing collieries developed, but there was an important influence on the siting of new collieries. For example, it was after borings had proved good quality thick seams nearer the surface than had been expected that it was decided in 1951 to sink a major new colliery, Lea Hall, near Rugeley in south Staffordshire. Other information was stored up for future development. Thus it was shown that reserves were substantial and generally workable in the eastward extension of the Yorkshire and East Midlands field, that coal continued south-east of Nottingham, where its presence had been doubted, and that the North Staffordshire field could be worked further south.[3]

Exploratory boring was very neglected in the sixties, and in 1963–4 and 1966–7 there was less than at any other time since nationalization. The original programme of offshore borings was kept going to completion in 1965–6 and some desultory attention was still given to exceptional prospects. Borings in the fifties had for the first time revealed

congress at the University of Newcastle upon Tyne, 1965, 2–11; NCB *Ann. Rep.* 1948, 20; 1951, 17–18; 1953, 39; 1956, I, 22–3.

[1] NCB *Ann. Rep.* 1953, 38–9.

[2] *Colliery Guardian*, vol. 231, no. 4 (Apr. 1983), 208.

[3] NCB *Ann. Reps.*, especially 1951, 18; 1956, I, 22; 1957, I, 22.

workable coal seams to the east of the northern edge of the Yorkshire field and had led to the sinking of Kellingley Colliery, the last major new project started before continuous contraction was recognized. In the belief that there might be a much greater extension to be found, five boreholes were drilled in 1964 north, east, and south of Selby. The two to the north found the Barnsley seam in thicknesses of over 2 metres at depths of not more than 850 metres, which was greater abundance than anyone could have expected. But, in the depressed conditions of the time, no detailed exploration followed until 1972.[1]

There was a gradual revival in drilling, and when the competitive position of coal changed very sharply in relation to oil a large new exploration programme was set in train. In 1973–4 the length of boreholes drilled exceeded the maximum of the fifties, and in 1976–7 a new maximum was reached, with about four times as much drilling as in 1954 or 1955. Thereafter there was some reduction, but the amount continued very high into the eighties.

In this phase of exploration the techniques used were much superior to those available twenty years earlier. The NCB got together with the drilling contractors after specifying a set of requirements and together they were able to overcome most of the obstacles which had previously prevented the full application to coal exploration of principles employed in other mineral exploration. One limited change was that offshore boring for coal was done through the bottom of specially designed ships instead of from towers that had to be floated out to each site. On land, drilling was done much more quickly than before. Open boreholes were drilled to the top of the coal measures in much the same way as in the oil industry, with satisfactory cores taken below that level as a result of improved choice of drilling bits and stricter control of the drilling mud. Better logging techniques, mainly designed by British Plaster Board Industries for the NCB, made it possible to obtain a greater variety of information from the sondes placed in the boreholes and record it digitally on magnetic tape for computer processing. Though there were some instances where satisfactory data were not obtained, there was a further great gain from a general success in developing a system of seismic surveying adapted to the needs of coal exploration and using much more sensitive geophones than before. All the information obtained in this way was fed into computers and related to the information obtained from boreholes and from present and past mine plans. It was also practicable to make underground seismic surveys

[1] Various Authors 1976, 98.

by firing small charges in horizontal holes and obtaining information about geological conditions up to 500 metres away. A further improvement came in 1981 when three-dimensional seismic surveying was first applied in the coal industry. Marine seismic surveys were also carried out, from a ship towing a line of hydrophones above an electric discharge. After smaller experiments for a year or two, seismic surveys began to be carried out regularly and on a substantial scale from 1974 onwards, and a very large programme of marine seismic surveying was undertaken in 1979–80 and the following year.[1]

The use made of these improved techniques is illustrated by the activities of 1976–7 when 144 boreholes over 2,000 feet (610 metres) deep, 41 betwen 305 and 610 metres, and 5 offshore holes were drilled and 390 line-km. of seismic survey were run. In this year exploration was completed both in the Selby field and the more recent discovery in north-east Leicestershire.[2] The exploration at this time was concerned with new mine prospects to a much greater degree than ever before: of 109 boreholes of more than 300 metres drilled in 1979–80 only 25 were for existing mines. This distribution of effort was partly to compensate for the lack of exploration in the sixties, partly to restore the known level of recoverable reserves (for which a new and more cautious definition was officially adopted in 1977), but most particularly to prepare for an expected increase in demand and output which could be met both in quantity and in cost terms only from new collieries, and to establish locations for these in the most prudent way.

The choice of new reserves for development proceeded through six stages. Potential prospects were first noted from prior evidence. Some of these were then given preliminary examination by exploration to determine the limits of the coal. The third or investigative stage was to determine the geological and chemical features that could affect the mining of the coal. The fourth was a feasibility study to determine possible sites for mining access and to assess potential output levels and the economics of extraction. All these four stages involved exploration in different degrees and different ways. Only prospects that passed all these reached the fifth stage of planning (including the preparation of a planning application), and the sixth stage of development if the planning application succeeded. In the late seventies projects for new mines at Royston, Betws, Kinsley, and in the Selby complex, and for what was

[1] *Coal's new burning challenge* (articles reprinted from *Achievement* magazine in book form, n.d. [1981?], 19–25; NCB, *High Technology in Coal*, 3–5; NCB *Ann. Rep.* 1975–6, 26; 1981–2, 12.

[2] NCB *Ann. Rep.* 1976–7, 13.

effectively a new mine absorbing and replacing Thorne Colliery, which had been closed for over 20 years, went through all six stages (though the Thorne project went into suspense because of low demand for coal). But inevitably most exploration was concentrated in the preliminary stages (because some prospects got no further) and the feasibility studies, because that was where it was needed most intensively. Sometimes the investigative and feasibility explorations indicated that reserves might best be worked, if they were worked at all, from an existing colliery, but mostly they pointed to an entirely new site. Deep boreholes were, however, drilled at existing collieries with a prospective long life, in order to guide the planning of future developments there.

By the early eighties, besides the North East Leicestershire Prospect which had reached the fifth stage but only in a reduced form the sixth, the feasibility stage had been reached in South Warwickshire, at Witham (between Newark and Lincoln), in East Yorkshire (where since 1977 substantial reserves had been proved near Snaith), and in the Hirst Basin, west of Longannet. Prospects which had been carried to this stage but for technical or commercial reasons were not currently being pursued were at Park (near Stafford), Margam in South Wales, and under the Firth of Forth near Musselburgh. Areas where coalmining had not been known and where at least preliminary exploration had found something included two districts of Oxfordshire (round Witney and Banbury), Till near Gainsborough, and the area north of York; and another area was the Canonbie field, straddling the Scottish–English border, which had never been worked on more than a very small scale and not at all for many years.[1]

It was a cautious approach, but it used an improved technology for exploration very intensively. By 1977 it had demonstrated the probable availability of well over 1,000 million tonnes of coal for new mines and the amount went on increasing. It is probable that never before had such a large and widely dispersed array of new prospects been demonstrated in so short a period. But there was a host of other factors determining whether, or how soon, exploration would lead to production.

## x. Technology and the characteristics of the coal industry

The most obvious general influence of the whole range of technological

[1] NCB *Ann. Rep.* 1979–80, 19–20; 1980–1, 20; 1981–2, 12–13. For activities at some existing collieries see NCB, *New mines for old* (n.d. [1979]). The prospect at Margam was the subject of renewed attention, with a possible simpler and cheaper scheme of development in view.

changes was that by the seventies the coal industry looked much more like a science-based, research-based industry, following rationally tested and standardized practices as far as it could, than it had done thirty years earlier. Yet that state of affairs had come about erratically, with many delays and interruptions. It was by no means clear that technological advances had been applied when they would have been physically and commercially most advantageous. In particular, when demand for coal was rising and the absolute level of demand was difficult to satisfy, the industry appeared slow in innovation. It may well be asked why this should have been so.

There were various contributory factors and their relative strength can be only a matter for speculation. One might be summed up as the inadaptability of old age. Britain had, and retained, a high proportion of old collieries. Although a lot of old capacity was closed, especially in the sixties, even in 1980 well over a quarter of the operating collieries had been sunk before 1880, and the hundred-year-old collieries exceeded in number those established since nationalization. The Reid Committee was alive to the handicap and declared that a regular programme of new sinkings would help to keep the industry abreast of the latest developments in technique and that 'no useful purpose is to be served by planning an unduly long life for a mine'.[1] But longevity remained common for collieries, though many old collieries were eventually drastically reconstructed. While the detailed underground layout of collieries was continually changing, the surface buildings, the shafts, the position and arrangement of the pit bottom, the direction and dimensions of the main roads, and the location of old workings all remained as permanent constraints. It was far harder to introduce new equipment into old structures than into those designed with the new equipment in mind. In the long run it was practicable to establish mechanized coal faces in nearly all mines that it was physically possible to keep in production, but many of the most sophisticated elements of high technology were never easily transferable to old mines that had not been thoroughly reconstructed. Old structures also encouraged the retention of old ideas and practices, and resistance to new, among both managers and miners. It is quite clear that in the earlier years of nationalization, several members of the national board believed that the slowness in getting greater mechanization at the face was due in part to widespread lack of interest among Area and colliery managers. It is, however, uncertain whether the evidence was sufficient to support this belief. A more encouraging

[1] Reid Report 1945, 123.

environment for innovation depended on the necessarily protracted physical reconstruction of collieries and on retirement, recruitment, and new training.

Another significant feature was the slow progress in the systematic organization of research and development, perhaps especially development. Under private ownership the coal industry had devoted only small amounts of money and effort to research, despite a better example from one or two firms, and the equipment manufacturers tended to rely on the inspiration of individuals for occasional important innovation rather than pursue it in a regular organized way. The NCB sought to do better. From the beginning the national board had a scientific member with his own department, though much of the department's work was in relation to day-to-day operations. In 1948 the NCB set up their own central research establishment at Stoke Orchard, near Cheltenham, but this had to spread itself very thinly over the whole technical range of the industry and in its early days probably gave more attention to coal quality and the mine environment than to the mechanical problems of mining.[1] By 1951 it was clear that it could not cope with all the demands upon it and in 1952 a separate Mining Research Establishment was set up at Isleworth to deal with underground mining problems and the application of mechanical and electrical engineering to the coal industry, and Stoke Orchard became the Coal Research Establishment. Initially the MRE was only half the size of the CRE, though it was intended from the start to build it up to similar size.[2] But even the new establishment did not meet pressing needs for the fruitful *application* of research. At the end of 1952 E. H. (later Sir Humphrey) Browne, the NCB's Director-General of Production, told the national board that more and more ideas were coming forward from a variety of sources, including the research establishments, and that 'if these ideas were not to go for nothing, the Board must themselves provide the resources to develop them, which the manufacturers could not'. He therefore proposed the setting up of a Central Engineering Establishment.[3] The board agreed in principle, but it was three years before the new establishment was set up, at Bretby in the South Derbyshire and Leicestershire coalfield, and 1956 before it came fully into operation.

This was a most influential change. The CEE came directly under the

[1] NCB *Ann. Rep.* 1948, 89–90; 1950, 41–2.

[2] Ibid. 1952, 46–8. The work of both establishments is discussed in R. W. Idris Jones, 'Research', in *Colliery Guardian*, *National Coal Board*, 1957, 83–90.

[3] NCB 358th meeting, min. 171, 5.12.52.

Production Department and from 1957 it was provided with testing facilities at Swadlincote Colliery. It tackled specific problems presented by those in the industry; it designed and developed machines to prototype stage, but if it then wanted to take them further it called in equipment manufacturers to share in developing them to the production stage; and it provided a testing ground for materials, components, and machines from their manufacturers, with whom it discussed modifications to make them more suited to users' needs.[1] Thus both user and manufacturers were made much more development-minded and kept in more continuous touch with each other's needs and problems. Measures to induce closer collaboration and better communication among users, researchers, designers, developers, and manufacturers created a more favourable environment for speedy and successful innovation than had ever existed before.

Once these conditions had been created there was a general wish to maintain and, if possible, strengthen them. For a long time any enhancement came from the informal effects of growing familiarity in collaboration. But in the longer run there were organizational changes which pushed in the same direction. From the late sixties there were a number of mergers and takeovers among mining equipment manufacturers which created some firms with access to larger resources for research and development. The NCB in 1969 closed their Isleworth establishment and merged it at Bretby with the CEE. The combined undertaking was named the Mining Research and Development Establishment. In 1974 it was reorganized to give it a still stronger practical influence. It received more staff and equipment and its responsibilities were extended to include the exploitation of research and development to the point at which they can be applied throughout the coal industry, and the training of management and men in new technologies and methods. Its work, which was given more emphasis on the remote control of existing mining methods, was divided into project groups, each supervised by a Major Development Committee chaired by an Area Director.[2]

The same sort of trend can be seen in the use made of operational research. A Field Investigation Group was set up by the NCB in 1948, but it had at first a staff of only six and its studies, which had to be highly specific, were necessarily few. But it gradually increased in size to become a full Operational Research Branch (and, later, Executive) with a

[1] NCB *Ann. Rep.* 1955, I, 20; 1956, I, 55–7.

[2] Ibid. 1969–70, I, 10–11; 1974–5, 20.

much wider range of enquiries, especially in relation to business management and planning, though it did not give up its enquiries into some detailed technical questions. It was, however, not until 1968–9 that it first offered a direct service to Areas so that there was then much more use of operational research at collieries and in Area management. The practice was adopted of each field of work having to have a sponsor, who typically might be the Director-General of a headquarters department or the Deputy Director of a deep mining Area. The stress, that is, came to be on having the user in a position of both authority and consultation to ensure that effort was concentrated on the problems which arose as part of the work for which he was responsible.[1]

Experience and organization were thus by the later sixties and seventies encouraging more research, and its development to production, in a way that was mostly lacking before the mid-fifties. It is also probable that some of the lines of development had their own momentum, speeding up after initial uncertainties and problems had been overcome. The sheer increase in research investigations helped, too. From a bigger pool of starters there was likely to be some absolute increase in the number of potential winners and a better basis of comparison to enable the earlier identification and abandonment of potential failures. So the slow build-up and then the sustaining of the influence of technological innovation are not surprising. It remains to be asked how much and in what way this influence altered the nature of the coal industry.

There was some reduction in the labour-intensity of the industry, because much of the innovation was to get machines to do in part what had once been done almost wholly by hand. Differences of definition make long-term comparisons slightly imprecise, but if labour costs are taken as wages plus wages charges (i.e. such items as national insurance, sick and injury pay, and holiday pay), then in 1947 they were 68 per cent of colliery costs, or 70 per cent if salaries are also included.[2] In 1981–2 they were 48 per cent on a basis as near as practicable to the former, and 51 per cent to the latter, but if certain 'social costs', covered by specific government grants, are excluded, these figures come down to 46 per cent and 49 per cent respectively.[3] There is some case for saying that a

[1] NCB *Ann. Rep.* 1948, 88; 1968–9, I, 30; Tomlinson (ed.) 1971, *passim*; R. C. Tomlinson, 'The Work of the Operational Research Executive', in *Colliery Guardian*, Jan. 1977 (article no. 13 in series *Structure of the National Coal Board*).

[2] Calculated from NCB *Ann. Rep.* 1947, 164.

[3] Calculated from MMC, *National Coal Board: a report on efficiency and costs* (1983), Cmnd. 8920, I, 31. These figures give the best comparison because they relate only to deep mining. Figures on

more realistic definition of labour costs would include some charge for overheads and services, and in that event all the figures would be higher, in 1981–2 by up to 8 percentage points, in 1947 almost certainly by much less. The fall in the proportion of labour to total costs is perhaps lower than might have been expected in view of the great emphasis on mechanization, but it took place against the counter-influence of a long-term rise in the price of labour relative to that of capital. Some of the effect of the greater input of capital is illustrated by the rise in labour productivity in physical terms. Here again problems of definition slightly distort long-term comparisons, but, on the definitions used at the time, overall OMS was 1.09 tonnes in 1947 and 2.40 tonnes in 1981–2, with a smaller rate of increase in output per man-year, which was 267 tonnes in 1947 and 497 tonnes in 1981–2. In both cases most of the increase was achieved by 1970–1.[1] This, too, might seem less remarkable than the descriptive evidence would lead one to expect.[2] It is an indication that the technological innovations were not all taken up comprehensively, especially the sophisticated electronics towards the end of the period, and it also indicates that the innovations were much less effective in some conditions than in others. Overall OMS in 1981–2 at new mines was almost double the national average.[3]

To a large extent the limitations and the contrasts arise from the tension between the new technology and the conditions in which it has to be applied. A modern coalmine, using the most advanced technology available, with a face being worked quickly to produce a regular large daily output, and coal moving all the way on conveyors from the face to the surface, looks much more like the continuous flow production industries familiar for decades in many types of manufacturing. In a sense, this is clearly the character towards which the coal industry was aspiring and moving. But it is a character that could not be wholly achieved. Quite apart from the hard physical conditions which must sometimes damage machinery and overstrain

cost structure in NCB *Ann. Reps.* relate to all mining and are therefore affected by changes in the proportion of opencast mining, in which labour inputs are low.

[1] NCB *Ann. Rep.* 1981–2, 32–3.

[2] Harlow 1977 has a useful chapter on 'Mechanised Mining and the Growth of Labour Productivity' (175–218), which concentrates almost entirely on coal face operations. It argues (213–18) that the favourable effect of face mechanization on labour productivity was restricted by the survival of small and badly laid out collieries, by failure to achieve comparable improvements in labour efficiency in underground work away from the face, and by inherent limits to the potential of machinery as a means of raising productivity at the face.

[3] NCB *Ann. Rep.* 1981–2, 12.

men, the operating conditions (unlike those in manufacturing) do not remain uniform and cannot be regulated into something near uniformity. The difference shows itself in the time when the running of machinery is interrupted in a coalmine and is not in an assembly line manufacturing industry. It also accounts for some of the continuing large need for labour in mining.

Technology could not change the geology which still determines the physical conditions in which coalmining takes place. The most it could achieve was to promote more continuously effective counteraction to the variable conditions and the random severe physical obstacles that threatened the continuity of production which had become otherwise attainable. This was what was being attempted from the mid-seventies.[1] To make equipment more robust so that it would be less likely to break down under the strain of conditions temporarily more severe than normal, and develop fewer minor faults under normal conditions, was one contribution and led to some of the potentially most productive mines being equipped with heavy duty faces. To gain fuller advance knowledge of geological variations, for example by seismic surveying, before operations reached them was another. To monitor operations automatically and to switch in automatic adjustments to recorded changes was a third; and the automatic monitoring of machine health so that components could be replaced before they failed during operations served the same purpose.

Only towards 1980 was comprehensive innovation of this sort available. It was costly and therefore appropriate only for large collieries with an expected long period of operation, and it needed to make a large improvement to the average rate of output if it was to justify its cost. In the early eighties, therefore, the technology of the coal industry was again in a somewhat transitional phase. A general technology directed towards continuity and speeding up of production through mechanization had been widely adopted, but the hostility of the environment had been only partly overcome and the continuity only very imperfectly achieved. Newer technology, intended to reduce these imperfections, had been introduced in only a minority of the industry and was still in process of proving its value.

[1] The thinking behind the choices then made in technological development is explained in N. Siddall, 'What R. & D. will contribute to coal's future', *Colliery Guardian*, Oct. 1975, 439–42.

PART B

# HISTORICAL PHASES

CHAPTER 4

# Launching the Nationalized Industry

Some of the important decisions about the running of the coal industry were taken even before nationalization had been enacted. The names of the members of the National Coal Board were announced in March 1946, with the deputy chairman's name coming a little later than the others because of the need to release him from his duties in the administration of occupied Germany. At this time the bill was still in progress through Parliament. It did not receive the Royal Assent until 12 July and the members of the board were formally appointed on 15 July. But they were already settled into their work. They met first on 17 April 1946, and altogether met six times before their formal appointment.[1]

All the members were full-time from the date of the appointment of the board and they were very much an executive body. The legislation merely required that the board should consist of a chairman and eight members, with one of the members appointed by the Minister as deputy chairman.[2] But the Minister decided that, except for the chairman and deputy chairman, each member should have a specific functional responsibility, two for production and one each for finance, manpower and welfare, labour relations, marketing, and science. There was some later criticism that the board members were insufficiently experienced in large-scale business management, but nobody had had to run a business of the scale and character of the NCB's and, in fact, the members brought together a long and varied experience of different aspects of industry, much of it directly concerned with coal. A more reasonable question might have been about the extent to which their talents were complementary and compatible.[3]

[1] The minutes show that they first described themselves as 'National Coal Board Designate', but after the first meeting they settled for 'National Coal Board Organising Committee'.

[2] CINA s. 2 (2) and s. 2 (5).

[3] Biographical details for the board members who are discussed in the following paragraphs can be found in the relevant volumes of *Who Was Who*, except for Lord Citrine and Sir Geoffrey Vickers, whose deaths are too recent but who are recorded in *Who's Who 1981*. There is a summary in Haynes 1953, 117–20. The present writer has had the benefit of conversations with some who worked with the first members of the board.

Lord Hyndley, the chairman, was 63 years old in 1946. His appointment was generally acclaimed, though a few complained that he knew much more about marketing than about production. He straddled perfectly the two worlds of colliery business and government control, and his dual experience was continually applied with telling effect in his new role. He had been a managing director, since 1931, of the largest colliery firm, Powell Duffryn; he had undertaken government service in relation to coal in both World Wars and in the Second, as Controller-General in the Ministry of Fuel and Power in 1942 and 1943, he had administered the wartime coal controls; and he had been Commercial Adviser to the Mines Department from 1918 to 1938. His colleagues recognized that he was first-rate in the conduct of business at meetings and he was skilled in reaching satisfactory arrangements with politicians. He took trouble with personal relations and was both respected and liked.

The deputy chairman, Sir Arthur Street, who, in practice, directed the administration of the NCB, was one of the ablest civil servants of his generation. He had come up to London from the Isle of Wight in 1907, at the age of 15, to enter the civil service as a boy clerk and had worked his way up from there. He was experienced in the work of public corporations as instruments of economic policy, for he had been involved in the agricultural marketing boards and the British Overseas Airways Corporation. He had been Permanent Under-Secretary of State at the Air Ministry from 1939 to 1945 and in 1945 he became Permanent Secretary of the Control Office for Germany and Austria. He was high principled, broad minded, and conscientious almost to a fault. Although he was very concerned to establish and maintain rational and economical administration, he had just as much interest in running the coal industry humanely and improving the lot of all who were engaged in it. Almost certainly he delegated too little, even though, to a large extent, he did the job of secretary as well as deputy chairman, as the NCB did not succeed in appointing a secretary until late in 1947. Yet he took trouble to know all the senior staff and to judge who should be kept in check and who should be allowed wide scope, and he was a chosen arbiter in disagreements who gave clear rulings without arousing resentment. In the end he (perhaps literally) worked himself to death. But despite that early death (in 1951), his was the greatest single personal influence in making the NCB a stable and orderly institution, organized on basic principles which survived throughout the period of this history.[1]

[1] 'Arthur Street', *Public Administration* (winter 1951), 303–15, an obituary article written anonymously by one of his private secretaries, adds to our knowledge and understanding of Street.

The board members were a most varied collection of individuals, probably too individual ever to form an ideal group. Sir Charles Reid, born in 1879, was the oldest member of the board and one of the two production members. He was an outstanding mining engineer and until 1942 had been manager and a director of the Fife Coal Co., a firm which operated in varied and sometimes difficult conditions and which was regarded as a model of efficiency. Since 1942 he had been Production Director of the Ministry of Fuel and Power and had chaired the committee which brilliantly analysed the technical problems of the coal industry and pointed the way to specific drastic reforms. He had clear ideas on the decisive role of management and was impatient of prolonged discussion and of a preference for edging round difficulties instead of tackling them head-on, when it seemed clear to him what needed to be done and to be done quickly. There was much in NCB procedure that he found continually frustrating. His production colleague, T. E. B. (later Sir Eric) Young, twelve years his junior and not matched to him in personality, was also a mining engineer who was managing director of a successful firm, the Bolsover Colliery Co., which operated in less varied and generally rather easier conditions than the Fife Coal Co. He, too, had served since 1942 in the production directorate of the Ministry of Fuel and Power. Though dissimilar in personality, Reid and Young appear to have co-operated easily.

The support of the mineworkers had been deliberately sought by the appointment of prominent trade union leaders to deal with all matters concerning labour, but the two chosen contrasted sharply in style. Sir Walter (later Lord) Citrine, the member for manpower and welfare, had been general secretary of the TUC for the preceding twenty years and both then and later he showed great ability as a policy-maker and an administrator. He never undervalued his talents and sharply resisted anything he saw as an encroachment on his own wide interpretation of his sphere within the NCB. With good reason, he found the boundary between his department and that of industrial relations rather odd and probably believed he should have had responsibility for both. Other business sometimes kept him from board meetings and he had a habit of rather sharply questioning what had been done in his absence. It made things easier in some respects that he departed as early as May 1947 to become chairman of the British Electricity Authority. He was replaced by Sir Joseph Hallsworth, a much less dominant trade unionist.

The member for industrial relations, Ebby Edwards, was a different sort of man from Citrine. He knew very well the objectives of labour policy which he believed were necessary for the good of the mineworkers and the industry, and could express them clearly and persuasively and pursue them persistently. He was fair-minded and full of common sense. He showed no interest in administrative procedures and administrative detail, to the frequent exasperation of his colleagues. He was regarded by a vast number of people (including the occasionally exasperated colleagues) with great affection and respect. His trade union credentials were impeccable. He had started work in the mines in 1896, at the age of 12, rose to trade union leadership in his native Northumberland, became for a time MP for the mining constituency of Morpeth, was elected to the executive committee of the MFGB in 1928 and became its vice-president in 1930 and president in 1931. He was general secretary of the MFGB from 1932 until it was replaced on 1 January 1945 by the NUM, of which he continued as general secretary until he was appointed to the NCB. He was one of the notable characters of the twentieth-century coal industry.[1]

The other board members were less widely known to the public. The finance member, L. H. H. (later Sir Lionel) Lowe, came from the leading accounting firm of Thomson, McLintock, where he had been in charge of all the firm's work for colliery companies. He had wartime experience as a finance director in both the Ministry of Food and the Ministry of Fuel and Power. Of all the original members of the board he was probably the least pushing. J. C. Gridley, the marketing member, was another Powell Duffryn man and, indeed, special arrangements had to be made for him to be seconded by the firm to the Ministry of Fuel and Power until the NCB could be legally constituted.[2] At 42 he was the youngest member of the board. He showed an immediate grasp of the formidable problems of coal distribution, but in circumstances where coal was marketed only within an allocation policy for which the Ministry, not the NCB, was responsible, he may have found his scope restricted. Sir Charles Ellis, scientific member, was Professor of Physics at King's College, London and had served for thirteen years on the Advisory Council on Scientific Research and Technical Development to the Ministry of Supply and its predecessors. Though he came as an outsider, he was well equipped to develop scientific control and research in

[1] There is a brief appreciative biographical note of Ebby Edwards in NCB *Ann. Rep.* 1953, 1–2. His career as a trade unionist can be followed more fully in R. Page Arnot's historical volumes on the MFGB, esp. the third, *The Miners in Crisis and War*, 1961, and the fourth, *The Miners: One Union, One Industry*, 1979.

[2] NCB Designate, 1st meeting, min. 9, 17.4.46.

an industry that had neglected them; but he had perhaps less to contribute to general policy making than he sought to offer.[1]

Though the board formally had a chairman and eight other members, they operated, in effect, with one more. In May 1946 the 'Organising Committee', as the prospective board members called themselves until their formal institution, appointed Sir Geoffrey Vickers to be the board's legal adviser and to head a legal department which he would have to set up. He was a member of a leading firm of City solicitors, Slaughter and May, who had some of the larger colliery companies among their clients; he was a part-time member of the London Passenger Transport Board; and during the Second World War he was in charge of economic intelligence as a deputy director at the Ministry of Economic Warfare. From the very beginning the NCB had a formidable amount of intricate legal work, and Vickers, unlike all the other salaried officers, attended all board meetings on the same terms as members. He had a lifelong interest in administrative theory and what might be called 'the philosophy of management', and took part in the board's discussions of all aspects of policy. In 1948 he was appointed a board member.

It is also appropriate to recall that, besides Vickers, there were in a few of the senior salaried posts, men who were a good deal younger than he and who, by ability, personality, and long service, were able to exert a powerful influence. In particular, Joseph (later Sir Joseph) Latham, the first Director-General of Finance, and E. H. (later Sir Humphrey) Browne, who was the first Production Director for the North Western Division but moved to headquarters in 1948 as Director-General of Production, were both brought in from one of the best-run large firms, Manchester Collieries, and each in turn rose to be NCB deputy chairman; and C. A. Roberts, who was recruited as assistant secretary in 1946, served as secretary of the board from 1951 until in 1960 he became a board member for seven years. Latham joined the NCB in his early forties, Browne and Roberts in their thirties.

The creation of the board obviously had an effect on the personnel as well as the outlook of the trade unions in the coal industry. There was a good deal of political encouragement to them to let some of their officers take labour relations posts with the NCB and, on the whole, the leading trade union figures, who had immense goodwill towards the setting up of the NCB, responded sympathetically. There were some

[1] K. Hutchison, J. A. Gray, and H. Massey, 'Charles Drummond Ellis 1895–1980', *Biographical Memoirs of Fellows of the Royal Society*, vol. 27 (Nov. 1981), deals (217–23) with Ellis's period on the NCB.

who took the view that, nationalized or not, bosses were bosses and to join them was a betrayal; and they bitterly criticized those who took NCB posts.[1] There was also a widespread belief that in the best interests of the miners many of their leaders must stay in their present positions. But the NUM assisted the board to recruit from trade union ranks. Indeed, Ebby Edwards went to the NCB, despite his own reluctance, as a result of a decision of the NUM executive, made in response to a government appeal,[2] and several appointments at lower levels were made from lists of people put forward by the union.

Ebby Edwards had been a key figure for a long time in union affairs and his replacement was a matter of importance. His successor, Arthur Horner, served as general secretary of the NUM from 1946 to 1959 and was at least as influential as his predecessor. He remained politically active in the Communist party and had some large blind spots about international politics wherever the USSR was involved.[3] But he always put the miners and the coal industry first and his political preferences second. He combined intellectual power with immense charm, and together they made him a very persuasive advocate. He was the more successful because he could be relied on to stick to what he had agreed and he was trusted by both his colleagues and the NCB. In 1947, when Citrine left the NCB for the British Electricity Authority, Horner was offered his place on the NCB but refused it, despite Citrine's own urging.[4] He proved right in thinking he could accomplish as much or more where he was, as well as staying more content. He made an honest judgement on the way he did his job when he wrote:

> We would have been fools, and indeed would have betrayed the interests of the men who elected us if we had not insisted on getting the best possible terms we could while our bargaining power was at its height. But we never used that power irresponsibly . . . We always related our demands to the needs of our men and the needs of our industry, but also against the background of the wider needs of the nation.[5]

[1] This general attitude is illustrated in Moffat 1965, esp. 86–94 and 276–8. Moffat, who was President of the NUM (Scottish Area) wrote (89): 'Appointments of this kind seemed to change the whole outlook of a considerable number of labour officers; hene their isolation from the trade unions.' But, though he refused to let himself be proposed for NCB appointment, he agreed to Ebby Edwards accepting appointment, and he did not resist the decision of the Scottish Miners' Union to put forward names for some appointments.

[2] Arnot 1979, 188–9.

[3] Consider, for example, the apologetics in his autobiography, Horner 1960, 212–20.

[4] Horner 1960, 182.

[5] Ibid., 191.

The miners did not always respond in the same spirit to his efforts, but his own fair dealing was unquestioned.

The NUM, indeed, entered the period of nationalization with some notable men at their head. The president was Will (afterwards Sir William) Lawther from Durham, and he had a lot of experience of local and national politics as well as union affairs. He had been a member of the Labour Party Executive Committee from 1923 to 1936, MP for Barnard Castle from 1929 to 1931, and a member of Durham County Council from 1925 to 1929. He became vice-president of the MFGB in 1934 and remained in that office until, after a short spell as acting president, he was elected president in 1939. From then on he was president of the MFGB and its successor, the NUM, until 1953. His long experience gave him considerable sway in the development of union policy and he sought to exert this in a balanced and responsible way, though that caused him to be accused by some of his colleagues of being too right wing.[1]

The vice-president, James (later Sir James) Bowman, established himself as one of the most influential men in the coal industry. He had been secretary of the Northumberland Miners since 1935, and in 1939, at the age of 41, he became the youngest vice-president in the history of the MFGB, and he continued after the formation of the NUM as both Northumberland secretary and national vice-president.[2] He moved into management in 1950 when he was appointed chairman of the Northern Division of the NCB, and he became deputy chairman of the national board in 1955 and chairman from 1956 to 1961. Down to 1950 he was a powerful trade union figure, both within the NUM and more widely as a member of the general council of the TUC. His qualities of patience and firmness, and his combination of constitutional precision with the political ability to mobilize support for his favoured policy were often shown. They are well illustrated by his handling of one piece of local business, the gradual closure of Newbiggin colliery and the transfer of its men to nearby Lynemouth colliery in 1949. This was a scheme to which the Newbiggin branch of the NUM suddenly objected several months after the men had apparently accepted it. Bowman gave the Newbiggin branch all the opportunities it could reasonably seek: a deputation to the Northumberland NUM Area Executive, a transmission of its complaints to the NCB divisional chairman, and a special

[1] See, for example, the attacks in Moffat 1965, 98–9, 116–19, and 266–8. There are brief biographical details in Arnot 1961, 129–30.

[2] Arnot 1961, 284.

NUM Area Council meeting. But, before that meeting, he drafted detailed conclusions which the Area Executive agreed to and which included the statement: 'The Newbiggin men will be employed in richer seams at more pleasant work, enjoy greater security and will continue to live in the same houses, shop at the same store, use the same Pub or worship in the same Church.' And, armed with those conclusions, he had no difficulty in getting the representatives of the other branches on the Area Council to defeat the Newbiggin branch by fifty-six to one.[1] He was clearly a man skilled and forceful in getting his own way, but still seen as acting fairly, and he had a wider appeal within the NUM because at this time he was regarded as also having views acceptable to the left wing.[2]

Political responsibility for the introduction and oversight of nationalization lay with the Minister of Fuel and Power, Emanuel (later Lord) Shinwell, and the Parliamentary Secretary, Hugh Gaitskell, a very ill-matched pair. Shinwell had had a long and robust career in the service of the labour movement and this gave him a good standing in his dealings with the miners. He could and did consider shrewdly some of the large and difficult questions brought to his notice, provided they came to him one by one, and he was ready to stir others vigorously to do something about them. But he showed little capacity for the coherent treatment of the problems of an industry as a whole. Gaitskell, who had a very difficult time with him, noted of him in 1947 that 'he has no conception at all of either organisation or planning or following up . . . He will always try and evade an unpopular decision, procrastinate, find a way round, etc.'[3] Shinwell made in 1946 and 1947 some rash public statements about the state of the coal industry and was much criticized. Late in 1947 he was shifted to become Secretary of State for War and, later, Minister of Defence.

Gaitskell, who succeeded Shinwell, never had his political advantage of popularity with the miners, but gained a great reputation with the NCB members and their staff. He was anxious to be fully informed of the whole range of coal industry problems, to hear the detailed pros and cons of alternative proposals for tackling them, and to make up his mind on the merits of the arguments. Officials were astonished to find that he

[1] NUM Northumberland Area, *Report on Special Council Meeting on Saturday, 24th September, 1949*. The quotation is from p. 22.

[2] Moffat 1965, 274. Moffat regarded Bowman as showing a change in 1948 when he supported the TUC line on wage restraint.

[3] Williams 1979, 135.

did not use the claims of political necessity to override their proposals if they could persuade him that the administrative or commercial or technical arguments in their favour were stronger than those against. This was perhaps a political weakness in a politician and, indeed, Sir Donald Fergusson, Permanent Secretary of the Ministry of Fuel and Power, privately expressed the view that Gaitskell behaved more like a civil servant than a politician, and disapproved.[1] But, while there were benefits from having nationalization brought in by a populist politician, by 1948 it was desirable to have governmental oversight exercised by people who, within the political context, could see things in a more businesslike way and evaluate the practicalities of what was attempted and proposed within the coal industry. This change of viewpoint and widening of understanding was provided by Gaitskell and the new Parliamentary Secretary, Alfred (later Lord) Robens.

The suitability of the men in authority was, in fact, conditional as much on the nature of the tasks confronting them as on their own personalities. A wave of political euphoria was not adequate for more than the pushing forward of legislation and the infusion of a little extra effort that comes from goodwill. The tasks to be accomplished were formidable and called for leadership, based on judgement and skill exercised in a great variety of ways. For everybody concerned the main objective had to be to get more coal produced and, if possible, produced more cheaply; and, first, as the most immediate means of securing this, to attract more men to work in the mines and to retain them there, and second, necessarily over a rather longer period, to co-operate in the reorganization and re-equipment of the collieries. But there were preliminary steps which different groups of interested people thought it so urgent to take that the central objectives were in danger of being pushed into the background, at least for a good many months.

Reference has already been made to the initial overriding concern of the NCB with the creation of a new administrative structure and the recruitment of staff for its senior posts.[2] The method of nationalization, which put out of the coal industry nearly all the institutions of which it had hitherto been composed, made that inevitable. Without an administrative structure the coal industry could only have lapsed into anarchy, so one had to be made. For many mineworkers nationalization was first and foremost a guarantee of a better working life and their immediate preoccupation was to harvest the first fruits of that promise in the form

[1] Ibid., 147.

[2] Chap. 1 above, pp. 31–4.

of better pay and shorter hours. It could rightly be argued that something of this sort was essential to the achievement of a larger and more productive labour force; but the first emphasis was on getting the improvements as ends in themselves. Their function in labour recruitment and coal output was, if stressed at all, treated as automatic and the detailed relation between the means and the ends was not worked out. The ministers, with their responsibility for general policy, might have been expected to be the most impartial and the least distracted by the pressures of the detailed business of management and labour. But their preoccupation was much more with getting the fuel and power industries nationalized than with the formulation of a policy for them to work to after they had been nationalized.

In fact, nationalization was implemented in the coal industry in disastrous circumstances which, so far as they were attributable to policy, suggested neglect and misjudgement rather than careful foresight by the Ministry of Fuel and Power. The coal industry was taken over by the NCB on 1 January 1947 and within weeks everyone in the country was acutely aware of being caught up in a fuel crisis. In reality, crisis conditions had existed earlier. The press first took up cases of industrial firms going on short time for lack of fuel in December 1946, and the NCB in their own initial survey of the supply position they had inherited referred to 'the coal crisis that is now upon us'.[1] The less publicized difficulties were present for many weeks before. But the general public was hit by what was happening from 23 January 1947, when there began a period of exceptionally cold weather, with repeated heavy snowfalls. This lasted without a break until mid-March. There were delays and interruptions in the movement of coal and few stocks to draw upon; most consumers, both industrial and domestic, got less coal than they needed; there were power cuts and other restrictions on the use of electricity; unemployment rose rapidly (though the increase did not last long), and much of industry had to work well below capacity.

The outstanding events of the crisis period may be quickly summarized,[2] and the quantitative features (including the contrasts with the previous winter) appear in Table 4.1. A few minor changes in transport, approved by the cabinet on 7 January, operated from mid-January to

[1] From early January 1947 the NCB Marketing Department was continually monitoring the solid fuel supply position and recording the detailed changes week by week. The papers produced for this purpose, which include a retrospective survey to indicate the causes of the current crisis, provide invaluable material for the present account.

[2] For a concise summary see NCB *Ann. Rep.* 1947, 2–4.

Table 4.1. *The course of the fuel crisis of 1947: weekly general statistics*
(coal figures in thousand tons, railway wagons in thousands)

| | Week ended | 25.1.47 | 1.2.47 | 8.2.47 | 15.2.47 | 22.2.47 | 1.3.47 | 8.3.47 | 9.2.46 | 23.2.46 |
|---|---|---|---|---|---|---|---|---|---|---|
| 1(a) | Deep mined coal output | 3,709.7 | 3,625.9 | 3,407.9 | 3,650.7 | 3,830.6 | 3,652.0 | 3,501.9 | 3,475.0 | 3,563.6 |
| 1(b) | Opencast disposals | 175.6 | 158.9 | 117.4 | 107.7 | 175.5 | 188.8 | 194.8 | 126.9 | 148.0 |
| 1 | Total additions to coal supply | 3,885.3 | 3,784.6 | 3,525.3 | 3,758.4 | 4,006.1 | 3,840.8 | 3,696.7 | 3,601.9 | 3,711.6 |
| 2 | Consumption (including shipments overseas) | 4,172.0 | 4,175.9 | 4,107.4 | 3,835.5 | 3,774.4 | 3.850.0 | 3,917.7 | 4,070.5 | 4,016.2 |
| 3 | Difference: (1) minus (2) | −297.7 | −391.3 | −582.1 | −77.1 | +231.7 | −9.2 | −221.0 | −468.6 | −304.6 |
| 4 | Distributed coal stocks | 6,746.3 | 6,194.3 | 5,475.4 | 5,256.2 | 5,525.7 | 5,535.5 | 5,342.1 | 8,935.1 | 8,320.2 |
| 5(a) | Railway wagons available | 1,093.3 | 1,095.2 | 1,095.0 | 1,094.5 | 1,095.7 | na | na | 1,138.1 | 1,140.2 |
| 5(b) | ditto under or awaiting repair | 142.5 | 139.9 | 139.8 | 140.2 | 138.7 | na | na | 120.7 | 118.6 |
| 5(c) | ditto repaired during week | 91.0 | 89.4 | 73.8 | 77.7 | 82.9 | na | na | 84.7 | 90.4 |
| 6(a) | Average max. day temp. (°F) | 40.4 | 33.5 | 34.4 | 32.9 | 32.2 | 35.3 | 36.4 | 50.1 | 47.4 |
| 6(b) | Average min. night temp. (°F) | 30.5 | 26.6 | 30.0 | 29.5 | 25.6 | 22.8 | 23.5 | 41.6 | 37.8 |

*Note*: na = not available.
*Source*: NCB.

ease the movement of coal but were soon more than offset by the severe weather. The government had already laid down that first priority in the supply of coal was to go to gasworks, power stations, waterworks, and ships' bunkers, with house coal depots and the railways spared any restrictions if possible. Iron and steel works and coking plants were next in order of priority, but subject to some cuts, and the main restrictions fell on the rest of industry. In practice, coal could not always be delivered to the intended users and there was not enough for most industrial users to receive anything near to what was suggested. Adequate coal supplies failed to get to power stations in London and the south-east, the Midlands, and north-western England. On 7 February the export of coal was forbidden and close control was instituted over ships' bunkers, with British vessels urged to bunker abroad for both homeward and outward voyages. From 10 February most industrial undertakings in the three regions where power stations were worst affected were required to stop using electricity, and interruptions of electricity supply, mainly from 9 a.m. to 12 noon and 2 p.m. to 4 p.m. applied to domestic consumers in many areas. There were restrictions on theatres and cinemas and reductions in broadcasting hours. The consumption of coal by power stations was immediately reduced very greatly by such measures and special routing arrangements were worked out to ensure continuity of coal supply to power stations and increase stocks at those in the greatest difficulty. Gas consumption was harder to control but there were reductions in pressure which reversed the recent surge in coal consumption by gasworks.

These restrictions did not immediately help other coal users. Freezing weather and shortage of railway capacity prevented some movement of coal from collieries to intended priority users. But loaded wagons had to be got out of the way and often had to go on particular routes which happened to be available. Consumers at the end of such routes were lucky, while those at the end of congested routes went short. This was why consumers in Scotland and the three northernmost counties of England were better supplied than the rest. However, the crisis in the public utilities was gradually overcome. Factories were again allowed to use electric power in the Midlands from 24 February and in London and the south-east and in north-west England from 3 March. From 8 March the system of coal allocations to industry was restored to what had been announced in mid-January, though with a lower level of deliveries to non-priority consumers. Additional coal was made available to industry from 14 April and, though supplies for non-priority industries were then still far below what they needed for full capacity working, fairly rapid progress was made thereafter towards normal industrial and domestic activity.

The causes of this bitter experience were various. The weather received most of the public blame and deserved a good deal, but by no means all, of it. Many of the power cuts were due to the insufficiency of generating capacity to meet new peaks of demand. But, essentially, there was a crisis because not enough coal was available to satisfy current demand. This was not just a simple matter of failures of production, for the output of coal had increased by over 7 million tonnes in 1946. Moreover, thanks in part to some special efforts by mineworkers (those in South Wales, for instance, volunteered for work on seven Sundays), the output of coal in the crisis months of February and March 1947 was 1.1 million tonnes higher than in the same months of 1946.[1] Just as important was the difficulty in making the optimum use of the output. The very low level of distributed stocks, which fell lower than ever during 1946,[2] was particularly influential because it made it hard to compensate for any temporary interruption in supply or to adjust to any sudden surges in demand at particular points. The problem was all the more serious because the physical rundown of the railways had slowed the speed of transport of coal,[3] and therefore created a need for higher rather than lower stocks, and because a relatively larger share of demand was coming from activities subject to large fluctuations in consumption, especially the power stations. All this was well known; the supply of coal to consumers was subject to administrative allocation and did not depend on purchasing power in a free market; and therefore it might have been expected that adjustments would have been made in good time, but they were not.

Of course, if the output of coal could have been increased even faster in 1946, the need for complex adjustments in allocations would have been lessened and there might have been enough coal in users' possession to scrape by without a crisis. It might be contended that a lot of labour was being released from the forces in 1946, that the greatest immediate need of the mines was for more men, and that if only the Ministry of Labour and National Service had made a special effort to get more ex-servicemen to take up mining, the extra few million tonnes of coal might have been produced. There may be a grain of truth in this,

[1] NCB *Ann. Rep.* 1947, 4.

[2] Chap. 1 above, p. 3.

[3] The time taken for coal to reach some southern power stations was three times as long at the beginning of 1947 as it had been a year earlier.

but the argument cannot be pushed far. Housing and training facilities, especially the latter, set limits to the numbers of recruits that could be absorbed. Much recruitment was offset by wastage from the somewhat elderly and decrepit labour force of the wartime mines: in 1946 34,000 men were lost by retirement, death, or ill health, double the number who left for these causes in 1947. And the NCB, after they became responsible on 1 April 1947 for recruitment of mining labour, did make a special drive to attract men from the armed forces, but in 1947 the number of ex-miners recruited from the forces was only 12,255 against 27,862 in 1946.[1] In any case, it was clear long before the end of 1946 what the manpower and output position was and that security of coal supply could be ensured only with careful control applied well in advance. It is delay in acting on this knowledge that exposes the government and, especially, the Ministry of Fuel and Power, to severe criticism, for the delay made a 1947 coal crisis probable in any circumstances and certain if the winter were hard.

A fundamental weakness was that, except for domestic users, wartime controls had never tried to restrict demand to the level of supply. In practice, the immediate concern was to establish what each business needed, give it an allocation not much less than that, and then ensure that as much of that allocation as possible was actually met, whether from current output or from stocks. By 1946 such a system, always weak, had become dangerous. The officials administering it thought that it was only with luck and ingenuity that a crisis had been avoided in the winter of 1945–6. It was pointed out during the 1947 crisis that:

> at Budget Meetings held in the summer of 1946, Coal Supplies Officers urged, even to the point of offence, that it was not a question *whether* but *when* short time working by industry would occur. They pressed with great emphasis that, if the Government insisted that priority in coal supplies should continue to be given to gas and electricity works, immediate cuts must be enforced in coal consumption by industry. They pointed out that the system of programming supplies was inherently bad inasmuch as it did nothing to equalise consumption and output of coal.

Yet six months before the crisis burst upon the nation Shinwell said in public that everyone was expecting a fuel crisis except the Minister of Fuel and Power and that he had advised the Prime Minister not to rely too much on statistics and forget the imponderables.[2] By the time the winter coal budget was supposed to come into operation, on

[1] NCB *Ann. Rep.* 1947, 45–6 and 240–1.

[2] Williams 1979, 135.

1 November 1946, firm decisions about urgent matters of shortage and priority and the transport of coal had not been taken. The cabinet on 19 November approved proposals by Shinwell to cut deliveries of coal to non-priority industries by 5 per cent in some cases and 10 per cent in others and presumably accepted his claim that this would reduce consumption by 2,450,000 tons (2,489,000 tonnes) in the four months from December to March.[1] This was nonsensical, for deliveries of graded coal and smalls (the two categories most used by industry) were already about 25 per cent below allocations. By early January industry was receiving only about 70 per cent of its allocations, and the government on 7 January adopted a plan proposed by the President of the Board of Trade, Sir Stafford Cripps, under which the power stations were to get 50,000 tons (50,800 tonnes) extra per week for the next six weeks, iron and steel and coke ovens were to have 80 per cent of their allocation, and non-priority industry 50 per cent. It was believed that this would enable the creation of a reserve of 150,000 tons (152,400 tonnes) a week for distribution to cases of specially urgent need.[2] Again, the figures proved wrong. Within days the iron and steel and coke oven allocation was lowered to 75 per cent and the target for a weekly reserve to 100,000 tons (101,600 tonnes). The priority given to this reserve caused industrial supplies to fall far below the official 50 per cent and, in fact, in the early months of 1947 industry received no more than about one-third of the allocations; yet the reserve of 100,000 tons was not achieved. Only by mid-March, after the drastic measures introduced in early February, had supplies been made secure enough to ensure a weekly reserve of 150,000 tons for distribution where specially needed.[3]

Delay and muddle by the government clearly deepened the crisis. Not until mid-February was there really effective control of what was happening and even then there is an arguable case that the government over-reacted. The denial of electricity to a large part of industry, and its consequent temporary closure, was announced without previous warning in a Friday afternoon statement by Shinwell in the House of Commons. (It was only that morning that he had put it to the cabinet, which the previous day had said it was not concerned to make new cuts

[1] PRO, CAB 129/14 and 128/6, paper CP(46)423, and discussion at CM(46)98.

[2] PRO, CAB 128/9 and 129/16, papers CP(47)6, 15, 17, and 18, discussion at CM(47)3. The consequential instructions sent to Coal Supplies Officers and Programming Authorities are noted in NCB minutes, 54th meeting, min. 639, 10.1.47.

[3] PRO, CAB 128/9, CM(47)16.

for industrial or domestic consumers.)[1] As the ban operated from the Monday following there was not time to make any adjustments. In view of the damage to industrial output and exports, it is also possible that there might have been some benefit from using for immediate industrial purposes part of the coal saved in the next few weeks instead of concentrating on the improvement of power station stocks, even though operations at some power stations were very precarious because stocks had fallen so low. But it would be unfair to blame those who decided to reduce as quickly as they could the operational dangers of inadequate stocks, which they had belatedly recognized.

There were some indications that, at a rather lower level also, the Ministry of Fuel and Power had too few staff with experience relevant to the problems involved, and that some expertise in the coal industry was at first ignored. The NCB had no formal responsibility for coal allocation but there were benefits when on 6 February Lord Hyndley offered help, and his proposal of a joint committee representing the Ministry of Fuel and Power and the NCB Marketing Department was accepted within 24 hours. At the height of the crisis this committee met daily. It was the NCB who guided the coastwise coal distributors to pool their equipment and resources for the duration of the crisis and thereby secured increased deliveries of seaborne coal. It was the NCB, in consultation with the Central Electricity Authority, who worked out all the details, including the choice of supplying collieries and transport routes, of a programme of deliveries to build up stocks at all British power stations to two weeks' consumption by 10 March and to maintain them at that level thereafter.[2] Without such help the crisis would have been resolved more slowly. One of the sequels was that from 1 May 1947 responsibility for executing supply policies was transferred from the Ministry of Fuel and Power to the NCB, though the Ministry continued to draw up the programmes.[3]

In various ways, indeed, the government got away from the neglect and delays with which the coal situation had been treated until February and sought to create conditions as quickly as possible in which the crisis would not recur. The increase of coal supplies to industry was gradual and carefully controlled. The use of gas and electricity for space heating was forbidden from May to the end of September. Extra work by the

[1] PRO, CAB 128/9, CM(47)17 and 18.

[2] NCB minutes, 62nd meeting, min. 750, 7.2.47.

[3] NCB minutes, 81st meeting, min. 970, 11.4.47. The change arose from proposals made by the NCB in March. Detailed discussions leading to the change are in PRO, POWE 17/58.

miners was negotiated in order to maintain an expansion of output, and in 1947 as a whole 8 million tonnes more were produced than in 1946.[1] Imports, which had been refused at the height of the crisis because continental Europe was in even more desperate need,[2] were obtained in the latter half of the year, 606,000 tons (616,000 tonnes) from the USA and 108,000 tons (110,000 tonnes) from Poland.[3] Detailed planning of the amount, and the colliery of origin, of deliveries to each industrial user in 1947–8 was made in good time. Even the luck with the weather turned. A marvellous summer was followed by a mild winter. So coal stocks went on rising, from their very low level of March, right into November, and were 16.15 million tons (16.41 million tonnes) at the end of the year, the best position for three years.[4] The NCB were able to start their report for 1948 with the words, 'On 1st January, 1948, few traces remained of the critical situation in which the National Coal Board had taken over twelve months before.'[5]

Nevertheless, there could hardly have been worse conditions in which to launch nationalization, and such a traumatic beginning could not be without a continuing influence. In one way it may have helped those running the industry. The government and the public had had a tremendous scare that encouraged them to support ways of improving coal production and distribution. The Cabinet had set up a new Fuel Committee on 12 February with the Prime Minister, Attlee, as chairman. On 20 March this committee considered a paper from him in which he said that 'unless our resources of fuel and power can be expanded, the whole tempo of our recovery will be gravely retarded'. The committee then agreed that priority must be given to the provision of equipment for electricity generation, coalmining, and gas-making, the conversion of some plants from coal to oil burning, and the provision of freight locomotives and wagons for the railways.[6]

But in other respects the influence was less helpful. A scare induced concern to try to ensure that it never happened again. One effect was that government control over coal supply to users was maintained for a very long time and, whenever there was any worry, exercised with great

[1] NCB *Ann. Rep.* 1947, 216–17.

[2] President Truman had offered to divert some American coal ships to the UK (PRO, CAB 128/9, CM(47)31 and 32).

[3] NCB *Ann. Rep.* 1947, 75–7. A further 32,000 tons (32,500 tonnes) was imported in January and February 1948 under the Polish contract (NCB *Ann. Rep.* 1948, 67).

[4] NCB *Ann. Rep.* 1947, 250–1. This figure was later revised to 16.7 million tonnes.

[5] Ibid. 1948, 1.

[6] PRO, CAB 129/17, CP(47)58 and 92.

caution. Another was to increase emphasis on getting the earliest practicable increase in supplies. That meant a readiness by the government to make up some shortages by imports, even at high cost, if there were no other quick way; and expensive imports were a charge on the finances of the NCB. It also made it seem desirable to concentrate on quick improvements more than on others which might be of more lasting benefit. At the moment of taking over the coal industry the NCB were urged by their own expert membership that they should defer any long-term schemes that were at an early stage and concentrate on what would stimulate output quickly. The need for long-term reconstruction, which the Reid Report had insisted was essential for an efficient coal industry, was not ignored; but the inevitable choices between what would give a little now and what would give a lot in the future but nothing now were already weighted in favour of the former. That was one of the cumulative effects of contracting wartime output. The events of the winter of 1946–7 unbalanced the weights a little more in the same direction.

While those events were unfolding the many other preoccupations of management and men in the industry still accompanied them. There were questions which urgently needed answers, and the answers to some of them had long-lasting effects. The need to bring into existence as quickly as possible an effective organization to run the coal industry has repeatedly been stressed. The characteristics and methods of operation of something necessarily so large and, in detail, so proliferating need separate and more extended consideration;[1] but it is essential to understanding to note at the outset both what was being devised and what gaps there were in the arrangements at the time when the NCB took over.

Because the CINA gave only the national board a statutory existence there is a sense in which the NCB had enormous freedom to choose their own organizational form. In practice, the most important general choices were largely pre-determined, though there was a lot of freedom about detail. The huge size of the coal industry made it impracticable for one central body to keep responsibility for detailed management in its own hands, so there had to be a large subordinate structure. The Reid Report, which was so influential, had argued persuasively that efficient working and development depended on the merging of the individual

[1] Chap. 11, section iii below. The most helpful introduction is P. M. D. Roberts, 'Getting to Grips', in *Colliery Guardian*, *National Coal Board*, 1957, 8–12.

companies in each area into 'one compact and unified command of manageable size, with full responsibility, financially and otherwise, for the development of the area'.[1] So it had come to be taken for granted that area organizations would be the new units of commercial management. The choices were how big and how numerous they should be, and what internal structure and powers they should have. The scale of activity still seemed to be too large for the area managements to be supervised directly from the centre and the government appears always to have assumed that there would be a regional organization. It was only because there was such a hurry to get coal nationalized and because thinking about details was so incomplete that the form of regional structure was omitted from the CINA. The Minister of Fuel and Power discussed the matter on the second reading of the bill and said he envisaged regional organizations on the lines of the National Coal Board itself, but the details could not be determined until after the appointment of the national board.[2] Moreover, the Ministry of Fuel and Power had operated the wartime control of the coal industry on a regional system, so that defined regions already existed. When the designated members of the NCB were called to their first meeting they were *told*, without any circulated formal document, that it was proposed to have eight regional boards, and the geographical limits of each were stated.[3] The only queries offered in response were that marketing arrangements might need differently drawn boundaries and that there might be difficulties in the three northern English counties. In fact, Cumberland was taken out of its proposed grouping with Lancashire, Cheshire, and North Wales and put with Northumberland and Durham.

So the board went ahead with the creation and staffing of a regional organization headed by what were entitled *divisional* boards, which were similar in structure to the national board. In August they agreed on the nomenclature of the divisions as (1) Scottish, (2) Northern, (3) North Eastern, (4) North Western, (5) East Midlands, (6) West Midlands, (7) South Western, (8) South Eastern.[4] In discussions until then (2) had been referred to as North-East Coast, but could hardly keep that name when Cumberland was added to it; (3) had been more intelligibly called Yorkshire, but now reverted to the curious wartime usage of the Ministry of Fuel and Power; and (5) had been known as North

[1] Reid Report 1945, 138.

[2] Chester 1975, 388–91.

[3] NCB Designate, 1st meeting, min. 4, 17.4.46.

[4] NCB 10th meeting, min. 100, 16.8.46; 11th meeting, min. 109, 20.8.46.

Midlands, as in the wartime regional structure. As nearly as possible there was a divisional organization for each coalfield or group of coalfields, with two for the largest coalfield east of the Pennines. The small detached Kent coalfield was treated as a division but with a less elaborate organization and lower paid senior posts.

Recruitment of divisional board members presented many difficulties. As far as possible, the chairmen were sought outside the coal industry, though this was not an absolute rule. Some men within the industry were considered and some of those appointed had a connection of some kind. For instance, Sir Hubert Houldsworth, chairman of the East Midlands Division, though neither a coalowner nor a mining engineer, was well versed in the industry from experience in the wartime regional control. Likewise, Lord Balfour, who had been wartime Regional Controller of the coal industry in Scotland, became chairman of the Scottish Division. But they were exceptional. No doubt there were advantages, in an industry with such an appalling history of labour–management relations, in avoiding any appearance that the old bosses were still in charge; but suitable outsiders who were willing and free to move into the coal industry were not plentiful. And their unfamiliarity with the industry gave an admitted urgency to the attraction of the best possible men within the industry to be deputy divisional chairmen. This in turn involved salary problems. There had to be revisions of ideas about these because recruitment difficulties threatened anomalies by which chairmen would be paid less than their deputies. Despite frantic efforts the first divisional boards (for the Scottish, Northern, West Midlands, and South Eastern Divisions) were not complete until 12 September 1946 and the last (the South Western) was announced only on 1 November and lacked a chairman until 1 December.[1]

This was important because the initial tasks of the divisional boards were mainly concerned with organization and structure. They had to propose the detailed area structure and, though the national board laid down the general principles of establishment and salaries, they had to find the staff, subject to approval of the most senior appointments by the national board. All this had to start rather late and be done hurriedly. Not until 15 October could the NCB consider divisional proposals for the delimitation of areas, by which time four of the eight divisions had schemes ready, and even for the first three of these it was 25 October before the details could be settled and approved.[2] Everyone

[1] NCB *Ann. Rep.* 1946, 5.

[2] NCB 24th meeting, min. 234, 4.10.46; 30th meeting, min. 314, 25.10.46.

knew that a complete organizational structure could not be in existence by 1 January 1947 and the agreement of the NCB to take over the industry then has to be seen as a recognition of the political dangers of delay and a sign of keenness to get on with the job. It was an agreement given despite apprehension about the gaps in control that were likely to appear.

In fact, at the beginning of 1947, in most of the country, the area organizations which were to be the main units of management had not been built up to the stage at which they could be operational. So temporary arrangements had to be made in advance for the headquarters of the old colliery companies to carry on for a time as before, with the title of 'Coal Board Units'. Each such unit was placed under a controller, who was responsible to the Area General Manager, if there was one, otherwise to the divisional board. This involved generous co-operation by the companies which had just been legislated out of the coal industry. It inevitably posed special problems for composite firms which had a continuing business to run in the iron and steel industry, but in one way and another continuity was maintained in the routine management of the coal industry.[1]

Below the area level there was less of a problem. There were, in existence or projected, sub-areas (in some places also called 'groups') supervising several collieries, either directly or through an intermediate set of smaller groups in a few cases. This was an arrangement like that under private enterprise when a colliery agent and his staff exercised a similar function. At the individual collieries things went on much as before, though, in the early months of 1947, many a colliery manager must have been puzzled to know who he was working for. Fortunately, there was good progress in completing the area organizations and the Coal Board Units steadily diminished in numbers in the first half of the year. By the end of the summer the NCB's five-tier structure—national headquarters, division, area, sub-area, colliery—was established in all but a few areas, and a proper start could be made in managing the industry on the lines intended.[2]

Less difficulty had been experienced than had been feared. The NCB had been so apprehensive in late 1946 that they had enquired whether the government would meet any losses arising from imperfections in financial control which would arise from agreement to an early vesting date. They had been told it was improbable that the government would

[1] NCB *Ann. Rep.* 1947, 5–6.

[2] Ibid., 7. In a few areas there were two tiers between the area and the colliery.

do this, but they nevertheless took the risk.[1] The accounting practices and reporting procedures instituted by vesting date proved adequate for effective control of day-to-day operations. In particular, there is nothing to suggest that the incompleteness of the arrangements at the time when ownership was transferred had any hampering effects on the management of the crisis in coal supplies. The serious problem was that detailed proposals for non-routine improvements in production were hard to make until the administrative structure was established.

Haste had also led to some ambiguity in managerial arrangements in the areas and these exacerbated that problem. The NCB had resisted suggestions (notably from Houldsworth in the East Midlands) that they should set up a board for each area and followed Reid's view that responsibility should be concentrated on one manager; but they left the Area General Manager with a double role and separated channels of responsibility. The men appointed to posts of this type were all mining engineers and were expected to give most of their time to technical matters, including colliery reconstruction, and report on all production matters direct to the Divisional Director of Production. In their minority time they had to look after the efficiency of the general management of the area, co-ordinate the activities of the specialist departments, and be directly accountable to the divisional board for all this. It did not take long to discover that these arrangements were unhelpfully confused, and by the end of 1947 the NCB had decided to change them as opportunity arose in each area. In future Area General Managers were to exercise full responsibility for all aspects of management and report on all their duties to the divisional board. Every Area General Manager was to have on his staff a Production Manager to take charge of the area's production activities.[2] This was an essential change if efficient management of current operations and the devising of both small and large schemes of improvement were to go on together, and it was a handicap that the managerial arrangements had not been on these lines from the beginning.

Until new arrangements had been introduced at every level it was inevitable that the effects of nationalization should appear to concern mainly

[1] NCB 32nd meeting, min. 340, 1.11.46.

[2] NCB *Ann. Rep.* 1947, 110–11. For earlier proposals and decisions about area managerial arrangements see NCB 29th meeting, min. 307, 22.10.46 and 30th meeting, min. 314, 25.10.46. Street remarked that it would be unfortunate if the industry got the reputation of being 'Board-ridden'.

those matters which were obviously urgent and could be decided centrally. Many of these were labour questions. The government and the NCB were desperately anxious to get more men into the mines. The unions had been demanding nationalization for many years and, now that it had come, they took it for granted that the role of labour in the industry, and its rewards, and the methods of determining both, should all be changed. This belief was supported by the statutory obligation laid on the NCB to establish joint machinery with the unions not only for the settlement of the terms and conditions of employment (including provision for reference to arbitration) but also for consultation on safety, health, and welfare and on the conduct of the operations in which their employees were engaged.[1] So in yet another respect the early preoccupations had to be mainly institutional.[2]

By the end of 1946 the NCB and the NUM (which represented most of the workers in the coal industry) had agreed on conciliation procedures. Questions concerning all collieries were to be dealt with by the national conciliation scheme consisting of a Joint National Negotiating Committee (on which the NUM could have up to fourteen members and NCB representation was the entire national board) and a National Reference Tribunal of three independent members to deal with cases of disagreement. This was a continuation of an existing scheme set up by the Mining Association and the Miners' Federation in 1943. Provision was also made for the continuation of all existing district conciliation schemes and before the vesting date there was agreement on all these except that for Cumberland. There were some difficulties of application because the NUM were unable to change the boundaries of their districts to match those of the NCB divisions, so one division might be a party in two or three district conciliation schemes. But much of the district business was done by informal negotiation between officials, with few or no formal meetings of the conciliation boards. The new element was the pit conciliation scheme adopted on 1 January 1947 by national agreement between the NCB and NUM. This provided for the discussion of all questions in dispute at a pit at successively higher levels, with a tight timetable for reference to the next higher level if a settlement was not reached at a lower. Unresolved disputes were to be referred, by the joint secretaries of the district conciliation scheme, to a Joint Disputes Committee and, if there was then still no settlement, they would go to an umpire whose decision was final, unless he decided

[1] CINA s. 46.

[2] Chap. 11, section ii below for further examination of the machinery of industrial relations.

that the question involved principles affecting the whole district or the whole country. It was apparently without irony that the NCB commented: 'if . . . everyone concerned keeps to the rules laid down, there should never be any need for a strike.'[1] That would have been a miracle and the miracle did not happen.

The NCB also had to set up negotiating arrangements with unions representing employees not in the NUM. Colliery managers had hitherto had only a professional association, but the British Association of Colliery Management (BACM) was now set up and during 1947 was recognized as a negotiating body for most managerial grades, though it was not until 6 June 1948 that formal conciliation machinery was agreed at national and divisional levels. The National Association of Colliery Overmen, Deputies and Shotfirers (NACODS) included practically all deputies and during 1947 the union and the NCB set about devising formal conciliation machinery for them, but this attempt remained unsuccessful for a long time because the union wanted it to cover overmen and shotfirers also, many of whom were in the NUM. The NCB agreed to negotiate jointly with the two unions about men in these grades, but not until early 1948, and then only in Scotland and even there only for overmen, did this lead to any permanent conciliation machinery involving all three parties. It was also impossible to set up permanent machinery for clerical staff, most of whom had previously worked for the private colliery owners. Both the Clerical and Administrative Workers Union and the NUM claimed the right to represent them, both declined to negotiate jointly, and the TUC was unable to recommend a choice between them when the NCB sought its help; so the NCB had to negotiate separately with each. There was more cooperation in providing for workers at coke oven and by-product plants, who, in different parts of the country, had previously been divided among three unions. One of these, the National Union of Blast-furnacemen, Ore Miners, Coke Workers and Kindred Trades, withdrew and a second, the National Union of General and Municipal Workers, agreed that the third, the NUM, should negotiate on behalf of the members of both unions. This made it possible for the NCB and the NUM to start devising permanent machinery modelled on the conciliation scheme for mineworkers, though it was only gradually during 1948 that this was fully established.[2]

The negotiating and conciliation machinery was to a considerable extent an adaptation of previous arrangements to new conditions where

[1] NCB *Ann. Rep.* 1947, 15–17. [2] Ibid., 20–2; 1948, 37–40.

there was a single employer with a changed administrative hierarchy. The schemes for consultation were more novel. A National Consultative Council was agreed on in 1946. The chairman of the NCB was its chairman and it had twenty-seven members, six appointed by the NCB, nine by the NUM, nine by the National Association of Colliery Managers (i.e. the professional body, *not* the BACM), and three by NACODS. It was precluded from dealing with wages or terms of employment (which were matters for the conciliation schemes) but it could, and did, range over anything else of importance in the operation of the coal industry. One of its earliest decisions was to establish Divisional Consultative Councils and to draw up a model constitution for them. Such councils were established in all divisions by July 1947 and many set up standing committees to keep particular subjects (e.g. safety and health, production, training) under review. The national council also wanted a complete set of area councils, but the divisional consultative councils preferred to delay until they had seen how their own arrangements worked. So it was only in the course of 1948 that area or sub-divisional councils became generally established, and even then Scotland, with only one area council, was an exception.

The Colliery Consultative Committees came more quickly and before the end of 1947 most collieries had one. They were described as the collieries' 'parish councils'. The colliery manager was chairman and nominated three other members, two of whom must be underground officials. The NCB mining agent, the area agent of the NUM, and the colliery lodge secretary of the NUM were members *ex officio*. One deputy was elected by secret ballot and six other members were elected by secret ballot—two by face workers and one each by underground haulage workers, contract workers (not employed at the face), surface workers, and tradesmen. The nature and purpose of Colliery Consultative Committees were not always clearly understood by the workers, and their early reception was mixed. Some met only apathy, some were used almost entirely for verbal sniping, some were initially welcomed and then resented when it was realized that they were purely consultative and had no executive authority. But others proved useful in promoting understanding (which often led to co-operation, though not always) by early notice and explanation of proposed changes in workplaces or practices, or more far-reaching changes such as colliery mergers. And the whole range of matters, large and small, affecting the working lives and housing of colliery employees could be brought into the open.[1]

[1] NCB, *Ann. Reps.* 1947, 24–7; 1948, 51–8.

As the new leaders of the industry started to use the emerging new institutional arrangements their early emphasis was heavily on labour questions. The NCB wanted a smooth transition to operation by public ownership and knew that it could not happen without the co-operation of the workers and their unions. The latter wanted quick implementation of what they regarded as the promise (at least implicit) that nationalization would give them better pay and conditions. All the strongest forces thus pushed in the same direction.

One of the potential difficulties of transition was the need to continue existing wage agreements for the time being. There was such a huge number of local differences, locally determined, that it was essential to have union co-operation in not disturbing or exploiting them and in making any minor adjustments that were unavoidable in individual cases, as a result of changed arrangements under a new administration. That co-operation was given. Whether it would have been given so fully without favourable treatment of other labour demands is very doubtful.

The two demands most strongly pressed were for payment for statutory holidays and for the five-day week without loss of pay. As Arthur Horner put it:

> Under the old owners we had six bank holidays a year, but they were not paid for. In the old days, Christmas Day and Boxing Day, without pay, meant a serious blow to the mining families. Obviously that could not be allowed to continue, and we had little difficulty in securing agreement from the Board to pay for each bank holiday.[1]

It had not been quite so easy as that. The agreement was signed on 21 November 1946, and the following day it was only out of loyalty to his colleagues that Reid withdrew a request to have his dissent formally recorded and he was subsequently supported by Young. They objected that the agreement would give bank holiday pay to 60,000 men who were not at the pit that week. The August bank holiday had been paid for in 1946, but only subject to conditions about attendance. Hyndley told Horner that, unless the same conditions applied, the NCB would not agree to payment for Christmas Day and Boxing Day without referring the matter to the Minister. Thereupon the NUM produced a letter from the Ministry of Fuel and Power which stated that the August conditions were without prejudice to arrangements for future bank holidays, and the NUM interpreted that to mean that the conditions

[1] Horner 1960, 199–200.

would not apply again.[1] It was an episode which revealed the beginning of strains among members of the board about the handling of labour questions and also a feeling among some of them that their position was weakened in advance by the Minister.

The same sort of thing appeared in discussions of the much larger issue of the five-day week. This was one of the demands in the Miners' Charter adopted by the national executive committee of the NUM in January 1946. In March of that year Shinwell wrote to the NUM to say that one of the principal objects of government policy in nationalizing the coal industry was to achieve 'the kind of far-reaching reform and improvements contained in the Charter'.[2] In June he said publicly that the government had no objection to the five-day week provided that output was safeguarded and he gave other indications of support. Not unnaturally the NUM attitude was that, as Ebby Edwards put it to his NCB colleagues,

> the Minister had promised them the five-day week and they were not prepared to continue with a six-day week; accordingly, whatever the terms and whatever the cost involved, they were insisting they should have the five-day week. The same attitude had been shown in regard to statutory holidays . . . They were relying entirely on the Minister.[3]

Members of the NCB had been involved in discussion of the five-day week since June (when it was first formally put to them by Shinwell)[4] and developed contrasting attitudes to it. Several were worried by the expectation that it would raise costs of production by an unknown but possibly substantial amount. All recognized that it was bound to come, but differed about timing and conditions. Whereas Reid, on one side, wanted to demand that the Minister should publicly declare his responsibility by issuing a formal direction to the board, Citrine wanted everything to be done spontaneously by the board as a demonstration of their goodwill. He even insisted on recording his dissent to the board's decision to seek fuller information from the divisions about the expected effect on costs and output, because this would involve delay. He commented that the Minister had said nothing about costs, a subject which only the board had raised, and he maintained it was not an

[1] NCB 37th–39th meetings, mins. 400, 419, and 430, 19.11.46, 22.11.46, and 26.11.46.

[2] Arnot 1979, 125–7.

[3] NCB 41st meeting, min. 461, 3.12.46. According to Arnot 1979, 125, it was Edwards who, while secretary of the NUM, had drawn up the Miners' Charter; which adds piquancy to the discussions.

[4] NCB Organising Committee, 3rd meeting, min. 1, 2.7.46.

argument that should be used to prevent the men getting the five-day week. Citrine and Shinwell were among those who thought the cost might be met from the economies of co-ordination and national ownership. Hyndley, more realistic, suggested that co-ordination would take time and that the benefits of nationalization would not show themselves at an early date. After Hyndley had seen Shinwell on 16 December 1946 he told his colleagues that the five-day week was coming even if it cost four or five shillings a ton and that Shinwell had passed on the information that the NUM wanted to be able to tell a delegate conference later that week that there was a good prospect of the five-day week being introduced on 1 May 1947. So the NCB told the NUM that this could be done if the latter gave a written assurance that they would get the acceptance in the industry of the conditions they had already suggested were needed in order to maintain output, and if (in the absence of agreement between them) they would accept arbitration on any other conditions put forward by the NCB.[1] Negotiations about the conditions went on for the next three months, during which the coal crisis caused growing apprehension in the NCB and some members of the board, especially Reid and Gridley, were severely critical that too weak a line was being taken.[2]

On the average about 350,000 tons of coal were produced every Saturday and the need was for annual output to be not merely maintained but increased. Agreement about conditions for the five-day week was reached in March and they both looked, and soon proved, to be insufficient for the achievement of this objective. Miners in future could be paid for one more shift than they worked, but better attendance was encouraged by agreement that the bonus would be paid only to those who had worked five shifts (subject to some concessions for unavoidable absence). The full length of each shift was to be worked and old customs of leaving as soon as a set 'stint' had been completed were to be abandoned; and there was provision for local negotiation of increased tasks. Pieceworkers (who included virtually all faceworkers) were to receive as a bonus 16 per cent of their earnings over five days, excluding overtime, so that they had an incentive to produce more and thus get a bigger bonus. Surface workers were to have, in most cases, a longer working day to make up a working week of 42½ hours in five days instead of the 46½ hours in six days which had been the commonest system.

[1] NCB 42nd, 44th, and 46th meetings, mins. 487, 499, and 533, 6.12.46, 10.12.46, and 17.12.46.

[2] See especially NCB 71st and 72nd meetings, mins. 845 and 849, 4.3.47 and 7.3.47.

Recruitment improved and there was some increase in total output in the months following the introduction of the five-day week, but this was the work of a bigger labour force, though there was a small increase in output per man. Absenteeism declined but much of this was the arithmetical consequence of the reduction in the number of shifts from which it was possible to be absent. There were numerous disputes about the application of the conditions to particular cases. In general, though there were many local exceptions, the response to the appeals by both the NCB and the NUM for the men to take on greater tasks was disappointing. So it soon became apparent that, at current five-day outputs, total coal production for the year would be below the target. As this was happening while recovery from the coal crisis was a major preoccupation, some means of getting more work done in the mines was soon being sought.[1]

By late July the Production Department of the NCB was considering whether financial penalties of some kind might be imposed for repeated absence and appeared to receive some support from the senior officers of the NUM, though they stressed that their executive might not agree. But, for several reasons, this line was not pursued. It was thought difficult to devise a system of penalties which would not risk infringing the Truck Acts. The NUM recognized that the workers had not fully carried out the undertakings given in return for the five-day week, and deplored it, but they were unable to get them to co-operate more fully any more quickly. But the main new factor was that the government had gone directly to the NUM and asked the miners to work longer hours. The NCB thought that the joint pursuit of new measures to get better attendance might be used by the NUM to argue that the government's new demand was unnecessary.[2]

The Prime Minister called a conference on coal production at 10 Downing Street on 30 July 1947. A group of ministers met representatives of the NCB and the NUM, and the Foreign Secretary, Ernest Bevin, asked that the working day should be increased from $7\frac{1}{2}$ to 8 hours for two years or, failing that, some alternative proposal for increasing output should be adopted.[3] The government, having thus opened negotiations, left their detailed conduct to the NCB, but expected the NCB to keep the government informed and to act within governmental guidance. The NUM were in a very strong position. They

[1] NCB *Ann. Rep.* 1947, 10–14.

[2] NCB 114th meeting, min. 1485, 7.8.47.

[3] NCB 113th meeting, min. 1440, 1.8.47.

soon made it plain that, in general, they preferred a Saturday shift to a longer working day (though the NCB made some proposals which included both, as additions to each other, not alternatives) and they insisted that additional hours worked must be voluntary, must be temporary for a specified period, and must be treated as outside the normal working week, not as an extension of it. The five-day week agreement was not to be infringed. The NCB were negotiating under difficulties. The government expected them to stick to the five-day week agreement and to get extra hours worked and to agree to nothing that would increase the cost of production. Governmental guidance occasionally seemed contradictory, but it also kept recurring to the restrictive point that it was the working day which must be extended. Some members of the board, Gridley most persistently, wanted at times to tell ministers that they were interfering in matters which should be left to the board, but passages of that kind were, in accordance with the majority view, omitted from the final version of communications. Negotiations dragged on for two months and successive proposals moved steadily nearer to the NUM wishes.

In mid-September Shinwell told Hyndley that he was prepared to leave an agreement to the board, provided they did nothing that would increase absenteeism. Yet within a week he told the board that the Cabinet would not allow negotiations to continue on the lines which the board were proposing and which the NUM were expected to accept. The Cabinet still demanded a longer working day and insisted that, if a Saturday shift was worked, no one should be eligible for the bonus of pay for one shift more than his total unless he had worked *six* shifts. At that point the board had to tell the Minister that either he would have to review his communication or accept, on behalf of the government, the responsibility for the breakdown of negotiations. The government backed down, renewed negotiations were authorized on 29 September, and a draft agreement was reached on 2 October for submission to a delegate conference of the NUM on 10 October.[1]

The agreement ran until 30 April 1948 and provided that NCB divisional boards and NUM area executives would be invited to arrange either an extra half-hour a day or as many extra Saturday shifts as possible and they were free either to make the choice themselves or to leave it to be made at individual collieries. The extra work would be paid

[1] The negotiations for extended hours were discussed at almost every NCB meeting from the 113th on 1.8.47 to the 135th on 10.10.47. Horner 1960, 181–2 gives an account which is seriously misleading as faulty memory seems to have led him to telescope the events of several weeks.

for at overtime rates and the working of overtime must not interfere with the cycle of operations on the five normal shifts or with the full carrying out of the five-day week agreement. An Order in Council,[1] under the Coal Mines Regulations Act 1908, temporarily allowed hours of work below ground to be increased from 7½ plus one winding time to 8 plus one winding time.[2] There were some tactful extensions of choice in the agreement. During the negotiations the divisional boards had expressed doubts about the line the NCB were taking, because they feared the imposition of arrangements ill suited to local conditions. In the outcome they were given the chance to make their own adjustments. Sir Charles Reid had suggested that choice of working arrangements should be made pit by pit, in the hope that the men would feel they had a genuine choice of their own and were not just having something imposed from on high. His suggestion had a lot of support and was at least partly met.

In the event, only Northumberland and some collieries in Durham opted for the extra half-hour a day. Elsewhere, most collieries worked either every Saturday or every other Saturday. Scotland sought to economize in the use of railway wagons by having half the collieries work one Saturday and half the next. In the last two months of 1947 about 2 million tonnes, 5 per cent of total output, were produced in the extra hours.[3] This was exactly in conformity with what had been forecast early in the negotiations, but which had then been considered insufficient and a reason for treating Saturday working as an inadequate approach to the solution of current problems. Nevertheless, the additional output was both welcome and necessary. If it was to continue, it seemed that the extra work would have to continue, too. Young had told his NCB colleagues that output could not be increased in a spectacular way by improvements here and there; 'in the transitional period it is necessary to rely on maximum effort from the men.'[4] So Saturday working, though less universal than before and always as voluntary overtime, was restored for some years to come as a feature of coalmining operations. But it was done under the terms of an agreement which had to be renegotiated every year. So there was a continuing risk of uncertainty and friction.

The other great immediate concern, in getting more output by getting

[1] SR&O no. 2405 of 1947.
[2] NCB *Ann. Rep.* 1947, 101.
[3] Ibid., 102.
[4] NCB 114th meeting, min. 1458, 7.8.47.

more work done, was with recruitment. The readiness of the NCB to meet demands for improvements in working hours and holiday pay was partly influenced by a desire to make the coal industry more attractive to workers outside and to men leaving the forces. But some of the efforts needed specific union help. At the beginning of April 1947, when the NCB took over full responsibility for recruitment, the NUM agreed to consider proposed changes in local recruiting arrangements. The greatest need for union help, however, was with the recruitment of foreigners. There were, in particular, large numbers of Poles who had served in their country's armed forces and who did not wish to return home after the war. The government was anxious to get as many as possible, especially those with mining experience, into the coal industry and left negotiations with the unions to the NCB. These were conducted at the end of 1946 and beginning of 1947. An agreement was notified to the NCB on 21 January 1947. Its terms were that (a) no Polish worker should be placed in employment without the agreement of the local NUM branch; (b) Polish workers must join the NUM; and (c) in the event of redundancy Polish workers would be the first to go. Poles volunteering for the coal industry were interviewed by panels representing the NCB, the NUM, and the Ministry of Labour and National Service. Out of 8,600 volunteers, over 6,000 were accepted and trained. But there was a lot of local resistance by union branches, despite pressure by exhortation from national officers, who decided in April that they need no longer participate in the selection interviews, with which they were quite content. By the end of May there were Poles whose training was complete and for whom no placements could be found where the men would accept them. The NCB made repeated appeals for help to the NUM. The latter agreed in August that in some areas they would have to have one official give his whole time to overcoming branch objections to the employment of Poles and they also put to a delegate conference a strong resolution in favour of the acceptance of Poles. In the end the Poles were found jobs and fitted in well, but they were very unevenly distributed. More went to the East Midlands Division than anywhere else.[1]

Even more difficulties were experienced with the scheme to recruit European Volunteer Workers which began, with full NUM agreement, in September 1947. By the end of the year 688 out of 1,230 so far recruited were at work in the collieries, but opposition in local NUM branches was very strong. Many recruits who had been put through a

[1] NCB 57th, 58th, 80th, 97th, 98th, 118th, 125th, and 126th meetings; NCB *Ann. Rep.* 1947, 46–7.

training course by the NCB could not be placed and had to seek jobs outside the coal industry. About 8,500 foreigners, from various sources, were absorbed into the coal industry employment in 1948, but the target had been 30,000 and the difficulties had been so great that the recruitment of foreigners gradually faded out in 1949.[1]

In 1947, when conditions for recruitment of British as well as foreign workers were a little easier than they became in the next year or two, the coal industry was not as successful as it had hoped in increasing its manpower. Just over 94,000 recruits were obtained, of whom 41,000 had previous experience in mining. But this was fewer than had been sought and wastage, though less than in 1946, was unexpectedly heavy. So, although there was a net gain of 26,000 workers, the total of 718,000 men on colliery books at the end of 1947 was 12,000 less than had been hoped for.[2] This was an indication of continuing difficulty, as it was inevitable that, for the next few years at least, increased output to the extent that was wanted would have to depend more on a greater input of labour than anything else.

The first twelve or eighteen months of the existence of the National Coal Board looked, in retrospect, mainly like a period of preparation and a struggle with day-to-day pressures. The innovations were in the creation of a new administrative framework, and new directions of policy had to wait until that was established. It is true that a start was made in putting industrial relations on a new footing, but even there the biggest changes were administrative. The one great change of substance was the five-day week, but otherwise the influential mass of detail continued as before. Production was kept going on familiar lines, and whatever local measures promised some early, though often small, increase in output were adopted as opportunity served.

It could hardly have been otherwise. Nationalization was introduced in a somewhat awkward way and in difficult circumstances. It caused a hiatus in the administration of the coal industry which had to be dealt with immediately. The work involved was on such a large scale and had to be done so quickly that it was likely to have some serious faults and it severely restricted the attention that could be given to other problems. The great crisis in coal supply at the beginning of 1947 added to the pressure and encouraged everyone to concentrate on immediate needs and put other things aside for the moment. It also caused such alarm in

[1] NCB *Ann. Rep.* 1947, 103; 1948, 4–5; 1949, 73.

[2] Ibid. 1947, 47.

the government that the coal industry looked for the time being as though it must be a major field of government policy. Thus the NCB experienced greater uncertainty about the extent of their freedom of action than they otherwise might have done. This could have been unsettling at a time when relations between government and nationalized industry were being worked out by experience. As the crisis receded into the past, however, the urge to take more into governmental hands also receded. The conduct of the negotiations for extended hours, which seemed at times to threaten a serious limitation to the authority of the NCB, was not, as it happened, followed by anything similar. The government thereafter ceased for some years to interfere much in questions of wages and industrial relations.

The coal crisis also illustrated vividly the enormous difficulty of the task which had been given to the NCB. Anything that involved so distressing an upheaval obviously had deep-seated causes that could be removed only by large changes which went far beyond the day-to-day routines. This was widely appreciated in the coal industry. To produce more coal year by year, without raising costs if possible, was seen as the objective, and it was recognized that it could not be attained for long unless piecemeal improvements were supplemented by more thorough reorganization and reconstruction. A wish to act on this recognition was present from the beginning. In the early months it could not be given a lot of practical scope. But the preliminary thinking was in progress and, if one turns aside from the immediate large preoccupations and looks at the great variety of activities which were being pursued on a smaller scale, many signs can be seen of preparation for a future of more fundamental change. During the next ten years some of these were themselves developed as major preoccupations in the effort to maintain a large coal industry with a restored efficiency.

CHAPTER 5

# The Maintenance of Maximum Output 1947–1957

## i. The availability and application of productive resources

For ten years after nationalization the strongest influence on policy in the coal industry was the high and generally rising demand. Economic recovery and growth required the use of more energy, and most of that had to come from coal because there was no alternative that was so suitable, cheap, and abundant. If, unexpectedly, the British coal industry could completely satisfy the home market, there were, especially in the first few years, plenty of export opportunities. So the urge all the time was to try to produce more coal, so long as this could be done without a large real increase in average costs.

Recruit more miners. Get a greater quantity of work, and more effective work, by winning such co-operation from the miners that it shows in industrial peace, reduced absence, and lower labour turnover. Reap the benefits of having both unified and public ownership by more rational deployment of productive resources, by standardization, and by the wider spread of best practices; and then enlarge those benefits by increased investment wherever good returns can be expected. Find and work the coal which can most readily and economically add to output. These were the kinds of proposition that seemed appropriate for the making of policy and for speeding its application. They made a drastic change for an industry which, from the beginning of the First World War to the end of the Second, had grown accustomed to generally smaller markets, constant internal strife, and low investment in a large proportion of the collieries. The extent of the turnaround required was one of the chief difficulties in achieving it, while clear perception of what was needed ensured at least well-aimed activity to reduce the difficulties. Much thought and effort went into the search for improvement in accordance with every one of the propositions just stated. In a few directions no more than small temporary advances were made, but in most there was some sustained progress, though it was usually slow. Only when each in turn has been considered is it possible to see how a high output was achieved and sustained and how, nevertheless, coal suffered some decline in its share of the fuel market.[1]

In view of the emphasis on maximum output, it was only to be expected that any easily attainable sources of productive capacity would be examined hopefully. Among the possibilities were additions by producers other than the NCB. The largest such additions came from opencast coalmining, which had helped to check the alarming decline of wartime output. It has already been noted that the NCB, unwilling to be burdened by an unpopular loss-maker, had declined an invitation in 1946 to take over opencast mining from the spring of 1947. They were not opposed to its continuing (though there were some locations where they thought it preferable to work shallow coal by their own drift mines) and, indeed, they welcomed its contribution to the task of meeting demand. But opencasting was treated as a temporary expedient and threatened several times with termination. In the autumn of 1946, when the coal crisis was imminent, Shinwell, with astonishing timing, announced that opencasting would not continue beyond 1948, and some of the senior personnel immediately sought other posts. Within a few weeks he changed his mind and said there would be opencast production for another five to ten years.[2] In June 1947 his ministry suggested that the opencast programme should be kept going until 1951 and then tailed off during the three following years.[3] When 1951 arrived there were renewed fears of a serious coal shortage and, despite adverse but ill-informed press criticism,[4] it was decided to maintain opencasting without reduction. In fact, though the NCB had remained hesitant about its prospects for several years, opencasting became steadily more valuable. The rate of losses declined and from the early fifties turned into gradually rising profits. The quality of the output improved and the reserves suitable for recovery by opencast methods increased. The opencast activities of the Ministry of Fuel and Power were taken over by the NCB from 1 April 1952 and became an important part of their business. Throughout the period of high and rising demand opencast

[1] For the nature and influence of the market, and coal's share in supplying it, see Chap. 2 above.

[2] NCB 28th meeting, min. 283, 18.10.46.

[3] NCB 108th meeting, min. 1,361, 15.7.47.

[4] The *Manchester Guardian* (2 Feb. 1951) described opencast mining on agricultural land as 'the expedient of the bankrupt who starts to sell his furniture', the *Daily Express* (2 Feb. 1951) called its continuation a 'shameful proposal [which] is a confession of failure'. The *Daily Graphic* (2 Feb. 1951) claimed that there were only 33 million tons of worthwhile coal that were winnable by opencasting and that much of the 50 million tons it was proposed to produce in the next five years could only be rubbish.

contributed more than 5 per cent of national output each year except 1954. Only after demand turned rapidly downwards from 1957 was a great diminution of opencasting planned.[1]

More speculative and much less successful were investigations into the possibility of mining lignite, which was outside the activities of the NCB.[2] Lignite deposits had long been known at Alum Bay in the Isle of Wight, where a new investigation in 1946 confirmed that they were commercially insignificant, and near Bovey Tracey, in Devon, where they had been worked for a long period in the late eighteenth and nineteenth centuries. These latter deposits had been bought up in 1945 and were again being worked by a succession of companies from 1946. Powell Duffryn worked small quantities by drift mining for a time, but most of the activity involved El Oro Mining and Railway Co. Ltd., subsequently British Lignite Products Ltd., which was taken over by George Wimpey and Co. in January 1948. The later mining was entirely by opencast and was undertaken mainly in order to establish an industry producing montan wax,[3] with fuel as a secondary product. However, there is no evidence that montan wax was produced. In the summer of 1947 from 350 to 500 tonnes of lignite a week were being mined as an emergency fuel to mix with coal or coke, and it appears that for a period in the autumn this output was greatly increased, with the suggestion that it could help the winter fuel position. At that stage the Central Electricity Board even considered a proposal to have one power station fired by lignite, but it came to nothing. Significant outputs of lignite were not maintained. By 1949, when British Lignite Products Ltd. was still seeking planning permission for further lignite mining, the Ministry of Fuel and Power refused support on the ground that it was of no economic value except in a severe coal crisis. That effectively ended the mining of lignite.[4] The quantities existing in Britain were simply too small to compensate for the low quality of the fuel, and it is difficult to think that lignite could have been taken seriously except in the desperate days of 1946 and 1947.

[1] Opencast mining is examined in detail in Chap. 8 below.

[2] CINA s. 63 (1) defines 'coal' in such a way as to exclude lignite, which is consequently not covered by the Act. However, lignite is included within the definition of coal by the European Communities (Treaty of Paris (ECSC Treaty), Annex 1). Lignite contains more moisture and less carbon and has a much lower calorific value than bituminous coal. The terms 'brown coal' and 'lignite' are both used loosely for a range of low-grade carboniferous fuels, superior to peat, but more precise usage distinguishes (in ascending order of carbon content) brown coal, brown lignite, and black lignite.

[3] Montan wax is an ester wax, hard in texture, used in polishes and candles.

[4] The papers on this episode are in PRO, POWE 28/6.

A useful supply of coal, which it was very necessary to keep up when every little helped, came from the large number of small mines[1] which did not need elaborate services, management, and equipment, and which the NCB had no wish to operate directly. Some of them had belonged to owners of a group of collieries of various sizes and, on nationalization, these owners went out of the business completely. The NCB then had to take charge of both the small and larger mines and they found themselves operating a significant number of small mines, mainly in Scotland and South Wales. But most small mines belonged to undertakings which wanted to continue them if possible. The NCB had power to grant licences to private operators of small mines[2] and the policy was to leave it to the divisions to deal with applications for licences. But at the outset it was clear that they would not have time to examine the details, so it was decided to grant temporary two-year licences to any existing operator of a small mine who applied,[3] and by the beginning of 1947 434 out of the 481 previously operating small mines had sought and been granted licences. The terms were settled after consultation with the Federation of Non-Associated Collieries. The plant and equipment of these mines now belonged to the NCB and the owners were entitled to compensation for them. Until the capital sum was paid they received 'interim income' in respect of it. The licences provided that the interim income would be repaid to the NCB as rent for plant and equipment and that the licensees would pay to the NCB the amount of their pre-nationalization royalty, adjusted upwards or downwards to reproduce the effect of the Coal Charges Account, which was now abolished. They also paid the NCB rent for surface land wherever their liabilities under the lease had passed to the NCB. The licences provided that at their expiry the NCB would buy at a fair price any additional plant and equipment installed by the operator with NCB approval, and that all agreements between the NCB and the NUM about wages, conditions of service, and the settlement of disputes would be binding on the operators. They also empowered the NCB to give directions about the markets in which the operators might sell their coal and the prices they might charge.[4]

[1] 'Small mine' has a statutory meaning. Under the Coal Mines Act 1911, which was the main regulating act until the Mines and Quarries Act 1954 was brought into force, it was defined as 'a mine in which the number of persons employed below ground does not exceed thirty', and the Mines Inspectorate could exempt it from the requirement to appoint a qualified manager.

[2] CINA s. 36 (2). The slightly more flexible qualification was that 'the number of persons to be employed therein below ground is at no time likely to exceed, or greatly to exceed, thirty.'

[3] NCB 35th meeting, min. 385, 12.11.46.

[4] NCB *Ann. Rep.* 1946, 10–11.

It was intended to move to a system of permanent licences as soon as practicable and, in the meantime, the royalties were adjusted in accordance with major changes in costs and proceeds. The system of substantive, instead of temporary, licences, agreed with the Federation, was introduced in April 1950 in England and Wales and three months later in Scotland. The substantive licences were usually for ten years. Licensees would buy back their plant and equipment from the NCB and thereafter pay a reduced royalty. Where the NCB had been making payments to a licensee, to replicate payments which would have been due under the Coal Charges Account, these would cease, and licensees would henceforward pay a fixed royalty.[1]

Licensing was a very helpful way of keeping up the output from small and separated deposits of coal. Many small mines had only a short life but the availability of a licence encouraged an operator who had exhausted one mine to start another, which could often be done quickly, and also attracted some new operators. Of the 340 substantive licences granted from mid-1950 to the end of 1951, 158 were for mines first opened or reopened since nationalization. The number and output of licensed mines fluctuated a little but, in general, they produced about 1 per cent of the national coal output. Their share tended to fall slightly in the early fifties and then to rise a little. By 1956 the number of licensed mines had risen to 540, the growth in numbers having been specially large in South Wales. In that year their output reached a new high level of 2.7 million tonnes, which varied little in the next few years.[2]

The greatest enlargement of productive capacity, however, was to be sought where most of the production was: in the deep mines operated by the NCB. Attention was given to this very early. In November 1946 Reid and Young produced a short-term programme, which was (a) to get Area General Managers to concentrate on introducing more conveyors so as to release pony-drivers and other haulage workers for upgrading to work at the face, and (b) an increase of power loading. But within days, after they had been to South Wales and seen the state of some of its collieries, they added a third proposal. This was that every division should survey its collieries with a view to closing some of low productivity and concentrating their men at neighbouring collieries of higher productivity. They knew how delicate a matter this was. They said no division should tackle more than one or two such cases, which must be carefully explained through the consultative machinery, and

[1] Ibid. 1947, 84; 1948, 82; 1951, 48–9.

[2] Ibid. 1951, 49; 1956, I, 12.

that initially there should be closures only where the reserves could be better worked from adjoining collieries, the justification for closure was very ample, and the men could be transferred without unreasonable inconvenience.[1]

Such a programme could make a modest contribution to the increase of capacity, but no more. The slow progress at this time in the design and introduction of new equipment, especially for power loading, has already been noted.[2] Shortage of conveyors had been expected to be a serious handicap, but more was achieved in the first few years than the pessimists had feared, partly because British manufacturers proved capable of bigger output than had been suggested, partly by the import of additional conveyor belting from the USA. In the two years 1947 and 1948 conveyor belting in NCB mines increased from 9 to 13 million feet (2.74 to 3.96 million metres).[3] The attempts to get more men working at the face, by saving on manpower elsewhere underground, went on, but in general, though there were local successes and fluctuations, the improvement was so slow as to have only slight significance. At the beginning of each year, the percentage of men on colliery books who were employed at the face was 40.6 in 1947, 40.7 in 1948, 40.8 in 1949, 41.3 in 1950, and 41.2 in 1951. By 1956 the annual average was back to 40.7, compared with 40.5 for 1947 and 41.3 for 1950.[4] Closure of uneconomic pits took time and was attempted on so small a scale that its contribution could at first be only very slight. In fact, the very first colliery to be closed by the NCB, Brierley in Yorkshire, in January 1947, had a good production record but had exhausted its reserves. Nevertheless, closures on economic grounds began in April 1947 with Standhills in West Midlands Division. Altogether in 1947 twenty-one collieries were closed, thirteen of them on economic grounds, and nine of these thirteen were in South Western Division. In 1948 the total of closures went up to thirty-five, of which twenty-one were on economic grounds, and again the largest number of the latter (ten) were in South Western Division, though one of the collieries closed there, Lucy Thomas Drift near Merthyr, was reopened two years later. Many inefficient collieries were kept open at a large financial loss, because it was impossible to replace their output quickly.

[1] NCB 34th and 35th meetings, mins. 367 and 377, 8.11.46 and 12.11.46.

[2] Chap. 3 above.

[3] NCB *Ann. Rep.* 1948, 3.

[4] Calculated from figures in NCB *Ann. Reps.* Definitions were changed in 1954 so that the 1956 figures are not exactly comparable. See particularly *Ann. Rep.* 1956, I, 67. On the old basis the figure for 1953 showed a rather freakish increase, which is not apparent on the new basis.

Replacement for closed collieries had to come mainly from new developments within existing collieries, for the only new collieries that the NCB could open quickly were fairly small drift mines. Most of these in the early years were in Scotland, where there were numerous locations in which, although there was no prospect of long-term working, small pockets of coal could be got for some years by new workings of simple construction. Every advantage was taken of such opportunities: for instance, the small and isolated Machrihanish coalfield, where production had ceased in the 1920s, was reopened in 1947 with the Argyll Colliery, a new drift mine. But in England, where new prospects existed, either they could be left to licensed operators or they were usually large enough for it to be clear that small schemes begun in haste were inappropriate: in 1947 the NCB did not open a single new colliery south of county Durham. The scope for short-term projects in existing mines was, however, very extensive in every coalfield. In 1947 550 were begun, and 340 of them completed before the end of the year. They involved the installation of more face machinery, the improvement of haulages, opening new coal faces, upgrading labour and taking in more, and rearrangement below ground so as to concentrate machinery and men on the more productive seams. With so many such schemes it was not difficult to find jobs nearby for men who were displaced. In 1947 2,680 men were involved in colliery closures and 2,136 of them were transferred to other collieries.[1] Projects of this kind continued in large numbers for the next few years. The NCB later commented that until about 1952 it was possible at many collieries for output to be increased without making radical physical alterations to the pit. But 'after this period, such improvements became more difficult to make: there is a limit to the number of small schemes that can be introduced to raise productivity; when these schemes have been carried through, further progress depends on more fundamental and time-consuming measures of reconstruction.'[2] That was the great change of the mid-fifties. Even some of the remaining smaller opportunities were apparently not as well taken as they had been. In 1953 the Director-General of Production, E. H. Browne, complained that a 4 per cent increase in face room had been planned and no increase achieved, and he attributed the failure to inadequate managerial control, particularly by divisional production directors, and to insufficient attention to development at divisional and lower levels.[3]

[1] NCB *Ann. Rep.* 1947, 39–43; 1948, 2. The figures for colliery closures have been corrected.
[2] Ibid. 1956, I, 10.
[3] NCB 367th meeting, min. 64, 5.6.53.

Another contribution to more effective operation was sought from standardization of equipment, which unified ownership encouraged. In this respect also, progress was slow and gradual. The main reason was the sheer magnitude of what had to be done. When the NCB took over the collieries there were 60 different underground rail gauges, 77 pit-tub wheel diameters, and 47 combinations of weights for tubs. It was something of a feat that within three years the number of rail gauges had been reduced to 4 and the number of wheel sizes to 16. But standard specifications were still not available for a large number of items, including wire ropes, electric cables, and conveyor rollers.[1] The large quantity of non-standard equipment, not all of it physically due for replacement; the slowness in achieving large-scale production of many items to improved designs; the prevalence of local purchase; all delayed the standardization which could help to keep output capacity more constantly available by regularizing maintenance procedures and simplifying replacements, as well as making operational methods more uniform and therefore less dependent on special local knowledge. Greater benefits were achieved only over many years and were far from complete by 1957 or, indeed, much later.

As it was known that production in the early years could be expanded only by short-term expedients that would have only a gradual effect on methods of working, it was all the more important to keep up the total quantity of productive resources, especially of labour. But all the attempts to make coalmining more attractive had rather little success. In the country generally jobs were so much more readily available than before the war that the labour market had become highly competitive and all the coalfields felt the effects, though not all equally. The greatest difficulties were experienced in the West Midlands, South Wales, Lancashire, and Yorkshire. The annual average number of men on colliery books rose from 697,000 in 1946 to 711,000 in 1947 and 724,000 in 1948, but then started to decline, and in 1950 it was back to 697,000. Renewed efforts achieved a second expansion to 716,000 in 1952, but then, each year, the total fell back a little and was 703,000 in 1956. A mild recession in the following year helped recruitment, particularly in the Midlands, and the average total for 1957 was 710,000.[2] Within these rather discouraging figures there were some details which contributed a little

[1] NCB 297th meeting, min. 53, 31.3.50.

[2] NCB *Ann. Rep.* 1956, I, 3, 9, 63–7; 1957, I, 1, 10, 39–40. The figures included men in licensed mines.

improvement. The most important was the recruitment and retention of more young workers, even though the wastage of young adults remained high. The recruitment of boys under 18 never reached 15,000 a year until 1951, but from 1951 to 1957 inclusive it always exceeded 19,000 except in 1957, when it was 18,439. The proportion of workers under 21 in the industry rose from 7 per cent in 1948 to 11 per cent at the end of 1956, and there was an increase also in the proportion aged 21–5, which was 8.4 per cent in 1956 and 8.9 per cent in 1957. But the average age remained between 40 and 41, and the coal industry continued to have many more workers over 50 than under 26.[1]

It was always likely to be difficult to keep up numbers as there were special influences leading to high wastage in the earlier years. Apart from the large number of departures arising from age, death, and ill-health in the twelve months after the end of the war, there were many thousands who were in the coal industry only because they could not get out, and who could not be kept indefinitely. In July 1943 the Ministry of Labour and National Service had begun a scheme whereby some of those becoming liable for national service were chosen by ballot to work in the mines instead of being placed in the armed forces, and some conscripts already in the forces were given, and took, the option of moving to the mines instead. At the end of the war the coalmining labour force included about 45,000 of these mining conscripts, popularly known as 'Bevin boys'. When the war in Europe ended in May 1945 the ballot for the mines also ended, though for the next few months those called up for military service were still given the option of going to the mines and a few hundred chose to do so. By that time there was increasing discontent among the Bevin boys because of delay in announcing a demobilization scheme for them and because of some adverse disparities between the scheme for them, when it was announced, and those for the armed services. There seems also to have been a decline in discipline among some of them, when prosecutions of Bevin boys for non-attendance ceased. It was thus neither practicable nor desirable to retain this element of compulsory labour very long, and it gradually disappeared in the three years after the end of the war. In 1947 and 1948 20,000 young men left the coal industry on completion of national service obligations and had to be replaced before anything could be done to expand the labour force.[2]

[1] NCB *Ann. Rep.* 1956,I, 64; 1957, I, 39–40 and II, 136–7.

[2] Figures in NCB *Ann. Reps.* For experience, attitudes, and demobilization arrangements at the end of the war see Agnew 1947, esp. chap. 11.

Wider in its effects was the Essential Work Order of 1941 which provided that anyone who had been working in the coal industry for more than three months could not leave it without permission from the Ministry of Labour and National Service. After the war that permission was by no means always refused. Nevertheless, after a lot of discussion, the NCB gradually came round to the view that the legal restriction of occupational mobility deterred potential entrants to coalmining and kept in the industry many discontented and unsatisfactory workers. So, by agreement, the 'ring fence' (as it was popularly described) was removed at the beginning of 1950, in the expectation that there would be a once-for-all increase in wastage, and the hope that it would be followed by an increase in recruitment.[1] That was more or less what happened, but the rise in recruitment was not permanently sustained.

One of the disappointments about recruitment concerned foreign workers. The delays in getting British miners to accept Poles in 1947, and the still greater difficulties in finding places for European Voluntary Workers later in 1947 and 1948 have already been described.[2] Nevertheless, the NCB, with the agreement of the NUM, kept on trying. In 1951 10,000 out of 18,000 foreign workers, recruited earlier, were still working in the mines, and in that year a new scheme was launched to recruit Italians; but, whatever their national officials said, many NUM local branches remained immovably opposed to it, even where labour was very short.[3] In 1951 900 Italians had been placed or were training as miners, but in 1952 only 400 were placed in British mines, and recruitment ceased in April 1952. The NCB had to make a special payment to all those (a majority) who could not be found jobs in British mines. Ironically, 387 of the recruits went to Belgian mines, which already employed many Italians.[4] Experience was no better when in December 1956 the NCB and the NUM agreed on the recruitment of Hungarian refugees in Austria and the UK. As in all the earlier schemes many local branches of the NUM exercised their right to refuse to work with foreign recruits. By the end of 1957 the NCB had recruited and trained 4,186 Hungarian volunteers, but had been able to place only 731 in their own employment, not all of them at collieries, and still retained another 370 in the hope of finding places

[1] NCB *Ann. Rep.* 1949, 74; 1950, 9–10.
[2] Chap. 4, above.
[3] NCB *Ann. Rep.* 1951, 12.
[4] Ibid. 1952, 14.

where they would be accepted. More than two-thirds of the recruits went to other industries.[1]

Insularity and obstinacy thus played a depressing part, though by 1957 anxieties about job security were becoming more understandable. But it was the strong competition for labour in conditions of sustained high employment that was the main limitation on obtaining the optimum size of labour force. There were prolonged advertising campaigns (with much of the cost borne by government), there was exemption for mineworkers from compulsory military service, and from 1951 there was a special increase in house building in mining areas. There were pay increases which kept miners among the higher-paid industrial workers.[2] But none of this appears to have been very effective for more than short periods. Coalmining still offered arduous conditions to most of its workers and seemed a rather strange and unattractive world to those outside. It had long been considered difficult to recruit men into the coal industry unless they had some experience of it in youth, though many returned to it later after some years in other jobs. And so it still was. The improvement in recruiting boys and youths helped to prevent the shortfall in total numbers becoming too serious, but could not do more than that; and the abundance of other jobs in many industrial areas caused the loss of men to be especially great among the naturally mobile young men in their twenties.

The other main element in labour supply was the regularity and frequency with which the employees worked. The coal industry had long been notorious for high levels of absence. The severity of its physical conditions and the crudity of much of its underground environment made it almost inevitable that there should be a lot of involuntary absence, but voluntary absence was a serious permanent drag on output, and the effects were greater because there was more absence among faceworkers than others. The hope that better status and rewards for labour would lead to better attendance was partly fulfilled. The very high levels of absence in the pre-nationalization years came down very quickly. The percentage of absence for all workers, which averaged 15.95 in 1946 fell to 12.43 in 1947.[3] But this proved to be a once-for-all gain, not the beginning of a sustained trend of improvement, even though 1948 was better than 1947. In the first five years of nationalization the annual

[1] Ibid. 1956, I, 67; 1957, I, 40.

[2] Ibid. 1951, 12.

[3] Ibid. 1947, 236–7. Absence is measured as a percentage of possible shifts which were available to be worked.

average percentage of absence never fell outside the limits of 11.5 and 12.4. In 1952 it was slightly better at 11.2, but then gradually rose and became worse than at any time since nationalization. The annual percentage reached 12.5 in 1955, 12.9 in 1956, and 13.8 in 1957. For face workers it was, as always, worse, and rose each year from 14.36 in 1954 to 16.15 in 1957, a serious matter when every one percentage point increase in absence among faceworkers meant the loss of about 2 million tonnes of output. Another way of looking at the regularity of working is to count the average number of shifts per man each year. This is affected by holidays, overtime, and strikes, as well as by absence. On the whole the adoption of the five-day week in 1947 was offset by these other influences, but the introduction of a second week of paid holiday in 1953 was not. The average number of shifts worked per man per year, 244 in 1947, changed little until 1950, after which it rose appreciably to 250 in 1951 and 249 in 1952. The additional holiday week caused a drop of 5 in 1953, and a slight downward trend set in. The figure was 240 in 1957, when for faceworkers it was only 219.[1]

The attempt to secure greater continuity of work by cutting out industrial disputes, which was one of the objectives of the conciliation schemes and the new consultative committees, was not very successful. In the short term the chances of success could not be high because the number of detailed negotiating points, on which disagreement was easy, was enormous, and because much of the coal industry had a tradition, deeply rooted over many decades, of quick resort to strikes when disagreement arose. The multiplication of potential points of conflict was caused mainly by the labyrinthine complexity of the wage system. Most faceworkers, tunnellers, and back rippers (about 40 per cent of the labour force) were paid piece-rates, and nearly all other workers received daywages. Piece-rates consisted of a basic contract rate plus a host of allowances for specific items such as roof conditions, height of roof, the amount of clay or stone to be cut, and many others. These items necessarily had to be negotiated locally and varied frequently, and were of great importance because they made up a substantial proportion (sometimes over 50 per cent) of a worker's pay. Day workers' wages were not uniform and even after nationalization their details were worked out in seventeen different districts, of which the boundaries coincided neither with those of NCB divisions nor those of NUM areas.

[1] Figures from NCB *Ann. Reps.* The new definitions of manpower introduced in 1954 caused the figure for number of shifts per man per year to fall by one, so the later figures look very slightly worse than they would on a strictly comparable basis.

The NUM were formally pledged both to the abolition of piece-rates and to the creation of a national wages structure, which was not defined but implied that the same job would be paid at the same rate everywhere. But there was such rich regional variation in the vocabulary of coalmining that nobody knew what was the same job: in 1952 more than 6,000 different job names were in use in the coal industry, although there were probably not more than 100 clearly differentiated occupations. It was thus not practicable for the NUM to pursue their proclaimed principles more than very slowly, and in many ways it was also not politic, because that pursuit tended to squeeze differentials which the best paid wanted to keep and it also threatened to diminish the degree of local autonomy, which was much cherished. So for a good many years the centralized wage bargaining of the NUM and the NCB produced agreements which, in essence, produced new national minima and national *additions* to wages, and some general agreements about the differential application of particular additions; but the final determination of what went into a worker's pay packet usually involved some further, and possibly disputatious, bargaining.[1] Moreover, some of the changes introduced early in other matters affecting labour created new causes of contention. The terms agreed for the introduction of the five-day week and then for additional working may have been satisfactory to the national officials of the NUM, but that did not make them immediately acceptable to all the men. The new arrangements for the machinery of industrial relations left some minorities dissatisfied about their representation. The adoption by the NCB of a policy of closing some uneconomic pits, however restrained its pursuit, altered the expectations of some workers and could arouse opposition.

It was, therefore, disappointing rather than surprising that, though there were no official strikes, unofficial strikes and restrictions of effort remained a permanent feature limiting the input of labour and the output of coal. In 1947 there were 1,635 stoppages and go-slows, losing an estimated 1.68 million tonnes of coal, compared with 1,329 losing 0.78 million tonnes in 1946. But on 30 June 1947 a ban, agreed in 1944, on changes in wage rates at individual collieries was lifted and this suddenly brought a backlog of contentious cases into the open, and the five-day week agreement brought its first disputes. Indeed, more than one-third of the loss of output came from one strike when men on four faces at Grimethorpe colliery refused to accept a reassessment of tasks (under

[1] Handy 1981 gives an excellent detailed analysis of these and other related problems; chaps. 2 and 3 are particularly relevant to the points raised here.

the terms of the agreement) worked out jointly and unanimously by representatives of the NCB and the Yorkshire Area NUM. The Grimethorpe men were on strike from 11 August to 15 September, and from 27 August they were joined by men from more and more collieries in the North Eastern Division, sixty-three collieries being affected at some time.[1]

In 1948, though wage disputes were the biggest cause of strikes, a new element appeared. Waleswood colliery, near Sheffield, was much the biggest pit so far scheduled for closure and miners there resisted the closure by twice engaging in stay-down strikes at the pit bottom. One of the anomalies there was that the surface workers were in the Yorkshire Area NUM (which was represented on the North Eastern Division Consultative Committee) and the underground workers were in the Derbyshire Area NUM (which was not so represented). In the end the strikes were prevented from spreading. The Derbyshire NUM, which had made a separate presentation of views to the North Eastern Division, reminded the strikers that it had done so and that the closure decision had been taken in a constitutionally correct way. The dispute was brought to an end in May and the loss of output kept down to about 10,000 tonnes, but it illustrated some of the present and prospective scope for further trouble.[2]

Disputes about union recognition and negotiating rights also took their toll. Not all winding enginemen were satisfied to be represented by the NUM, though most belonged to and remained in that union. In 1947 some of them formed the National Union of Colliery Winding Engine-

[1] There is a fairly detailed account of the Grimethorpe strike in NCB *Ann. Rep.* 1947, 18–20. It was clearly called in defiance of agreed disputes procedures and in defence of practices which had long enabled the men to work appreciably less time than the full shift for which they were paid, but the way it was handled by the North Eastern Divisional Board and their chairman, Gen. Holmes, was adversely criticized, especially by the NUM (Horner 1960, 196–7), but also by others. The NCB noted that the divisional announcement that the men's contracts of employment would be terminated had been made without consulting HQ legal department (NCB 120th meeting, min. 1534, 29.8.47). The national board (121st meeting, min. 1544, 2.9.47) took the view that the dispute should be firmly handled and brought to a successful conclusion as 'a preliminary to securing a change in the attitude of the mineworkers at Grimethorpe, which had been very unsatisfactory for a long time past'. But, though the men went back agreeing to work the new tasks as soon as a fact-finding committee declared that conditions were back to normal, all sorts of excuses were found for postponement, and throughout 1948 progress in settling new tasks under the five-day week agreement was very slow in North Eastern Division (NCB *Ann. Rep.* 1948, 11). Horner (1960, 197) thought the strike remained 'a big blow against the whole policy of fixing new stints to make effective the introduction of mechanization'.

[2] The Waleswood dispute was discussed at NCB 176th, 177th, 178th, 179th, and 180th meetings (the first sit-in strike), 203rd and 204th meetings (the second sit-in strike).

men, which the NCB found insufficiently representative for recognition, and the Durham members went on strike as soon as the five-day week was introduced. The next year another new body, the Colliery Winders' Federation of Great Britain, was formed, and in 1949 another strike of winding enginemen, in Lancashire and Yorkshire, cost 374,000 tonnes. Not until 1951 did the Federation settle its inter-union disagreements and accept absorption into the NUM. And in 1948 the biggest loss of output came from two strikes about the wages of shotfirers, affecting forty collieries in North Eastern Division. This was more difficult to deal with because there was no permanent conciliation scheme for shotfirers, as a result of their representation being split between the NUM and NACODS.[1]

For the first five years the number of stoppages and go-slows remained steady, between 1,528 in 1948 and 1,637 in 1951. The loss of output varied between 0.9 per cent and 0.5 per cent of the year's total. After that the number of disputes was appreciably greater and rose to 3,771 in 1956 and 1957. The loss of output was greatest in 1955, when it was well over 3 million tonnes. Nearly a third of this loss came from one strike, about coal fillers' piece-rates, which began in Doncaster Area at a time when a revision of fillers' price lists, pit by pit, was under discussion. It spread to eighty-four pits and involved nearly 80,000 men. In the five years 1947–51, 6.5 million tonnes were lost by disputes, in the next five 1952–6, 10.1 million tonnes, and a further 1.9 million tonnes in 1957.[2] It was not that the coal industry was in perpetual unrest or that the machinery for resolving disputes was in a constant state of idleness or breakdown. In 1952, when the number of stoppages suddenly leaped up, 11,685 disputes were settled by the conciliation machinery; and in 1954, when the number of stoppages again rose sharply, half the pits had none at all.[3] Year after year, more than four-fifths of the tonnage lost was in Scottish, North Eastern, and South Western Divisions, and even there most of it was located in only ten of their twenty-five areas. Difficult labour relations were not new there and it seems impossible to explain the contrasts without reference to deep-rooted communal attitudes which long antedated nationalization.[4] But it was particularly

[1] NCB *Ann. Rep.* 1947, 21; 1948, 36; 1951, 71.

[2] Ibid. 1952, 51–2; 1953, 14–15; 1954, 14–15; 1956, I, 78–9; 1957, I, 48.

[3] Ibid. 1953, 15; 1954, 14.

[4] Cf. Ebby Edwards's comments to his colleagues about the Waleswood strike: it 'is evidence of a lack of discipline at the pit of a type that is not infrequent in Yorkshire'. (NCB 180th meeting, min. 277, 20.2.47.) This chimes in with the views expressed in the board about Grimethorpe (p. 168, n. 1, above). The contrast between the ease of labour relations in what, before

exasperating that losses of output from stoppages were particularly high in parts of South Wales producing special coals of which there was a serious shortage, and in South Yorkshire where costs were lower and productivity higher than in most of the industry. The coal industry had not returned to its notorious unrest of the inter-war years, but labour stoppages remained an obstacle to the achievement of maximum input and output in the years of coal scarcity.

The remaining element in the input of labour was that of managerial and administrative staff. In this case it is impossible to define and measure a shortage in the sense of identifying quantities of necessary posts which were established and left unfilled, but there were various symptoms of an unsatisfactory state of affairs. One was that the number of staff appointed, and their functional allocation, were directly affected by the restrictive influence of the available supply. The pre-nationalization coal industry was very backward in its managerial techniques. Most colliery companies had not even made a beginning with the use of standard costs, budgetary control, method study, and planned maintenance. The NCB sought to do something about this, but progress was slow and patchy. This was partly attributable to indecisiveness by the national board, which refused to coerce the lower tiers on such matters and left even their experimental adoption to the discretion of each division. But it was also due to the shortage of staff who understood or were readily capable of learning such techniques, and a lack of training facilities to impart them. Likewise, after ten years of nationalization, the NCB recognized that 'notwithstanding the increase in the number of engineers, there are still not enough for the efficient and speedy carrying through of the Board's reconstruction programme simultaneously with the maintenance of the highest possible level of current production'.[1] There was also an apparent tendency to try to keep down the size of the managerial and administrative establishment, almost regardless of the range of things that had to be done. This was doubtless the result of a defensive attitude inculcated by the hostile criticism of the numbers of staff, chiefly from politicians and journalists, but also expressed in sniping from collieries and areas at those above them. Whatever the sources of the attitude, it helped to prevent the shortage in the supply of managers from appearing in the statistics.

nationalization, was the West Yorkshire District and their difficulty in much (though not all) of what was the South Yorkshire District has continued to be noted and clearly has historical causes traceable back into the early twentieth or even the nineteenth century.

[1] NCB *Ann. Rep.* 1956, I, 88–9.

But the most serious symptom of shortage was apparent whenever there was close examination: the large number of posts occupied by people who were not able enough or sufficiently well qualified to fill them satisfactorily. The Fleck Committee, which reported in 1955 on the industry's organization, found this theme recurring in many of the problems it uncovered. Its comments were numerous and sharp. Here are some:

> One of the industry's greatest needs is better management at all levels. . . . At every level of the Board's organisation, and in most departments, there is a serious shortage of able people equipped with the right qualifications and experience. . . . Quick promotion has come to more people than is good for them or the industry. . . . While in some appointments—we have instanced Colliery Managers—the turnover is too rapid, in many others it is too slow. Too many men settle down early to a career in the department in which they started. . . . Once people have been put in a post, there is little planned training for them and virtually no scheme designed to teach them the art of management or widen their general administrative experience. . . . The nationalised industry inherited more than a few people who were not then, and are not now, able to fill responsible positions satisfactorily. A policy for the retirement of these people is needed and must be vigorously pursued.[1]

Such criticism, made after more than eight years of nationalization, suggested that serious inadequacies in managerial input had been much too slowly reduced.

The Fleck Report indicated that, when managerial needs and quality had been reassessed, there would have to be some increase in the number of non-industrial staff. Such an increase quickly followed the report, but started earlier. The long-term trend of staff numbers cannot be established with any precision. At the end of 1946 the number of administrative and non-industrial staff of the coal industry was reckoned to be 30,123, but this did not include the chairmen and non-executive directors of colliery companies, a total of more than 1,000 people. At the end of 1947 NCB staff totalled 34,535, of whom more than two-thirds were at the collieries; but the NCB were using their own staff to do things previously contracted out by colliery companies, such as legal, accounting, and insurance work, and they also had taken over various institutions which were not previously included in the official coal industry statistics.[2] Between 1948 and 1949 there was a reclassification of the statistics, which transferred about 1,350 people to the

[1] NCB, *Report of the Advisory Committee on Organisation* (Fleck Report 1955), 23–5.

[2] NCB *Ann. Rep.* 1947, 32–4.

non-industrial category without any additional recruitment, and another reclassification between 1952 and 1953 probably added a further 750 in a similar way. In 1952 the NCB took over the Opencast Executive from the Ministry of Fuel and Power, and most of the personnel (over 1,000) were non-industrial. So only from 1953 were there figures on a uniform basis. The total number of staff rose from 43,659 (of whom 28,300 were in clerical and equivalent grades, and 9,688 were mining and other engineers and technical staff) in 1953 to 51,200 in 1956 and 55,300 in 1957, including 34,400 in clerical and equivalent grades and 12,700 mining and other engineers and technical staff. In the latter year over 80 per cent were at collieries, colliery groups, and areas, and not quite 6 per cent at national headquarters and the NCB research and engineering establishments.[1]

The NCB had, before this, begun to consider the better use of their staff and ways of improving quality. One problem was that area management was responsible not only for collieries but for ancillary activities. In 1951, when the Director-General of Finance, Latham, drew the board's attention to the riskiness of their carbonization business, there was a long discussion of the possibility of letting some of it go to the Iron and Steel Corporation and having a joint board with the gas industry to deal with domestic coke and coke oven gas. One of the elements in the debate was the distraction to management, which was deprecated by several board members, including the chairman, and senior officials. Lord Hyndley remarked that the standard of management was hardly good enough to cope with the immense problems of coal production, that extensions of activities were bound to reduce still further the effectiveness of control, and the time devoted by key managers to ancillaries would lessen their effectiveness in coal production. In the end that line of argument did not prevail and other reasons were found for keeping all the carbonization activities, so, for the time being, one possible source of greater concentration of managerial effort was forgone.[2] A year later Area General Managers were relieved of responsibility for coke oven and briquette plants, which passed to newly created divisional directors of carbonization. This used managers more sensibly but did not cure shortages.

Schemes which sought to attract more people into non-industrial posts and train them to better standards were gradually started. In 1949

[1] NCB *Ann. Rep.* 1956, I, 89; 1957, I, 41–2.

[2] NCB 316th, 317th, and 329th meetings, mins. 38, 49, and 155, 16.2.51, 2.3.51, and 21.9.51. The main final discussion and recommended decision was by the general purposes committee.

'Directed Practical Training' was introduced for engineering graduates and for men who had obtained engineering qualifications at a technical college while employed in the coal industry. Those accepted were given up to three years of practical experience to supplement their technical education. The original aim was to recruit 200 people a year, and the scheme began with only 30 in 1949, but the objective was later increased and 360 people were recruited in 1956, by which time the scheme had produced 31 colliery managers and 116 under-managers, as well as people in group, area, and divisional posts. The scheme was backed up by the NCB scholarships to universities, of which 100 a year were offered from 1948 onwards, open to young men both in and outside the industry; but there were never enough good candidates and 1956 was the first year in which as many as 80 scholarships were awarded. A few non-technical graduates were recruited from the beginning, in an unsystematic way, and in 1950 an Administrative Assistant Scheme was introduced for them, and was thrown open in 1952 to young people of similar calibre who were already in the industry. They were given up to three years' planned experience in several departments at all levels from colliery to national headquarters. It was hoped that this scheme would be the recruiting-ground for future occupants of senior positions in the non-technical departments and in general management.

Mainly for lower managerial positions as well as for craftsmen the NCB tried to develop more of the ability of the existing workforce by introducing the 'Ladder Plan'. This allowed young employees in the industry to study for one day a week at a technical college for various qualifications, such as those leading to jobs as craftsmen or under-officials (shotfirers and deputies and, with more experience, overmen). From the mid-fifties some who obtained the Higher National Certificate were selected for full-time one-year courses leading to a National Diploma, after which they might qualify as colliery managers. From the mid-fifties also the NCB began to do more to train people in the human aspects of management. Courses were included in the training of some prospective under-officials. In 1955 a few scholarships were offered each year to enable men with at least five years' experience below ground to go to university to take a social science degree, with a view to employment in industrial relations posts. In general, progress was hampered by a shortage of management training facilities. The idea of having a staff college was favourably discussed for several years, but it was not until 1956 that the college, at Chalfont St Giles, was in existence, and 1957 was its first full year of operation.[1]

[1] NCB *Ann. Rep.* 1949, 76 and 185–94; 1954, 33–7; 1956, I, 94–8; 1957, I, 56–8; J. O.

Thus a good deal was begun in the fifties, but it had not had time to produce very large results, and much still remained to be done. The first products of the best of the NCB's schemes were nearly all still too junior to make a lot of difference to management, and some necessary forms of training had started so late that they had not yet had any products. The Fleck Committee, while commending the Directed Practical Training and the Administrative Assistant schemes, remarked that 'the staff problem is still far from solution. Indeed, the right atmosphere for its solution has yet to be created.' It was particularly critical of the lack of training in personnel management, and told the NCB they ought to stop relying exclusively on recruitment from the NUM for their senior industrial relations posts. Instead, they should seek the unions' co-operation, but draw on the whole body of their own employees and train them for responsible positions in the personnel field.[1] It also pointed to the continued unattractiveness of senior positions to suitable outsiders, and suggested that one reason was that board members were underpaid and that this resulted in their chief officials' salaries comparing poorly with those in other industries. And it emphasized the further weakness that able people in the board's employment were reluctant to take headquarters posts in London because they thought that the rewards, expenses, and career prospects compared unfavourably with those in the coalfields.[2] For a variety of reasons, then, the supply of managerial resources, which for most of the early years was not over-abundant in numerical terms, needs to be weighted downwards for quality, and this was one of the restrictive factors under which the coal industry operated.

The remaining important element in the supply of resources was finance. There was not a serious problem in this respect. Many difficult questions arise about financial structure and about the sources of funds for investment. The NCB were not statutorily required to do more than break even financially, and even had they sought to improve on this, government policy on prices did not permit them to do so. As they were committed to enlarge, revivify, and reconstruct the industry, they were bound to need more investment than could be set aside for normal depreciation. Under such financial constraints the extra capital could be obtained only by borrowing from the government. This was a practice

Blair-Cunynghame, 'Careers in the Coal Industry', in *Colliery Guardian*, *National Coal Board*, 1957, 69–73.

[1] Fleck Report 1955, 24–6.

[2] Ibid., 17 and 33.

which could be risky and could become burdensome, as everything was financed by interest-bearing loans and not by equity capital. In the not very long run some of the risks turned out badly and the burdens became all too heavy. But these penalties were not incurred to any significant extent until after the years of expansion. The coal industry got all the finance it could use and, to begin with, it got it cheaply, for the coal industry was nationalized before the end of the era of cheap money. Indeed, one of the problems in the fifties was that physical investment could not quite keep up with the finance available for it. For the more forward-looking managers, unless they had come from one of the handful of prosperous and progressive companies, easier financial conditions were among the most welcome changes brought by nationalization. If they had a well thought-out project for improving productive capacity, their hopes were no longer dashed by the non-availability of finance, as they often had been in the past. In this period, when the determinants of productive capacity and efficiency are sought, the answers have to be found mainly in physical, technical, and behavioural factors, and only in a few limited and special ways in finance.

## ii. The price structure

One way in which the most rational use of available productive resources might be encouraged was through the relative prices of coal of different types and from different sources. As coal remained subject to allocation by quantity and every consumer was entitled to a specified quantity within each year's programme, there was not a great deal that could be done through the general average price level, particularly as price changes needed the approval of the Ministry of Fuel and Power and it was government policy not to let scarcity lead to prices which might be well above average costs. But each consumer had to purchase what he wanted, up to the maximum of his allocation, and there was the possibility of keeping down demand for high-cost and less accessible coals and raising that for less popular types by differential pricing. It was also desirable to improve the working of the market (and the confidence of the customer in the supplier) by ensuring that both parties to a sale knew the quality and characteristics of what they were dealing in. With something as heterogeneous as coal, this need could be met only by a system of classification and grading by criteria based on the qualities that most affected the utility of the product to the user. But any price structure needed to avoid too great complication if it was to be understood, accepted as equitable, and administratively workable.

The price structure which existed in 1947 was not conducive to rational use of resources. It had been moulded by the various selling schemes set up under the Coal Mines Act 1930, and not much progress had been made in working out a sensible relationship between the schemes of different districts. During the war a further irrationality was introduced because all coal price increases were by lump sums, not percentages, applied equally to all coals. The longer this went on, and the greater the number of increases, the more distorted the price structure became, because the effect was to make the better coals relatively cheaper and the inferior coals relatively dearer. For lack of time to work out any alternative, the practice of general flat-rate increases continued in the early years of nationalization, but it was the intention of the NCB from the beginning to devise a new price structure, and they set up a working party which was able to produce a scheme on consistent principles, which remained permanently in use. Before this could be done, coals had to be reclassified by quality. An important first step was to agree on a national system of description by size. The British Colliery Owners' Research Association had produced recommendations about this in July 1946. The NCB formally adopted these proposals in April 1947 and put them into practice later in the year.[1] There still remained the task of analysing about 8,000 different sorts of coal on a uniform basis. The most important general factor for all users was the calorific value, but allowance was made also for ash content, size, and whether or not the coal had been treated in a preparation plant. For carbonization coals the yield of gas and coke was a major factor. All these characteristics could be incorporated in formulae to establish relative prices on the basis of quality in different classes of use. The comprehensive analyses made it possible for basic prices, established in this way, to be adjusted in certain cases for special factors such as the sulphur content of the coal.

To establish the scientific data and apply them quantitatively in formulae appropriate for price calculation took a good deal of time. Numerous large consumers were consulted about their needs, and the

[1] NCB 24th meeting, min. 250, 9.10.46 and 84th meeting, min. 1019, 23.4.47; it was indicated that the BCORA grades and sizes would be used in the new price structure. The scheme described 'large coals' in terms of the aperture of the screen over which they are made, 'graded coals' in seven size groups with standard names, and 'smalls' in terms of their treatment, the aperture through which they are made, and the proportion of fines (i.e. coal passing through a one-eighth inch square mesh) in their content. It also provided a definition of 'slurry'.

views of the Coal Consumers' Councils were sought. There were other sources of delay which were equally serious. Although the devising of a new price structure was a task decided on and carried out by the NCB, the Ministry of Fuel and Power took a great interest in it and joined in continual discussions. The Ministry eventually decided that the price structure needed ministerial approval, as it could alter the general price level of coal, even though its introduction was proposed in a way intended to alter only relativities.

Ministerial concern led to Treasury concern. The NCB proposals went to an Interdepartmental Working Party on Prices Policy of Socialised Industries, which failed in three years to produce a single report on any substantive issue. That working party was provided with memoranda by the Economic Section of the Cabinet Office, which was totally unsympathetic to the NCB proposals and caused much irritation in both the NCB and the Ministry of Fuel and Power. The Economic Section was advocating pricing practices which implicitly assumed that every consumer had perfect knowledge of the marginal costs and revenues resulting from every choice in the purchase of coal that was open to him. That body asserted in 1949 that the NCB should not attempt any valuation of the technical characteristics of coals because all that mattered was relative value at the margin of substitution. A year later it suggested a draft report which included the statement that:

> ... it would be necessary to adjust output at each pit so that the difference between (1) the additional cost of producing, and (2) the additional net proceeds from selling, a small additional output should be *the same at all pits*. The total proceeds of the National Coal Board would depend on the extent of this constant difference between marginal costs and proceeds, and could be adjusted accordingly to secure the desired overall balance.[1]

The writer seemed to think the rate of colliery output depended on turning a tap. Fortunately, it was possible in the end to by-pass such unworldly innocence, but the delays did not help the pursuit of rational industrial operation.

The NCB had hoped to introduce a new price structure by the autumn of 1948 or the beginning of 1949, but this proved impossible,[2] so in 1948 and 1949 two *ad hoc* interim adjustments were made, with the object of widening the gap between the prices of the best and worst quality coals. This was desirable because, despite the general shortage of

[1] PRO, POWE 37/29 and 37/31. The italics are in the original document.
[2] NCB 199th meeting, min. 587, 23.4.48.

coal, there had been difficulty in selling some of the lowest quality. The new price structure was introduced in two stages in 1951, that for house coal in June and that for industrial and carbonization coals in December. Even then there was some controversy about relativities. The Area Gas Boards, in particular, made loud public protests because the new scheme raised the relative prices of carbonization coals, though these were more expensive to mine and, by their special qualities, contributed to higher-priced products. The Gas Boards got little sympathy from the Industrial Coal Consumers' Council or the Ministry. The Chancellor of the Exchequer also threatened to delay the introduction of the scheme by referring it for cabinet consideration, because of the amount of the reduction in house coal relative to industrial coal prices. So, to avoid further delay, the NCB increased the proposed house coal prices before they had been announced, in order to permit a lessening of the increase in the price of industrial coals.[1] This was probably unfair to captive customers, for it seems likely that, since some time during the war, domestic consumers had had poorer value for money than anybody else.

The house coal scheme was novel in fixing delivered prices at the merchant's depot. All house coals were classified in grades numbered in descending order of quality from 1 to 8. The whole country was divided into sixty-four zones (a few of which were further sub-divided), and a price for grade 4 coal was fixed for each zone and sub-division. The price *differential* between grades was identical in all zones. The zonal price covered the cost of rail transport (with variations for other modes of transport), and attempted some rough averaging of the pithead costs of the different sources from which merchants in a particular zone were accustomed to get their coal. As the supplying collieries often varied from week to week, frequent fluctuations in the price of coals of similar quality were avoided by using a delivered price basis. The highest prices were in Cornwall and Devon, the lowest in the Cudworth sub-division of Doncaster zone. From December 1951 an additional unnumbered grade was sold under the name 'nutty slack'. At first it was cheaper than grade 8, but from July 1953 it was priced between grades 7 and 8. From May 1954 nutty slack and grade 8 were both withdrawn. From then on there were seven grades, and the price of grade 5, was used as the datum for each zone. This was part of the policy of discouraging demand for the better qualities, which were becoming scarcer and more expensive to supply. From the summer of 1951 to the summer of 1957 the price differential per ton between the best and worst grades available increased

[1] PRO, POWE 21/70.

from 26*s.* 8*d.* (£1.33) to 85*s.* 8*d.* (£4.28), and the biggest increases were at the top of the scale.

The NCB contemplated the introduction of delivered prices for all coal,[1] but found it more difficult and less urgent to do this for industrial users, who were more accustomed to buy large quantities from the same source over a period. So the scheme for industrial and carbonization coals embodied pithead prices and, despite indications that it would be changed to a delivered price basis when experience from the house coal scheme could be studied and applied, it was left unchanged in this respect. Prices were based on calorific values, modified to take account of the various factors already mentioned, and there were separate formulae for industrial and for carbonization coals. One other ingredient of the scheme was the 'coalfield addition'. Coal drawn from some coalfields (particularly those in regions which were not self-sufficient in coal) was priced higher by specified amounts, partly to take account of higher costs of production and partly to match the transport cost of similar coal bought more cheaply in other coalfields. Thus Lancashire coal was priced more highly than that from Yorkshire and the East Midlands, which was used in large quantities in Lancashire, and the coalfield addition was greatest of all for the Kent coalfield. Some later adjustments to the price schemes were made in order to improve the matching with relative costs and to try to divert some demand away from coals of increasing scarcity. So in May 1954 there were selective additions to prices of coking coals from Durham and South Wales, coals produced in Scotland, and all large coal.[2]

The price schemes appear to have been generally acceptable to consumers (though, of course, they did not all approve of price *levels* within them), and they were judged to have worked well, in the sense that they permitted a smoothly operating market in which transactions were based on adequate information. Whether they promoted the best practicable application of resources[3] depends to a large extent on the closeness with which the size of price differential matched differences in the cost of supplying the various classes of coal and in the value of the products to which their use contributed. Table 5.1 illustrates the position in late 1957.

Railways and ships used high proportions of large coal. Coke ovens

[1] NCB *Ann. Rep.* 1951, 37.

[2] R. J. Moffat, 'Demand and Supplies', in *Colliery Guardian*, *National Coal Board*, 1957, 37, and NCB *Ann. Rep.* 1956, I, 39–42, give clear, concise summaries of the price structures.

[3] NCB *Ann. Rep.* 1957, I, 32–7 discusses the problems of adapting demand for and supply of particular classes of coal in mutually advantageous ways.

Table 5.1. *Pithead price relativities for major classes of consumers, 1957*

| | | | |
|---|---|---|---|
| Railways | 102 | Coking plants: operated by NCB | 97 |
| Bunkers, coastwise and trawler | 101 | Coking plants: operated by others | 95 |
| Domestic consumers | 101 | Iron and steel (excluding coke ovens) | 92 |
| Gasworks | 100 | Electricity works | 72 |
| | | Other inland consumers | 86 |

*Source*: Calculated from NCB *Ann. Rep.* 1957, I, 32.

operated by the iron and steel industry saved a little on coal prices by buying some untreated coals and preparing them in their own washeries. Nevertheless, the figures suggest that differentiation among the better-quality coals was rather narrower than might have been expected and that supplies to coke ovens, which included coal of exceptional quality and some that was very expensive to mine, were underpriced, though the carbonization price scheme was probably doing part of its job by diverting some coke oven consumption to the cheaper and more plentiful coking coals of Yorkshire. The price schemes, of course, covered a lot of averaging among both collieries and coalfields. So they could not encourage a comprehensive redistribution of resources among them. But that was impracticable and, within fairly wide limits, undesirable in a period when the maintenance of national economic growth needed a supply of coal which could not be met without some production at a marginal cost well above the average price at which consumers were willing to buy the whole output.

The price schemes appeared to be operating in the right direction, though they needed frequent adjustment in detail and had probably not been adjusted quite enough by 1957. They remained as the permanent basis of coal pricing, though later experience suggested that in the form of their early years they were administratively more complex than they need be. This was particularly so with the house coal scheme. The foundation of pricing policy laid down in the fifties thus proved lasting, but in the longer term the trend was to have fewer categories, to operate the principles in a simpler way, and to leave the exceptions to be dealt with by individual arrangements.

### iii. Doubts, criticisms, and changes

So much had to be done so hurriedly in the first two years after the passage of the CINA that it would have been surprising if it had all been done in the best possible way and if doubts had not arisen about the merits of some of the measures taken. Some frustration was likely because the overriding objectives were to increase the output of coal and improve the efficiency of the coal industry, and most of the early activity addressed itself only indirectly to these objectives, as the creation of administrative machinery and attention to the pent-up demands of the unions made inescapable claims on initial time and effort. And neither morale nor the public repute of the NCB were enhanced by the recording of a substantial financial loss on the first year of nationalized operation. Even though the biggest item in the loss was the payment of interim income to the owners of the still unvalued assets that had been acquired, and the loss was swollen by the costs of coal imports which were a sequel to the crisis of supply which the NCB inherited, it had to be recognized that colliery costs had risen sharply and there was an appreciable loss on colliery operations. All these conditions helped to provoke a good deal of criticism of the NCB which was being expressed by 1948.

Some of the criticism was directed at the men at the top, who were alleged to be insufficiently businesslike. Much of it concerned the organization. It was said that because the national board was entirely composed (except for the chairman and deputy chairman) of members with functional responsibilities, each board member fought for his own field and there was too little attention to disinterested policy making. It was also said that the whole organization was much too centralized, with the twofold adverse consequence that costs were inflated by taking on hordes of officials, and that those at the point of production felt utterly remote from the upper hierarchy and were unable to get the prompt decisions and actions they needed in order to do their job well.

These criticisms were mostly voiced by outsiders, but some came internally. The national board was not a quarrelsome body (partly because both Hyndley and Street took so much care to soothe any ruffled feelings), but there were some repeated differences, particularly between those most immediately concerned with getting more coal produced and marketed and their colleagues. The former felt that both specific projects and general policy for improving production were hampered by long consultative delays and excessive concessions to labour, even (perhaps especially) where labour was being most unco-operative. Quite apart from any disagreements at national level, the production members of the national board found it frustrating that Divisional Labour Directors (who had been recruited from the NUM) had so much say on new production projects which were drawn up by areas for their collieries. They seem at times to have felt that projects which were matters for managerial decision were held up and perhaps modified by long consultations in which both sides gave priority to union interests. Divisional boards sometimes complained of having too small a role. Complaints about this from Sir Hubert Houldsworth, the East Midlands chairman, began even before the vesting date;[1] and they came at times, though less frequently, from others. But it appears, though it was not put explicitly in this form, that divisional boards did not think they were greatly inhibited by control from above, but rather that they (and Houldsworth in particular) thought they were too little consulted about, and had too little influence on, the formulation of *national* policy. At area level and below, the feeling of remoteness from the top administration probably was widespread, as the critics maintained. Before nationalization most colliery agents and colliery managers had worked for firms small enough for them to have direct access to the boards of directors who were the ultimate decision-makers. Now they no longer had it. The enormous size of the nationalized industry made a much longer line of management necessary, and the problems of rapid communication and decision making along it had not all been solved. But most of the practical questions of management of collieries and colliery groups were settled without having to be referred up and up through the hierarchy. How much initiative the area managements were free to take varied from one division to another. Some divisions, the Scottish for example, appear to have kept a lot in their own hands, whereas others, such as the North Eastern, gave their areas more scope.[2]

The national board probably got the reputation of keeping too much in its own hands simply because in the first 18 months so much had to be done *for the first time*, and the setting of precedents was a matter with national policy implications.[3] But there was neither an intention nor a

[1] NCB 50th meeting, min. 589, 27.12.46. The national board held monthly conferences with the divisional chairmen.

[2] As one illustration see Abe Moffat (1965, 92) on his complaint to Scottish Divisional Board about the lack of authority given to colliery managers to settle disputes on colliery issues. The contrast between divisions which left more initiative to areas and those which left less appeared in the investment projects put forward through the fifties.

[3] For instance, it might seem excessive that the national board had long discussions at several

practice that the national board should thereafter involve itself in the detailed cases following the precedents. In general, the national board reserved for itself only five subjects: the determination of national policy, national negotiation with the unions on pay and conditions of service, the approval of capital expenditure over £100,000 on any one colliery scheme, the operation of some common services, and the supervision of coal distribution.[1] There were occasions when the board was too inclusive in what it classed as 'national policy', but these were isolated, not habitual.

Pressure for change built up quickly from the time Gaitskell became Minister of Fuel and Power in November 1947. He was opposed to functional boards and wanted the full-time members supplemented by a number of part-timers, and he thought the existence of functional board members made it harder to establish a clear chain of responsibility from the top to the bottom. He got a lot of support from the Ministerial Committee on the Socialisation of Industries.[2] Several changes soon followed. Gridley, the NCB member for marketing who had not been particularly happy with the way some things were going, resigned without fuss in February 1948 and soon afterwards rejoined Powell Duffryn as a director. Gaitskell replaced him by a non-functional part-time member, Sir Robert Burrows, the chairman of the London Midland and Scottish Railway Company until the nationalization of the railways, and former chairman of Lancashire Associated Collieries and of Manchester Collieries. The NCB relieved two of their members, Reid and Hallsworth, of their functional responsibilities, and on 4 May 1948 it was agreed that the national board need meet less frequently and that much of the business should be taken by a new General Purposes Committee, made up of the chairman and deputy chairman and the non-functional members, Reid, Hallsworth, and Burrows. At the same meeting it was also agreed to set up a committee to examine NCB organization.[3] Burrows's appointment to the board had taken effect on

meetings about the continuation or termination of the contributions formerly made by many colliery companies to religious bodies. But the practice was so widely dispersed that it was a national rather than a divisional matter, and the subject was taken up by the Archbishop of York who was not happy to deal with anyone of lower rank than the NCB chairman. After a temporary respite, the churches lost.

[1] C. A. Roberts, 'The Development of the Organisation', in *Colliery Guardian*, *National Coal Board*, 1957, 13. £100,000 was the capital expenditure limit in 1948. It gradually increased and was £250,000 in 1956.

[2] Chester 1975, 551–4.

[3] NCB 203rd meeting, mins. 239 and 240, 4.5.48.

1 May but he had not yet attended a meeting. He was made chairman of this committee on organization, and his colleagues were two industrialists, who remained outsiders, Sir Charles Renold and Sir Mark Hodgson.

On 13 May Sir Charles Reid, who had attended a meeting of the national board two days earlier without any indication of any impending change, suddenly announced his resignation from the NCB in a press statement which attacked NCB organization, over-staffing, and over-centralization. The NCB responded next day with a brief statement recognizing that they had been unable to accept his views about the retention of some features of organization which had existed under private ownership, but pointing out that many of the things to which he was now objecting resulted from decisions in which he had shared.[1] Reid's action and statement provided increased ammunition for the critics of the NCB, many of whom found their prejudices, as well as their knowledge, reinforced. Material for the attack was further increased in November when Reid summarized his views in three articles in *The Times*.[2] He very properly drew attention to the failure of attendance in the pits to get back to pre-war levels and of OMS to rise above them. When he turned to remedies he concentrated mainly on organization. He attacked the NCB for being incapable of taking a broad view because of the functional preoccupations of board members, for having inflated staffs at national and divisional headquarters, for stifling initiative by excessive centralization, and for being remote from managers and men in the collieries so that there was nothing to replace the loyalty which these had given to their companies before the war. He insisted that the profit motive should not be despised, and remarked sensibly that 'the nation demands a profit in the shape of cheaper coal'. His positive plan was to replace the existing divisions and areas by twenty-six area-based corporations, each with its own capital, financial responsibilities, and accounts, an executive managing director, a general mining manager, a sales manager, and a board otherwise made up of part-time local members, one of whom might be chairman. The national board would become, in effect, a holding company with direct responsibility only for general policy (including the appointment of corporation boards), overall finance, and wages policy.

This plan attracted the more comment because of its similarity to the proposals of Col. C. G. Lancaster in a pamphlet published in the same

[1] NCB 206th meeting, min. 712, 14.5.48.

[2] Sir Charles Reid, 'The Problem of Coal', *The Times*, 22, 23, and 24 Nov. 1948.

week by the Conservative Political Centre. Col. Lancaster, a former coalowner, was, of all MPs, the one most active and vocal in discussions of the coal industry. He favoured having twenty to twenty-five enlarged areas, whose general managers would have executive power for production and development. The national board would have no executive power on these matters, but would be responsible for wages and prices and research, and would offer general guidance on expenditure and development.[1] Schemes such as these quite rightly put an emphasis, that had been inadequately present, on commercial drive and initiative. At a different stage in the coal industry's history they might, with modifications, have worked. But in 1948 the number and size of the proposed corporations was too big for the coal industry's scanty supply of top-class managers, and they were so big that they could not have run all the collieries without some intermediate level of management. Moreover, though most of the members would have been part-time, the coal industry under either of these schemes would have had more boards than in the existing NCB organization. So none of these proposals, despite all the discussion they provoked, had any positive results.

In fact, more use was made of Reid's recriminations than of his proposals, which was unfair to him and a sad ending to a great career. Reid may have been influenced by two events shortly before his resignation. One was a decision by the national board (after consultation with the divisions) to give guidance on the procedure for upgrading men in the collieries. Reid objected strongly and said it was a matter that should be left to the good sense of colliery managers, but he was in a minority.[2] This was an occasion when Reid was probably right and the national board was too centralist in deciding what was a subject of national policy. But it was too exceptional to be the basis of a general charge. The other matter was a request by Lord Hyndley to his colleagues not to take on any additional headquarters staff.[3] Reid may have regarded this as support for his view about swollen staffs, but Lord Hyndley was not suggesting that there was no need for extra staff. He was pointing out that the NCB's financial position was such that they should not incur any extra costs on staff. Reid drew attention to some genuine weaknesses. For example, though he underestimated the great extent to which the balance of power in negotiations had tipped towards the unions, he was justified in criticizing the board's conduct of industrial

[1] *The Times*, 23 Nov. 1948 had a separate article on Col. Lancaster's proposals.

[2] NCB 196th and 197th meetings, mins. 538 and 555, 13.4.48 and 16.4.48.

[3] NCB 198th meeting, min. 578, 20.4.48.

relations. The NCB had a habit (as in the negotiations for an extension of hours agreement) of presenting certain conditions as indispensable and then retreating from them, and this threatened to make their bargaining position weaker than it need be. But Reid's criticisms as a whole were made in a way that was likely to lessen the receptivity of his ex-colleagues to them. Some of his generalizations were unsustainable and, uncharacteristically, he was not altogether frank in one of his statements. He claimed that he had continually pressed for the most vital changes in NCB organization. He had certainly spoken and been outvoted on some specific detailed questions, but on comprehensive structural reform he had proceeded only in private. In November 1947 he suggested to Lord Hyndley in a letter that board members should be relieved of departmental duties and that the seven divisions other than South Eastern (which should lose its divisional board) should each be converted into a separate corporation with its own managing director and a board of part-time members. But he asked Hyndley not to discuss the proposals with others until the principles had been agreed with him and Street. As Reid did not convince the other two about the scheme for divisional corporations, his colleagues did not hear of his proposals until his post-resignation statement to the press.[1] This was a way to ruffle feelings unnecessarily, though the board nevertheless agreed to consider his proposals after the Burrows Committee had made its recommendations.

November 1948 was a busy month for organizational argument and recrimination, for it was then that the Burrows Committee added its contribution to those of Reid and Lancaster. On balance the Burrows Report probably did more harm than good, and its influence on the subsequent running of the coal industry was slight. There were three reasons for its poor reception. First, it included attacks on individuals and on classes of employees and thus embarrassed the NCB when they considered publication. In the end, though the recommendations and the NCB comments on them were published, the complete report was not.[2] Secondly, parts of the report were unclear. The board had to go back to the committee members and ask them what changes of practice some of the proposals were meant to imply. Thirdly, some of the proposals displayed muddled thinking about managerial responsibility. In particular, the committee wanted all the chairmen of the divisional boards to be members of the national board, and the area general

[1] NCB 207th meeting, min. 723, 18.5.48.

[2] NCB 240th meeting, min. 1216, 12.11.48; NCB *Ann. Rep.* 1948, 233–6.

managers to be members of their divisional board. Besides increasing the boards to an unwieldy size, such arrangements would have greatly increased the proportion of members who were concerned to further the interests of particular constituencies, and would have given rise to many potential conflicts of responsibility. The NCB totally rejected these proposals. The Burrows Committee had a few better ideas, but its general clumsiness reduced their effect, except where it was pushing at an open door. One recommendation was that the creation of a separate supplies or purchasing department was worth examination, and the board agreed, but did not proceed from examination to action. This was a mistake because the lack of such a department was a serious and continuing source of waste. The other most helpful recommendation was that board members who headed departments should leave executive action to their chief officials. The lack of definition of the respective responsibilities of board members and directors-general of departments was the chief weakness of a functional board—the regular charge that functional members could not or would not pull their weight in dealing with general policy is contradicted by the record—and it was a good thing to remove this weakness.

The Burrows recommendation most closely in line with subsequent revised practice was that the national board should be enlarged by the addition of a second deputy chairman and up to three part-time members. The first clause of the Coal Industry Act 1949 altered the composition of the board so that there must be a chairman and not less than eight nor more than eleven other members, of whom not more than eight might be full-time. The Minister was empowered but not obliged to appoint a second deputy chairman from among the members.[1] Mr Robens, introducing the bill in November 1948, said this clause was there to give effect to the Burrows Committee's recommendation, but a change on these lines had been under discussion inside and outside government for many months before then. The bill, which also empowered the NCB to conduct some business outside Great Britain, was debated with much hefty swiping at the NCB, helped by numerous quotations from Sir Charles Reid, but with nothing that was new, little that was constructive, and much that was the reverse.

The NCB, for the most part, let the barrage pass by, though Street, in particular, tried to ensure that those within the industry were given some chance to understand what was really happening. Probably the most ignorant and unhelpful of the constant reiterations was that about

[1] 12 & 13 Geo. VI, c. 53, s. 1.

vast headquarters staffs. The true need was to have more, not less, and to be able to attract talent, not to have it repelled by morale-damaging attacks. Street eventually, but with too little publicity, put the position quite plainly, when he pointed out that the NCB proportion of non-industrial staff, 5 per cent, was low by the standards of private enterprise, and that as only $3\frac{1}{2}$ per cent of the staff were at the London Headquarters, the numbers there were extraordinarily few. He remarked that he knew of no other concern operating on a national scale which had less than six times as many staff in proportion in its head office.[1] There was a good deal in the NCB that might have been improved by careful analysis and criticism from outsiders. One of the few bodies attempting to provide this in a dispassionate frame of mind was the Fabian Society. The investigations made by the members did not go very deep, but they did produce some sensible comments on communication, training, and personnel policy, and showed a recognition that most of the worst features of the coal industry were deeply rooted in its history and not mainly attributable to current follies of the NCB or anybody else.[2] But most of the wave of criticism in 1948 and 1949 did little good to anyone.

In October 1949 the national board was revised in conformity with the new act. Sir Geoffrey Vickers had been appointed to the board within days of Reid's resignation and had been made responsible for manpower and welfare, and Young had been relieved of his functional responsibilities. So there had continued to be four functional members, for labour relations, manpower and welfare, finance, and science. Hallsworth, who had not been in good health, resigned on 30 April 1949 and Burrows on 28 October. They were both replaced by part-time members, and three other part-timers were added, so the board now consisted of a chairman, six full-time members (of whom four were functional), and five part-timers. The power to appoint a second deputy chairman was not used. The new members included three prominent businessmen, J. H. Hambro of Hambro's Bank, Sir Geoffrey Heyworth of Unilever, and Sir Godfrey Mitchell of George Wimpey and Co., and two trade unionists, Alderman Sydney Jones of the NUM and Gavin Martin of the Confederation of Shipbuilding and Engineering Unions.[3] The NCB did not fundamentally alter their mode of operation, though

[1] Street 1950, 13–15.

[2] See, in particular, M. I. Cole 1949, esp. 16–23. Other comments are in G. D. H. Cole 1949.

[3] Chester 1975, 554–6 discusses the changes in the structure of the board and the arguments preceding them. See also Gaitskell's notes in Williams 1979, 191–2.

the national board acted with a less crowded timetable and more delegation to the standing committees, but from this time onward they were better supplied with well-informed advice. Heyworth and Mitchell, in particular, were quietly helpful in many ways. And the board were left freer to get on with urgent work without the distraction of public clamour about organization.

Two matters in particular needed very full attention. One was a switch of emphasis from short-term expedients to the long-term reconstruction of the industry. The period of detailed enquiry and preparation was coming to an end and the NCB were preparing to unveil their plans for a revivified and expanded industry and to start implementing them. The other matter had the opposite character. In 1950 and again in 1951 there were fears of a repetition of the 1947 crisis of coal supply. Every likely counter-measure had to be looked at well in advance and applied where practicable.

Both these matters were dealt with. The reconstruction scheme was launched in 1950 under the title *Plan for Coal*, though, of course, its success or failure depended on actions during a good many years after that.[1] And the winter months of 1950–1 and 1951–2 were got through without disaster. The difficulties in 1950 arose from a sharp increase of internal consumption, by about 7 million tonnes, at a time when falling manpower in the mines was making the growth of output very slow. The increased consumption came mainly from expanded industrial activity, and so kept summer stockbuilding below planned levels. A cold autumn and early winter followed and put up the demand for space heating. Coal stocks remained much above the desperate levels of 1946 and early 1947, but memories of experience then were vivid and the question of coal supplies became a cabinet as well as a ministry matter. The government in November 1950 authorized the import of coal and the NCB bought American coal in the open market. From December 1950 to October 1951 1.21 million tonnes were brought in and there were also very small shipments from Canada, India, and Nigeria. From 1 January to 28 February 1951 householders were allowed to buy only 8 cwt. (406 kg) in the north and 6 cwt. in the south, although their total winter allocations were 30 cwt. and 24 cwt. respectively. The latter half of the winter was relatively mild, and distributed stocks at the end of April were rather higher than they had been a year earlier. But home consumption was still rising—it went up by another 7 million tonnes in 1951—and the government remained very worried, even though an

[1] pp. 199–205 below.

increase in Saturday working was helping output. Those worries were repeated in the NCB, whose official view was that distributed stocks for normal conditions needed to be 16–17 million tons at the end of every October, and that only distributed stocks above that level provided a reserve against worse than average winter weather. This was a rather timid view, but is perhaps understandable in the light of previous experience.

In the summer of 1951 the government was urging the NCB to produce marginal tonnage almost regardless of cost, and from September the government, at the request of the NCB, authorized another programme of coal imports, at first from India and later from the USA as well. Householders' nominal winter allocation was unchanged, but in the first three winter months they were allowed to buy only 12 cwt. (610 kg) in the north and 10 cwt. in the south. In fact, distributed stocks at the end of October were back to the best levels of recent years and the winter brought no special problems. Some of the import contracts were cancelled, though just over a third of a million tonnes were brought in during the winter. Industry suffered a recession in 1952 so the upward trend in coal consumption was interrupted, while there was some increase in supply. The result was that, at the end of the summer, distributed stocks, at not quite 20 million tonnes, were higher than at any time since 1943.

A period of intense official anxiety, and of some pressure on domestic consumers, had been safely surmounted, with the help of imports.[1] It is, however, reasonable to ask whether the anxiety need have been so great and whether the imports (except for some large coal for railways and householders) were strictly necessary. A recourse to imports in two successive winters was not helpful to the public reputation of the NCB, particularly as the cost of imported coal was about three times that of home produced and the difference in cost was a loss that had to be borne by the NCB.

A good reputation was much to be sought at a time when a newly reconstructed board was being tried out, and all nationalized boards had to start dealing for the first time with a government from the Conservative party, which had opposed their creation. The original members of the National Coal Board had been appointed for five years and their periods of office were due to expire on 14 July 1951. Sir Eric Young claimed to have evidence that the government was determined not to reappoint him, and he resigned with effect from 28 February 1951, and

[1] The basic facts and figures are in NCB *Ann. Rep.* 1950, 3 and 22–5; 1951, 31–5; 1952, 1 and 33–7.

four days earlier Sir Arthur Street died. So for several months the board had only four full-time members besides the chairman, and included neither a deputy chairman nor any mining engineer. The terms of the retiring members were extended to the end of July and from 1 August the board was restored to its full complement. Lord Hyndley and Sir Lionel Lowe retired in accordance with their own wishes. Sir Hubert Houldsworth, previously chairman of East Midlands Division, was appointed chairman of the board, and for the first time there were two deputy chairmen: one, W. J. (later Sir Walter) Drummond, from within the industry and one, Sir Eric Coates (a former Indian civil servant), from outside. The other newcomer was Sir Andrew Bryan who, until this time, had been HM Chief Inspector of Mines. The other members were reappointed. The part-timers had been appointed in October 1949 for three years. One of them, Sir Godfrey Mitchell, resigned on 31 January 1952, when the NCB were preparing to take over the Opencast Executive. As his firm, George Wimpey & Co., regularly took contracts for opencast mining, he avoided any possible conflict of interest. He was not replaced. In October 1952 the other part-time members were reappointd except for Sydney Jones, who was replaced by another mining trade unionist, Alderman W. Bayliss. By this time the national board had come to have a large preponderance of non-functional members.

Under the new board, methods of running the industry changed in significant ways. To some the changes amounted primarily to a removal of the practices which had produced the accusations of over-centralization. To others they involved slackness and the creation of uncertainty about where authority lay. The national board, in deliberately leaving more to the divisional boards, went so far on occasion as not to insist on divisional boards carrying out policies laid down by the national board, and not to call the lower formations firmly to account. Likewise, the divisional boards were encouraged to lighten their supervision of area management and, in so doing, they often failed to exercise effective control over budgets and forecasts. The relations between departmental officers of the formations at different levels also changed, often much to the distaste of senior officials. It was, on occasion, accepted that a specialist at divisional level might treat what issued from the corresponding specialist at national level not as instruction which he *must* follow, but as advice which he *might* follow if he chose; and this even though what was involved was the implementation of nationally determined policy. And it was also left increasingly to the lower formations to decide whether they would inform or consult the specialists in higher formations, even on matters which clearly had much more than local relevance.[1]

At first these relaxations were pursued informally, but there had long been a felt need to codify NCB practice, and attempts to do this led to the issue in October 1953 of a General Directive.[2] This document gave written authority to many of these new practices. Among numerous examples are these. Heads of departments at headquarters were instructed to maintain continuous contact with the corresponding departments at divisions, and in these contacts their position was to be *primus inter pares*—no more authority than that. Individual members of divisional boards were stated to be responsible only to their own boards; the only responsibility to the national board was the collective one of the divisional board as a whole. Divisional boards were told explicitly that they were bound to consider, but not necessarily to accept, the suggestions of headquarters departments. Divisional boards were told that they should act towards areas in a supervisory and co-ordinating capacity but should only minimally be managerial. Area General Managers were stated not to be subordinate to any individual member of a divisional board.

The General Directive was issued (after numerous amendments and compromises) with the agreement of the national board,[3] but the special features of the kind just illustrated were very much asociated with the chairman, who had developed ideas of this style of management when he was a divisional chairman. There were many (particularly, but not exclusively, officials in the headquarters departments) who intensely disliked these features and wanted to see a quite different style of management adopted. It was not only that they felt their own authority was being weakened, but also that the new style created 'grey areas' in which it was uncertain who was responsible, and that grey areas allowed scope for intrigue. Disagreement about administrative principles and methods was exacerbated by personal disagreements. Houldsworth brought an unwelcome change of personality to the head of affairs. One of his most charitable colleagues said that whereas Hyndley conducted cabinet government, Houldsworth's style was presidential.[4] But the difficulties went deeper and wider. Houldsworth was able to remain on

[1] C. A. Roberts, in *Colliery Guardian*, *National Coal Board*, 1957, 14.
[2] Fleck Report 1955, App. D, 95–100 is a reprint of this directive.
[3] NCB 373rd meeting, min. 116, 16.10.53.
[4] Sir Geoffrey Vickers, in conversation with the author.

consistently good terms with probably no more than a couple of his colleagues on the national board, and had uneasy relations with some of his senior officials. There was a spread of mistrust, and there were one or two others who had separate disputes among themselves. In particular, Coates (one of the two deputy chairmen) was involved in various disagreements. The necessary work of the board went on, but this was an unhappy period in their history and the atmosphere was not helpful to optimal achievement.

The main instrument in changing this unsatisfactory state of affairs was the Fleck Committee. An enquiry into NCB organization was discussed as a corollary to the preparation of the General Directive and the idea found favour. It was originally intended to be a board enquiry, perhaps with some outside help, but suggestions that it should be conducted mainly by officials met with objections, notably from Houldsworth, who said there were too few officials who were suitable (only one in the whole of headquarters, he maintained) and who would not be put in a potentially embarrassing position. So when an enquiry was approved, in December 1953, it was left to a committee of outsiders.[1] They were Dr A. (later Sir Alexander) Fleck, H. A. (later Sir Henry and then Lord) Benson, Sir William Lawther, Sir Herbert Merrett, and Sir Godfrey Mitchell. The committee was free to elect its own chairman and chose Dr Fleck, the chairman of Imperial Chemical Industries Ltd. The member most strongly influencing the form and substance of the report (which was unanimous) was Mr Benson, a leading accountant. The national board gave the committee wide terms of reference: 'To consider the organisation of the National Coal Board and to make recommendations to the Board'. But they also put twenty-two specific questions for answer. These were prepared mainly by Sir Eric Coates and were a mixed bag, but some of them obviously envisaged the possibility of relaxing even more the authority of higher over lower formations, and some important aspects of management were not covered at all by the questions. The committee, however, did not overlook these aspects, and dismissed some of the questions with an abruptness which suggested a low opinion of their significance.

The Fleck Committee spent more than a year on its work, received evidence from national, divisional, and area levels, from some individuals on the staff, and from trade unions and professional associations connected with the coal industry. There was probably a hope, perhaps an expectation, that a supportive report might endorse the recent

[1] NCB 373rd, 374th, and 376th meetings, mins. 116, 124, and 149, 16.10.53, 6.11.53, and 4.12.53.

changes of style and practice and enable them to become thoroughly established; but the committee heard all sides (it talked to members of the national board separately rather than collectively) and found the opponents of these practices much more convincing than their supporters. The Fleck Report was a concise, lucid, comprehensive, and definite document, which pounced on anything that appeared fudged, and proceeded to unfudge it. The basic structure of the NCB organization was judged to be sound, though there were proposals to simplify and make uniform the administrative formations between area and colliery. But the national board was thought to be in urgent need of reform. The committee took the view that full-time members needed specific as well as general knowledge and experience. Therefore most of them should be drawn from the coal industry and all of them, except the chairman and deputy chairman (of whom one was sufficient) should have a special field of responsibility, though they would leave to the officials who were heads of departments the executive duty of day-to-day management. So there should be a board consisting of a chairman, a deputy chairman, six functional full-time members, and four part-time members. This could not come about without new appointments to the board. The committee added a reminder that was significant in relation to previous, and especially recent, experience: that in making appointments the Minister should give special attention to the necessity for members to work together as a team. He should also secure continuity by arranging for members to retire at different dates.[1] There was also a need to rearrange the departments, in particular by a more rational organization to deal with personnel, and by the creation (as the Burrows Report had hinted) of a purchasing department. At divisional level, too, specific knowledge needed to be increased, and chairmen, as well as deputy chairmen, of divisional boards should normally be drawn from the coal industry.

Perhaps the most incisive of all the comments and proposals of the Fleck Committee were those relating to the methods, style, and personnel of management. The NCB had too few managerial staff and their average quality was too low. Therefore they ought to get rid of the worst, increase the staff, and train everyone to a better standard. The committee added that because the industry had inherited backward management, 'the positive direction and guidance which must emanate from the top must be very much greater than would normally be the case in a competitive industry whose pattern and traditions of manage-

[1] Fleck Report 1955, 11–17. The report was published in full.

ment had been established for a long time'.[1] Again and again, in one context after another, the committee insisted that direction from above and accountability from below must be more thorough and more firmly enforced than they had been, and there was a reminder that this could not be achieved unless the board gave to headquarters departments the support they needed and had not consistently received.[2] A key passage is this:[3]

> No organisation will work well unless the people who give the instructions and frame the policies also ensure that they are carried out. This should be done, not by interference in detailed or day-to-day matters, but by modern management techniques of approved programmes followed by periodical reports and reviews of progress and by physical inspections. The Board appear to have assumed that decentralisation means that they should not, or need not, impose their will on Divisions and Areas. We do not agree with this policy.

The general lightness of managerial control from above had been common in the NCB from the beginning and so had some of the particular features which the Fleck Report condemned. But some of the practices and, most strikingly, the confusion as to whether the relation of higher to lower departments was one of direction, or merely one of advice when solicited, were more recent.[4] In particular, it is impossible to see the committee's severe commentary on the General Directive of 1953[5] as other than a rejection of the Houldsworth style and method of management.

The recommendations of the Fleck Committee were accepted with very little modification, and the Minister agreed to those which could not be implemented without his intervention. The proposed new structure of the board was dependent on new appointments and, in any case, though some continuity was needed, it would hardly have been appropriate for all the former members to be responsible for the elimination of methods which they had promoted. There had already been some change among the older members. Ebby Edwards retired in September 1953 and was replaced by W. H. Sales, and in 1954 Sir Geoffrey Vickers indicated his intention to resign from 28 February 1955. In that month, in order to permit the implementation of the Fleck recommendations, all the full-time board members placed their resignations in the Minister's hands, and two of the part-time members, Heyworth and

[1] Ibid., 22. [2] Ibid., 20. [3] Ibid., 58. [4] Ibid., 21.

[5] Ibid., 69–71. A new General Directive on the lines approved by the Fleck Committee was issued in July 1955. It is reprinted in NCB *Ann. Rep.* 1955, I, 54–8.

Hambro, chose to go at the same time and were not replaced. The other two part-timers stayed until the end of the year, when their periods of office expired. In the reshuffle of February 1955 only Bryan and Sales were reappointed, and there were five new full-time members. One was the former trade union leader and chairman of the NCB Northern Division, James Bowman. He was made deputy chairman, with an expectation that he would succeed to the chairmanship when Houldsworth's term finished in the summer of 1956. Another new appointee was Joseph Latham, who had been the NCB's Director-General of Finance ever since nationalization. Houldsworth died on 1 February 1956 and Bowman succeeded him then, with Latham moving up to be deputy chairman and leaving the functional membership one below complement. Four new part-time members had been appointed at the beginning of the year.

The post-Fleck reconstitution of the national board took place in stages, however, and could not be said to be complete until the new chairman and deputy chairman were settled in office and supplemented by other new appointees. Further changes took place in 1957, most of them early in the year. H. W. Hembry came in to fill the functional vacancy left by the promotion of Latham. The two survivors of the pre-Fleck board ended their membership. Sir Andrew Bryan retired after fifty years in the mining industry and was replaced by J. O. Blair-Cunynghame, who had joined the NCB as Director-General of Staff as part of the implementation of the Fleck recommendations on the reorganization of staffing matters. W. H. Sales left to become Chairman of North Eastern Division, and J. Crawford, a trade unionist who was one of the new part-time members appointed a year earlier, became full-time to replace Sales as labour relations member, though from this time onward the chairman and deputy chairman usually took personal charge of major negotiations with labour. There was also a change on the production side. Dr William Reid, who had been appointed to the board only in 1955, became Chairman of Durham Division. He was replaced by H. E. Collins.

Thus by the summer of 1957 there was a new board, very different from that of the Houldsworth era; stronger at the top, not necessarily stronger throughout in personality and ability, but with more professional knowledge and more harmony. A fairly rapid implementation of most of the Fleck recommendations went ahead, in consultation with the divisions and the unions, with not too many or too serious signs of strain or resistance, though there were some.[1] The Fleck Committee had

[1] C. A. Roberts, in *Colliery Guardian*, *National Coal Board*, 1957, 16.

warned that no immediate significant results should be expected.[1] Indeed, the development of detailed procedures of supervision and accountability, which had previously been neglected, was bound to take time and involve some trial and error. The early changes, in fact, were probably not drastic enough, and time was needed for amendments in the light of experience. But there was at least the opportunity to prepare more effectively for future problems which, it soon turned out, were going to be very different from those that had constantly pressed, year after year. And there was the benefit of being rid of much of the persistent distraction by argument and uncertainty about organization. The passage in the Fleck Report which probably aroused the most general and heartfelt agreement was this:

> It has been put to us that, since it was created, the National Coal Board's organisation has been too often subjected on the one hand to special examination and on the other to ill-informed criticism without proper enquiry. Therefore, it has been suggested, the present review should be the last of its kind for some time so that the organisation may have a chance to settle down and get on with its job. We agree.[2]

Criticism did not suddenly melt into total silence, but that particular hope was realized.

### iv. Long-term reconstruction

It had always been intended that a major investment effort should go into the long-term development of the collieries, and a start was made before the end of 1947 with a few schemes and an examination of the possibilities of others. At the time of nationalization, projects for two new collieries, thirty-seven major reconstructions, twenty-one partial reconstructions, and four new drift mines had been prepared by the former owners and sponsored by the Ministry of Fuel and Power, and all these proposals were taken over by the NCB. It was practicable to start implementing a few of these fairly quickly, but it would have been unwise to adopt the lot as an initial programme. They all needed re-examination in a new context, for the list was likely to indicate the relative enterprise and resources of different undertakings more than the relative urgency of needs within the industry, and planning was no longer restricted by the boundaries of the old colliery leaseholds. In practice, not a lot was done for a few years about most of the larger

[1] Fleck Report 1955, 79.

[2] Ibid., 3.

schemes, though even in 1947 work (interrupted by the war) was resumed in sinking a new colliery at Calverton in East Midlands Division, and a start was made on the new Rothes colliery, which had been planned by the Fife Coal Co. The latter colliery, closely related to the building of the new town of Glenrothes, was one of Scotland's greatest economic hopes, but was to prove one of its great disappointments.[1]

The need to re-examine earlier schemes was not the only influence holding back long-term development. The sheer pressure of work involved in increasing immediate output; the need for new project planning to be done *in the coalfields* and the impossibility of doing this until the new area organizations were fully established; the shortage of engineers competent and willing to specialize in mine planning—all these had a retarding effect. Just as important was the abundance of small schemes which for five years or so (as already noted) could be carried out as useful piecemeal contributions to the creation of capacity.

But the need for a large reconstruction remained. Nobody could forget the demonstration in the Reid Report that big increases in productivity could not be achieved without redesigning the layout of, as well as re-equipping, many of the collieries. That was mainly the argument for reconstruction in order to use capacity more efficiently. There was also the argument based on the need for a continuous increase in total productive capacity. The evidence for that was found in the experience of coal shortage and currently rising consumption. All plans for long-term investment in the coal industry were related to necessarily hazardous exercises in forecasting the trend of future demand. All the forecasts indicated a rise in energy consumption continuing as far ahead as one cared to look, and throughout the NCB's first ten years it was taken for granted that by far the greatest proportion of that energy must come from coal. For most of that period there were few indications in relative prices and ease of supply that this assumption was erroneous, and, as yet, no awareness of irony could attach itself to the coal industry's many expressions of welcome to oil as a means of filling the gaps which the utmost attainable expansion of coal output must still leave. But by the middle fifties there were signs which might have induced some doubts.

Reconstruction was supported by one other argument which was well known, but the full weight of which was not realized for several years. Mining has special additional forms of depreciation which do not occur in manufacturing. A coal mine has the costly characteristic that in

[1] NCB *Ann. Rep.* 1947, 43–4 and 111–15.

the process of production it consumes its own workplaces and the access to them. These have to be continuously replaced, and much of the day-to-day routine development work of a mine is directed to this end. A new coal face is prepared, with appropriate access roads and equipment, ready for use as soon as an existing coal face nearby is worked out, and so on. But, sooner or later, in every mine, this is not enough. Entirely new districts in a mine have to be opened up with all their necessary services, and at some point they may be so far from the pit bottom that they have to have new access to the surface. And eventually a mine will become commercially unworkable and another will be needed. Thus reconstruction is necessary to replace capacity consumed. What is involved is no different from what is involved in creating additional capacity and, so far as concerns this kind of activity, it becomes impracticable to distinguish clearly between gross and net investment. In a time of declining demand the need for full replacement of capacity ceases to be urgent and, as it is costly, is likely to be neglected. It seems certain that, for many years before 1947, conditions in the coal industry had been such that many firms had only partly made good the capacity consumed. There was thus no recent experience to guide judgement about the extent of provision necessary for this purpose. In the fifties it came to be reckoned that investment had to be provided to replace at least 2 per cent of the industry's productive capacity consumed every year. (This, of course, is additional to the normal depreciation of buildings and plant.) Only after that does investment contribute to a net enlargement of productive capacity. Until the experience of the fifties, investment had been expected to expand the industry faster than it did.[1]

In 1948 and 1949, under the guidance of the NCB's Director-General of Production, E. H. Browne, there was a detailed examination of the condition and potentiality of every coalfield and every NCB colliery. This was related to expectations not only of total demand but of demand in different markets and for different types of coal. The results were brought together in a plan for the development and reconstruction of the entire British coal industry. Because the establishment of new collieries and the thorough reconstruction of others takes so many years, it was thought appropriate to extend the plan over fifteen years. The proposals, which incorporated projects already begun but still incomplete, were published in October 1950 under the title *Plan for Coal*. This was not presented as a rigid programme but as a guide which

[1] This aspect of reconstruction investment is discussed in E. F. Schumacher, 'Plan for Coal', in *Colliery Guardian*, *National Coal Board*, 1957, 61–4.

would necessarily be affected by market changes and by changes in technology that could not be accurately foreseen. It would be modified in execution as circumstances altered. On that understanding about flexibility it was formally approved by the Minister of Fuel and Power in the following year.

*Plan for Coal* noted that all the reconstruction projects so far begun belonged to a 'hard core' which would be wanted even if there were no expansion. Schemes to provide for expansion would have to be started soon if they were to make an adequate contribution to need, and reconstruction was also wanted in an attempt to check the rising trend of costs, in 'real' terms, which had been observable for at least seventy years. The plan was based on the estimate that in the period 1961–5 about 240 million tons (244 million tonnes) of coal should be produced and sold annually, of which 210 million tons (213 million tonnes) would go to the home market and the rest to exports and bunkers. The planned output was 18 per cent more than the actual output of deep mines in 1949, which is the appropriate basis of comparison, as it was assumed that opencast mining would have ceased before 1961–5. Manpower was projected to fall by 11 per cent from mid-1950, with gradual, but evenly spread, improvements in productivity more than offsetting that decrease. *Plan for Coal* was expected to cost (at 1949 prices) £520 million for collieries and £115 million for other establishments, mainly carbonization plants. The rate of investment was intended to be at its highest in the quinquennium 1951–5, when £190 million was suggested as the investment in collieries.

The plan had a broad strategy for reaching these objectives. For additional output it relied mainly on the reconstruction of existing collieries and only to a small extent on new collieries, 'because detailed surveys of production possibilities have shown that the greater part of future production must come from areas already explored and much worked in the past'; and also because many reconstructions were expected to give a better return on capital than new collieries. 67 major and 192 minor reconstructions, 22 new large collieries, and 53 drift mines were indicated. The expectation was that, when the schemes had been completed, the collieries with major reconstructions would supply 30 per cent of total output, those with minor reconstructions 40 per cent, new large collieries and drift mines 10 per cent, and about 250 collieries continuing with little change 20 per cent. Between 350 and 400 collieries, mostly small, would disappear, mostly by closure, though about 90 would probably be merged with neighbouring pits into larger units.

Partly because of the overriding concern to maximize total output and to maintain the supply of coals of special quality, the strategy did not seek to achieve the greatest possible increase in efficiency. Indeed, it was remarked that much greater reductions of cost could be achieved if the NCB sought only to maintain output at its existing level. Such a strategy would have given the largest possible share of investment to the coalfields which had, or were expected to achieve, high productivity and low costs, i.e. the East Midlands and Yorkshire and also (it was suggested) parts of Scotland, though the latter suggestion appears to have rested on hopes about East Fife and the Lothians that were over-optimistic. As it was, all these, along with the highly productive North Staffordshire, were among the regions scheduled for expansion, but so, too, for the sake of their special coals, were the generally high cost coalfields of East Durham, South Wales, and Kent. It is remarkable that the biggest percentage expansion suggested was in the anthracite areas of South Wales, where both mining conditions and financial results had been very poor, though it was made plain that this suggestion was more tentative than almost all others and its adoption dependent on an improvement in the appalling labour relations. Some features of a strategy of reconstruction that might have seemed ideal were also modified for external reasons. For instance, the degree of expansion proposed for the East Midlands and Yorkshire was reduced because of congestion on the transpennine railway routes and in the Humber ports, which suggests that national needs might have been better served by appropriate investment in transport rather than the diversion of funds to less attractive colliery projects. And the planned rundown of coalmining in Cumberland and West Durham was slowed because of the social damage it was likely to cause. This was an indication of an emerging conflict of aims that had great future significance. Despite these departures from strict economic criteria, however, it was believed that the measures of reconstruction should bring down the cost of producing coal by about 7 shillings per ton, though it was assumed that some of the saving would be swallowed up by higher wages.[1]

The adoption of *Plan for Coal* led to preparations to start implementing many more projects, but progress at first was disappointingly slow. It had been intended to build up investment quickly to a maximum to be attained by 1952. But actual capital expenditure at collieries was in 1950

[1] The foregoing paragraphs are based on NCB *Plan for Coal* (1950). The main elements are set out in 3–9 and the production proposals are further discussed in 25–30. Three tables on 10–12 summarize the basic figures.

only 76 per cent of that planned, in 1951 57 per cent, and in 1952 64 per cent. In 1952 it was only at about the same level, in real terms, as in 1949, before *Plan for Coal* had been brought in.[1] There was a good deal of anxiety that year, for a government appointed enquiry[2] suggested that the probable demand for coal in the early sixties had been underestimated and that 260 million tons (264 million tonnes) was a more realistic target than 240 million tons. At the same time recruitment was improving and collieries generally were finding that they could no longer provide any more jobs than there were men forthcoming to fill them. Until then it had been assumed that increased recruitment would *always* help to increase output. Now a limit had been reached beyond which that rule could not apply until more productive capacity had been created. Something had got to be done. Shortage of planning engineers and imperfections in managerial and administrative arrangements were delaying the detailed working out of projects. After this, more engineers were attracted into mine planning and there was time for even slowly prepared schemes to come forward. Higher levels of investment began to be regularly achieved, but the carrying out of *Plan for Coal* still lagged behind what had been intended. Of the 281 principal schemes (i.e. excluding the small drift mines) in *Plan for Coal*, only 20 had been completed by the end of 1955, and 147 were then in progress. On all purposes, including ancillaries, £353 million was invested from 1950 to 1955 inclusive, which was equivalent to £248 million at mid-1949 prices, only £20 million less than the investment proposed for these years in *Plan for Coal*.[3] So the deficiency in making planned investment was not very large, even though so few schemes had been completed. This indicated that projects were proving more costly than had been expected. The annual amounts of capital investment at current prices remained much higher after 1952,[4] as indicated in Table 5.2. By this time the severity of the problem of replacing consumed capacity had been recognized, and it had come to be adopted as a rough approximation that the whole colliery investment other than major schemes should be regarded as having been used up for this purpose. So only a very modest contribution to expansion had been made.

*Plan for Coal* was reviewed and updated in 1955, and revised figures

[1] NCB *Ann. Rep.* 1952, 17–18.

[2] *Report of the Committee on National Policy for the Use of Fuel and Power Resources* (Ridley Report 1952), Cmd. 8647.

[3] NCB *Investing in Coal: Progress and Prospects under the Plan for Coal* (1956), 9 and 12.

[4] NCB *Ann. Rep.* 1957, I, 25.

Table 5.2. *Capital expenditure, 1947-1957*
(£ million, current prices)

| | 1947 | 1948 | 1949 | 1950 | 1951 | 1952 | 1953 | 1954 | 1955 | 1956 | 1957 | Total |
|---|---|---|---|---|---|---|---|---|---|---|---|---|
| Collieries: | | | | | | | | | | | | |
| major schemes | 2 | 3 | 4 | 7 | 9 | 15 | 21 | 32 | 39 | 42 | 44 | 218 |
| other schemes | 13 | 18 | 23 | 18 | 18 | 23 | 31 | 36 | 35 | 34 | 40 | 289 |
| Ancillaries | 4 | 4 | 4 | 4 | 5 | 10 | 13 | 16 | 19 | 19 | 19 | 117 |
| House building | — | — | — | — | — | 2 | 17 | 15 | 2 | 1 | — | 37 |
| TOTAL | 19 | 25 | 31 | 29 | 32 | 50 | 82 | 99 | 95 | 96 | 103 | 661 |

*Source*: NCB *Ann. Rep.* 1957, I. 25.

were published with a brief commentary the following spring. Little change was foreseen in total demand, though the amount wanted for export was likely to be less, but the target of 240 million tons (244 million tonnes) of deep mined coal in 1965 was seen to be out of reach. It was thought that 230 million tons (234 million tonnes) might be achieved and the gap filled by continuing opencast mining, which had turned out to have a much longer life than had been thought possible and had become consistently profitable. Even the reduced deep mining target for 1965 was thought attainable only with some fairly large changes. These included the retention of 54,000 more men than had originally been proposed, the completion before 1965 of all the schemes already in progress and the completion of 100 more which still had to be started, as well as a start on additional schemes which would be incomplete in 1965 and might enable 250 million tons to be produced by 1970. The need for more men arose because the hoped-for reduction in absence had not occurred, and because saleable output was a declining proportion of what was raised and weighed from the collieries. All the extra schemes, on top of rising prices, required increased expenditure. It was estimated that, in addition to the £353 million spent in 1950–5, an investment of a further £1,000 million, at current prices, would be needed, £590 million of it in 1956–60 and £410 million in 1961–5. Of this total £860 million was for collieries and £140 million for ancillary activities.[1]

This was almost a last look for nearly twenty years at a prospect of expansion. The NCB observed that the use of oil would grow rapidly and there would be some contribution from nuclear power after 1960, but concluded that 'even in the longer term the problems of overproduction for the coal industry can scarcely arise.'[2] But the next revision of *Plan for Coal* had to be downwards because the market required it.[3] For the moment, however, the need was to press on with the completion of the many uncompleted projects and to try not to repeat the delays so far experienced. In 1956 the NCB set up a Reconstruction Department at headquarters, which had exactly this responsibility for the next few years, and corresponding departments were established in Scottish Division in 1956 and North Eastern Division early in 1957.[4] It was at this time that an attempt was made to improve both technical and

[1] NCB *Investing in Coal*, 13–17.

[2] Ibid., 13. Assumptions about the rate of growth of the use of other fuels are in ibid., 4.

[3] The basic details are summarized in NCB, *Revised Plan for Coal* (1959), 22–3.

[4] NCB *Ann. Rep.* 1956, I, 20.

managerial methods in shaft sinking and tunnelling. In 1957 revised arrangements were adopted for the approval of major projects and the control of their progress, so as to avoid delay and to establish quickly when anything was going wrong and apply a remedy. There was also an attempt to save time and reduce cost by preparing standard designs of colliery layout, major equipment, and certain classes of buildings.[1] By the end of 1957, 62 major colliery schemes had been completed, which was 30 more than a year earlier, but there were 166 which had been approved but were still incomplete.[2]

It was important for the health of the industry to bring more of the investment to fruition, for collieries which had not undergone major reconstructions were beginning to find it difficult even to maintain their output,[3] as appears from Table 5.3.

As far as it had been completed by 1957 it was looking as though *Plan for Coal* had been a safeguard against decline rather than a successful recipe for expansion. Of course, once the uncompleted majority of the projects had been finished, the means of some expansion would be there. But that was too late. After 1957 the market conditions for expansion were never present. So the reconstruction investment had to justify itself on rather different grounds, viz. in terms of improved efficiency and cost-saving alone. And that, if achieved, was likely to make most of the unimproved sector of the industry look dispensable. It was a weakness that, while the market was still growing, the proportion of the industry that had been so improved by reconstruction as to look indispensable was not as large as had been intended.

## v. The later years of high demand

The increasing emphasis on long-term reconstruction suggested that there was no longer such great sensitivity to the pressures constantly exerted by the urge for a bigger coal output here and now. It was not that the pressures disappeared, but there was an awareness that they could be contained. Threats of renewed crises in coal supply in both 1950 and 1951 were met without even near disaster and seemed to have been exaggerated. Things were unlikely to get so much worse in any

[1] Ibid. 1957, I, 22–3.

[2] Ibid., I, 24.

[3] Schumacher, in *Colliery Guardian*, *National Coal Board*, 1957, 63. Even in 1957 new collieries and drift mines contributed only just over one-eighth of the output of the first group in Table 5.3.

Table 5.3. *Output changes of collieries by groups, 1947-1957*

| | 1951 output as percentage of 1947 output | 1952 output as percentage of 1947 output | 1955 output as percentage of 1951 output | 1957 output as percentage of 1952 output |
|---|---|---|---|---|
| New collieries, drift mines and completed major reconstructions | 125 | 128 | 126 | 122 |
| Collieries scheduled for major reconstructions | 116 | 118 | 102 | 99 |
| Collieries continuing without major reconstruction | 122 | 121 | 100 | 97 |
| Collieries likely to close within fifteen years | 95 | 93 | 88 | 75 |
| All collieries | 114 | 115 | 100 | 98 |

*Source*: NCB *Ann. Rep.* 1957, I, 24.

future year that they could not be coped with, and they were unlikely to get much better without a lot of reconstruction schemes. This was particularly so because hopes of higher achievement from better attendance and fewer labour disputes had been deferred for so long that they were ceasing to be hopes, and because highly productive small improvements in the collieries were becoming harder to find.

Many and varied as were the daily activities, the heart of the coal industry's task was still to try successfully to balance supply against demand. In the half dozen years after the alarms of 1951 this was accomplished without enormous difficulty but also without any glory, particularly as it appeared that the industry and the country were settling down to use imports as a permanent cushion against any immediate shortages. From 1951 to 1957 inclusive, coal output remained on a plateau, always above 225 million tonnes but never surpassing the maximum of 228.4 million tonnes reached in 1952. Gradual increases in the productivity of faceworkers, interrupted only in 1952, were accompanied by a more fluctuating performance from other workers; and the slight improvement in OMS overall (which was not sustained between 1954 and 1957) was generally offset by the effects of increased holidays, absence, and disputes, so that the output per man-year reached in 1951 was not exceeded and from 1954 the trend was slightly downwards.[1] On the other hand, the home consumption of coal rose steadily in every year to 1956 except 1952, though more slowly than it had been doing down to 1951.[2] In 1957 home consumption was 5.3 million tonnes less than in 1956, a reduction which was identified at the time as a temporary fluctuation attributable to the first industrial recession since early 1952 and to the exceptional warmth of the year, when average temperature was 2.1 °F higher than in 1956: the reduced space heating needed by domestic and other consumers could explain about 60 per cent of the drop in consumption.[3] It may well be that this diagnosis was correct for 1957, and that it was fortuitous that demand fell then rather than a little later, but the turning point in that year proved to be the beginning of a long downward trend, not a brief fluctuation.

Until 1957 the discrepancy between the trends of output and home consumption was covered by variations in the amounts exported, and by imports. No risks were taken with stocks. In 1953, when about 0.6 million tonnes were imported (more than half from France), the reason was to insure against a possible shortage of large coal,[4] and this was always a factor encouraging a willingness to import. But in other years the shortfall in total supply was even more influential. Conditions were particularly bad in 1955, when losses of output from industrial disputes were greatest, and imports, at 11.8 million tonnes, were nearly as much as exports and threatened a new sort of crisis, because the ports were not designed for much coal import and could only just about cope with that amount. The NCB and the government decided that this sort of thing could not continue, but the needs of 1955 had led to the signing of some contracts which ran on, with the result that 5.3 million tonnes were imported in 1956. Even in 1957, when a surplus of some kinds of coal began to emerge, imports were still nearly 3 million tonnes, in order to meet the shortage of large coal; but the government then

[1] NCB *Ann. Rep.* 1957, I, 1.
[2] Ibid. 1956, I, 34.
[3] Ibid. 1957, I, 28.
[4] Ibid. 1953, 29.

decided that imports should not continue after deliveries under the 1957 programme had been completed.[1] Altogether 25.9 million tonnes of coal were imported from 1947 to 1957 inclusive, more than three-quarters of it after the beginning of 1955. Roughly half came from the USA and nearly one-fifth from France, with Belgium as the third largest supplier.[2]

This policy was good for neither the public image nor the finances of the NCB. It certainly looked odd that deficiencies in the UK should be made up from a country so much more poorly endowed with coal as France. Finances were injured both indirectly and directly: indirectly because of an immediate reduction in exports, which were more profitable than home sales, and a developing loss of export markets because of unreliability of supply; directly because imports cost more than home production but were sold at the same price. British coal was produced more cheaply than that in continental western Europe, and transatlantic freight charges were so high as to put the landed cost of American coal far above the cost of the home product. From 1947 to 1957 the NCB had a loss of £69.3 million on imported coal, which was more than double the total accumulated financial deficit for the period.[3]

The obvious defence of this state of affairs is that it was bound to happen unless either the demand for coal was forced down or greater production was achieved. The government had no wish to reduce economic activity by a denial of fuel, and the coal allocation system was operated as unrestrictively as possible for all except householders and coal exporters. The idea of using the price mechanism to reduce demand was not pursued very far, though, as has been seen, the NCB used it (probably not quite enough) to switch demand from scarcer to more abundant types of coal. The main consideration of possible switches in demand was associated with the Ridley Committee of 1952. The committee required some of those who gave evidence to look at the subject, as one of the purposes was to try to optimize the use of all the main competing sources of energy. In fact, the committee became rather enmeshed in a mass of detail, and the members could not agree among themselves on some of the fundamentals of price policy. The final recommendations had a miscellaneous character and involved some additional bureaucracy and additional expenditure, which did not

[1] NCB *Ann. Rep.* 1955, II, 140; 1956, I, 44; 1957, I, 31 and 33.

[2] Ibid. 1956, I, 45; 1957, II, 148.

[3] Ibid. 1957, II, 2–3 and 8–9. From 18 July 1955 the NCB ceased to treat losses on imported coal as a separate item in their profit and loss account and instead simply deducted the amount from the total income derived from sales of coal, but the figure given in the text includes the losses incurred from 18 July 1955 onward.

commend them to either the government or the nationalized industries. So they did not have much lasting influence. But the exchanges between the committee and the NCB are revealing on a few points.

In preparing evidence the NCB were anxious to show that in order to meet the expected increase in demand it was much cheaper to invest in the reconstruction of the coal industry than to import the additional requirements, and to show that in certain respects the cheapness of coal militated against economy in its use.[1] On the last point the NCB do not appear to have been altogether consistent, for, though they were anxious that electricity tariffs should not encourage greater off-peak consumption or more use for space heating, they were markedly unenthusiastic about a large rise in the price of coal. Their main evidence was given in a written memorandum,[2] but their representatives also gave oral evidence. Having been asked whether the board had considered the desirability of raising coal prices, they expressed doubt as to whether this could be reconciled with the statutory requirement to have regard to the national interest. They noted that if average prices were raised by £1 a ton the NCB would make a profit of £200 million a year. This, they claimed, was much too large to put to reserve because it far exceeded annual capital expenditure. Moreover, easy profits would encourage inefficiency and lead to large wage demands from the miners to have their share. So the result might be to reduce output. In their report the members of the Ridley Committee were evenly divided about coal pricing policy. Half the members wanted to move towards marginal cost pricing and to make a start by adding £1 a ton to average cost price, with the resulting surplus treated as a royalty payable to the state. Only thus, they argued, would consumers pay an economic price for coal. The rest of the committee would have none of this. They maintained that such a rise would not call forth an increase in the output of coal, and all the burden of adjustment would be placed on consumers.[3] Such a disagreement is a sure way for a committee to get itself ignored. Nothing was done to modify policy about the general price level of coal (as distinct from the internal relativities of the price structure) in order to influence demand. It was a question that might have deserved a little more attention in a different context from that of the Ridley Committee, but for the time being it was effectively buried.

[1] NCB 334th and 337th meetings, mins. 184 and 200, 2.11.51 and 7.12.51.

[2] Printed in full in Ridley Report 1952, 116–28. The oral evidence was not published but is included in PRO, POWE, 28/197.

[3] Ridley Report 1952, 14–19.

One of the less prominent recommendations of the committee was that an attempt should be made to make more coal available to the public by paying the miners to give up their concessionary coal.[1] This was a proposal easily set aside because the report was made to the Minister of Fuel and Power who could rightly say that this was not a matter for him but for the NCB. The latter did not want to touch it, not merely because the miners were most unlikely to agree to it, but because it was so complicated that any negotiations would be almost endlessly detailed, and because it was a very delicate element in industrial relations. Arrangements about concessionary coal varied from district to district, not only in the amount of the entitlement but in the agreements about who qualified for it and who did not. When the subject came up in negotiations anywhere it was never because anyone wanted to have less coal and more money, but usually because some excluded group sought to receive concessionary coal, or because some change was sought in the differences between the entitlements of different groups. The familiar problem for the NCB was how to maintain or achieve equity without incurring higher costs.

It was still one of the ultimate aims to try to improve output by lessening the number of potential causes of dispute, and few people wanted to risk increasing this number by starting arguments about concessionary coal. This was particularly so in 1952 when, after years of unsystematic wage bargaining in despite of the professed aim of establishing a national wages structure, something was at last being attempted to make the wage system more rational. Late that year pressure on the NUM and the NCB to complete this task was deliberately increased by the coal industry's National Reference Tribunal, which declined to consider a union wage claim lest it jeopardize progress on the revision of the wages structure.[2] Both sides took the hint, and it was believed that a daywage structure could be established by the end of 1953, though no date could be forecast for piecewages.[3] In fact, even when a determined effort had begun, the complications were too great to permit speed, and at the end of 1953 wage claims still had to be settled without reference to a new structure. At that time, indeed, the NUM chose a new target and urged the NCB to revoke the 'Gentlemen's Agreement' on coal prices. The union argued that the agreement caused industry to be subsidized by low coal prices and also denied the NCB

[1] Ridley Report 1952, recommendation 15.

[2] Handy 1981, 38.

[3] NCB 357th meeting, min. 164, 21.11.52.

revenue to make adequate wage increases. The board were not free to deal with the price agreement in that way, but were reminded by one of their trade unionist part-time members (Gavin Martin) that they would be subject to that sort of wages pressure as long as they failed to evolve a rational wages structure.[1]

It was only in 1954 that, daywagemen's work having at last been classed under 300 well-defined job names which were then divided into groups, colliery managers could be asked to allocate their workers under the new classification. This trial exercise revealed that for 300,000 workers there were 10,000 different wage rates in operation, and the scatter of wage rates was haphazard in relation to the skill and responsibility in the new job categories. All this had to be sorted out without large numbers of workers being alienated from new proposals. At the same time the problems of the piecework wages were still being closely examined. The NCB hoped that a revised wages structure could be evolved for pieceworkers in order to give some control over the persistent upward 'creep' of earnings, which was said to be adding 4*d.* a ton to costs of production every year.[2]

The problems of rationalizing the piecewage structure proved too difficult then and for many years after, but in 1955 agreement was reached on a national daywage structure and applied from April to the 400,000 men concerned. Such a structure needed clear differentials between jobs and grades, such as had not previously existed, and it was bound to conflict with many existing local relativities and so to run the risk of provoking unrest. An agreement was reached whereby no one's wage rate was reduced, and 95 per cent had an increase of at least 1*s.* (5p) per shift. Differentials were fixed of 1*s.* per shift between each of the ten grades (five underground and five on the surface) of mineworker, and 2*s.* 6*d.* (12½p) per shift between each of the three grades of craftsmen. Other wages questions were settled at the same time: the national minimum rates were raised by 1*s.* 11*d.* (9½p) per shift, and proportionate adjustments were made in the wage structure for youths under 21. The deal added £14 million to the NCB's annual costs. In view of the complexity of what had been done, the new structure was accepted with remarkable smoothness. The union shared with management in deciding the appropriate job classification for the work of each individual. The whole exercise was completed in three weeks and only 6,000 of the 400,000 workers disputed their classification. The disputed cases were

[1] Ibid. 375th meeting, min. 136, 20.11.53.

[2] Ibid. 392nd meeting, min. 125, 1.10.54.

settled by a national grading committee composed of NCB and NUM representatives.[1] It was a worthwhile, though belated, exercise, for it not only revealed in advance more of the implications of any wage bargaining, it also permitted a much clearer managerial view of the rationality of manpower deployment throughout the industry. It could not bring an outbreak of industrial peace, but it did greatly reduce both the number and the obscurity of potential points of conflict. As it happened, 1955 was a very bad year for disputes, but the stoppages which caused the great loss of time and output were among pieceworkers.

The introduction of the national daywage structure necessarily had repercussions on those in supervisory positions. For colliery under-officials there was a new wage agreement in January 1956 which came much nearer than before to a national wage system. For weekly paid industrial staff, whose wages had been settled locally, a separate exercise in identifying similar jobs among a multiplicity of local names and then defining and grading the reduced list of job names was carried out. It led to the adoption of a uniform wages structure from August 1956.[2] So the main daywage agreement produced some additional useful tidying up.

Other major agreements between the NCB and the unions were less encouraging. Apart from the annual claims for wage rises the most prominent question concerned the qualifications for the five-day week bonus. Except where absence was for certain defined acceptable reasons (and the list of these had been extended from time to time) a man lost the bonus (i.e. pay for one shift more than he had worked) unless he worked the full five shifts of a normal week. In October 1955 the NUM adopted a new Miners' Charter which included the removal of this disqualification as one of the principal aims. In 1956 the union pressed this demand more strongly than any other. The NCB resisted the claim on the ground that the concession would encourage more absence, which was bad enough already. The board's resistance was maintained for many months, but there was a reluctance to push matters to an open breach, particularly because the board wanted to get another annual renewal of the agreement for Saturday working, so various modifications were offered. One offer was to remove the disqualification for all men with a satisfactory attendance record over a period of several

[1] NCB *Ann. Rep.* 1955, I, 38–40. For details of remaining anomalies (in particular 'personal rates' arising from the undertaking not to reduce anyone's wage rate) and the slowness with which they were reduced, see Handy 1981, 39–41 and 87–90.

[2] Handy 1981, 44–6. Weekly paid industrial staff held supervisory jobs on the surface at collieries or in central workshops. The number of separate job names among them was reduced from 837 to 65.

immediately preceding weeks, but this scheme, though acceptable to the NUM executive, was rejected by a coalfield ballot. So was a later offer to remove the disqualification in return for union support for a campaign to fill all vacancies, with British workers where available but otherwise with foreigners. (This was the time when Hungarian refugees were seeking placements.) In the end, with union leaders saying 'Trust us' and undertaking to try to ensure there would be no significant loss of output, and in the light of improved output in the early months of 1957, the NCB gave way. From 1 June 1957 the only disqualification remaining was the denial of the bonus to a man who lost one or more shifts because of a strike at his place of employment.[1] But by September the NCB were complaining that the sequel was already a decline in attendance and output; and in December Bowman remarked to his colleagues that, although the NUM had pledged their honour that output would not be prejudiced by the concession, their spokesmen were now publicly challenging his assessment when he started a campaign for better attendance.[2]

In 1957 it did not look as though improved arrangements for labour relations were achieving much for output and efficiency. For that it seemed better to place hopes on the recent reforms of management and extensions of training, on the coming to fruition of more of the protracted reconstruction projects, and on the recent achievement of much more fundamental advances in technology, especially in mechanized mining.[3]

## vi. International relations

To anyone in the industry the international relations of coal meant essentially one thing: exports. Coal was traditionally a large export industry and, even though the market had weakened in the inter-war period, exports and foreign bunkers were still taking just over a fifth of British coal output in the later thirties. In the last year of peace, 1938, they came to 47 million tonnes. Though exports almost ceased during the war, it was taken for granted that, as the post-war scene returned to 'normality', the coal industry could and should seek to restore its export

[1] The matter was discussed at practically every NCB meeting from the 415th on 3.8.56 to the 426th on 15.3.57. There is a summary account in NCB *Ann. Rep.* 1957, I, 43–4.

[2] NCB *Ann. Rep.* 1957, I, 9–11, suggests the result by presenting the relevant statistics before and after 1 June and removing bias by separating out those figures affected by the influenza epidemic in September and October.

[3] On the timing of technological advances see Chap. 3 above, especially section iii.

trade, though it was realized that the demand for bunkers had shrunk and would go on shrinking. Some exports were fairly quickly restored, but the quantities were nowhere near those of the thirties; for a time, as has been seen, the feature of international trade in coal was not an all-out export drive but the balance between exports and imports; and the industry's international relations became political and institutional as well as commercial.

The great shortages of coal, particularly in Europe and to some extent in other parts of the world, were for about three years from the end of the war; but in that period Britain had very little available for export, and for much of 1947 coal exports stopped. Coal supplies were allocated by the government, and the allocation system included a programme for exports, drawn up after the receipt of advice from the NCB and taking account of any international agreements entered into by the government. Small amounts outside the programmes were also exported if and when they were available. In 1946, 1947, and much of 1948 the gap in west European coal supplies was filled, so far as it was filled at all, mainly by the USA and, to a lesser extent, Poland, but by 1948 western Germany and Britain were also supplying substantial quantities and there were significant exports from France. By 1949 the European coal output was almost back to pre-war levels, and the European demand for coal imports was below the pre-war level. This was mainly because the experience of recent years had persuaded several countries in continental Europe to turn to oil and hydro-electric power to a much greater extent than Britain had done; and France, the largest importer of coal, had also developed some additional coal production at home. In 1949, nevertheless, British coal exports increased appreciably, partly because coal output in the USA was greatly reduced by a long strike. In that year exports and foreign bunkers together reached 19.5 million tonnes, but that total was never attained again. Sales for foreign bunkers declined in each successive year after 1948, but cargo exports fluctuated. After falling heavily in 1951, during the scare about the level of home supplies, they picked up again and in 1953 and 1954 were only a little less than in 1949, but then they declined once more.

The strongest reason for the fluctuations was the variation in the tightness of supply at home, as demand irregularly increased and production failed to keep pace with it. Both the NCB and the government reduced exports only reluctantly. They wanted the revenue, and they believed that in the long run they would be able to supply an export market reliably and they needed to preserve it until then. With this in mind they sometimes refrained from pushing a short-term advantage very hard. The NCB, in general, followed a policy of pricing exports commercially, but chose to follow rather than lead the market. In 1949, when American coal was scarce for several months, they refrained from putting up export prices to the maximum that was then attainable because they believed that loss of goodwill would damage their long-term interests. In the late forties and early fifties coal was sold on average for about £1 a ton more for export than at home, and by the mid-fifties this premium had risen to about £1.25. But the extra profit on coal exported was much less than the loss sustained in importing a similar quantity, so the experience of 1954 and 1955, when imports built up rapidly, caused doubt about the good sense of exporting any appreciable amount, and export programmes were reduced. The recession in home demand in 1957 suggested in the course of that year that export programmes might be expanded again, but by then it was too late. The export market was showing signs of contraction, and competition to supply it was growing; and the difficulties two or three years earlier had caused some of the larger foreign importers to place long-term contracts for coal from the USA and these were still running. So it was impossible to arrest the downward trend in British coal exports. Altogether, from 1947 to 1957 inclusive, exports totalled 113 million tonnes, and foreign bunkers 37 million tonnes.[1]

For the whole of this period most of the export market was in Europe. The only large markets elsewhere were Argentina in the earlier post-war years and Canada, which was a rather special case because it had a strong demand for anthracite which the British government was desperate to satisfy as a means of earning hard currency. Indeed, early in 1948 the NCB gave way to pressure from the government to sell anthracite to Canada even if it involved a loss (for which there would be no government compensation); and later that year the Chancellor of the Exchequer went to Canada and promised 350,000 tons of anthracite without checking that such a quantity was available. It was not until early the next year, after some increase in output, that it was decided that the target could be met, and the NCB made the minimum stipulation that the f.o.b. price must not be below the home market price—not a strong commercial position, as anthracite was invariably sold at a loss

[1] NCB *Ann. Rep.* 1949, 52–5; 1956, I, 44–5; 1957, I, 31. The Ministry of Fuel and Power and the NCB argued at length in late 1954 the case for and against substantial exports if imports were also necessary. NCB 391st to 394th and 396th meetings, mins. 116, 123, 133, 141, and 161, 17.9.54, 1.10.54, 15.10.54, 5.11.54, and 17.12.54.

at home.[1] A few years later even the Canadian demand for anthracite diminished, though it did not disappear, and the European share of British coal exports became even greater. Anthracite was a fuel of which there remained a shortage in various countries besides Britain, and in 1956 and 1957 about a third of the British anthracite output was exported (even some of the tiny Scottish output went abroad).[2] But the market for other coals, especially for general purpose coals, narrowed. In 1949, partly from normal commercial considerations, and partly as a result of internationally agreed allocations in a time of scarcity, five countries (Eire, Denmark, France, Italy, and Sweden) each bought over a million tons of British coal, and Argentina and Spain not a lot less. By the mid-fifties only Eire and Denmark remained as large customers. The Netherlands, France, and Gemany came next, but their combined demand was slightly less than that of Denmark alone.[3]

This was a disappointing outcome after so much political emphasis had been put on coal exports in the immediate post-war years, but it was not altogether surprising. A succession of international political organizations had tried to sort out problems of the supply and allocation of fuel, so as to lessen the effects of scarcity and to reduce the scarcity as soon as possible; and these efforts had been supplemented by bilateral agreements between governments. In all these activities Britain was fully involved. In general, British commitments were met and occasionally surpassed, but Britain was usually unable to offer unanticipated additions at moments of crisis, though in 1948 more British coal was supplied to countries participating in the Marshall Plan for European Recovery than had been promised.[4]

At the end of the war a European Coal Organization was set up, and this was absorbed in 1947 by the Economic Commission for Europe (ECE) which established a coal committee. The European Coal Organization drew up a plan for supplies and their distribution, which put impracticably high demands on the British coal industry. The plan proposed that in 1951 British output should be 249 million tonnes, of which 29 million tonnes would be exported, but reality came nowhere near that.[5] In 1947 the Marshall Plan was launched, and Britain was heavily involved in that and in the Organization for European

[1] NCB 182nd, 189th, and 263rd meetings, mins. 303, 421, and 229, 27.2.48, 19.3.48, and 29.4.49.

[2] NCB *Ann. Rep.* 1957, II, 146–7.

[3] Ibid., II, 144–5.

[4] Ibid. 1948, 63.

[5] Jensen 1967, 12 and 16.

Economic Co-operation (OEEC) which grew out of it and which also had a coal committee. The NCB noted as early as February 1948 that coal shortages in Europe were turning out to be less severe than had been expected,[1] and the NCB and the UK government were able to participate successfully at a realistic level in coal supply under the Marshall Plan. In the early fifties the coal committees of the ECE (in Geneva) and the OEEC (in Paris) made an effort to co-ordinate their activities. The main task was to secure agreement on the international allocation of imported coal, mainly from the USA and Germany, but occasionally there were more specific pressures and tentative essays in wider international proposals which were regarded with slight irritation in Britain. In 1951 the ECE, for instance, asked for extra coal exports in the third quarter, and it was urged in the Ministry of Fuel and Power that, as this was quite impossible in view of British shortages and as Britain was already supplying all that the government had promised, the Foreign Office should simply say 'no', with the minimum of explanation.[2] Lord Hyndley, in particular, had been rather suspicious of the ECE coal committee, fearing at one point that it might take the first steps towards the creation of an international coal cartel which would be against British interests while Britain had little surplus available for export, and wondering whether the committee ought not to be wound up. But some of his colleagues valued the contacts they maintained there and the opportunity to take a full part in the discussion of such subjects as the prices to be paid for internationally traded coal.[3]

Both the UK government and the NCB thus had plenty of experience of international bodies dealing with coal in the immediate post-war years, and had found them extremely useful but sometimes inclined to make impractical and unwelcome suggestions about the British coal industry and trade. No doubt this experience had some effect on British reaction when, in February 1950, Robert Schuman proposed the pooling of the coal and steel industries of France and Federal Germany and of any other European countries that wished to join. This was the origin of the European Coal and Steel Community of France, Federal Germany, Belgium, the Netherlands, Luxembourg, and Italy, which came into existence in July 1952. The decision of the British government to send only an observer to the negotiations in Paris for a draft treaty, and not to join the community, was taken on general political grounds. There have

[1] NCB 177th meeting, min. 237(b), 13.2.48.

[2] PRO, POWE 41/20.

[3] NCB 263rd and 279th meetings, mins. 230 and 457, 29.4.49 and 16.9.49.

been suggestions, which may be unfounded, that France deliberately put its proposals, which required participants to commit themselves in advance to the principle of placing their coal and steel industries under a common authority, in a form that would make them unacceptable to the UK,[1] but there were many signs that the participating countries wanted a close association with the UK. The UK Cabinet had set up a working party of officials in April 1951 (i.e. when the Schuman treaty was being signed, but long before it was ratified), with representatives from the Iron and Steel Corporation of Great Britain and from the NCB, who sent as one of their two members the young Derek Ezra who, years later, was to be the board's longest-serving chairman. That working party concluded that full membership would have positive market advantages and no unwelcome policy restrictions for coal, but fewer advantages and more risks for steel. The members recommended that the UK should seek at least formal association and possibly limited membership of the ECSC.[2]

The Cabinet in July 1951 agreed unanimously to enter into a relationship with the ECSC, but wanted, at least to begin with, something less definite than the formal association which was the least close of the options recommended by the working party. The NCB were acquiescent but, at board level, feelings never appeared more than lukewarm. Their view was expressed to the Ministry of Fuel and Power in the statement 'that without accepting any prior commitments, we should put ourselves in a position of being able to come to reasonable understandings with the Community from time to time on particular subjects'.[3] The UK government decided to set up a permanent delegation to the ECSC at Luxembourg. Its leader, Sir Cecil Weir, presented his credentials on 1 September 1952. There was provision for both the NCB and the NUM to nominate one adviser to the UK delegation, and their respective choices were Derek Ezra and Hugo Street. The President of the High Authority of the Community, Jean Monnet, made it clear that he wanted the UK connection to be as close as possible, with the UK representatives participating fully in most of the early discussions.[4] Even then the NCB remained somewhat suspicious. They feared the ECSC members would control the regular coal trade and leave for the NCB only the fluctuating residuum, they thought the community's working

[1] Jensen 1967, 27.
[2] PRO, POWE 41/29.
[3] PRO, POWE 41/33.
[4] NCB 352nd meeting, min. 122, 5.9.52.

parties would be unfruitful, and they were reluctant to supply detailed information.[1] When it became known that the government was contemplating the making of an agreement of association with the ECSC, they not only complained that they and the steel industry had not been consulted, they stated repeatedly that they thought there should not be a treaty, and that if the government, for wider political reasons, insisted on having one, they wanted specific safeguards.[2]

In the end the government signed an agreement of association with the ECSC on 21 December 1954,[3] and ratified it on 17 June 1955. It continued in force until the UK joined the European Communities at the beginning of 1973. Both the NUM and the NCB declared themselves satisfied—Houldsworth was one of the three signatories on behalf of the UK—and congratulated themselves that nothing had been done to circumscribe British freedom of action in the last resort. They need not have worried. The judgement of the 1951 working party that the British coal industry had nothing to fear from the Community's system of regulation was almost certainly sound, especially as coalmining was the one economic activity in which Britain could have been the big fish in the Community pond. And in any case, the regulating system was not included in the agreement of association. The agreement set up a Standing Council of Association between the UK and the High Authority of the ECSC and provided for the continuous exchange of information and consultation about most aspects of production of, and trade in, coal and steel; and consultation could be about the possibility of co-ordinated action. In the case of action to deal with declines in demand, or with shortages, such co-ordination should, not merely could, be considered, if possible.

It might be asked whether the agreement brought any benefits, even if it did no harm. One commentator has described the formal arrangements between the UK and the ECSC as elaborate, but only symbolic.[4] The association agreement certainly did not make an enormous difference to the parties, but it probably had two advantages for the British. One was the supply of information, not just on technical matters, but on business practices and on government and High Authority policies in detail and during their formation. If, for instance, there was a new policy on coal

[1] NCB 354th and 357th meetings, mins. 136 and 163, 3.10.52 and 21.11.52.

[2] This subject came up at most NCB meetings from the 371st on 4.9.53 to the 383rd on 2.4.54. Agreement on a satisfactory draft is recorded at 394th meeting, min. 140, 5.11.54.

[3] Cmd. 9346.

[4] Lister 1960, 352. This work gives a clear summary of the formation of the ECSC (3–4 and 7–18) and of the development of the formal relations of the UK with it (350–7).

prices, the NCB representative in Luxembourg knew about every step in its formulation, and it was easier to adjust to changing provisions in such circumstances. The other advantage was that, although the British coal industry did not have to keep to ECSC rules and in some respects acted rather differently (e.g. in relation to its price lists, which had to be more complicated because Britain produced a greater variety of coals than any ECSC member), the NCB always knew how those rules were developing and what adaptations would have to be made in order to conform to them. In the coal industry British and Community practices were never far apart, and the ways of bridging the small gaps were continuously known. Thus the association was, in effect, helping to keep one option open. As far as the coal industry was concerned, the necessary adaptations were known and could be easily made if at any time it became policy for the UK to join the Community.

What the association did not do was to open a large export market to British coal. That was not because of restrictiveness in the constitution of the ECSC. It was primarily because the west European market for imported coal had contracted and because, after 1949, Britain rarely had enough coal available when western Europe did have need of it. So by the mid-fifties only about 10 per cent of the coal imports of the ECSC were coming from Britain. This was a symptom of the weakness of the international position of British coal, which became all the more damaging as other trends in demand turned downwards. But it was a weakness which sprang from industrial and commercial causes rather than from external institutions.[1]

## vii. Performance

Disappointment with exports suggests that the objectives which presented themselves in 1947 had not been fully achieved. It would have been surprising if they had, for the magnitude of the tasks to be carried out was mainly determined by the huge difficulties confronting the coal industry. The tasks can be briefly listed: to bring under common public ownership and efficient common administration an industry in obvious

[1] If there was any adverse influence from European institutions it probably came through the opportunity for oil to increase its share of the energy market, thus damaging ECSC coal producers even more than outsiders, and did not operate until the end of the period. The Treaty of Rome, signed in March 1957, established the European Economic Community from the beginning of 1958 and had somewhat more flexible provisions on the publication of standard price lists than did the ECSC. Oil came under the Treaty of Rome and could therefore adjust prices in order to capture additional business more easily than coal could.

decline, made up of hundreds of units that were diverse in almost every respect, many of them seriously affected by neglect and obsolescence; to turn the industry round immediately from decline to expansion, so as to ensure that the national economy would not go short of energy when expanding at levels of employment higher than had been known for many decades; to restore the colliery labour force in morale, numbers, and efficiency, and to gain the co-operation of the workers in increasing the output and productivity of the industry; to undertake the physical and organizational reconstruction of the industry on the lines indicated by the Reid Report, incorporating the best practices already known and providing for the discovery and application of continual improvements in technology; to accomplish all these things without running into financial loss or burdening consumers with high prices. Some of these subjects are best dealt with descriptively. The state of achievement in others can be at least partly summarized in figures. Both treatments draw attention to some wider questions and also suggest some of the answers.

The need to create a common administrative system had been met quickly and comprehensively, and there is no reason to think that, on the whole, it had not been met as well as could reasonably be expected, even though at times the system received much abusive criticism. The verdict of the Fleck Committee tells strongly in its favour. Within the administrative system, practices necessary for basic competence and control, such as uniform accounting methods and provision for standard preparation of statistical and other records, were also promptly established. The remarkable abundance and clarity of detailed information in the NCB Annual Reports and Accounts is public evidence of this. But in the use of up-to-date techniques for detailed management and accountability, and in the development of training for management, things were much less healthy. The NCB compared poorly with many other large business organizations in these respects. The problems were partly recognized, but progress was slow in the first few years, and after 1951 managerial practices retrogressed in some ways. It was only with the reforms that followed the Fleck Report that management began to take on the detailed characteristics it needed. By 1957 the NCB were equipped for better management than they had ever had before, but until then there were weaknesses which were liable to have had some limiting effect on many aspects of performance.

The aims that can be indicated quantitatively were achieved with partial success, but various qualifications have to be added, and progress was in most respects better in the first few years of nationalization than later.[1] The British economy, except in early 1947, got the coal it needed as its needs increased, but not always in the qualities preferred, though in this respect householders were the chief sufferers, not business.[2] But in the middle fifties the supply had to be marginally topped up with imports. Deep mined output rose steadily in the first five years of nationalization. In 1951 it was $13\frac{1}{3}$ per cent more than in 1947, but thereafter there was no sustained improvement, just slight fluctuation around a similar level. The highest deep mined output was 217.0 million tonnes in 1954, but in each of 1955, 1956, and 1957 deep mined output was slightly lower than in 1951. The maintenance of opencast output helped to keep up total supplies in a way that had not been expected in 1947. After a reduction in 1954 it gradually increased again, and in 1957 was 13.8 million tonnes, the highest output so far achieved. This was a success story to offset some of the other recent checks.

The halt in the growth of deep mined output was associated with the trends in manpower, attendance, and productivity. All these were more favourable in the earlier than the later years. The fluctuations in manpower, with no clear sustained trend after the rises of the first two years; the initial reduction in absence, maintained but not significantly improved on until, after 1952, it was reversed; and the rise and subsequent decline in the average number of shifts worked in a year, with 1951 and 1952 as the high points, have all been noted. The record of productivity has something in common. Overall OMS rose each year except 1952 until 1954, and then fluctuated slightly round a steady level. In 1951 it was nearly 13 per cent higher than in 1947, but in 1957 it was only just over 0.5 per cent higher than in 1953. For faceworkers the improvement (again with 1952 as the aberrant year) was more sustained, though at a declining rate. The percentage increases in the same two sub-periods were 11.2 and 4.6. Between 1953 and 1957 there was actually a small decline in productivity away from the face.[3]

The financial results were not greatly out of line with what was statutorily required. The collieries as a whole showed an operating profit in every year except 1947, and the ancillary activities in every year. Open-

[1] NCB *Ann. Rep.* 1957, I, 1 is the most convenient statistical summary.

[2] Probably the iron and steel industry complained most about quality because of the decline in the proportion of prime coking coal available for making metallurgical coke. It was a matter of controversy just how much of a handicap was caused by the blending of more of what were good, but not the very best, quality coking coals.

[3] Comparisons bridging the gap either side of 1952 would be slightly inaccurate because of a change in definitions.

cast mining, which the Ministry of Fuel and Power operated at a loss except in 1950–1 and 1951–2, was profitable to the NCB except in the takeover year, 1952. There were outgoings to offset these results, chiefly the payment of interest, and at the end of 1957 the profit and loss account had an accumulated deficit of £29.2 million, which was mainly attributable to the results of 1947 and 1955. The NCB had a surplus in three of their first five years and two of the next six. In the continuous period from 1948 to 1954 inclusive the board were very successful in balancing good and bad years. They had a surplus in four and a deficit in three, and over the whole seven years a small surplus of just over £6 million.[1]

These financial results were, however, dependent on repeated price increases to match increases in costs. The average price of coal (a concept open to question) was 103.9 per cent higher in 1957 than in 1947, and the rise was fairly even over the period: the level in 1952 was 42.2 per cent above that in 1947, the level in 1957 43.4 per cent above that of 1952. These figures may be a little unfair to the earlier years because, though there could be controversy about the appropriate index to use in order to convert coal prices to 'real terms', general prices on any measure were rising faster to the end of 1951 (especially from 1949 to 1951 inclusive, when sterling devaluation and the Korean War were strong inflationary influences) than they were from 1952 to 1957. For a brief interlude at the end of the forties coal was, indeed, probably becoming slightly cheaper in real terms, but it soon resumed its secular upward cost trend. That interlude was remarkable when it is borne in mind that the prices of some of the materials used by the coal industry were rising faster than the general average. On the other hand, coal prices had risen very fast in 1947, and even then lagged behind the rise in costs of production. If comparison is made with prices at the beginning of 1947 instead of with the average for the year, the price rise from nationalization to 1957 goes up to 117 per cent, i.e. about 8 per cent per annum, around double the general rate of inflation. That is not in line with the proclaimed expectations at the time of nationalization.

Some of those expectations, however, were mutually contradictory, especially those which looked for both cheaper coal and much higher pay for miners. Despite some economies in the use of labour, wages and wages charges, which were much the biggest element in the cost of producing coal, rose by 74.7 per cent from 1947 to 1957, with most of the increase in the later years: from 1947 to 1951 it was only 15.2 per

[1] NCB *Ann. Rep.* 1957, II, 2–3 and 9.

cent.[1] Average earnings per shift for faceworkers rose by 98 per cent from 1947 to 1957, and the rise was fairly evenly spread over time: the figure in 1952 was 43 per cent above the 1947 level.[2] The national minimum weekly wage for adult mineworkers underground increased from 1947 to 1957 by 90 per cent, with a rise of 40.5 per cent in the first half of the period and 35.2 per cent in the second half. Because of changes in the structure of the labour force and in wage differentials, the average weekly earnings of all mineworkers moved a little differently. These rose by 127 per cent from 1947 to 1957, with the biggest rise between 1947 and 1948. The rise was 62.2 per cent in the first half and 35.5 per cent in the second half of the period.[3] These figures give some indication of representative movements in the complex of wages in the coal industry. They suggest that (except in 1947 when earnings outpaced coal prices) earnings increased nearly but not quite as fast as the price and cost of coal, and improved significantly in real terms. In relation to those in other industries the relative position of adult male earnings in the coal industry changed little between 1948 and 1956, but on most measures improved slightly. Coal remained one of the best paid industries, with hourly earnings more than 40 per cent and weekly earnings around 25 per cent above the average for all industries.[4] So in material terms the workers in the coal industry had gained, though they had probably not advanced their relative position as much as they had expected at the time of nationalization. This was mainly because the general high level of employment had so greatly increased the opportunities and bargaining power of workers elsewhere. These relativities may have reduced the satisfaction of workers in the coal industry. Whatever the reasons, the response in terms of attendance, industrial peace, and productivity did not come up to the hopes of 1947.

When one turns to consider progress in reconstructing the industry, the record by 1957 is very incomplete. Unavoidable concentration on the pressing needs of the moment, and a proper care for detailed preliminary examination before heavy investment was planned, meant that not much was begun before 1950. Thereafter a thoroughgoing programme of reconstruction was always in train, but its progress was disappointingly slow. It had been intended that 1952 would be the peak

[1] The figures from which the foregoing calculations are made are in NCB *Ann. Rep.* 1956, I, 16 and 1957, I, 14 and II, 2–3.

[2] NCB *Ann. Rep.* 1956, I, 69; 1957, I, 43.

[3] Ibid. 1956, I, 69; 1957, I, 42; 1963–4, II, 108–9. The big jump in average weekly earnings from 1947 to 1948 was partly due to the extended hours agreement made late in 1947.

[4] Handy 1981, 172.

year for investment in reconstruction, but it was not until 1953 that the annual rate of capital expenditure was pushed up to a level expected to provide reconstruction at the planned rate. Even then it was found that a given level of investment was providing a smaller net increase in capacity than had been forecast in 1950. One result was that there was a serious lag in the physical restoration of the coalmining industry. It was good going that more than half the projects in the 15-year *Plan for Coal* had been started by 1955, but only 20 had been finished. The speed-up during 1956 has been noted, and much more came to fruition in 1957, but even at the end of that year only a little over a quarter of the projects begun had reached completion.

Thus, throughout the period of expanding demand, coal had to be supplied by what, in physical terms, was essentially the old pre-nationalization industry with a lot of small adaptations and improvements, not by an industry which had acquired a large renovated and modernized sector. Old capacity which, in different circumstances, might have been given up, had to be kept going and patched as necessary, even though there was no way of thoroughly curing the decrepitude and obsolescence of some of it. The total number of producing collieries operated by the NCB had fallen from 970 at the time of nationalization to 822 at the end of 1957,[1] with the highest rate of closures in the first three years, when many small collieries were found to be worked out. This might look like a fairly large closure of old capacity, but these collieries accounted for only a very small proportion of total output. All but 25 of the collieries closed had an annual output below 100,000 tons, in many cases very far below that figure. The only large colliery closed was Thorne, near Doncaster, which was made unusable by flooding in 1956. That colliery had produced over 600,000 tons a year, but no other closed colliery had produced as much as a quarter of a million tons. Of large new collieries only Rothes, in East Fife Area, and Calverton, in Nottingham Area (both projected before nationalization), had been brought into production. Most of the completed major colliery reconstructions, which by the end of 1957 were contributing a valuable but small minority of modernized capacity, had then been in full use for only two years or less. Moreover, they absorbed only 12 per cent of the gross investment in collieries since 1947; £157 million was still tied up in uncompleted schemes. So the generalization about

[1] Classification of some collieries as separate and producing has changed, so the historical statistics differ slightly from those which appeared at the time in *Ann. Reps.* From the end of 1950 to the end of 1954 the number of producing collieries fell only from 901 to 867.

reliance mainly on slightly adapted old capacity remains true.[1] This had been known to be inevitable in the early years, but had been expected to be less extensive than it was ten years later.

The fact that a given amount of investment was discovered to provide less net additional capacity than had been planned had some financial consequences, too. It had been intended that about three-quarters of the cost of reconstruction would be met from depreciation provisions, so that only a minority would require additional loans from the government.[2] The need for higher capital expenditure in order to achieve the same projected output, together with the adverse effect of inflation on provisions for depreciation, created a need for more borrowing, since coal prices were not set at a level designed to provide internal financing for capital expansion. Between 1950 and 1955 inclusive the NCB had a gross capital expenditure of £401 million (£353 million of it on projects in *Plan for Coal*), and £217 million of this was financed by loans from the government.[3] In 1956 over 80 per cent of the capital expenditure was financed from internal sources,[4] but in 1957 there was a return to dependence mainly on borrowing. It was mainly the heavier than expected reliance on loans to finance long-term investment in reconstruction that caused the NCB to have outstanding capital liabilities of £349 million by the end of 1957, apart from the liabilities relating to the assets vested at nationalization. It was also of some importance that the greater recourse to loans came at a time when the trend of interest rates was slowly upwards. In 1947 the NCB had been able to borrow from the government at $2\frac{1}{2}$ per cent. In 1957 the average rate on their new loans was 5.675 per cent.[5]

Whether many of the frustrations of the reconstruction programme were avoidable is an open question. It is not implausible to speculate that there might have been some advantage in starting fewer projects at the same time, concentrating a bigger proportion of the available resources on each, and trying to get them finished more quickly. But it is far from certain that much would have been gained. There are serious physical limits to the quantity of labour and materials that can be absorbed by one project in a colliery where the necessity of maintaining current output is an ever-present constraint. That limitation should not,

[1] NCB *Ann. Rep.* 1957, I, 26.
[2] NCB, *Plan for Coal*, 5.
[3] NCB, *Investing in Coal*, 12.
[4] NCB *Ann. Rep.* 1956, I, 8.
[5] Ibid. 1957, II, 22–3.

however, have been so severe for the construction of the small number of entirely new large collieries, though these had the additional hazard of encountering the unknown in a higher proportion of the work involved. There was a handicap that the largest relevant technical advances, for example in shaft sinking and tunnelling, did not come early enough for this reconstruction programme. It was also recognized that a shortage of able and appropriately trained planning engineers and project managers was a delaying factor. It is arguable that this was remediable, and that more should have been done to remedy it earlier.

Another question is whether effort was sufficiently concentrated on projects likely to produce the best returns. Some examples that gave rise to doubt can be cited. One was Solway colliery in Cumberland, where, by the end of 1957, £1.4 million of capital expenditure had been authorized, and that was barely two-thirds of what was expected to be ultimately required.[1] The work there was first for a new sinking, and then for a reorganization to prove possible undersea reserves and work them if they were satisfactory. This was known to be a very speculative undertaking, especially because no satisfactory techniques for undersea exploration were available when the work began, and the project turned out to be a costly disappointment. It was justified at the time on the ground that the northern part of the Cumberland coalfield was doomed, with serious social consequences, if a workable undersea field were not proved.[2] Another speculation was the investment put into the revival and expansion of anthracite production in the west of the South Wales coalfield. Two of the projects for new collieries of substantial size were located there. It was understandable that such an investment had attractions. The government wanted more anthracite sold in hard currency markets, and demand at home as well as abroad was so strong that some NCB members believed it offered the chance to earn large profits, although the costs of anthracite mining had been so high that a price that covered them (if anybody had been willing to fix one) might well have choked off much of the demand. Moreover, the physical arrangement of the existing mines was so ruinous that a fresh start seemed necessary. Nevertheless, Young declared in 1949 that any large-scale developments in the anthracite field were doomed to failure unless the men there changed their attitude. But the new developments went ahead, although there were great differences of opinion within local, as well as national, management. This was particularly so in the case of the

[1] Ibid., II, 58–9.

[2] NCB 386th meeting, min. 72, 21.5.54.

new Gwendraeth valley colliery (renamed Cynheidre at an early stage in its construction), where there was disagreement about both the size of the reserves accessible from the site and the economics of working them.[1]

Both Cumberland and the Welsh anthracite field seemed to offer the choice 'renew or die', and nobody was prepared to cause avoidable doom when there was a general problem of physical shortage of output, to which no end could be foreseen. So a few large projects were undertaken where there was a lot of doubt about the achievement of long-term profit, though in any of these the investment was projected, at worst, to improve on previous rates of loss. But such projects were not representative. It is obviously not practicable to measure the financial results of the reconstruction programme by 1957, because most of the schemes were still not completed then; and the eventual results of the projects then still in progress were affected by the shrinkage of markets below the levels assumed when the start of the projects was authorized, though this need not have been a very powerful influence, as the main impact could be expected to fall on the unimproved collieries. One independent study suggested that about half the pit reconstructions in Scottish and South Western Divisions were financial failures, but that in the rest of the country 90 per cent of the reconstruction investment was at pits which were profitable in 1951 or increased their profits in the next ten years, and that 70 per cent was at pits which met both criteria. The same study calculated that by 1961 the reconstruction investment was yielding a direct marginal return of 8 per cent for the coal industry as a whole, or 12 per cent if Scottish and South Western Divisions were excluded.[2]

All the indications are that the tasks that existed at the time of nationalization were performed with, in general, a moderate degree of success but a lack of completeness. Most things moved in the desired direction, but more slowly than was needed. The great disappointments

[1] NCB 263rd meeting, min. 229 and 348th meeting, min. 93(a), 29.4.49 and 6.6.52.

[2] Shepherd 1965, 115–17. Shepherd argued (70–1) that, if comparable accounts were prepared, the NCB from 1948 to 1961 achieved a much higher rate of profit on their total net fixed assets than the gas and electricity industries, and a rate (before interest) slightly better than manufacturing and distribution as a whole. But some of the adjustments made in the attempt to produce standardized accounts—they included, for example, the omission of all NCB losses on imported coal, presumably on the ground that they were outside NCB business as a producer—might seem unduly favourable to the NCB. It is not clear from the information given whether any comparable adjustments were made to colliery profits in calculating marginal returns on *additional* investment, but it seems improbable that there was anything likely to exaggerate the returns.

were the inability to maintain a substantial increase in the input of labour, the failure to achieve a permanent reversal of the long upward trend in coalmining costs, and the frustration of seeing so many target dates recede. It is also noticeable that steady improvement over a wide range is more characteristic of the first quinquennium than the second, and this must raise the question whether a momentum, once gained, had been lost.

The reasons for the generally better performance to 1951 cannot be suggested more than tentatively. It could be that there was a better board, giving clearer direction to the industry. This was probably an important influence, but it cannot be the whole story. In the first few years there was positive help from residual goodwill. Within the industry most people wanted the great new experiment of nationalization to work, even if there were many who too easily let a grievance interrupt their goodwill. By the fifties the novelty had worn off. There was no longer much that seemed rather special about the coal industry to call forth, even intermittently, some extra response. Within the first few years, too, much was done to use the benefits of common ownership by diffusing more widely the better practices long applied within familiar techniques. This helped, though it was by no means solely responsible for, the execution of a large number of small schemes to create a little more productive capacity within collieries that underwent no fundamental change. By 1952 there was much less scope left for this sort of thing. The conditions of expansion and higher productivity had become more difficult. They involved bigger projects, with more possibilities of delay, and these created distractions, not strongly present before, from the task of maintaining current output on familiar lines. Continued improvement was also increasingly dependent on technological innovation, which did not come quickly. Some innovations of great importance were in gestation, but they did not greatly affect output until the late fifties. Even then there was some initial uncertainty about their value. In 1957 there was great worry in the NCB about the decline in profits which was attributable to the degradation of the coal, particularly by the use of shearer loaders,[1] and a slowing down in the rate of introduction of shearer loaders followed.[2] So in the fifties, though the sheer weight of the burdens of directing the coal industry was less than in the late forties, there were some new complexities that were hard to deal with.

[1] The case was argued that degradation from shearer loaders had to be accepted, but that types of power loader could be used more selectively in accordance with differing face conditions.

[2] Chap. 3, pp. 84 and 103 above.

Whatever the explanation of the apparent reduction in momentum, there were several features in the results achieved down to 1957 that had serious implications for the future. The delays in completing reconstruction projects and the recent smallness in improvements in efficiency had appeared to involve only postponement of the day when a clearly defined goal would be achieved. But the goal was defined by assumptions which included continuous expansion of the demand for coal throughout a period to which no end could be foreseen. The basis of those assumptions was beginning to change, so the goal to which current activities and future projections were related was losing some of its relevance.

There had been indications of such a change for several years. After 1951–2 the usual experience, except during the Suez crisis, was that more than 40 per cent (sometimes much more) of each year's increase in the UK consumption of energy was supplied by primary fuels other than coal; in fact, almost entirely by oil. In 1957 oil supplies were still restricted until the end of May, but oil consumption thereafter resumed its growth as fast as before the Suez crisis. Down to 1951 the share of oil in the increments of energy consumption had been much slower.[1] In the mid-fifties it was still customary in the coal industry to welcome the availability of oil as a relief from the burdens of attempting faster expansion of coal production. It seems to have been too little appreciated that oil was changing from an auxiliary to a competitor, still far behind but coming up fast. The NCB had been warned by their staff as early as 1950 that, except for steam raising in power station and industrial boilers, oil was competitive in price for many uses, and even in steam raising coal had no advantage in convenience, but discussion then concentrated on the loss of the bunker market for coal.[2] By 1957 convenience, price, and security of supply were all helping to lessen the competitive strength of coal.

There is a case for arguing that coal was sold too cheaply for an appreciable period after nationalization. Demand was regulated by allocation as much as by price. Despite the return to rising real costs of coal production, British costs remained lower than those elsewhere in western Europe and, when transport costs are added, British coal retained a huge cost advantage over imported. If more of that advantage had been reflected in home prices, some British demand for coal might have been diverted, the maximum target for output (which was proving

[1] NCB *Ann. Rep.* 1956, I, 34; 1957, I, 28.

[2] NCB 295th meeting, min. 34, 3.3.50.

so difficult to attain) could have been lowered, and the coal industry could have built up bigger financial reserves out of which to provide more self-finance for reconstruction, which would have remained essential for efficiency even if future output projections had been lowered. But by the mid-fifties that sort of relaxed price policy, which would have posed many problems for the economy in general, could no longer have been helpful in any way to the coal industry. Its competitors were becoming much stronger and it needed to exploit any competitive advantages of its own which it could find.

Those competitive advantages were more narrowly based than they used to be, because of a wider regional dispersion of costs about the national mean. For a very long time there had been a tendency for district wage rates to show a rough and ready accommodation to differences in productivity and profitability in the different coalfields, though there were a few striking exceptions, such as the relatively high wages in Kent. The government policy of allowing flat-rate cost of living increases in wage rates in wartime seriously disturbed these arrangements. Whereas in 1938 district wage rates ranged from 15.8 per cent below the unweighted average to 26.3 per cent above it, the range by 1946 was only from 10.6 per cent below to 17.1 per cent above.[1] As wages were the biggest element in costs it was more difficult to make the lower productivity coalfields profitable or keep their losses under control. This was evident in the difficulties of the Coal Charges Account, which tried to offset the profits of some coalfields against the losses of others, but found itself unable to stay solvent without an Exchequer subsidy. This was a problem inherited by the NCB. They could do nothing to reverse recent policy, they were enjoined to avoid losses, and they could not attain the desired output without continuing to use the coalfields which relatively higher costs had driven into permanent loss. The Director-General of Production emphasized in 1950 that it was a major difficulty in the way of satisfactory results. He said that 'in 1938 the difference between the most profitable and the least profitable coalfield was only 2*s.* 6*d.* In 1948 it was 36*s.* 10*d.* In that year 111 million tons were profitable and 84 million tons were unprofitable.'[2] Wage movements continued to push in the same direction. Even though the NUM did not try to carry to completion their proposed national wages

[1] Handy 1981, 32–4 and 140–5.

[2] NCB 291st meeting, min. 8, 13.1.50. For *coalfields* it is not clear how the figure of 36*s.* 10*d.* (£1.84) was reached. The difference between the most heavily losing NCB Area (Swansea) and the most profitable (Edwinstowe) was 30*s.* 7.4*d.* (£1.53) per ton in 1948.

policy, nationally agreed wage additions were biased in favour of lower-paid districts and workers, and locally negotiated elements of the wages structure did not do a lot to redress the balance. By 1954 the range of district rates had narrowed further and then ran from 8.4 per cent below the average to 11.6 per cent above it.

In terms of total costs, things did not get much worse on the whole, though the relativities between localities changed somewhat. In 1948 the costs in the worst NCB area were not quite two and a quarter times those in the best. In 1952 they were just over, and in 1957 just under, two and a half times, but the difference was due more to boundary changes than anything else.[1] Some regional changes in costs could be matched wholly or partly by regional changes in prices, and it is, indeed, suggested that cross-subsidies as a proportion of total costs changed little,[2] though there were more changes in the destinations of the cross-subsidies.

The significant point was that wide differences in costs and profits had been built into the coal industry before it was nationalized and it was difficult or impossible to improve most of them. The NCB regularly made money in East Midlands, West Midlands, and North Eastern Divisions and lost it nearly everywhere else. Between 1947 and 1957 there was not a single year in which a profit (after interest and apportionment of losses on imported coal) was made by Durham[3] or North Western or South Western Division, and Northern and South Eastern Divisions achieved a profit on this basis only once each. On colliery activities alone 24 areas made an operating loss and 24 an operating profit in 1948, and in 1957 there were 28 losses and 23 profits. East Midlands and North Eastern were the only divisions which never had an area on the short list of very high loss-makers. Of the loss-making regions the best that can be said is that the trend of colliery losses was generally downward in Northern, Durham, and South Western Divisions. Indeed, the first-named moved into operating profit in 1956 and 1957, and South Western, which had very bad results in the anthracite district and in the small Forest of Dean and Bristol and Somerset fields, also included the Aberdare Area which, after making losses, became one of the most profitable.

[1] After 1948 the NCB area boundaries in South Wales were redrawn. A new Neath Area consisted of very high cost anthracite mines to a greater degree than any previous area and remained thereafter the highest cost area in the industry.

[2] Shepherd 1965, 102.

[3] The Durham Division was made at the beginning of 1950 by dividing the Northern Division into two.

The most alarming changes were in Scotland. Scottish Division was profitable (even after interest) down to the end of 1949 and was regarded as one of the best potential regions for heavy investment. It was known that there was not much more to be got out of the central coalfields, but there were great hopes of development elsewhere, especially in central and East Fife and the Lothians. But physical conditions in the newer coalfields often turned out much worse than expected, and did not help the notoriously difficult labour relations, which got no better. Scottish costs and losses mounted. In 1947 costs per ton in Scottish Division were practically the same as in North Eastern Division, but in 1957 they were 28 per cent higher. In the latter year they were higher than costs in Durham and only a little below those in South Western Division, although these two divisions produced a high proportion of special coals which commanded high prices, whereas Scotland produced mainly general-purpose coals. In 1948 Fife and Clackmannan had the fifth-highest profit rate per ton among the forty-eight NCB areas. In 1957 East Fife, West Fife, and Alloa, which covered much the same territory, were all among the six areas with the highest losses per ton. Indeed, Alloa was worse than all others except the Forest of Dean, and five of Scottish Division's eight areas were among the eleven in the country losing over 7*s.* 6*d.* (37½p) on every ton produced (Alloa's rate of loss was four times as great as this).[1] It was a disastrous transformation, which turned Scotland from one of the great hopes to one of the great problems of the coal industry.

The commercial rationale of the wide diversity of costs and financial results was not readily apparent. The saving on transport costs in meeting local demand from the output of nearby pits did something to redress the adverse balance in North Western and Scottish Divisions. Large and growing demands for special high-quality fuels, such as prime coking coal and naturally smokeless coals, encouraged the maintenance of output in Durham, South Western, and South Eastern Divisions. But policies here were concessions to the 'public interest' rather than direct services to the commercial interest of the NCB. Essentially, the line that was followed was that maximum output, with national break-even pricing, was the overriding objective, more important than the level of costs of individual producing units. Provided that revenue as a whole could be maintained near break-even point, any group of collieries that could be kept contributing to maximum output, without costs becoming astronomical, seemed worth keeping for the time being, although

[1] All the figures in this paragraph are based on the published annual accounts.

simple arithmetic ensured that this involved permanent cross-subsidy. Every coalfield was, in fact, retained in production, even the most costly.[1] But this was a policy conceived in relation not only to a particular set of market conditions but also to an expectation that those conditions would continue indefinitely. Once that expectation began to weaken and then to disappear, there was an emergent need to reconsider the size and geographical distribution of the coal industry. The experience of the first decade of nationalization suggested that there was a substantial part of the industry which could not go on as it had done if price competition rather than maximization of output was given scope to be the dominant force in the market. And that is what was beginning to look likely in 1957.

[1] In the Bristol coalfield, within the Bristol and Somerset Area, production ceased with the closure of Coalpit Heath colliery in 1949, but was resumed in 1954 when the new Harry Stoke drift mine was opened. Production was continuously maintained in the Somerset coalfield. The NCB never operated in the Brora coalfield, but there was a licensed mine there.

CHAPTER 6

# The Industry in Contraction 1957–1973

## i. The changed commercial setting

The worries which were caused by the unexpected fall in coal consumption in 1957, even though there were obvious temporary influences that could be responsible, were soon much increased. In the next two years the drop was far greater and was not attributable to any immediate pressure on the general demand for energy. Energy use was increasing, but cheapening oil was competing much more strongly with coal than ever before. In 1959 inland consumption of coal was 28.1 million tons (28.5 million tonnes) less than in the peak year of 1956; and total consumption was 33.5 million tons (34.0 million tonnes) down, a drop of 15.0 per cent in three years. In the same three years undistributed stocks of coal rose just over twelvefold, and total stocks much more than doubled, to reach 50.0 million tons (50.8 million tonnes). These sudden changes were appalling to an industry that had been trained by recent experience to regard over-production as an impossibility and, with all the urging of official policy, had been striving for years to expand productive capacity to match a forecast demand of 240 million tons (244 million tonnes) for 1965.

The timing of the change seemed almost as cruel as its magnitude, for, after several years of difficulty with output, the coal industry was poised for further expansion. In 1957 most of the projects of *Plan for Coal* had still not matured, but the number coming to completion had begun to increase fairly quickly, and an appreciable enlargement of modernized capacity could be expected to become available between then and 1965. In addition, the long period of disappointment with attempts to increase mechanized mining had come to an end. Since 1955 the proportion of power-loaded output had been leaping upwards. It passed 20 per cent by 1957, 40 per cent early in 1961, and 60 per cent by 1963. The compensation was that such changes helped with costs as well as with the level of output, and, if total sales were to be determined by price competition, they gave a chance to fight back in the market. But for a coal industry with the size and structure that existed in 1957, the likelihood of over-production remained.

A fairly rapid adjustment to changed conditions was obviously called for, but just how big the adjustment needed to be was not at all clear for several years. In the late fifties hardly anyone guessed how great the market pressures on coal were going to be ten years ahead. Even before 1960 some permanent losses seemed inescapable. Technological developments showed that the railways were following shipping in the transition to oil-burning. Export opportunities were shrinking because the switch to oil was going on faster in other industrial countries than it was in the UK, and the consequent sharpening of competition for international coal markets led the cheaper producers and shippers to offer price and delivery terms which British coal could seldom match. But there were two or three years to wait before it was evident that the technology of gas production was eroding another important market for coal. Provided that an all-out effort could be made to limit to small amounts the use of oil and nuclear fuel for electricity generation, it seemed for the moment that the future, though difficult, need not be catastrophic. In fact, coal consumption recovered a little and, without returning to the level of the mid-fifties, it remained above that of 1959 until 1964–5. That had once (when reconstruction began) been set as the target year when the coal industry must try to meet a demand for 240 million tons, but the actual consumption then was only 192.6 million tons (195.7 million tonnes), a shortfall of 19.75 per cent on the old forecast which, however, had been abandoned in the late fifties. Thereafter, the prospects for coal sales slid down and down as cheap supplies of alternative fuels increased and government policy turned to push down the use of coal still further. The later sixties thus became the most difficult time ever known in the history of the coal industry. From 1964–5 to 1970–1, despite a brief respite in 1968–9, coal consumption fell by 21.4 per cent, to 151.3 million tons (153.7 million tonnes), and at that point the government permitted the resumption of coal imports. In the next year the great strike of early 1972 contributed to a further huge drop in consumption, as supplies were not fully available even to the extent that they were wanted, and there was only a slight recovery in 1972–3, when total consumption was 130.4 million tons (132.5 million tonnes). Since the high point in 1956 coal consumption had fallen by 42.6 per cent, with the losses concentrated in the years 1957 to 1959, 1964–5 to 1967–8, and 1969–70 to 1971–2 inclusive. In the first of these sub-periods much of the fall in consumption was offset by the building up of stocks, but stocks were never again allowed to stay at the level of 1959, though (by previous standards) they remained high for the next eight years. But between 1967–8 and 1970–1 they were more than halved, to 19.8 million tons (20.1 million tonnes).[1] So the later phases of the contraction of market demand had to be more than fully matched by a squeeze on output.

Not only market conditions but government reaction to them indicated that drastic changes in the policy of the NCB were unavoidable. By the beginning of 1958 it was evident that what had happened in the previous twelve months was more than a brief interruption of past trends. Some economies had already been put in hand in the preceding autumn, particularly in non-operational expenditures. Latham, reviewing in February 1958 the financial prospects for the year, insisted that improvement would have to come from internal measures to cut costs, not from price rises. Attention was given to the reduction of overtime working, the cessation of recruitment except in special circumstances, the reduction of stocks of materials and stores, and an attempt to increase proceeds by checking the decline in the proportion of large coal in total output.

The changed attitude of government, based on a belief that it was no longer necessary to industrial security to maximize capacity for coal production, was made plain when the national board met the Minister of Power, Lord Mills, and his Permanent Secretary, Sir John Maud, in April. Sir Henry Wilson-Smith, a businessman who was a part-time member of the board, told the Minister how impressed he had been by the speed and commercial manner the board had shown in tackling their problems, and Bowman supplied the detail of what had been, and was being, done. But it is doubtful whether Lord Mills, who was always found by the NCB to be the most negative minister with whom they had to deal,[2] took much notice. He said the board was a business

[1] For further details of the influences on demand, see above, Chap. 2 *passim*. The timing of new developments in technology is discussed in detail in Chap. 3 above. The figures of consumption (which includes exports) and stocks are from NCB *Ann. Rep.* 1971–2, II, 84–5; 1972–3, 9; and 1982–3, 20–1. In order to permit accurate comparison the figures used for 1972–3 are for 52 weeks, although there were 53 weeks in that financial year. The dates used change in the sixties because in 1964 the NCB changed their financial year, to terminate at the end of March instead of the end of December. The stock figures in the text differ from those in NCB *Ann. Rep.* 1982–3. This is because later definitions of stocks (which have been applied to historical, as well as contemporary, statistics) are on a narrower basis than those used in the reports down to 1971–2.

[2] It was at one of Lord Mills's meetings with board members that his contributions to the discussion provoked the retort, 'If I were to say to you, "Night follows day", you would say, "I can't agree to that".'

undertaking, but one which had to meet public demands at a price the public was prepared to pay, and at the same time face competition like any other business. This was a reasonable statement in itself, but ignored the fact that the board had never been allowed to charge competitive market prices at home when demand was high and rising. He complained that of £500 million spent on reconstruction, only £100 million was spent on positive improvement, figures which would not have been accepted by others as even an approximation, and which showed no recognition that new replacement capacity is more efficient than the old capacity that has been withdrawn. He did not respond directly to either Bowman's reminder of a relation between the number of colliery closures and the morale of the workers, or Wilson-Smith's point that the coal industry was left with a difficult problem of public relations until the government defined a clear and consistent long term fuel policy. And he departed with the remark that 'it would be a useful exercise' for the NCB to consider keeping coal prices unchanged, despite a 3 per cent rise in the cost of living.

Other and later ministers often showed more understanding than Lord Mills, and there were important ways in which the government assisted the adaptation of the coal industry, in particular by some increase in borrowing powers at a time when additional costs were imposed by the keeping of high stocks, and, later, by the reduction of interest payments through financial reconstruction. But, in general, the pressure of policy was towards a stricter application of commercial criteria, when these made life more difficult for the coal industry, than in earlier years when they would have made its life easier. For instance, a white paper of 1961[1] had laid down that nationalized industries should observe the principle that, in a five-year period, surpluses on revenue account must at least cover deficits; and from 1963 the application to the coal industry was a target of earning £10 million a year, after payment of interest and depreciation, as a contribution to the excess of the replacement cost of fixed assets over their historic cost.[2] That provision was not insisted on after the first fifteen months, for the point was then reached that a financial reconstruction, to reduce the burden of interest, was unavoidable. But this did not mean a relaxation of pressure for the coal industry to adapt to a contracting market. Indeed, on the whole[3] the

[1] Cmnd. 1337.

[2] NCB *Ann. Rep.* 1962, I, 4. This financial objective was lower than for other nationalized industries except the British Railways Board and the London Transport Board.

[3] At variance with this strand of policy was the introduction of an excise duty on fuel oil in

opposite was true, as the government helped to contract the market further and pursued with the NCB ways and means for the industry to follow suit. It was in 1965 that the government produced the National Plan and a white paper on Fuel Policy, both of which projected substantial reductions in the use of coal in the next five years, and revived for 1966–7 the objective of earning £10 million towards fixed assets replacement (though, in the event, this was not enforced). It was in 1967 that the government brought out its new white paper on Fuel Policy, which allotted to coal the lowest position for new developments in a four fuel economy, and started to implement the policy, even though it was never debated in the House of Commons.[1] At that stage the corresponding contraction of the industry was rapid indeed.

Though there was sometimes resentment about the nature and speed of the changes that the government sought, which in some cases seemed to go beyond what the logic of a competitive market required, the people in the coal industry did not seek to avoid a commercial response to an increasingly difficult commercial situation. The NCB continued to pursue, and achieve, an operating profit, and many of the workers in the industry were so responsive to the changing comparison of job opportunities that, from time to time, especially in the later sixties, particular coalfields found themselves short of some much needed classes of labour.

Among the leaders of the industry there was much determination to make the changes which would keep it viable in reduced circumstances. The board of the late fifties had plenty of managerial and financial skill at the top and plenty of business advice among their part-time members, to whom had been added in 1956 S. P. Chambers, who became chairman of Imperial Chemical Industries. The appreciation by his colleague, Sir Henry Wilson-Smith, of the speed with which the board applied commercial remedies to their difficulties has already been noted. The board could claim a fair degree of success. At the end of 1959 they had made an operating profit in each year of contraction, and in the last year had not only kept the cost of coal production below the trend rate of inflation, but had reduced it in cash terms; they had shed 70,000 men in three years, yet in the last year only 1,300 from 53 closed collieries had been left without alternative employment; and, despite the reductions, they

1961, ostensibly to raise additional revenue, and its retention after its protective value for coal was admitted in 1963.

[1] Cmnd. 3438. See the fuller discussion in Chap. 2 above, pp. 53–6. See also Robens 1972, Chap. 10.

had kept intact enough capacity to meet the slightly higher demand of the next few years.[1]

The later years of difficulty were mostly in different hands. In 1960 the deputy chairman, Sir Joseph Latham, asked not to have his appointment to the board renewed, and he moved to the private sector. He was replaced by E. H. (later Sir Humphrey) Browne, his colleague at Manchester Collieries before nationalization, and latterly, since his service as Director-General of Production, chairman of West Midlands Division. Later in the same year the government appointed the leading Labour party politician and former minister, Alfred (later Lord) Robens, to succeed Sir James Bowman as chairman when the latter retired on 31 January 1961, and he joined the board in October 1960, with a four-month running-in period as a second deputy chairman. Lord Robens was a formidable figure who added commercial drive to a readiness to tackle politicians both publicly and privately in whatever way seemed most likely to serve the interests of the coal industry. He was chairman for a little over ten years, and throughout that period was the dominant personality in the industry. Sir Humphrey Browne left the coal industry in 1967 and was replaced as deputy chairman by D. J. (later, successively, Sir Derek and Lord) Ezra, who had a long experience of NCB activities, particularly in marketing and international matters. The latter succeeded to the chairmanship in 1971, with W. V. Sheppard, who had joined the board from the production side in 1964, as the new deputy chairman. By this time the board had a preponderance of men whose careers had been spent wholly or mainly in the nationalized coal industry, although there were still part-time members to bring experience from both management and unions in other industries.

The new conditions also presented difficult problems of judgement and leadership for the unions. The first impact of contraction came while the leadership was still in experienced hands. Arthur Horner remained general secretary of the NUM, and the president, Ernest Jones, a cautious man of moderating outlook, had been in office since Sir William Lawther retired in 1953. They were much aided and guided by the Durham miners' leader, Sam Watson, a skilled and reasonable advocate whose abilities and integrity were greatly respected even by those who disagreed with him.[2] All these had easy relations with the

[1] NCB *Ann. Rep.* 1959, I, 3.

[2] Cf. Moffat 1965, 290–5. Moffat devotes several pages to what he regarded as Watson's errors in politics and wage policy, but describes him as 'more astute and more capable than any of the right-wing miners' officials or other members of the Executive'; and says that 'no one could

NCB who, of course, still had a former union leader as chairman. Bowman was often strongly critical of the miners and their attitudes, but he and his colleagues knew that the union leaders would negotiate always in good faith and that it was possible to reach with them understandings and agreements that would be kept, even though they would not all be free of some undercurrent of local, unofficial disputes, which the union could never completely control.

There were worries that after 1959, when retirements would bring changes, the relations that could maintain sensible and reliable compromises between management and unions might disappear, but these worries proved unfounded. Horner was succeeded in 1959 by another Communist, Will Paynter, from South Wales, who was regarded by some as a firebrand. Not long before his election he published a review of coal industry questions which included such passages as:

> The doctrine of 'sacrifice in the national interest' is for the miners—but not for the Tory leaders or big business. It is the advice of the burglar to his intended victim. . . . The Coal Board has rejected our demands and in doing so has indicated its attitude to all claims. They have set their course for 1958, as the servants of capitalist big business and the Tories. . . . Once again, as so often in the past, the banner of struggle is in the hands of the miners, for struggle we must if we are to avert the catastrophe of widespread unemployment in our industry and throughout the country.[1]

In fact, for the next nine years Will Paynter was the dominant personality in mining unionism and he negotiated hard and skilfully to get the best deal he could for the workers of a shrinking industry, but always within the constraints of reality as he perceived them, never tilting at windmills. His supporter, Joe Gormley, records arguing with him that he was too reasonable in accepting small increases in money wages as a means of keeping more pits open.[2] Paynter became so trusted by the NCB that at one stage Lord Robens urged their negotiators to impossible speed to try to conclude what became the Third National Daywage Agreement before Paynter retired, as he did in December 1968. There were other contributions to continuity and pragmatism. A hiatus occurred in the presidency of the NUM because Ernest Jones's elected successor in 1960, Alwyn Machen, died before taking office, but in 1961

underestimate his ability and the influence he had in the Durham coalfield, in the British National Union of Mineworkers and the whole Labour and Trade Union movement.'

[1] Paynter n.d. [1958], 10, 11, and 15. For more detail on his work and views see his autobiography (Paynter 1972).

[2] Gormley 1982, 62.

Sidney (later Sir Sidney) Ford, whom Paynter had defeated in the election for secretary, became president and remained until he retired because of ill-health in 1971. He was very unusual in that he had never been a miner but was a member of the COSA (Colliery Officials Staffs Association) affiliate of the NUM. He had none of the gifts of the popular orator, but had long experience as a member of the NUM headquarters staff, and was a quietly competent chairman well versed in the mass of routine union business. Sam Watson remained as an influential figure until his retirement in 1963. After that his advice and help in coping with continuing difficulties were not lost, for he was immediately appointed a part-time member of the NCB and retained that position until he died in May 1967.

Only towards the end of the years of contraction was there another general change in union leadership. There were curious echoes of previous events. In 1968, once again (and this time rather unexpectedly) the politically more radical candidate, Lawrence Daly, secretary of the Scottish Area NUM, won the general secretaryship. But in 1971, as ten years earlier, his defeated opponent won the election for president. The new president was Joe (later Lord) Gormley, who had been secretary of the Lancashire Area NUM since 1961.[1] This time, however, the president had a robust personality and there was no question of him being overshadowed by the general secretary. Both achieved a lot of prominence in the next few years.

Whatever the character and abilities of the leaders of both management and labour, the achievements that were open to them were more constrained than before, not only by economic conditions but by the increased political pressures which economic conditions made it easier to exert. The nature and effect of those pressures appears at most times to have depended only a little on the individuals most immediately responsible for exerting them. For a time the case for adjustments between government policy and industrial need seemed to be better appreciated. The NCB certainly found relations with government easier under Lord Mills's successor, Richard Wood, who was Minister of Power from 1959 to 1963, and who, in all the long line of sponsoring ministers, probably ranked with Gaitskell as most willing and able to understand the problems of the coal industry. But generally it was more significant that too few ministers stayed long enough in one post to produce initiatives based on knowledge—Robens complained that he had to deal with ten ministers in ten years—though they did not

[1] Gormley 1982, 72–3 and 76–7.

always move on fast enough to avoid making initiatives out of ignorance.

It was also important that changes of departmental organization and of general government policy led to some influential decisions being taken rather more remotely. It had made no difference in the mid-fifties when the Ministry of Fuel and Power dropped the reference to fuel from its title; but in October 1969 the Ministry was abolished and governmental supervision of the coal industry passed to a somewhat amorphous Ministry of Technology under, first, Anthony Wedgwood Benn and then, for a short time after the 1970 election, Geoffrey Rippon. But then the coal industry was transferred to the responsibility of another new amalgam, the Department of Trade and Industry, headed by John Davies, a businessman newly elected to Parliament and given high office almost immediately. Though such changes did not necessarily alter relations with civil servants very drastically, they ensured that high level decisions had to be referred to people to whom the coal industry was far less central than had formerly been the case. And before this, the adoption of special financial obligations and provisions, the application of a series of incomes policies, and the need to move miners to other occupations, gave the Treasury and Employment Ministers more direct influence over what could be done in the coal industry.[1]

So although the new setting in which the coal industry had to operate was moulded mainly by new market conditions, there was also an important political aspect, which was increasingly influential. The lines of policy open to the industry were to seek and try to retain sales, to improve the quality and reduce the quantity of productive capacity, to cut costs in any way that could be found that was not damaging to efficiency, to deal with labour in such a way as to ensure that the industry could avoid overmanning without creating human scrap heaps. Each of these activities needs more particular examination, but, as they are examined, it has to be borne in mind that both the dimensions of the task to be attempted under each heading, and the degree to which it could be satisfactorily accomplished, were gradually modified, not always in intended ways, by political decisions.

## ii. Looking for sales

There were many features of the market, adverse to coal, which the coal industry could not control. What it could do was to make more of itself,

[1] Robens 1972, Chap. 8 is called 'Mixing it with Ministers' and deals with some of these points, with illustrations from specific episodes and comments on the role of the Prime Minister.

and seek to please the customer better, in every field where a chance of effective competition still remained, and to see whether there were new fields which it could capture.

The NCB had always had to keep a close eye on costs and prices, because of government reluctance to allow more than the smallest practicable price increases, and they had had to respond as well as they could to the demands of groups of users and of the two Coal Consumers' Councils about quality, but they had never had to make a major sales effort to dispose of their coal. Ever since the outbreak of the Second World War coal had been rationed for domestic consumers and individually allocated by government order to industrial users. By 1958 numerous industrial users were no longer buying all the coal available for allocation to them, so in the summer of that year the Ministry of Power ended the statutory control of coal supplies. From then on, householders could order as much coal as they wished from whomsoever they wished, and merchants could fix their own prices and profit margins, though the NCB continued unchanged their system of zone delivered prices to merchants. Industrial users could buy coal like any other commodity, though the NCB continued the pithead price formulae for industrial and carbonization coals with no change other than an increase in the price of doubles and larger sizes relative to that of singles. This was done in order to have a bigger proportion of the larger sizes available to attract the domestic purchaser.[1]

Price and quality were more important than ever, but there was also much more need to persuade the customer that coal was worth while, to simplify his problems of identifying what he needed in order to be sure of getting the best value for money, and then to ensure that this was what he got. Prices, over an extended period were raised generally at only infrequent intervals and by only small amounts in relation to the general rise in the cost of living, a practice made possible through the improvement of productive efficiency by technological innovation. From July 1957 there was no general increase in coal prices for a little over three years, but in the autumn of 1960 increases averaging 7*s.* per ton were applied.[2] Thereafter, for several years, only selective price increases were made, for particular types of coal and in particular places. In 1962, for instance, the price of all large coal was increased by 10*s.* per ton, but general price increases were made only for coal pro-

[1] NCB *Ann. Rep.* 1958, I, 31–2. Singles were made between a 2-inch and 1-inch screen, doubles and larger sizes over a 2-inch screen.

[2] Ibid. 1960, I, 6.

duced in Scottish and North Western Divisions, where costs and losses were high.[1] In fact, for four and a half years to March 1965, while the retail price index rose 17½ per cent, pithead prices remained stable for three-quarters of all coal output, but in 1965 costs of production were beginning to outpace productivity per man. The government refused a request for price increases from September 1965 and arranged for any resulting financial losses to be written off. Only from 1 April 1966 (1 May for domestic fuels) was there again a general increase in coal prices, averaging 12 per cent. Even then there was no increase in the prices of industrial coals from the East Midlands, and a much less than average increase in the price of those from Yorkshire; and the price rise for carbonization coals (which followed a price reduction to coke ovens in September 1964) was restricted in order to limit the excess of their prices over those of other industrial coals. Indeed, some coals supplied to coke ovens remained lower in price than they were at the end of 1960.[2] The NCB had to recognize that it was only for their lower-priced coals that demand was increasing, and to use further increases of productivity to stabilize the prices charged for the more expensive coals. Not until October 1969 were there further selective increases in coal prices, followed by a general increase in January 1970. The two increases together averaged 12 per cent, and the October changes included the first price increases for nine years on industrial coal produced in the East Midlands.[3] Higher railway charges had also added a little more to delivered prices.

Thereafter, it was becoming more difficult to achieve extended periods of price stability, as most major items of cost (by no means only wages costs) rose rapidly. An average increase of 16 per cent was applied to carbonization coals on 1 September 1970 and to other industrial coals on 1 November, though coal (but not manufactured smokeless fuels) for the domestic market remained unchanged in price.[4] Another increase, averaging 8 per cent, was put on industrial and carbonization coals on 13 April 1971, and domestic fuels went up by an average of 7 per cent on 1 July 1971. The 1972 strike and its settlement brought a great increase in costs and financial losses, and a general price increase of 7½ per cent was introduced on 26 March 1972 as a gesture towards the recovery of a little of these losses.[5] Clearly, the policy of keeping prices nearly stable, which had been operated with a good deal of success in the sixties, fell apart in the early seventies.

[1] Ibid. 1961, I, 4; 1962, I, 2.
[2] Ibid. 1965–6, I, 4–5 and 17.
[3] Ibid. 1969–70, I, 1–2.
[4] Ibid. 1970–1, I, 1, 4, and 18.
[5] Ibid. 1971–2, I, 28.

The average proceeds per ton saleable show the effects of both the level of prices and changes in the relative proportions of different classes of coal in total sales. The latter influence helped to depress proceeds because of the rising proportion of the cheaper general purpose coals, especially for burning in power stations. The figures show clearly the fairly stable conditions until the end of the sixties and then the change to soaring prices. Of the thirteen financial periods from 1957 to 1969–70 (1963–4 covered 15 months, the rest were financial years), five had average proceeds per ton that were less than in the preceding period. Proceeds in 1965–6 were at a lower average than in 1962, and in 1968–9 were below the level of 1966–7. The chief jumps appeared in 1961, when the price increases of late 1960 first operated for a full year, and in 1966–7. From 1957 to 1961 average proceeds per ton rose from £4.10 to £4.54, a rise of 10.7 per cent. From 1961 to 1966–7 the rise was 10.8 per cent to £5.03, and from 1966–7 to 1969–70 it was 1.8 per cent, to £5.12. The level of 1969–70 was 24.9 per cent above that of 1957, twelve and a quarter years earlier, a rise well below the general level of inflation. But from 1969–70 to 1972–3 average proceeds per ton rose by 36.3 per cent in three years, to £6.98.[1]

The attraction of better quality was not easy to provide consistently, but many efforts were made on both the production and marketing sides. It was at the end of the fifties and in the first half of the sixties that the proportion of trepanners in the supply of coal cutting equipment was raised to its maximum, in an effort to have large coal available for all who wanted it; and more attention was given to coal preparation. A particular difficulty came from the increasing influence of the Clean Air Act of 1956, under which more and more smokeless zones were established each year. Householders and commercial premises in such zones could no longer burn smoky coal and, as the supply of naturally smokeless coals was severely limited, a significant market was under threat. Efforts to develop improved smokeless fuels, manufactured from lower grades of bituminous coal, were slow to produce satisfactory results, though the problems were eventually overcome. There was also no rapid progress with new designs of stoves and other appliances to burn bituminous coal smokelessly. Something was achieved in keeping business through the provision of substitutes. For example, NCB sales of blast furnace coke fell away sharply from 1957, and from 1960 the attempt was made to dispose of more of the surplus of hard coke by selling it on the domestic market under the standard brand name

[1] NCB *Ann. Rep.* 1971–2, II, 4; 1972–3, 36–7.

'Sunbrite',[1] and at the same time it was decided to extend the productive capacity for 'Phurnacite', the long-established premium grade smokeless fuel for closed stoves, which had been inherited from Powell Duffryn. But these were devices for arresting a decline and did not add to coal consumption.

Changes in methods of marketing were also adopted in an effort to provide better service and attract more customers. In 1961 a large-scale sales drive was launched. More showrooms and service and advisory centres were opened. Loan schemes for solid fuel central heating were developed in association with Forward Trust Ltd., and the NCB arranged credit facilities wherever they sold solid fuel direct, either on contract or through their retail subsidiaries. Pre-packing facilities for relatively small quantities of solid fuel were increased. There was much more attention to publicity. From 1 May 1962, in association with various representative organizations of merchants, the NCB sponsored an 'Approved Coal Merchants Scheme', under which basic standards of service were defined. Coal merchants had to meet these in order to obtain recognition.[2] At the same time the price and grading scheme for house coals was simplified, with only five groups instead of seven and a standard price difference of 10*s.* per ton between groups.[3] So as to be able to sell heating installations to interested enquirers, instead of having no means of directly following up advice sought and given, the NCB from the beginning of 1965 became the majority shareholder in a firm of builders' merchants, J. H. Sankey and Son Ltd.[4]

Other efforts were devoted to rationalizing and cheapening distribution arrangements. To encourage a more even flow of coal sales the NCB had for long charged lower prices to domestic consumers in the six months from the beginning of May to the end of October, but this still led to a heavy concentration of sales in early autumn. So in 1964–5 the period of reduced summer prices was extended to seven months, with greater reductions in prices in the first three.[5] Other changes were technical and physical. Arrangements, in association with the railways and local coal merchants, were begun in 1963 to concentrate the handling of coal at fewer, larger, and better equipped depots. A change in numbers

[1] Ibid. 1960, I, 23.

[2] Ibid. 1961, I, 6–7.

[3] Ibid. 1962, I, 16–17.

[4] Ibid. 1964–5, I, 17 and II, 27–8; 1965–6, II, 24; Robens 1972, 314–15 gives the background, including the difficulties which private builders' merchants made for the NCB. On the whole marketing effort at this time there is additional material in Robens 1972, Chap. 4.

[5] NCB *Ann. Rep.* 1964–5, I, 17.

was inevitable because so many railway stations were being closed, and an improvement in quality was a sensible accompaniment. The change began with the opening of the fully mechanized West Drayton coal concentration depot in December 1963. Most depots were not as large as that, but the programme provided for the replacement of over 4,000 depots and yards by only 500, and rapid progress was made in carrying it out. At the same time, to help the most important market for industrial coal, the NCB, CEGB, and British Railways Board were developing a new type of high-capacity wagon for use in permanently coupled trains for 'merry-go-round' working between selected collieries and power stations.[1]

All these were useful measures. None of them individually could make a lot of difference, but together they almost certainly helped to create extra sales, though in the medium term this may have meant no more than the slowing of the rate of decline of total sales. More assurance about consumption might have been obtained if long term contracts could have been made with large users, but not much proved possible. Between 1962 and the spring of 1964 the NCB negotiated an agreement with the CEGB which ran (with some modification for the retrospective year) from 1 April 1963 to 31 March 1966. It gave the NCB an assurance of delay in the conversion of two power stations to oil-firing and a guarantee of penalty payments by the CEGB if specified minimum quantities of coal were not taken. It also provided for the CEGB to receive rebates, varying from one NCB division to another, on specified tonnages taken, and to pay to the NCB fixed charges calculated on a formula which would offset most, but not all, of the rebates. There was some talk of trying to go on to a long term agreement, but the NCB could see no advantage in such an agreement on any terms that were likely to be obtainable. Even the short term agreement was not continued. Subsequently, bargaining with the government about provision for coal burning in power stations had to be the main way in which to try to lessen insecurity of sales. Lord Robens was very active throughout his chairmanship in seeking to protect the coal industry in this way, though his battles were particularly difficult after the adoption of the policy which the government set out in the white paper on Fuel Policy in 1967.[2]

Long term contracts with private users were seldom practicable, but the NCB did negotiate one in 1968 with Alcan (United Kingdom) Ltd.,

[1] NCB *Ann. Rep.* 1963–4, I, 11; 1964–5, I, 19–20.

[2] Robens 1972, esp. Chaps. 8–10.

which was planning to set up a new aluminium smelter, with its own power station, and which decided that a coal-fired station was a better proposition than a nuclear station. The NCB offered long-term supplies based on anticipated costs at particular supplying collieries, and in May 1968 the government agreed to support the smelter, to be sited on the Northumberland coast, beside Lynemouth and Ellington collieries.[1] This arrangement captured a market which originally had appeared unobtainable, but it was ultimately not without problems because the trend of colliery costs after 1970 was much more unfavourable than had been anticipated when the agreement was made.

Despite the occasional successes, no marketing effort, however vigorous and well conceived, was able to do more than slow down the decline of coal consumption. The adverse influences, political as well as commercial, were too strong. Sometimes it looked as though it might be better to try to turn the adverse influences to profit, as when, from the end of 1966, the NCB, having lost most of the coal sales formerly made to the gas industry, associated themselves with various oil companies in drilling for natural gas under the North Sea. This was a successful effort that found gas in the Viking field, from which deliveries began in 1972.[2] But the NCB's main business was with coal, and the inability to maintain the level of sales meant that the major adjustment had to come from the reduction of output.

### iii. Production and productive capacity

When it became apparent that the growth of demand for coal had ceased, it was clearly unnecessary to go on struggling for every bit of extra output. Some changes were practicable without seriously affecting the capacity for future production. The agreement for Saturday working had been renewed annually since it was first made in 1947, but there was no point in paying out extra money for coal produced outside the normal week when the only result was to add to unsold stocks. The NCB were anxious to end Saturday working on their own terms and with goodwill, and not to have it withdrawn by the NUM as an act of opposition to unpopular measures, such as closure of collieries on economic grounds, so timing and method were important. The formal procedure was that, by agreement, Saturday working was suspended from the end

[1] Ibid., 77–80; NCB *Ann. Rep.* 1967–8, I, 20.
[2] NCB *Ann. Rep.* 1966–7, I, 26; 1971–2, I, 17.

of April 1958.[1] This was nominally for a trial period, but it was effectively the end of Saturday working. Some agreements were made from time to time in the sixties for Saturday shifts in particular areas in order to produce types of coal (e.g. large house coal) of which there was an immediate shortage; but there was no return to the general system of agreed voluntary Saturday working, though, of course, normal week-end maintenance work, which was outside the old agreement, continued.

Other steps taken early in 1958 included the reduction of the amount of overtime working, which had been costing £40 million a year, an amount deemed no longer justifiable. There were also drastic restrictions on the recruitment of workers. The numbers recruited were allowed to fall well below natural wastage, as the beginning of a new policy which, in broad principle, became permanent. In 1958 the recruitment of anyone but craftsmen ceased, except at 120 undermanned collieries. In future years recruitment policy had to be continuously adjusted, both to fit what the unions would accept as reasonable (though they were very co-operative on this subject), and to meet changing circumstances and varying local conditions.[2] In some respects it had to be more stringent in order to keep jobs available for men transferring from closing collieries, and it was reinforced by agreements with the NUM for mineworkers not to continue beyond the age of 65. But in detail it often had to be relaxed. There was a need to keep on recruiting youths in order to try not to have an ageing labour force with a distorted age structure. Collieries in Yorkshire, East Midlands, and West Midlands (the three most successful divisions financially) and in South Wales were often threatened with labour shortage, which had to be corrected. And there were times in the sixties when prospects in the coal industry looked so uncertain that natural wastage caused problems by becoming too rapid. But though there were tricky problems in varying the rate, reduced recruitment was a major element in the adjustment of the coal industry to its new circumstances. In 1957, 70,711 recruits joined the industry, in 1971–2 only 13,722; and the average age of the labour force had risen from 40.5 years to 43.7 years.[3] What had begun as an aid to a limited curtailment of current output had become part of a huge reduction in productive capacity.

There were other possible ways of making small reductions in output

[1] NCB *Ann. Rep.* 1958, I, 4.

[2] Ibid., I, 3 and 35; 1959, I, 32–4; 1960, I, 28–9.

[3] Ibid. 1957, II, 136–7; 1971–2, II, 92 and 94.

fairly quickly. One was to stop opencast mining, which could be kept up only by constantly moving to new sites. This was a change for which the NUM pressed hard, as a means of preserving jobs in deep mining, but the NCB, and especially Latham while he was deputy chairman, were reluctant to concede this because opencast mining had become so profitable. Nevertheless, simply in order to reduce social hardship, the NCB agreed in 1959 to run down opencasting as existing contracts expired, but to keep it going fully for the production of scarce coals. This referred mainly to the naturally smokeless coals of South Wales, of which a high proportion came from opencast sites.[1] It took a little time for this decision to be fully effective as existing sites had to be worked out, but opencast output, which in 1958 had reached a maximum of 6.5 per cent of total output, fell to 2.8 per cent in 1963–4, and for the rest of the sixties was held between 3 and 4 per cent. In absolute terms it remained between 6 and 7 million tonnes per year, i.e. less than half the amount produced in 1958.

Another possibility was to restrict the small mines operated privately under licence from the NCB. These had attained their highest output (2.8 million tonnes) in 1959, and the NCB started to negotiate with the Federation of Small Mines of Great Britain an amendment of licences, which would provide for less output. This was practicable as the ten-year licences introduced in 1950 came up for renewal in 1960. Agreement was reached on new licences which included a basic annual quota with the possibility of additions variable in the light of circumstances. Voluntary restrictions on output were accepted by many, but not all, of those whose licences, taken out since 1950, were not yet due for renewal.[2] The result was a small but significant drop in output, and the downward trend continued as the difficulties of the industry led to a reduction in the number of licensed mines operating. In general in the late sixties the annual output of licensed mines, at just over a million tonnes, was only about 40 per cent of that in the late fifties.[3]

The effect of this change was, however, neutralized by the increased supply of non-vested coal, over which the NCB has no control. Non-vested coal production is defined as 'indigenous worked coal originating in workings of coal which by virtue of s.36(1) of the Coal Industry

[1] Ibid. 1959, I, 14–15; 1960, I, 12.

[2] Ibid. 1959, I, 6; 1960, I, 10–11.

[3] A small amount of the difference arises because there were a few licensed opencast sites, the output of which before 1962 was classified under 'licensed mines' and thereafter under 'opencast', but the quantity of coal involved is too small to affect the trend significantly.

Nationalisation Act 1946 did not vest in NCB'.[1] Though the term would cover production from any of the minute proportion of coal reserves alienated by the Coal Commission to private ownership between 1942 and 1946, most non-vested coal production is not 'mined' in any strict sense. It includes coal washed in by the sea and gathered on the beaches of north-east England, and its principal constituent is coal extracted by private operators from old colliery tips and slurry ponds. Down to 1955 the supply of non-vested coal was negligible, but it appears to have got up to a million tonnes a year by 1959. In the sixties, as more and more collieries were closed and, where practicable, their sites were sold, there were more and more sources from which non-vested coal could be extracted. The years from 1967 to 1973 were a peak period of supply, with the annual amount of non-vested production generally keeping between $2\frac{1}{2}$ and 3 million tonnes, which was much more than the licensed mines were producing.[2]

One other possible source of surplus supply was eliminated. No new import contracts were made after 1956, and it was just a matter of waiting for the earlier contracts to expire.[3] The last small deliveries under them came early in 1959. Thereafter the entire coal supply was home-produced until 1970–1 when the government granted a licence for private importation and 1.2 million tonnes came in, with a jump to 5.3 and 3.4 million tonnes in the two following years.[4]

These various changes in supply were small in proportion to the changes in demand. The main adjustments had to be in the amount of productive capacity kept available and the use made of it. An early necessity was to reconsider the existing plans for reconstruction and expansion. These had been made on the assumption that all the coal that could be produced within any feasible plan for growth would be sold,

[1] The term, in fact, was also used to cover the production of individual free miners of the Forest of Dean, whose activities were exempted by s. 63 (2) of CINA, even though the coal they work is legally vested in the NCB (NCB *Ann. Rep.* 1964–5, I, 3).

[2] From 1947 to 1955 inclusive the recorded totals of output from NCB and licensed mines and opencast sites equal the recorded totals of consumption, stock changes, and the excess of exports over imports (though there is not always an equality for each twelve-month period). After 1955 there is always a shortfall in recorded output, to be made up by non-vested production. There is curiously little attention to non-vested production. The NCB allowed for it in the sixties when making their own internal estimates of prospective sales to power stations and drew attention to its existence in *Ann. Rep.* 1964–5. Otherwise it gets scarcely a public mention. In the sixties the NCB began to find it increasingly profitable to have some coal extracted from their own tips, so it is not surprising that the working of private tips became more attractive.

[3] NCB *Ann. Rep.* 1958, I, 33.

[4] Ibid. 1982–3, 20–1.

and by early 1958 this assumption had obviously become false. There were contrary pulls on planning. Cheaper and more efficient capacity was needed in order to make coal more competitive and, in any case, many reconstruction schemes were so far advanced that it would have been wasteful not to complete them; but the total capacity needed to satisfy expected demand was less than already existed. On many grounds it appeared that the best course was to go on modernizing as much capacity as had already been proposed and to shut down the least efficient capacity much earlier than had been intended. But there were doubts whether it was financially wise, or even practicable, to maintain or increase the investment programme of a contracting industry, and there were physical and administrative limits to the rate at which collieries could be closed, as well as severe social consequences to be taken into account.

An attempt was made to steer between these conflicting hazards, and a *Revised Plan for Coal* was produced and adopted in 1959. Pessimism about the future market was still not very deep, and the new target was to produce 200 to 215 million tons (203 to 218 million tonnes) by 1965, with 2 million tons of the total coming from opencast. This replaced the last target of 230 million tons deep mined and 10 million tons from opencast. Investment plans were reduced, with a projected investment from 1960 to 1965 of £535 million at mid-1959 prices (£440 million for collieries, £71 million for ancillaries, and £24 million for contingencies). This was in addition to the £886 million (at current prices) already invested. The number of collieries to be closed, which in 1950 had been projected to be 350 to 400 by 1965, and had already reached 260, was now assumed to rise to 430 for certain, plus from 35 to 70 others, according to the realized level of demand. With a lower output target, some of the earlier inhibitions about switching output from one coalfield to another were abandoned. Proportionately more was now to be sought from the three low-cost divisions (North Eastern, i.e. Yorkshire, East Midlands, and West Midlands), and from South Wales, the chief source of special coals. All other divisions, except Scottish, were scheduled for contraction. It was, indeed, recognized within the NCB that so much was now being asked of Yorkshire, East Midlands, and South Wales that it might not be completely attainable, in which case some contractions elsewhere might have to be delayed.[1]

Even the reduced plan looked in some respects over-optimistic. Not

[1] Ibid. 1959, I, 45–7 summarized the main points. *Revised Plan for Coal*, published in October 1959, adds detail and discusses the reasons for the revisions.

all the smaller schemes were carried out. The completion of major colliery reconstructions was still wanted in order to increase efficiency and, in general, this objective was steadily pursued, though the rate of investment in some projects was slower than it might have been in more encouraging conditions. But the idea of providing entirely new capacity was given up. The projected new collieries were completed, but no more were planned. Kellingley colliery, in the north-east of the Yorkshire coalfield, was for the time being the last entirely new project and came into full production in 1965. Capital expenditure as a whole was kept somewhat below the level proposed in *Revised Plan for Coal*. By the end of March 1965 total capital expenditure since the vesting date had risen, at current prices, to £1,344 million. Of this total, £536 million had gone to major colliery reconstructions and new collieries, £573 million on other colliery schemes, £42 million on housing for mineworkers, and £190 million on ancillary activities, which included a small amount for opencast mining. 258 major colliery projects (including coal preparation plants and new collieries) had been completed, out of 328 approved. Most of those still to be completed were in the East Midlands and South Western divisions.[1] By then the coal industry had enough primary productive capacity for all foreseeable needs, but it still needed investment to increase its efficiency and competitive power.

Few large projects were started thereafter, though work continued on most of those still incomplete. But there were many new small projects to save manpower, particularly by improved transport underground and better surface layouts.[2] Gross capital expenditure at current prices amounted in 1965–6 to £88.0 million (£73.8 million on collieries), in 1966–7 to £89.9 million (£77.3 million on collieries), in 1967–8 to £81.7 million (£72.2 million on collieries), in 1968–9 to £58.8 million (£49.8 million on collieries), and in 1969–70 to £59.4 million (£46.2 million on collieries). This was a total of £377.8 million in five years, which was a very large reduction, in real terms, on what had become customary since 1950. Within the five-year total only £77.1 million went on major colliery projects. In the later years, in fact, the biggest category of colliery capital expenditure was 'coalface machinery and related equipment', which indicated the concentration on fairly quick returns at the point of production. There was some slight recovery of capital investment after this, with a total of £230.8 million in the three following years, with £180.5 million going on collieries.[3] But in 1972–3 it was

[1] NCB *Ann. Rep.* 1964–5, I, 8.

[2] Ibid. 1965–6, I, 9–10.

[3] The figures relate to the provision of fixed assets and are taken from NCB *Ann. Reps.* They

noted that expenditure on major colliery schemes was, in real terms, only just over one-tenth of the amount a decade earlier, and that this was inadequate if the industry was not to go on declining indefinitely.[1] Even though reduced investment was appropriate for a smaller industry, the paucity of major schemes would eventually cause a failure to provide enough replacement capacity.

This concern about the adequacy of future capacity was a newly emerging worry. Until then the market had been forcing a continual reduction of capacity, and the switch of investment to an almost exclusive emphasis on efficiency rather than quantity was not by itself a sufficient answer to the problem. The main change was to close a large number of collieries and to let the completed capital schemes replace only a small proportion of their capacity. At first this did not seem to require a very drastic policy. The successive versions of *Plan for Coal* had all recognized that there were many collieries which would exhaust their reserves within a few years, and the reduction of demand suggested that their inevitable closure might be carried out more quickly. But market pressure soon began to call into question the survival of many of the loss-making collieries that had been kept going because all output had been needed. It did more. The need to keep prices as nearly stable as possible, which could be met because productivity was rising quickly in much of the industry, worsened the financial position of the less efficient collieries. Some which had been economic ceased to be consistently so. Some produced coal which it was becoming difficult to sell at all. Financial pressure on the NCB to cover their costs reinforced the physical need for reduced capacity. So closure policy had to put increasing emphasis on the elimination of loss-makers rather than deal mainly with the closure of collieries approaching exhaustion. An indication of what was happening is given in Table 6.1.

It should be emphasized that any simple tabulation exaggerates the sharpness of the distinction between categories of closure. A colliery was closed only after consideration of all aspects of its performance and prospects. A summary statement of the grounds for closure could concentrate only on the principal reason, when there was often a combination of circumstances to be taken into account. Moreover, the categories used had some degree of overlap. Apart from the small number of closures caused by serious accidents or geological difficulties, most

are comparable from year to year except for the last year, 1972–3, where the total (but not the figure for colliery investment) is more comprehensive than before.

[1] NCB *Ann. Rep.* 1972–3, 7.

Table 6.1. *Number of colliery closures, 1958 to 1972-3*

| | 1958 | 1959 | 1960 | 1961 | 1962 | 1963–4 | 1964–5 | 1965–6 | 1966–7 | 1967–8 | 1968–9 | 1969–70 | 1970–1 | 1971–2 | 1972–3 |
|---|---|---|---|---|---|---|---|---|---|---|---|---|---|---|---|
| Closed on economic grounds | 13 | 36 | 6 | 4 | 15 | 15 | 13 | 21 | 25 | 31 | 38 | 10 | 2 | 0 | 0 |
| Closed for other reasons | 15 | 17 | 29 | 25 | 37 | 25 | 27 | 31 | 21 | 20 | 17 | 9 | 4 | 3 | 8 |
| Total closures | 28 | 53 | 35 | 29 | 52 | 40 | 40 | 52 | 46 | 51 | 55 | 19 | 6 | 3 | 8 |
| Mergers | 4 | 7 | 9 | 3 | 5 | 4 | 5 | 2 | 1 | 11 | 4 | 0 | 1 | 0 | 0 |

*Note*: The table includes mergers, which usually meant that one colliery swallowed up its neighbour. The latter thus lost its independent existence rather than survived as a joint equal.

*Source*: NCB Statistics Dept.

closures were given one of four descriptions: 'economic', 'exhaustion', 'exhaustion of realistic reserves', or 'manpower reservoir'. The third description involved some elements of both the first two, though the near approach of exhaustion was a predominant factor. A colliery described as a 'manpower reservoir' was usually an unprofitable undertaking which had been kept going in order to preserve in being a labour force that was going to be employable somewhere else in the near future, and the colliery was closed when the new employment opportunities became available. So 'manpower reservoir' was a description partly related to 'economic'. In Table 6.1 only closures where the description 'economic' was given at the time as the principal reason for closure have been counted as 'closed on economic grounds'. In one way this must appear to understate the 'economic' closures. But it is also relevant that an uneconomic colliery was often in this condition because it had started to run down towards physical exhaustion (and might in a few years have been described as experiencing 'exhaustion of realistic reserves') or because it was afflicted by some significant degree of geological difficulty. So there is a margin of error in both directions. If the appropriate qualifications are remembered, the figures in the table are not likely to be seriously misleading, and the trend towards greater influence of economic factors was genuine.

The figures do, indeed, show how much stronger was the economic pressure for closures in the latter half of the sixties, especially in the first two years of the new fuel policy inaugurated by the White Paper of 1967. But they do not fully convey how much more drastic things had then become, because they do not indicate the increase in the average size of the closed collieries and in the number of men involved. It might seem as though 1959 saw action as drastic as any later year, but most of the collieries closed then were of quite modest size. Half of them (and more than half of those closed primarily for economic reasons) were in Scotland, where not one of the collieries closed had been producing as much as 100,000 tons a year.[1] These closures were, indeed, a partial corrective for the great financial deterioration of Scottish mining since the late forties. The only really large closure in 1959 was Maypole colliery near Wigan, which had 1,287 men on its books and was closed on economic grounds. Only one colliery of comparable size was closed for economic reasons between then and 1965–6. This largest and most disappointing example was the new Rothes colliery, which had

[1] The largest of them, Jenny Gray colliery at Lochgelly, would meet this criterion if metric tons were used.

encountered a continual series of unpredicted physical difficulties that kept it hopelessly uneconomic. It closed in 1962, when its labour force, previously much larger, was 944.[1] But the closures for economic reasons in 1965–6 included three collieries each with a labour force of over 1,000. There were two more of this size the following year, and seven in 1967–8, including Mosley Common colliery in Lancashire, which had 2,892 men on its books.[2] Three more such large collieries, of which the biggest, Kirkby in South Nottinghamshire, had nearly 2,000 men, were among the economic closures of 1968–9. There were two more in 1969–70 and one in 1970–1. And in the late sixties there were many more with a labour force of between 500 and 1,000 each. A large proportion of these collieries had annual outputs exceeding a quarter of a million tons.

To carry out such a large number of closures required very careful management. If a vast amount of capacity had to be sacrificed, it was necessary to ensure that what survived was what was most needed, in terms of types of coal, continuing access to large reserves, convenient location for continuing markets, and profitability. There was an obligation to deal fairly with those whose livelihood was involved. That meant the provision of adequate warning of the risk of closure, and consultation about possible ways to avoid it and about the way it would be carried out if it was not avoided. It meant also the greatest care to provide alternative employment when collieries closed, which was a consideration that set a limit to the numbers of collieries that could be closed in a given period; and there had to be compensation for those who could not get other jobs. Such provision for labour relations was needed not only as a protection against social distress. It served a direct economic purpose. The coal industry needed to control the rate at which its labour force ran down, both nationally and locally, and it still needed to attract and retain young and skilful workers. Most of its most efficient production was in regions where there was strong competition for labour. There was no way of preventing the industry's contraction damaging the belief in the security of employment which it could offer. But at the very least, the contraction could be conducted in a way that might persuade people that there still were secure jobs worth having and that employment risks were countered by the assurance of getting a fair deal.

[1] Robens 1972, 145–6 discusses the trauma of closing Rothes colliery. The other large colliery to close was Upton, in Yorkshire, in 1964–5, but this was the result of a fire and explosion.

[2] The events leading to this closure, and the failure of the miners to respond to opportunities to improve performance, are described in Gormley 1982, 52–7. One more pit with over 1,400 men, Mid Cannock, closed in 1967–8, but this was due to exhaustion.

For most of the sixties, though there were numerous other issues which absorbed continual administrative attention, closure policy was one preoccupation at every level.[1] The scale of closures needed each year, and their approximate distribution, was a headquarters decision, but the choice of specific collieries for closure was, in the first place, a matter for divisions or, from 1967, for areas. The divisional proposals were scrutinized in detail at headquarters, with particular attention to the likely consequences for marketing and industrial relations, and, above all, to the continuous record of each colliery. It was the practice at headquarters, when closures were being considered, to review the physical and financial performance of each pit in each of the four preceding years and each quarter of the current year. The usual result was that most of the proposals were approved, but headquarters often suggested a few additions or variations. A division that disagreed with one of these suggestions had to put up a detailed case against it, or put forward a preferable alternative. This was sometimes successfully done, but most of the additional proposals were implemented, though often with varying delays that made the outcome seem rather less systematic than the proposals. There was a tendency for the lower formations to be slightly less drastic than national policy required.

It was, of course, fairly widely known where closures might first be sought. If a colliery did not have access to enough reserves for an indefinite future, there was a record of when it was likely to run out of coal. That, with modifications for known geological problems and financial record, gave an indication of expected remaining life. These considerations governed a system of classification which applied to all collieries and related to additional investment in them. The more uncertain a colliery's expected life, the more restricted was the authority delegated to lower formations to incur capital expenditure on it. There were three categories, and the allocations of collieries to them were frequently updated. In general, the first category covered long-life collieries, the second those with an expected life of over five but less than ten years, the third those unlikely to be viable for more than five years. When closures were being considered, they were sought mainly among collieries in the second and third categories, especially the third, though any colliery in the first category that repeatedly made losses so large that it contributed nothing to overheads and depreciation was certain to get questioning scrutiny.

[1] It would be impracticable to give detailed references for this. The paragraphs which follow are based on a large body of papers, which were prepared at frequent intervals over a number of years and discussed by the NCB General Purposes Committee and the National Board.

Late in 1965 it was agreed that future discussions with the unions and public authorities about closures should be on the basis of a list for each division, with every colliery classified as A, B, or C. 'A' referred to continuing collieries, 'B' to those whose future was 'doubtful' on geological or financial grounds, 'C' to those expected to close within five years. There was clearly some similarity to the classification for delegated capital expenditure, but the two were not identical. For example, 'A' collieries included some that were expected to exhaust their reserves within ten years, and that was why the term 'continuing' was used and 'assured long life' was not. It should be stressed that, although this classification was a good indicator of the destiny of most collieries, it was never final. New discoveries and developments could improve a colliery's prospects, and a persistently bad financial performance could worsen them. There were a few collieries classified 'C' in 1965 which were able to operate into the nineteen-eighties. Frances colliery at Dysart and Denby Grange, south-west of Wakefield, are examples. On the other hand, a few category 'A' collieries, with Mosley Common as the most prominent example, were closed before 1970.

When a colliery's performance was under review it was discussed with the unions through the pit consultative arrangements and at higher levels. Schemes for possible improvements were examined and, if tried, their results were made known and discussed. Definite proposals for closure never came out of the blue, though there were always some workers who seemed to be taken by surprise. It was, indeed, a common experience when a pit was in jeopardy for quite a lot of workers (especially those in their twenties and thirties) to start looking for other jobs. When a closure was decided, as many mineworkers as possible were offered alternative jobs in the industry. There was the chance to move to coalfields with better prospects, but most mineworkers preferred to go on living in their familiar district. Because natural wastage was high there were many opportunities to redeploy men to nearby pits, and there were schemes to preserve wages for a period if a man had to move to a lower grade job. Until the late sixties, when financial provisions had been improved, very few workers at closed collieries were made redundant. Most of these were over the age of 55 and affected by some disability which made it difficult to find alternative work for them. To try both to retain a willing labour force and to lessen hardship the NCB planned the timing of closures as carefully as they could. Where new developments were in progress it was the regular practice to keep open a few unprofitable, low productivity pits as 'manpower reservoirs' until new jobs which their men could take became available. The combination of this care to preserve mining jobs for those who wanted them with the generally buoyant levels of employment in the economy as a whole had a beneficial effect in reducing the personal hardships and ill-feeling which a massive closure programme could cause. In this respect the transition appears to have gone better than in the coalfields of Belgium, France, and the Ruhr.[1] Of the 678,000 people who left a mining job at some time between 1957 and 1967 inclusive, only 3.8 per cent did so because they had been made redundant.[2]

An effort was made to carry out the closure programme in such a way as to leave the coal industry with a bigger proportion of its capacity in the more productive and profitable central coalfields of Yorkshire, East Midlands, and West Midlands. This was likely to happen anyway because those coalfields (save for part of the West Midlands division) had fewer collieries approaching exhaustion or making large losses, and were still bringing into use new or reconstructed capacity started under *Plan for Coal*. From 1958 to 1965, when colliery manpower fell by almost one-third, it was reduced by less than a quarter only in Yorkshire and East Midlands divisions. But this was not an easy policy to sustain because competition for labour was more intense in the central coalfields than in any other. As a partial counter to this difficulty, attempts were made to induce miners from the coalfields with poorer prospects to transfer to the central coalfields, and also to parts of South Wales which had recruitment difficulties.

There were two schemes. In 1962 an Inter-Divisional Transfer Scheme was introduced and made a promising beginning, with over 6,000 men transferring in the first three years, though this was fewer than the importing divisions would have liked. The amount and terms of various forms of financial assistance to transferees were improved from time to time after negotiation with the NUM and NACODS.[3] Additional efforts were made from 1966 by means of the 'Pick Your Pit'

[1] House and Knight 1967, 1.

[2] Knight 1968, 3–4. Several detailed local studies of pit closures and their effects were made in this period. The most telling are probably those concerned with the northern coalfields because workers displaced there were helped less than those elsewhere by the intensity of competition for labour from other industries. Besides the works cited in this and the preceding footnote, which relate to Cumberland and Durham respectively, there is Department of Employment and Productivity, *Ryhope: a pit closes* (1970), which is particularly illuminating. All identify some types of hardship and material loss, but all indicate a high degree of successful adaptation to the circumstances of mining contraction. Rather, more difficulty is indicated in some of the studies of South Wales. These include Sewell 1975 and Town 1978.

[3] NCB *Ann. Rep.* 1963–4, I, 31; 1964–5, I, 33; 1965–6, I, 37.

arrangement. Men at closing collieries in Scotland, Northumberland, Durham, and Cumberland were interviewed by mobile teams and given details of jobs available at particular long-life collieries, and about housing and removal arrangements provided by the NCB. Miners and their wives were offered free visits to collieries and nearby communities to which they might consider moving. If they decided on a move they could be recruited for a specific job at a designated colliery. The same sorts of provisions were made available for transfers from one part of the South Wales coalfield to another.[1] The second scheme was an attempt to get ex-miners in the northern coalfields who had left the industry to rejoin and transfer to other coalfields. This scheme, known as the Long Distance Re-entrant Scheme (not a title likely to rouse fervour in recruitment offices!) was started in January 1964. Recruits were paid transfer allowances, by the Ministry of Labour if they were unemployed, by the NCB if they came from another industry.[2] Both schemes were reinforced by the provision of additional housing where recruits were most needed. This was done by arrangement with local authorities or, to a lesser extent, by the Coal Industry Housing Association. The NCB paid local authorities a subsidy additional to that which they could get from central government under the Housing Acts.[3]

These schemes were modestly successful in the sense that most of the men who moved under them settled down in their new jobs and localities. But most mineworkers had stronger ties to their home community than to their industry, and the number of long distance transfers was not large. By the end of March 1971, when the great wave of closures was over, 14,974 men had taken new jobs under the two schemes,[4] and most of the movement had been in the earlier years (the total was already about 8,500 by March 1966). Indeed, in the later years, when higher compensation was available, the effects of closure were dealt with more by redundancy than redeployment. For instance, the collieries closed in 1968–9 had employed 29,300 men, and 20,367 were declared redundant.[5] This was a change of policy resulting mainly from a greater government recognition of the national causes and consequences of contracting colliery employment, and the provision of more government money to enable the NCB to make various supplementary payments to mineworkers aged 55 and over if they were declared redundant.[6]

[1] NCB *Ann. Rep.* 1966–7, I, 35.
[2] Ibid. 1963–4, I, 31.
[3] Ibid. 1966–7, I, 36.
[4] Ibid. 1970–1, I, 34.
[5] Ibid. 1968–9, I, 35–6.
[6] Coal Industry Act 1967, s. 3 and Redundant Mineworkers (Payments Scheme) Order, 1968.

Thus long distance labour transfer had practically ceased to contribute significantly to a policy of shifting the balance among the nation's coalfields. There had always been other elements in this policy. The early concentration of closures in Scotland has already been mentioned, and it had the effect of cutting out some of the sources of heavy financial loss. The earlier years of the closure policy also saw a particularly severe rundown in North Western division, but there was, too, an above-average contraction in West Midlands division, which was not ideal. A particularly difficult problem of reshaping was provided by South Western division, which was wanted as the nation's largest source of special quality coals, yet was an obvious candidate for large cuts because it had difficulties of labour recruitment and numerous collieries with very high costs and losses. Until 1965 it was treated more gently than other divisions because, despite the general difficulty of keeping an adequate number of mineworkers under 40 years old, there were many pits in isolated districts with much more than average proportions of older and partly disabled men, whose alternative prospects were bleak if their collieries closed. But by 1965 it seemed impossible to maintain this restraint. By then the division had since nationalization accumulated a deficit larger than that of the NCB as a whole; it had received proportionately 50 per cent more investment than the average for the whole industry, but its OMS remained only just over two-thirds of the national average; and it was losing output through labour disputes at well over double the national average rate. So a working party, under Sir Humphrey Browne, was appointed 'to produce a plan to make the South Western Division a viable coalfield'. This body within five months worked out proposals which involved closing half the division's 91 collieries before 1970. It was even prepared to supply Llanwern steelworks partly with Yorkshire coking coal if the South Wales collieries failed to produce enough after this surgery. These proposals were adopted as the basis of future policy for the division. In fact, they were not completely implemented. There were 35 closures and 2 mergers in the division in the relevant five years. The financial performance was improved, and for a brief moment (the year 1967–8) South Wales was brought just into profit, but drastic closures had not got rid of all the centres of poor performance.

Whatever the tempering of the harshest winds when severe proposals were made, the closures were applied unequally in different coalfields and approximately in accordance with economic criteria. The result was a substantial shift in the geographical balance of the industry, as can be

Table 6.2. *Regional distribution of output and manpower, 1957 and 1972-3* (percentages)

| | 1957 | | 1972–3 | |
|---|---|---|---|---|
| | Output | Men | Output | Men |
| Scotland | 10.1 | 12.2 | 8.7 | 10.2 |
| Northumberland & Durham | 18.0 | 20.5 | 13.9 | 16.5 |
| Yorkshire | 20.9 | 19.6 | 28.4 | 25.8 |
| North Western | 7.7 | 9.0 | 4.3 | 5.5 |
| East Midlands | 18.8 | 12.6 | 23.6 | 17.9 |
| West and South Midlands | 12.2 | 10.4 | 11.8 | 9.8 |
| South Western | 11.4 | 14.8 | 8.5 | 13.1 |
| Kent | 0.8 | 1.0 | 0.8 | 1.0 |

*Note*: The contemporary divisional statistics of 1957 have been reworked so as to make the regions identical for both years. Because of rounding to one decimal place not all the columns total exactly 100.

*Source*: Calculated from NCB *Ann. Reps.*

seen in Table 6.2. The three regions making up the central coalfields, which showed so much better productivity than the rest, produced just over half the coal in 1957 but well over three-fifths in 1972–3. Most of the other coalfields, by shedding much high cost capacity, had at least not worsened their relative inferiority in productivity, and in some cases, notably in Scotland, had improved it. The great exception was the South West which by this time was almost the same as South Wales. South Wales now counted for less but, after all the surgery of the sixties, in the contrast between its share of output and its share of labour, its physical performance relative to that of the rest of the coal industry looked worse than before. But even South Wales offered some compensations because its share of the highest priced coals had increased and consequently the relative financial position had not deteriorated to quite the same extent.

But, however well the closure programme may have moved in the direction indicated by some important criteria of economic efficiency, it also imposed both hardships and real costs. Colliery output had been reduced by over 37 per cent, colliery manpower by nearly 62 per cent, and the number of NCB collieries by nearly 65 per cent. The NCB had not only ceased to mine coal in some of the small coalfields, as they had in Machrihanish, Forest of Dean, and Bristol, and at the end of 1972–3 were preparing to do in Somerset, but had also withdrawn from operations in numerous districts in the major coalfields. In some of these districts social adaptation had not been too difficult, in others the local communities had been much damaged.[1] But wherever operations ceased in part of a coalfield there were economic effects over a wider area. Reserves of coal that were part of the national endowment of natural resources were to some extent sterilized: i.e. some coal was left where it might become irrecoverable, and future access to other coal was made more difficult and costly. It may be that the amount of damage done in this way by the closures of the sixties was not very extensive, as many of the closed mines were old and not very deep, so that they were unlikely to have a lot of unworked coal in the vicinity and what there was might well be attainable, if required in the future, by opencasting or new drift mines. But there were exceptions, and they were at least a small source of capital loss. There were other incidental changes which added to the loss, sometimes in ways not at all precisely measurable. Many of the collieries closed were obsolescent, but not all. Some capacity and items of equipment were scrapped before their time, without replacement. Reserves of specialized skill, created by training and experience, were left unused. Perhaps most important, the concentration on closures meant that by the later sixties there were hardly any starts on new investment projects of appreciable size. An extractive industry needs a continual creation of new capacity in order to maintain output and efficiency. A policy that was all closures and no openings was injecting into the industry an inability to respond to new opportunities, except after a delay of several years from the time of their perception and the start of attempts to seize them.

The coal industry of the early seventies was left with a much reduced capacity. The average quality of what remained was much higher than it had been, because the most inefficient capacity had been shed and the projects planned in the fifties had come to maturity and added something better. But the losses have to be taken into account on the other side. By the mid-seventies it was becoming apparent how much harder these had made the task of halting and reversing the downward slide when there was a need to do so.

[1] Cf. the contrasts in Durham noted in House and Knight 1967 and summarized by them on 144.

### iv. The adaptation of the administration

A long period of sharp contraction in an industry is almost bound to affect administration in two opposite ways. The management of contraction is immensely demanding on administrative time and effort, and therefore makes it difficult to prevent administrative staff and costs becoming a larger proportion of the total, at least until the industry finds a period of stability at a smaller size. On the other hand, there must be a search for economies of every kind, which creates pressure for non-operational savings; and an industry of greatly reduced size needs to be organized and administered differently, which usually means in a simpler and cheaper way.

The coal industry conformed to these expectations. As early as September 1957 a freeze was put on non-operational expenditure, and just over a year later a 5 per cent cut in non-industrial staff was ordered for 1959. This was the beginning of a reversal of the trend, initiated by the Fleck Report, to increase administrative staff both absolutely and proportionally. From time to time changes in practice were adopted in an effort to get things done just as well with fewer people, especially by trying to ensure that things were not looked at twice, if once was enough, and so on. Thus in 1961 the Fleck doctrine of central direction was slightly reinterpreted. Essentially, what was done was to delegate more authority to lower formations to take decisions in accordance with established principles, without seeking prior confirmation from above, and to compensate for this by tightening the procedures for accountability. Another step in the same direction, made at Robens's insistence despite the doubts of some of his board members, was the decision in 1962 that estimates, once agreed, were no longer matters of mere guidance which might come back for further discussion on a different basis, but must be firm budgets on which the lower formations were strictly accountable. There were also changes in various aspects of headquarters organization and practice so as to concentrate functions more effectively in fewer bodies and to end separate provision for functions of declining importance. For instance, the Reconstruction Department had been set up to accelerate the completion of capital schemes when maximum expansion was wanted, but the end of expansion ended that urgency and in 1961 the department was absorbed into the Production Department.

A much more influential example of concentration was that early in 1963 the national board decided (though only on the chairman's casting

1. TIP RESTORATION. Pye Hill Colliery, South Nottinghamshire Area. An example of the progressive restoration of a colliery tip. Agricultural land, on which cattle graze, is provided on the reclaimed area of the tip.

2. OPENCAST SITE. Westfield Site, Fife. Excavations began in 1957, and production commenced in 1961, with output frequently exceeding 1m. tonnes per year. Reproduced by kind permission of Costain Mining Limited.

3. 'GEDLING'S RECORD BREAKERS'. Painting by D. Wharton of miners at Gedling Colliery, South Nottinghamshire Area celebrating the first occasion when two faces cut over 1,000 metres on the same shift, 6 March 1979. 1st prize in CISWO 'paint a pit' competition, 1980.

4. COAL FACE MECHANIZATION. Ranging Drum Shearer, Armoured Face Conveyor and Powered Roof Supports in the Red Vein seam at Betws Colliery, South Wales Area, a new anthracite mine near Ammanford. Production commenced in 1978.

vote) to abolish all its standing committees except the General Purposes Committee, and to have the board and that committee meet in alternate weeks. One major consequence was to give the chairman a greater ascendancy in the conduct of the board's business.

There were also reductions in the number of separate bodies at lower levels as a result of fusions, which became easier and advantageous as the number of collieries and employees diminished. In 1957 the NCB had fifty-one areas, but by 1966–7 the total was down to thirty-five,[1] and there were many fewer groups (the level between area and colliery) than there had been. There had also been some rearrangement of the divisions. Cumberland was taken out of the Northern (N&C) Division on 1 July 1963 and placed in the North Western Division, and from 1 January 1964 the former Northern Division and Durham Division were merged in a new Northumberland and Durham Division.[2] Northumberland and Durham had lost 50 collieries by closure in five years before the merger, so the change of scale alone suggested a strong case for simpler organization, though the alteration at this stage was not drastic. Indeed, with these pieces of redistribution, and the renaming of North Eastern Division as Yorkshire Division, which took effect on 1 January 1963,[3] the divisional pattern had returned very much to what was suggested in 1946.

Occasionally the trend towards fewer administrative units was apparently reversed, in special circumstances, in the expectation that a greater concentration on a more limited range of functions would lead to both economy and greater efficiency. The NCB divisions, whose main concern was with the deep mining of coal, had also to supervise all NCB brickworks and carbonization plants within their territory, though technical supervision of brickworks was supplied by headquarters. These duties were taken away from the divisions, which were left free to deal with coalmining. A separate Brickworks Executive, with headquarters in the Midlands, was established from 1 April 1962. A Coal Products Division, organized in three regions, came into existence on 1 January 1963 and, after a six month transitional period, took over full responsibility not only for production, but (in the case of metallurgical coke, coke oven gas, methane, and by-products) for marketing as well.

[1] These figures count South Eastern (or Kent, as it was officially designated from 1 Oct. 1964) as an area. It lost divisional status at the same time as its name was changed, but it remained responsible to national HQ and thus had a different position from an area.

[2] NCB *Ann. Rep.* 1963–4, I, 26.

[3] Ibid. 1962, I, 1.

The NCB Marketing Department and its regional sales organization retained responsibility for selling domestic and industrial coke and manufactured smokeless fuel. Another change of the same kind, though smaller in scale, was to give departmental status, from 1 August 1963, to what had been the Housing and Surface Estates Branch of the Secretary's Department.[1] In all these cases the NCB were not adding to the administrative edifice. It was a matter of placing more direct responsibility, with a functionally more appropriate line of management, on the specialists who were already doing the job.

The effect of these various changes on the general search for economy in non-industrial staff was fluctuating. For one thing, there were offsetting influences from the pressures of the market. The need for more action to persuade customers to buy coal led to the recruitment of more sales and technical staff, and the same sort of pressure led to the engagement of more staff in a new Process Development Department, to try to create improved types of manufactured smokeless fuels. In 1957 the implementation of the Fleck proposals for stronger administrative control was still incomplete, and the numbers of staff did not start to fall until the imposition of a positive cut at the end of the next year. Then, in the two years to the end of 1960, there was a reduction of more than 10 per cent, from 56,500 to 50,600. After that, reductions were much slower. It took more than five years to achieve a further drop in staff numbers equal to that which occurred in 1959 and 1960. This was partly because of the extra tasks that had to be undertaken, partly because consumption of coal remained above the level of 1959, partly because any obvious 'fat' had been removed. Nevertheless, the effect was that non-industrial staff were forming an increasing proportion of the total number of employees. In the late sixties and early seventies the slimming process went a good deal further, with a particularly large drop (4,300) in the total of non-industrial staff in 1968–9 (i.e. a reduction of over 10 per cent in one year), and one of nearly 2,800 in 1972–3. The totals, at 31 March each year, were 45,100 in 1966; 42,200 in 1968; 36,150 in 1970; and 31,600 in 1973.[2] From the peak of 1958 the number of non-industrial staff had fallen by 44 per cent, which was more than proportionate to the drop in output, but appreciably less than the reduction in the numbers of men on colliery books. It should be remembered, however, that the subsidiary activities of the NCB used a higher proportion of non-industrial to industrial staff, and not all had shrunk as coalmining

[1] NCB *Ann. Rep.* 1961, I, 26; 1962, I, 21 and 25; 1963–4, I, 21, 26, and 27.

[2] Figures from NCB *Ann. Reps.* for years cited.

had done. From 1957 to September 1970, after which the total again *rose* slightly, NCB *industrial* employment at establishments other than collieries fell by only 30.4 per cent.[1]

In five years before March 1966 numbers of non-industrial staff fell by just over 5,000, in five years after March 1966 they fell by just over 9,000. The contrast is attributable mainly to two influences. One is the sharper contraction of the coal industry after 1966, especially the greater reduction in the number of operating collieries. The other is a drastic simplification of the industry's organization. Neither had an instantaneous effect on numbers, but, once they had been not merely introduced but thoroughly established, which took almost eighteen months, both changes made many former jobs unnecessary.

The case for a reconsideration of the NCB's organization was put forward in January 1964 by Lord Robens. He argued that, since the Fleck Report, there had been some large changes in the way the board ran their activities, with little revision of the line and staff structure to take account of them. He urged that the question should be faced, whether the present organization helped or hindered the main management task of keeping the price of products stable, and he also suggested that any reappraisal should be made internally by people with experience of the different levels of management they were reviewing. So a working party was appointed to examine organization in the light of developments in the industry and its future requirements. Its chairman was H. E. Collins, a board member who was a technical man with a long and varied administrative experience in the coal industry.

The Collins Report was completed in February 1965 and received very thorough discussion from April to November of that year. In essence its proposals were directed towards a move from a five-tier to a three-tier organization, through the abolition of divisions and groups; but the scheme did not completely meet the needs which Lord Robens had emphasized, for in some respects the recommendations lacked precision and there were some odd omissions. There were detailed proposals for strengthening the management of the individual collieries, but it was proposed that every colliery should have a Colliery General Manager (a higher status than the ordinary manager), and that a 'colliery' might be a single large mine or two or more smaller mines, and some people objected that this meant the retention of some groups under another name. The area was to remain the principal business unit of the industry, but in future it would be the only formation between the national board and the

[1] NCB *Ann. Rep.* 1958, II, 135; 1971–2, II, 97.

colliery. Yet the report, though it seems to have taken for granted that there would be further mergers of areas, said nothing about the future number, size, and location of areas. The working party noted that the managerial responsibilities of divisional boards had become increasingly concentrated in their chairmen and that divisional departments had become less useful because areas had stronger service departments of their own, and much of the rest of what they needed could be supplied by national headquarters. But instead of going the whole hog and seeking to abolish the divisions, the report insisted that an intermediate authority between national headquarters and area was essential. It therefore proposed that divisional boards should be abolished but divisional chairmen should be retained, with responsibility for the performance of their areas; and that services which could not economically be provided at each area should be provided in each division at a regional service centre, which would be a headquarters responsibility but would report to the divisional chairman on matters of staffing and discipline.

Some of the proposals seemed to blur the lines of responsibility. One severe criticism made in board discussion was that 'a new organisation in which the Area General Manager is uncertain whether he is accountable to the Divisional Chairman or to the National Board is damned from the start.' Much of the detailed examination of the next few months was devoted to the removal of the uncertainties and confusions, while retaining the basic framework proposed in the report. What emerged was a much clearer version in which the three-tier structure was unambiguous and there was no doubt about the line of responsibility. The divisional chairmen were retained, but were made part of the national headquarters establishment and attended meetings of the national board. Each one had to help to guide a number of areas towards the attainment of their objectives and to help to ensure that the needs and performance of those areas were properly understood at headquarters. But the accountability of areas was direct to the national board, not the divisional chairman. Colliery management was kept on an individual mine basis. Only in the rare cases where adjoining mines formed a physically interlocked complex was a colliery general manager to be responsible for more than one mine. Only very large collieries were given a colliery general manager, and the others still came under an agent manager or a manager. The areas were remodelled in size, function, and staff establishment. A scheme was prepared to divide the industry into seventeen areas (plus Kent as a smaller unit which continued to have more detailed direction from headquarters), and it was originally intended to move to this structure through a phased series of mergers, which would be completed in 1970. But it was recognized that, despite the risks from a more drastic upheaval, it would be helpful to introduce the entire new organization at once and avoid an awkward period of transition. So the new areas were scheduled to start at the same time as the divisions disappeared. The new areas took over most of the surviving administrative functions of the divisions as well as those of the previous, smaller areas. They were big enough to provide most of the services which they each needed. But some specialist services, including computing, were provided by headquarters through outstations. Very detailed arrangements were made which specified the location and function of each outstation and which areas were to use its services.

Decisions on most of these matters were reached at the end of 1965, and 1966 was used for preparation, for training in the operation of the new system, for discussions with staff and their unions, especially about those whose jobs and workplaces would disappear, and for the beginning of schemes of redeployment. Natural wastage and retirement accounted for much of the planned reduction in staffs, but there were redundancies also and there had to be plans to phase them and to reduce hardship.[1]

The new organization came into force on 26 March 1967, the start of the financial year. Its details and the lines of responsibility and accountability involved were set out, for the understanding of all concerned, in an entirely new General Directive. It was accompanied by two other changes. One was the introduction of management by objectives, applied to area planning and control. The necessary practices were detailed in an additional directive, with the introductory direction, 'Those set out in this Directive shall be mandatory'. The list of practices included those relating to the plans, forecasts, and accountability of collieries.[2] The other change concerned physical location. The abolition of two tiers of management left the NCB with a surplus of office buildings, some of which could be disposed of and some used for new purposes. The biggest change made possible was in the decentralization of some headquarters functions, which had no special dependence on London connections. There were, indeed, some supply and production

[1] The features of the new organization are summarized in NCB *Ann. Rep.* 1965–6, I, 31–2 and 1966–7, I, 30. The working out of its details came in a series of internal memoranda and reports. Robens 1972, 119–27 gives a personal account.

[2] There is a brief summary in NCB *Ann. Rep.* 1966–7, I, 30–1.

requirements which could probably be better met in the coalfields. So the headquarters Purchasing and Stores Department and many branches of the Production Department and Finance Department were transferred from London to Doncaster, where there was a large new office building which had been intended for the now disbanded Yorkshire Division. In London various units had been housed in widely scattered small offices. It was now possible to bring nearly all these into the main headquarters building, Hobart House. The few others still remaining in London went to a smaller new office block in Harrow.[1]

It had been stated before the new organization was introduced that it would be reviewed in 1968, and a reconstituted working party, with W. V. Sheppard as chairman, was set up in May of that year. This body was given some additional specific terms of reference, including the principles of management by objectives and the role of headquarters, and it was given authority to use management consultants, though it did not have time to do so. The outcome was a report which did not just consider how the new system was working out but went over the whole ground again and produced fifty-five detailed recommendations. The report insisted that organization should be built from the colliery upwards, not from headquarters downwards, and it proclaimed three axioms: (i) that decisions should be made at the lowest level consistent with the board's overall responsibility; (ii) that standard patterns of organization should not be imposed for the sake of uniformity, and that greater flexibility to take account of local requirements need not lead to loss of central control; (iii) that the organization should be capable of ready modification to meet changes. In many ways these axioms were particularly significant. The report did not attempt to bring about any fundamental changes in the new organizational structure, though it did propose changes in the method of management by objectives to 'a simpler and more effective procedure that is capable of being enforced', and it suggested various changes of detail. But it initiated a much more detailed examination of the way in which, within the existing structure, many specific functions were performed; its accepted recommendations provided for new arrangements for continuous policy planning; and they also provided for continuous future review of the organization, so that it would not be set in a mould but would evolve in response to changing circumstances. What the working party sought was a combination of firmer line management with greater adjustment to the varying

[1] NCB *Ann. Rep.* 1965–6, I, 32; 1966–7, I, 30.

conditions of different places and times, and a constant commitment to economy.[1]

To ensure that something approaching this was achieved, a steering committee was established, which used various working parties on particular subjects, including regional organization, computer services, marketing, provisioning, headquarters oganization, and management by objectives. All through 1969 and 1970 reports were being completed and modifications made in the details of practice and staffing. In 1969 a new and continuing scheme was adopted for dealing with cases of serious shortcomings in performance by particular formations. Performance Improvement Teams (PIT) were to be set up *ad hoc* to carry out efficiency appraisals when required, and a list of twenty to thirty senior officials, from whom teams could be made up, was prepared.

It might seem that one or two desirable reforms had been neglected. For instance, once areas had been enlarged in scale and function, for the reorganization of 1967, the role left for divisional chairmen (who were soon renamed regional chairmen) looked tenuous, but the Sheppard Report insisted that they were useful and ought to be kept. This looked like a recognition more of worthy individuals than of a genuine functional requirement, and it was remarkable how little mention of regional chairmen the subsequent working party on regional organization found occasion to make. In fact, the passage of time took care of this problem, as some regional chairmen were appointed as board members and others retired. From August 1971 a watching brief over named areas and other formations was assigned to individual members of the national board, and there was no further place for regional chairmen. Some additional mergers might also have seemed helpful, but were not necessarily easy to accomplish. There was a fairly strong wish in the NCB to merge the Staff and Industrial Relations Departments, but it was thought not worth while to do this because of the objections by the unions. And the Sheppard Report rejected any further mergers of areas for the time being, although the reduction in colliery numbers since 1966–7 suggested that this might be appropriate. However, by 1972 plans were being prepared for a whole series of mergers, so as to have only a single area management for each coalfield, except in Yorkshire and the Midlands. The first of these mergers, to create a unified South Wales Area, took place in July 1972.

[1] The working party's suggestions about HQ departments illustrate the approach to the combination of efficiency and economy, e.g. it proposed to have efficiency audits carried out by small teams and to have departments assist areas through individual consultants or 'flying squads' rather than 'standing armies'.

By this time the NCB had achieved a slimmer organization, with greater delegated discretion about the way in which defined objectives might be achieved, but greater respect for defined lines of communication and responsibility, and stricter procedures for accountability. They also had clearer arrangements for formulating policy and adapting it to changing conditions. The principles were summarized in these terms in 1971:

The basic objective of the Board's organisation ... is to ensure that at each level the functions and responsibilities are clearly outlined.

The tasks of the Board themselves are threefold:— (i) corporate responsibility for policy; (ii) setting objectives and determining organisation to achieve policy; (iii) accountability and supervision.

The tasks of departments at Headquarters and at Regional level are twofold:— (i) to provide the Board with advice and assist in the formulation and implementation of policy; (ii) to provide guidance and services to operating units.

The task of the operating units (i.e. Areas, Opencast, Coal Products Division and Ancillaries) is to achieve the results contained in the budgets agreed with the Board.

It was not an organization which was regarded as having reached a final form, but rather was expected to go on adapting in detail. A year later the chairman spoke of 'a longer-term strategy of fewer self-contained management units, producing and selling their product, and enjoying greater authority from the Board, with strict accountability, which would entail corresponding changes in the function and weight of work at Headquarters'.

Whatever the degree of finality, the economies had been urgent, and stricter care to define functions precisely and to establish best practice for their performance was highly desirable. The coal industry had run into hard times, both politically and financially. Organizational changes were not necessarily going to show themselves in better financial results, but they were needed in order to cope with current problems and as a demonstration that current difficulties were neither created nor magnified by administrative shortcomings.

## v. Financial adaptation

In the late fifties the NCB found themselves in a worsening financial situation. Operating profits were still being made, but were static or declining. This was because the level of sales had been reduced and the pressure of the market kept down price increases to an extent that made it impossible to cover all the rises in costs, even though productivity began a steady rise in 1958 after being almost stationary; because profitable opencast working was reduced in order to protect sales from deep mines; and because the carbonization plants, also affected by falling sales, ran into loss.[1] At the same time interest charges were rising rapidly. The outcome was that, after payment of interest, there were five successive years of deficit, starting with 1957. The deficit of £24.0 million in 1959 was the largest in any year so far,[2] and things were not much better in 1960. Corrective action was taken on the cost of operations and persistently rising productivity was achieved in both deep mining and opencast, so that operating profits rose appreciably from £13.1 million in 1959 to £45.4 million in 1962. But the burden of interest charges went on increasing. They were twice as high in 1962 as in 1956,[3] and threatened to go on weakening the finances of the NCB.

The questions were why there was this mounting problem and what could and should be done about it. The main difficulties originated in the peculiar capital structure and pricing policies of the nationalized industries, and were increased by some features of the history of the coal industry since nationalization. All capital was loan capital and therefore commanded fixed payments each year. Revenue was required to be sufficient to cover current costs but not to build up funds to provide for net additions to capital, and prices were kept down accordingly. Additional capital was provided not mainly by funds generated internally but by additional borrowing from the government, within the statutory limits from time to time laid down; and this was true not only of fixed assets but also of increases in working capital and the financing of any accumulated deficit. These are the conditions underlying the rise in the interest burden from the mid-fifties, and especially between 1958 and 1961 inclusive. The investment programme under the successive versions of *Plan for Coal* was both more expensive and less covered by depreciation provisions than had been expected, and therefore needed more borrowing. The need for working capital increased rapidly, especially while there were mounting stocks of unsold coal to be financed—the value of stocks of products increased two and a half times in 1957 alone—and the

[1] The NCB's non-mining activities made an operating profit in every year until 1957, but in only one year of the next six.

[2] In real terms (i.e. with allowance for the reduced value of money) it was not as big as the 1947 deficit.

[3] NCB *Ann. Rep.* 1971–2, II, 4–5.

accumulated deficiency, which had to be covered, trebled between the end of 1957 and the end of 1961. And the bigger borrowings were at higher rates of interest than before. Moreover, the recent heavier borrowings were added to earlier capital provisions which had one serious weakness. Down to 1956 all loans from the government (except short-term advances) were funded by annuities repayable over 50 years. Thus the outstanding principal diminished too slowly, only slight attention had to be paid in current business to the repayment of loans, and funded liabilities had a much longer life than many of the physical assets they were used to finance. Only from 1957 was the much more realistic repayment period of 15 years adopted.

The cumulative figures show how much the financial burdens had increased. Since nationalization the total advances by the government to the NCB had reached £353.8 million by the end of 1957 and £726.1 million by the end of 1963–4. The total amounts still outstanding on these loans were £349.2 million at the first date and £588.4 million at the second. And these amounts were on top of the liability for the vested assets, which was being repaid so slowly that it still totalled £339.1 million at the end of 1963–4. The inappropriateness of this figure is evident when it is noted that the vested fixed assets had originally cost the NCB £357.4 million (i.e. fixed assets accounted for over 90 per cent of the compensation payment for nationalization), and, through depreciation and disposals, had been written down to £123.9 million.[1] Thus, for this one large item the outstanding loan was about two and a half times as large as the book value of the assets it financed.

That was one respect in which over-capitalization was obvious. In others it was less clear. Fixed assets acquired since nationalization were still an increasing item in the balance sheet. Their written down value rose from £489.1 million at the end of 1957 to £722.5 million at the end of March 1964.[2] Yet the earning ability of these assets and the productive capacity of the industry were not expanding correspondingly. The discrepancy arose because the completion of many investment projects begun in the fifties added large new assets which, as yet, had been little depreciated, while most of the assets disposed of had already been written down so far that they had only a rather small effect on the balance sheet. There were also many assets that had not been disposed of and remained on the books at a value above zero, but had ceased to earn anything. Indeed, as many of them were at closed collieries where

[1] NCB *Ann. Rep.* 1957, II, 8–9 and 22–3; 1963–4, II, 12–13, 20–1, and 78–9.

[2] Ibid. 1957, II, 12–13; 1963–4, II, 16–17.

some costs were unavoidably incurred, their earning ability had become negative.

There was no doubt that the NCB were over-capitalized by the early sixties and, whatever their structure and status, would have had to make some drastic financial adjustments because of the need to operate on a reduced scale in a smaller market. But the form and manner of adjustment were dependent on their status and were highly political. If the NCB down to 1956 had been able to operate as a purely commercial undertaking, charging world market prices, they could have financed internally most or all of the new investment (which need not have aimed for such a high ultimate output) and could have paid off most of the compensation liabilities incurred on nationalization. The loan burden would then have been too slight to worry about. The book value of the assets could have been written down against financial reserves and, had the NCB been endowed mainly with equity rather than loan capital, the outcome would have been a reduction in the market valuation of the equity. Such a scenario was never politically feasible. Adjustment could come only through legislation which enabled the government to reduce the financial obligations of the NCB.

The possibility of a financial reconstruction was offered by the government late in 1961 at a time when they were trying to dissuade the NCB from implementing proposals to cut out the worst remaining loss-makers in Scottish coalmining—apparent threats to any major Scottish industry were always politically sensitive. As all this was happening just after the publication of the white paper on the obligations of nationalized industries,[1] it affords an interesting illustration of governmental ability to combine public advocacy of more commercial behaviour with private willingness to discourage it. The NCB, however, wanted the Scottish coalfield reforms and did not want the financial reconstruction. They had already consulted the accounting firm of Cooper Brothers and accepted their view that 'if the Board agreed to a capital reconstruction it would be popularly interpreted as a sign of bankruptcy'.[2] This was clearly a consideration that affected the choice of occasion on which to seek a capital reconstruction, but equally clearly the subject had to remain on the agenda. Not all the members of the board remained against it, but the chairman strongly favoured postponement and his view prevailed.

[1] Cmnd. 1337.

[2] Robens indicated that the board regarded government acceptance of the Scottish proposals 'as a test of the sincerity of the policy laid down in the White Paper'.

It was not until November 1964 that the NCB decided to seek a capital reconstruction. The need then was urgent and the time appeared suitable. The NCB had improved their image with better results since 1961. Prices had been held almost stable, operating profits steadily increased, and even the greatly increased interest burden just covered. But the effects of inflation were making it necessary to plan price increases, which could be significantly smaller if interest charges were reduced, and were threatening the reappearance of deficits. As the railways had already been given a substantial capital reconstruction, and the possibility of cancelling much of the accumulated deficit of BOAC was under consideration, the NCB were not seeking something which could be represented as reprehensible because novel. They decided to ask that, at the end of the financial year, £310 million of loans plus the whole of the accumulated deficit on revenue account should be cancelled. They also asked that, for the future, the Ministry of Power should adopt a coordinated fuel policy on agreed lines.[1]

By the end of the financial year the liabilities for advances from the government had reached £960.1 million, including £625.6 million for advances since vesting date. It took rather longer to get the legislation completed than had been hoped, but the provisions were retrospective to the end of the 1964–5 financial year. A white paper of November 1965[2] summarized the proposals, which were that the accumulated deficit at 27 March 1965 should be written off and that much of the capital debt should be transferred to a reserve fund, against which unremunerative colliery and coke oven assets should be progressively written off over a period. The government also insisted on delaying the price rises requested, and indicated that any consequential deficit, up to a maximum of £25 million, could be written off against the reserve fund. The proposals were embodied in the Coal Industry Act 1965. The NCB were relieved of £415 million of capital debt, transferred to reserve, and the remaining £545 million was divided into £18.1 million of temporary debt and £527 million of funded debt, which was to be repayable over 34 years with interest at $4\frac{1}{8}$ per cent. During 1965–6 £223.1 million of the reserve fund was applied to write-offs, £90.8 million being the accumulated deficit at 27 March 1965, £24.8 million the subsequent deficit attributed to the enforced delay in increasing prices, and £107.5 million fixed assets. The reconstruction resulted in a diminution of interest payments by £21.5 million and depreciation charges by £14.1 million in 1965–6.[3]

[1] Robens 1972, 160–3. [2] *The Finances of the Coal Industry*, Cmnd. 2805.

[3] NCB *Ann. Rep.* 1965–6, I, 1 and II, 29–30.

The implementation of the capital reconstruction was virtually completed in the next two years. By the end of 1967–8, £404.3 million had been written off against the reserve fund.[1]

Despite the NCB's improving financial record before the reconstruction, the measure left them, both at the time and afterwards, subjected to charges of financial inadequacy and of having become a protected dependant of a generous government. The Minister of Power, Fred Lee, helped to foster this view by publicly treating the reconstruction as some sort of government gift to the miners.[2] It is hard to find much evidential basis for these suggestions. The financial reconstruction did no more (indeed, probably did less) than cancel liabilities which would not have existed if the NCB had been free to act in a strictly commercial fashion in the past. The one deserved criticism was the artificial reduction of the annual repayments as a result of the adoption of the inappropriately long period of 34 years for the loans still outstanding.

More open to controversy were one or two other provisions of the 1965 act and a further Coal Industry Act in 1967. One was a power for the government to compel the NCB to defer the intended closure of some collieries and compensate the NCB for the costs incurred. This power was exercised in 1967 when the closure of sixteen pits was delayed for a few months at the request of the Prime Minister, Harold Wilson.[3] As the government were urging the NCB to act on strict financial criteria and then, for policy reasons of their own, preventing them doing so, it does not seem that government payment of the cost was the bestowal of a privilege. Another provision made in 1965 was for the government to pay part of the social costs of colliery closures, which included redundancy payments, travel allowances, and transfer payments for workers whose jobs disappeared when a colliery was closed. The government originally agreed to pay half of all such costs above £3.8 million a year, up to a maximum of £30 million over a period of five years. In 1967 the government agreed to pay two-thirds of all the 'relevant' expenditure, though in specific cases the NCB made some social payments which the government declined to treat as relevant. A related provision in the 1967 Act concerned additional payments to men made redundant when over the age of 55. Their unemployment benefit was to be made up for three years to 90 per cent of their previous net

[1] Ibid. 1967–8, I, 1.

[2] Robens 1972, 162–3. For a clear (and highly tendentious) expression of the anti-NCB view of this matter, see Kelf-Cohen 1973, 27–8.

[3] NCB *Ann. Rep.* 1967–8, I, 41.

earnings and, if eligible for a mineworker's pension, they could start to draw it at the end of the three years. Those who were over 60 when made redundant could, if they were qualified for a mineworker's pension at 65, start to draw it immediately. The entire cost of this scheme, which originally applied to the period to 28 March 1971, was to be reimbursed to the NCB by the government.[1]

Some people interpreted these various social cost payments as an element of subsidy to the coal industry. Others maintained that the costs did not arise solely from the activities or shortcomings of the coal industry, but from a general economic situation and economic policy, and were therefore properly borne, in part, by the government as a collective responsibility.[2] Lord Robens had affirmed, fairly enough, in 1965 that the coal industry, which had been substantially subsidized before nationalization, had not cost the taxpayer a penny since 1947.[3] Henceforward, critics were going to say that this no longer applied, though there would be counter-argument. The criticism increased because another provision of the Coal Industry Act 1967 offered government money up to a maximum of £45 million for the period up to 1970–1 to cover the additional costs of the gas and electricity industries in burning more coal, in preference to other fuels, than they had planned.[4] On this matter also the position was not clear cut and there was counter-argument that the coal industry gained less in this way than it lost from the government's support to uncompetitive nuclear power stations.[5] The extra amount of coal used in the relevant period was 6 million tons in each of the first two years, 2 million in the third, and none in the fourth. This provision was not renewed in the Coal Industry Act 1971. This measure did, however, continue for three years a government contribution to social costs, but at a tapering rate with a maximum total of £24 million for the three years; and it also continued the provision of extra payments to redundant mineworkers, though with power for the Secretary of State to vary the age ranges for eligibility.[6]

[1] NCB *Ann. Rep.* 1967–8, I, 42.

[2] Cf. the conclusion of Department of Employment and Productivity, *Ryhope*, 94, which, on the basis of the Ryhope study, suggests that the redundant men present the most serious national problem arising from closures, and comments: 'This problem goes far beyond what can be met by a nationalised industry, which has to organise its operations profitably within the terms of the Government's fuel policy, or by Government agencies, such as the Employment Exchange Service. With the best will in the world such organisations can assist only within a specified framework and their efforts must be limited.'

[3] Robens 1972, 161.

[4] NCB *Ann. Rep.* 1967–8, I, 6.

[5] Cf. Robens 1972, chap. 9.

[6] NCB *Ann. Rep.* 1970–1, I, 3–4.

After the capital reconstruction the financial position, though obviously easier than it had been, was by no means comfortable. There were significant increases in costs in 1965 and 1966, which left margins very narrow, and though until 1969 the stability of prices was matched by a similar success in holding down operating costs, the position was precariously balanced. Opencast remained profitable and coal products had been restored to profit, but it was the financial performance of deep mining which was decisive. The difficulties appeared in various ways. To keep up sales in a declining market the NCB tried to get the maximum advantage from the lower cost coalfields, with years of price stability in the East Midlands and below average increases in Yorkshire. But the pressure of this policy was so tight as to reduce East Midlands profits to a low level and push Yorkshire (where very good mining conditions were less widespread and there were more pits with difficult industrial relations) into a small overall loss (i.e. after interest charges). For the industry as a whole there had to be some new investment to maintain efficient replacement capacity, and not enough cash was generated to finance all this internally, so there had to be new borrowing from the government. New loans were at higher rates of interest (up to 7½ per cent and just over) and were repayable over 15 years. So the level of interest charges began to creep up: it was half as high again in 1968–9 as in 1965–6, though still well below what it had been before reconstruction. All this was happening while the physical capacity of the coal industry was being reduced by closures. At 30 March 1968, when the write-offs under the 1965 Act were almost complete, the fixed assets of the NCB had a book value of £642.0 million. Three years later the figure was down to £633.0 million, and the total loans from the government still outstanding were then £675.1 million, which was not far out of line when other items are taken into account. But the trend was adverse and it is not surprising that Lord Robens foresaw the likelihood of another capital reconstruction by 1975, to take account of the reduced value of fixed assets.[1]

Modest operating profits were made, though in 1969–70 deep mining failed to contribute to them, but by 1968–9 these were insufficient to cover interest payments. A correction was achieved in 1970–1, partly as a result of the large price increases of 1969 and 1970, partly through a doubling of opencast profits. In that year the NCB had an overall surplus of £0.5 million, and at the end of the financial year the accumulated deficit, most of it resulting from 1969–70, was £33.8 million. This was a

[1] Robens 1972, 163. The figures are from NCB *Ann. Reps.*, II.

manageable state of affairs, especially as the first half of 1971–2 indicated that another surplus was likely, but the precariousness caused by the pressure of rising costs was recognized by the Coal Industry Act 1971. This raised the legal maximum for the accumulated deficit, set at £50 million in 1967, to £75 million, with power for the government to put it up to £100 million without additional legislation. It also allowed the NCB, subject to government consent, to raise loans in foreign currency.[1]

In fact, the financial prospect was worse than it looked because the rapid rise in costs, which set in during 1969, proved impossible to check and continued until it was greatly increased by the national strike early in 1972 and its settlement. For three years the rise in costs came mainly from sources over which the NCB had little influence, especially the price of materials and of a great variety of supplies and services bought in from outside. Wage increases had some influence, but not a proportionate one, and wages and wages costs continued to form a declining proportion of total costs.[2] But an important change was appearing. For a decade the productivity of labour had been increasing faster than labour costs, and this had made it possible to absorb the inflationary pressure of rising external costs. Now the upward trend of productivity slowed down, while external costs were rising faster, and thus the most important aid for cost control was much reduced. There were signs, in serious unofficial strikes in 1969 and 1970, that the co-operation of labour was diminishing. In 1971–2 these portents were followed by wage demands which were much larger than any ever previously conceded. They were supported first by an overtime ban and then by a seven-week national strike, which together played havoc with output, costs, and revenue from sales, and, after a government-appointed enquiry, the demands were met almost in full.[3] These events contributed to a further great rise in costs. In the short term, in 1971–2, the continuing costs had to be borne by a much reduced output for the year, and labour costs were increased by the sharp reduction of OMS which accompanied these upheavals. But the main structure of costs was not immediately altered very much, for even the unprecedented rise in wage rates did not do much more than match the continued growth of other items. Wages and wages costs were only a fractionally greater proportion of colliery operating costs in 1972–3 than in 1970–1: 49 per cent to the nearest

[1] NCB *Ann. Rep.* 1970–1, I, 4.

[2] Cf. the broad analyses of the main items in costs in NCB *Ann. Rep.* 1968–9, I, 2 and 1970–1, I, 2.

[3] See section vii of this chapter.

whole number in both years.[1] The great changes were that the rise in costs was much larger than the rise in revenue from sales, and that the increase in wages and related items was not matched by productivity increases, though OMS in 1972–3 was higher than before.[2]

The financial effects of the overtime ban and strike in 1971–2 were disastrous. In addition to the rise in costs and loss of output and revenue, the shortage of cash flow led to higher borrowings and interest payments. There was a deficit of £157 million (after payment of interest) on the year's working. This would have taken the accumulated deficit to £190.8 million, which was far above the legal limit which the Minister was permitted to finance. So on 24 March 1972 the government made a grant of £100 million in order to keep the accumulated financial deficit within the statutory limit. The rest of the problems were left to be sorted out in the following financial year.[3]

The Wilberforce Tribunal, which proposed terms for the settlement of the strike, said that if the costs could not be met from the NCB revenue account, the government ought to accept the charge. The NCB immediately set about exploring the various ways of seeking assistance to cover the costs, any of which would require new legislation. A first rough estimate suggested that at the new level of costs rather more than two-thirds of all collieries would fail to cover their operating costs, and that the NCB would have a continuing additional cost of about £70 million a year which could not be met out of revenue. There was much discussion of the extent to which they should seek a capital write-off and specific or general subventions of varying kinds, including assistance which would be covered by ECSC rules after the UK joined the Community. During the spring and early summer proposals were worked out with the Department of Trade and Industry, which accepted the NCB submission as 'fair and reasonable'. In June 1972 the Department went so far as to ask the NCB not to close collieries on economic grounds as a contribution to the solving of their problems. After this progress there was alarm when, in August, without any prior warning, the government suddenly refused to make the expected public statement before Parliament rose for the summer recess and privately demanded that the NCB and the unions should reach agreement on 'the measures that were practicable for the industry to help itself and its competitive position by containing costs and improving productivity'.

[1] NCB *Ann. Rep.* 1972–3, 7.

[2] Ibid., 36.

[3] Ibid. 1971–2, I, 1–2 and II, 12 and 26.

However, the NCB drew up a programme of twenty points, ten for action by the industry and ten which the NCB and the unions might jointly request the government to implement. The latter ten were more or less what had already been thrashed out with the Department of Trade and Industry. The unions agreed to a joint statement and by October a set of proposals for legislation was agreed at the official level, though there was still some doubt whether the politicians would want substantial changes. In the end, what was enacted was close to what had been agreed in the long series of discussions.

The Coal Industry Act which was passed early in 1973 immediately reduced the book value of NCB assets and wrote off accumulated deficits, with the object of saving the NCB about £40 million a year by the lowering of interest and depreciation charges. Because the provision had immediate effect there was a benefit to the 1972–3 accounts. What was done was, first, to reduce the book value of the fixed assets by £275 million. The main items in this reduction were £190 million for mines, surface works, plant, machinery, and equipment no longer productive, and £51.6 million which was the depreciated value of the compensation originally paid to the Coal Commission for the transfer of ownership of the nation's coal reserves to the NCB. The latter change had the peculiar consequence that the unworked minerals, which are in reality a huge asset, thenceforward ceased to appear in the NCB accounts. The rest of the writing down concerned miscellaneous items of property, of which £6.6 million related to houses owned by the Coal Industry Housing Association, a wholly owned subsidiary of the NCB. Then the entire NCB liability for borrowings from the government, with accrued interest, was extinguished at 31 March 1973 and replaced by a new loan. This had a capital value equal to the former loans minus £275 million (to match the write-off of fixed assets) and minus a further £174.6 million, which was the accumulated revenue deficit at 31 March 1973. Under this scheme all the capital liabilities were consolidated in one loan of £412.7 million, bearing interest at 5.519 per cent and repayable over 20 years.[1]

The other provisions of the Coal Industry Act 1973 operated only from 1 April. The government agreed to continue to meet social costs arising from colliery closures and mineworkers' pensions, up to a maximum of £85 million in the next three years (with the possibility of extending the maximum to £140 million over five years), and to extend the Redundant Mineworkers Payments Scheme for three years, with a

[1] NCB *Ann. Rep.* 1972–3, 22–6, 29, and 31.

limit of £60 million (and a possible extension to £100 million over five years). The government provided grants of up to £50 million for the cost of extra coal burned at power stations in the next three years, with power to increase the sum by Order to not more than £100 million, and grants of up to £210 million over the next three years to help the NCB to moderate the contraction of the coal industry. This corresponded to the £70 million annually which the NCB estimated as the continuing adjustment arising from the Wilberforce award. There were also grants to cover coking coal production and the stocking of coal and coke above certain limits. These were limited to a maximum of £85 million over three years, which might by Order be raised to £145 million over five years. The NCB's limit for total borrowing was set at £550 million (which could be raised by Order to £700 million), and for the accumulated deficit at £50 million (which could be increased by Order to £100 million).[1]

The maximum assistance made possible was substantial in relation to NCB turnover, which in 1972–3 totalled £1,033.1 million. It is at this point that the financing of the coal industry clearly departed from commercial lines, at least in the sense of involving special privileges rather than special handicaps. The coal industry was not unique in its receipt of assistance and the government sought to justify the provisions in terms of safeguarding future energy supplies, the current market conditions which made large increases in coal prices impracticable, and the intolerable social consequences of a large programme of colliery closures. However, the level of assistance that was found necessary inevitably exposed the industry to a good deal of public criticism.

## vi. The Aberfan disaster

No other happening in the coal industry in the sixties cast so dark a shadow as the disaster at Aberfan. Merthyr Vale, the local colliery, had for many years deposited its waste on a complex of tips on the slopes of the hill known as Merthyr Mountain, which overlooked the village. On Friday 21 October 1966, at about 9.15 a.m., a large part of the tip in current use broke away and slid down the slope of Merthyr Mountain, partly obscured by mist, as an avalanche of colliery waste. A crust of solid material was carried on top of a flow of saturated waste and sludge. The avalanche engulfed the Hafod-Tanglwys-Uchaf Farm, killing the occupants, crossed a disused canal, surmounted a railway embankment,

[1] Ibid., 3–4.

and engulfed the Pantglas junior school, where lessons had just begun, and eighteen houses. It damaged the Pantglas senior school, which fortunately did not start work until 9.30 a.m. and where children were not yet gathered in numbers, and it damaged a number of other houses before coming to a halt. Not one person who was buried by the slide was rescued alive. Altogether 144 people were killed, 116 of them children (mostly aged 7 to 10). Of these children 109 were in the Pantglas junior school, where five teachers also died. In addition twenty-eight children and six adults were injured, some of them badly. Besides the houses destroyed, sixteen more were damaged by sludge, sixty had to be temporarily evacuated, and others were damaged to some extent in the course of rescue and clearance operations.[1]

Such a tragedy aroused widespread horror and indignation, alarm lest it could be repeated, and a search for causes (and culprits). The Secretary of State for Wales on 26 October appointed a Tribunal of Inquiry, with Lord Justice Edmund Davies, a native of the district, as chairman, and two other members. Its report appeared in July 1967. The NCB, who, through a press statement by their chairman on the day after the disaster, acknowledged full responsibility,[2] had an urgent need to take immediate precautions and to decide on practices that would remove the risk of any further accidents of this kind, which no one had thought to be possible. The first duty was to secure the stability of the Aberfan tip from which the slide had occurred, and this was done by immediate emergency measures, followed by continuous remedial work. There was also a need to look at possibilities elsewhere. The NCB owned 1,753 colliery tips, of which 477 were in current use. Every tip was inspected within a week of the Aberfan disaster and where, in the light of knowledge available at the time (which was not as full as it needed to be), there seemed any risk of immediate hazard, emergency action was taken to deal with it. There were only a few such cases. Then, while detailed investigations into the causes of the tip slide at Aberfan were in progress, comprehensive information about all NCB tips was assembled (for the first time) and policies were devised for controlling them. These policies dealt with technical methods, the more detailed definition of individual duties

[1] *Report of the Tribunal appointed to inquire into the Disaster at Aberfan* (HL 316, HC 553, of 1966–7) (subsequently cited as *Aberfan Tribunal Report* 1967), 26–30.

[2] Robens 1972, 249. The Tribunal in its report (81, 87, and 92) tried to maintain that the NCB refused, until a late stage in the enquiry, to accept any responsibility. It is possible, when time has softened the passions immediately aroused, to note that the report, which reflects understandable feelings of indignation, in various parts shows less rigour and impartiality in its treatment of evidence than might have been expected.

and lines of responsibility than had been centrally laid down before, and the arrangement of instructional courses. All this was well in hand by early 1967. One of the things that became apparent as a result of all this work was the great variety of physical characteristics at different tips, and this in itself suggested that it was desirable to keep a closer watch for signs of change than had been usual.

The report of the tribunal was very critical of the NCB and some of their employees, so much so that Lord Robens offered his resignation from the chairmanship, an offer which the Minister, after a month during which the board's detailed reply to the Report was being prepared and presented, refused to accept.[1] Some parts of the Report were privately deplored as inaccurate and unfair. In particular, the naming of a few individuals for detailed censure was regretted as unjustifiably selective. It was felt that either this exercise should have been omitted or the list of names should have been more comprehensive. But nothing could compensate for the grief that had been caused, and there was a wish to say nothing that anyone could misinterpret as an attempt to evade responsibility. So any doubts and resentments were swallowed, though Lord Robens put some of them on record several years later.[2] The NCB accepted the lessons and recommendations set out in the Report and gave the Minister of Power an immediate account of what they were doing to remedy shortcomings. He accepted this as satisfactory.[3]

The important consequences of the Aberfan disaster and the tribunal's report were national as well as local. The NCB had concluded that the tip slide at Aberfan was caused by geological conditions which it was very rare to find in combination. The tribunal dismissed this conclusion in scathing words and sought to show that similar slides, which fortunately had not gone as far, were well known, similarly caused, and provided an adequate, unheeded indication of what was liable to happen.[4] On the evidence it seems clear that, on this point, the NCB were right and the tribunal mistaken. The Report includes a detailed account of the causes and cumulative development of the disaster, which is admirably clear and indicates almost total agreement among the expert witnesses retained by different interests.[5] The conditions described could not be regarded as other than extremely rare.

[1] Robens 1972, 260–2. [2] Ibid., chap. 12, *passim*.

[3] NCB *Ann. Rep.* 1967–8, I, 16.

[4] *Aberfan Tribunal Report* 1967, 84–6.

[5] Ibid., 113–20.

But this disagreement is not the most important point. It was more significant that the tribunal was absolutely right in pointing to the absence of a clear policy on tipping and a lack of attention to the condition of tips. And it was right in suggesting that if things had been different in that respect there would have been no Aberfan disaster. Not only the NCB but the legislature, the Mines Inspectorate, and foreign governments and mining companies all treated colliery tips as of only slight importance. In the UK there was no legislation dealing specifically with tip safety, nor were there regulations under more general mining or safety statutes. The Mines Inspectorate had specialist senior posts in mechanical and electrical engineering but none at this time in civil engineering.[1] Elsewhere it was much the same. No governments except those of South Africa and North Rhine-Westphalia had any laws dealing with tip safety.[2] This is not very surprising. Mining has many hazards which preoccupy those concerned with making and supervising safety regulations, and colliery tips have not been reckoned among them. In the entire history of British mining the Aberfan disaster is the only recorded fatal accident caused by tip movement, but that one event was so appalling that things had to be very different in the future.

At Aberfan itself remedial work was continuous. New drains and boreholes were installed at the tips for extra drainage, and the hidden spring, which had been the main source of saturation of the tip material, was confined within a permanent structure. Large parts of the tips were moved and a scheme for recontouring the remainder in a completely stable structure, and grassing and landscaping the rest, was prepared. But the local people preferred to be rid of all permanent reminder of what had happened and have the tips completely removed. This was done at an additional cost of £3 million, shared between the NCB, the government, and local interests (the normal costs, of course, were borne by the NCB), and the work was largely completed in 1971.[3]

Nationally, there were important changes in practice and in legal requirements. The new procedures which the NCB began to apply immediately after the disaster were developed into a standard national system which was applied throughout the industry from the beginning of 1968, and, as recommended by the tribunal, a Code of Practice for

[1] The staff list in *Guide to the Coalfields*, 1967, 5–18 illustrates this point.

[2] *Aberfan Tribunal Report*, 36.

[3] NCB *Ann. Rep.* 1967–8, I, 17; 1968–9, I, 13; 1971–2, I, 8; Robens 1972, 262–3. Robens points out that it was the government who insisted on having part of the extra cost of complete tip removal met from the disaster fund to which the public had subscribed.

spoil heaps and lagoons was drawn up as an interim measure and a technical handbook was published in 1970 to supersede it. The NCB increased their civil engineering staffing both at headquarters and at areas, and in 1968–9 initiated a programme of research into fundamental factors affecting tip safety. The law was first changed by the issue of the Mines and Quarries (Notification of Dangerous Occurrences) (Amendment) Order 1967. Until then the Mines Inspectorate had never required any information about what happened to the condition of tips, but reports were made henceforward to the District Inspectors of Mines. The Ministry of Power in 1968 set up an Advisory Committee on Tip Safety, which received all the latest information on the revised practices and new findings of the NCB. A new measure, the Mines and Quarries (Tips) Act 1969, gave authority for comprehensive regulation and supervision, which came into force with the issue of the Mines and Quarries (Tips) Regulations 1971 and the Mines and Quarries (Tipping Plans) Rules 1971. The Mines Inspectorate equipped itself for its extra responsibilities by establishing new senior posts for civil engineers. So what had probably been the least regarded element in colliery operation and safety came to be a subject of continual regulation and special investigation. Probably more was then done to alter the structure and shaping of colliery tips, and to regulate the disposal of various forms of wet waste, than ever before.[1]

## vii. Labour relations

Work in the mines went on in a context that had unfamiliar features. Two influences demanded an exceptional amount of adaptability. The acceleration of technological change meant that many jobs, especially those at and near the coal face, had to be done in new ways. The shrinking of the market and the resultant huge number of colliery closures and job losses created insecurity among workers and altered both the detailed objectives and the bargaining strength of trade unions. These conditions introduced additional occasions for frustration, resentment, and dispute. Possibilities of this kind were all the greater because the constituents of the wages of faceworkers and some others underground were so varied, complicated, and tied to the detailed methods and conditions of working. Every change in the latter could cause the disagreements that come from a change in the size of the pay packet and the way its contents are calculated. Comparisons of past and present, and

[1] NCB *Ann. Rep.* 1967–8, I, 16–17; 1968–9, I, 13–14; 1969–70, I, 12; 1970–1, I, 12; 1971–2, I, 8.

comparisons with more prosperous industries, could add to resentment, for no other large industry came under such great and sustained pressure to contract. As coalmining had a tradition, so old as to be apparently ineradicable, of quick resort to small-scale action in cases of disagreement, it might perhaps have been expected that the changed conditions would make labour relations more awkward and disputatious. In fact, this proved not to be so. Not until the very end of the sixties were there signs of serious deterioration.

There were various reasons why the stresses did not have more serious effects. Comparison works two ways, and many men, especially among those at the peak of their earning ability, got out into other industries rather than stay in coalmining and struggle for something better there. Those who stayed, and who felt they must stay, because of age or because of lack of local alternatives, could see the pressures on the industry, and knew that their chances were unlikely to be helped by pushing for large improvements in pay and conditions, or by resort to frequent disruption. Certainly this was the attitude of the NUM, who made annual pay demands but were prepared to settle at modest levels and, in a general way, were willing to see some kind of balancing of small wage rises against a mitigation of job losses either in numbers or in the terms of compensation. The caution of the union seems also to have been accompanied by similar, though less complete, caution in the behaviour of individuals and of groups who might choose, or refrain from, unofficial action. There was also an important influence from the great care taken to carry out colliery closures as humanely as practicable, with redundancies at first kept very few and offers of other jobs for nearly all displaced workers, and later with substantial improvements in payments to those made redundant, especially older men who retired early. Indeed, the preference of some workers to take redundancy money rather than press for the continuation of their jobs was sometimes so marked as to exasperate their union officials.[1]

Changes in institutional arrangements were probably less influential, and it is unlikely that they all worked in the same direction. The most important were new agreements about the nature of the wage system for large categories of workers whose pay arrangements had not previously been reformed. The attempt was made to adapt the basis of wage determination to contemporary working practices. The general effect was to move towards greater uniformity, to reduce drastically the number of specific additions to basic contract rates, and so to move nearer to

[1] Cf. Gormley 1982, 56 and 58.

complete centralization of pay negotiations. It was hoped that, with fewer local variations and fewer items to be separately decided, the reduction in the number of potential points of argument would be translated into a reduction of actual disputes. The outcome seems to have varied from year to year and needs further consideration when particular disputes are examined.[1]

Other institutional changes were heterogeneous in both character and effect. A claim by the NUM, put forward in November 1957, for a sick pay scheme was rejected by the NCB, but the claim then went to the National Reference Tribunal, who awarded it and laid down the main conditions. The scheme, which came into operation on 28 September 1958, provided that men with at least one year's service could receive sick pay for up to 60 days in a year in respect of periods of absence of more than 7 days which were due to sickness or non-industrial injury. This provision was clearly relevant to levels of absence, and the NCB commented apprehensively at the time, 'the general experience in other industries is that sick pay schemes lead to higher absence.'[2] A different kind of change of practice was the amendment of the National Conciliation Scheme in July 1961 at the request of the NUM. Henceforward, unsettled national questions could no longer be sent to the National Reference Tribunal if either party objected. At most times this made little practical difference, but it was eventually found that, in the case of very serious disagreements, this change closed one avenue that might have helped to secure a settlement.[3]

A few changes in union representation had only a small effect, but made industrial relations easier to conduct in the limited groups which they concerned. From June 1966 practically all the NCB's management staff were represented by a single union, the BACM, as recognition was, by agreement, withdrawn from the Association of Scientific Workers (to which most of the scientific technical officers had belonged) and from the NCB Labour Staff Association, which had represented the industrial relations officials.[4] The overlapping and rivalry of the NUM and NACODS was also at last brought to an end by agreement in 1970. From that year all grades of underofficials were represented solely by NACODS, and all other coalmining weekly paid industrial staff (i.e. those in supervisory positions on the surface) solely by the NUM. One

[1] Below, pp. 301–14.
[2] NCB *Ann. Rep.* 1958, I, 37.
[3] Ibid. 1961, I, 29.
[4] Ibid. 1966–7, I, 39.

immediate effect was the negotiation, for the first time, of a single consolidated and simplified agreement for all underground 'officials', as the underofficials were then renamed.[1]

Most of the important changes in the wage system came in two agreements. One, made in 1966, concerned workers on mechanized faces. The other, completed in 1971, applied to most of the other underground workers who had been paid piece rates or task rates. Both agreements were regarded as further steps in the policy begun by the national daywage agreement of 1955. To the NUM they were major contributions to the establishment of national wages, to which they had been committed in principle since the foundation of the union, but which they had approached only slowly and erratically because of conflicts of interest and attitude among many of their members. To the NCB they were part of a simplifying and rationalizing process which could remove the enormous number of anomalies in wage differentials, relate pay more closely to the type of work done, get rid of the incessant pressures for local adjustments on particular points, and make the financial outcome of any wage settlement more predictable and controllable.

The hope that the rationalizing exercise for daywagemen, completed in 1955, could soon be followed by a similar one for pieceworkers had proved vain. The complexity of the task was daunting; the proportion of pieceworkers who benefited from some anomaly or other, and would have been reluctant to lose that benefit, was large; and method study, which was an obvious way to start trying to establish a more objective assessment of what was and could be contributed by particular jobs done in particular ways, was regarded with widespread suspicion by the men. Yet the continuation of existing piecework arrangements involved management in distraction by a vast amount of fine detail on wages questions, it reduced the central authority of the NUM, and it made it difficult for the NUM to maintain consistency in wage policy.

But for several years very little was done to change things. Under a national agreement of 1953, district officials were supposed to submit their proposals for revised district agreements to their national officials for approval. As a result of this procedure, some proposals that would have worsened existing anomalies, or introduced new ones, were removed, mostly at the instigation of NCB rather than NUM officials;[2] but that was about all. Two factors brought about a change. One was the discovery by experience that it was possible to work out national wage

[1] NCB *Ann. Rep.* 1970–1, I, 38.

[2] Handy 1981, 71.

systems for complex operations in locations with varying wage differentials. This happened because of the experiments in the early sixties with remotely operated longwall faces, which were intended to minimize human effort. The NCB, who then hoped to see the ROLF system spread further, did not want its use hampered by the adoption of a series of different local practices on pay, and a national agreement for a daywage system, with shift rates initially varying between districts, was concluded in November 1965.[1] The use of ROLF did not spread as had been hoped, and the number of men affected was small, but the agreement set an acknowledged precedent. The other factor was the very rapid spread of power loading. By the mid-sixties most coal faces were being operated in ways which had ceased to correspond closely to the arrangements embodied in existing wage agreements. So in 1965 a renewed attempt was made to construct a national wage system for all workers on power loaded faces, and agreement was reached on 6 June 1966. It applied to about 100,000 men and came into force immediately. All new faces opened after this date were operated under the new agreement, which also applied to existing faces in Scotland and Durham, where similar arrangements had been adopted some time earlier. By the end of the year 1966–7, 55 per cent of all mechanized faces were being worked on the new terms.[2]

The National Power Loading Agreement involved something of a revolution in the wages system. It applied a daywage to workers who had always been paid piece rates or task rates. The wage was an inclusive rate, unlike what had been known before. A member of a team who was appointed chargeman was to have an extra 4*s.* (20p) a shift. First aid workers retained the extra payments which had been agreed nationally, and any existing district agreements for extra payments for carrying safety lamps, carrying powder canisters, and working in wet conditions were continued. Otherwise no additions to the shift rate were allowed. The provisions of the agreement were much more egalitarian than past practice. The shift rate was the same for every member of a face team, which was carefully defined and included men working in any advanced heading or ripping worked in conjunction with a coal face. Some craftsmen came within the definition, and this improved their relative wage position. The shift rate was also the same for every face in each wages district. Different shift rates were laid down initially for each wages district, in accordance with the average earnings per manshift of faceworkers in the previous six months. They ranged from 75*s.* (£3.75) in

[1] Ibid., 71–3. [2] NCB *Ann. Rep.* 1966–7, I, 36–7.

Scotland, Durham, Cumberland, North Wales, South Staffordshire and Shropshire, and South Wales to 89*s.* 5*d.* (£4.47) in Kent. But the preamble of the agreement announced that 'the Board and the Union have agreed to introduce a uniform national shift rate for men employed on power loading faces not later than 31st December, 1971'. This was to be achieved by giving greater increases to lower paid than to higher paid districts in each future wage revision until the differences disappeared. Even more striking was the statement in the series of questions and answers which the NCB and the NUM jointly supplied for the information of the men: 'the aim of this Agreement is to narrow the disparity in wages between face and other workers in the pits.'[1]

Justification for such drastic changes was offered in the agreement itself, in various public statements, and in the information given to the men. The latter were told that a national agreement was needed because existing local agreements had produced wide differentials which were unjustified because, with power loading, 'output performance is determined *more* by the efficiency of the machines than by the labour effort of the men'; and that there had to be a daywage because only thus could there be national uniform wages related to the content of the job instead of to the pit or coalfield where it is performed. It was stated that piecework was no longer appropriate because it originated when output was determined mainly by the physical effort of the workman or group of workmen, which was no longer the case. It was added that the machine had also reduced the disparity in work effort between faceworkers and others, and hence justified a reduction in wage differentials between them.[2]

In all this there were some dubious assumptions and risky innovations, even if the agreement was the biggest single step towards implementing principles which the NUM had long proclaimed but only erratically supported, and even if there was the disadvantage of assuring the men of regular rather than fluctuating earnings. The dependence of output on machine efficiency was not nearly so exclusive and automatic as was proclaimed. If faceworkers felt there was inadequate reward for their skill and effort, they could easily relax a little, and productive work

[1] NCB and NUM, *National Power Loading Agreement* (1966), esp. preamble; s. 3; schedule ss. 1–3, 7–9; questions and answers no. 8. Craftsmen were not fully interchangeable with other members of the face team and were paid NPLA rates only when they worked on production shifts or, on other shifts, with members of their own face team. For any other work they were paid at the national standard rate, which was lower. There were rosters to try to share out NPLA work among craftsmen.

[2] Ibid., questions and answers nos. 1–8.

would then have less sustained continuity than it might have had, with adverse effects on the level and growth of productivity. Some men's earnings were immediately reduced, and it was asking a great deal of men in the better-paid districts (which, except for Kent, were mostly the more productive) that for several years they should do without much wage increase so that others could catch up. Such a requirement might have been acceptable if miners' wages generally had been advancing steadily. But in the late sixties miners were not faring well in relation either to prices or to the movement of wages in manufacturing industry. So the National Power Loading Agreement (NPLA) contained food for discontent, despite its important advantages. It might have been regarded as a warning that when the NUM put the agreement to the vote it was ratified by only the narrow margin of 269,000 to 226,000, and that all the higher paid districts had a majority against it.[1]

For a couple of years or so, however, it seemed that only the advantages were being reaped. With interchangeability within face teams there was more flexibility of working, productivity went on rising, there were far fewer matters to bargain about, and production was less interrupted by disputes. Moreover, the NCB found that the new agreement appeared to have practically stopped wages 'creep', whereas this continued to be serious, after every wage increase, for pieceworkers who were outside the agreement.

It seemed sensible to try to complete the rationalization of the wages structure by seeking an agreement to cover those who came under neither the 1955 daywage structure nor the NPLA. There were about 70,000 of these men, who were pieceworkers or taskworkers underground, mainly away from the face, but including those who worked the remaining unmechanized faces and a few who worked on power-loaded faces without being classified members of the face team. It was expected to be difficult to get agreement because there were no standard rates for the jobs concerned, even within the same colliery, and because many of the men were paid above the NPLA shift rate, which was thought to be the maximum that could be prudently embodied in a new agreement. Indeed the NCB calculated that if they could get the sort of agreement they had in mind, they would save over £1 million a year on the wages of the men concerned, and they intended to apply this saving to a reduction of the gap between those under the 1955 daywage structure and the higher paid men. In 1968 the NCB surveyed a 25 per cent sample of the

[1] Handy 1981, 78. Handy, in 73–8 and Chap. 9, discusses the context, content, and effects of the NPLA in some detail. For relative wage movements, see Handy 1981, 171–80.

jobs concerned and decided that, on the basis of work content, the jobs could be classified in three grades. They prepared an outline scheme on this basis and the NUM agreed to use it as the starting-point for negotiation. These negotiations began in November 1968 and ran into some difficult points of detail. They were far more protracted than those leading to the NPLA, but, though there were important differences of detail, the agreement eventually reached followed closely the general lines originally suggested by the NCB. A proxy vote of NUM areas first rejected the agreement, in March 1971, but after it was slightly amended, it was accepted in June. It was applied on 1 August, with effect from 1 June.[1] By this time the number of men affected had fallen to about 50,000.

This agreement, known as the Third National Daywage Structure, esentially completed the transfer of the coal industry to a system of national daywages. It was treated with fewer doubts than the NPLA had been, for a pithead ballot supported by 240,000 to 73,000 the approval given by a delegate conference. In classifying jobs it put in the top grade (A) such activities as ripping and dinting on roadways, cutting, getting, and loading on non-mechanized faces, installing or withdrawing equipment on power-loaded faces. Men doing such jobs were to be paid the NPLA rate, which meant that they got at this time a weekly minimum of £30. Rates for the B and C grades were to be the average paid throughout the country for all the jobs in each grade. This exercise, on the basis of six-shift pay for a five-shift week, gave £24.925 a week for grade B men and £23.20 a week for grade C men. Not all anomalies could be removed. Some men had been doing task work which did not fit any of the job definitions in the new agreement, but which did fit jobs covered by the 1955 daywage agreement. These men were spared downgrading and the wage reduction that would have gone with it, and they kept their existing pay as personal rates. Taskworkers on rates above the new schedule were also allowed to keep their old rates for the time being, but pieceworkers whose shift rates had been averaging more than the new schedules were transferred to the new rates and therefore suffered some reduction in pay. It could be expected, however, that personal rates would gradually be overtaken by general wage rises and that most of the anomalies would disappear within a very few years.[2]

There had also been increasing anomalies in the pay of craftsmen, who were a steadily increasing proportion of the colliery labour force (15 per cent by 1966). Their curious dual position in relation to the NPLA was

[1] NCB *Ann. Rep.* 1971–2, I, 36.

[2] Handy 1981, 83–4.

only one small example. As their numbers and occupational specialization increased, so did the variety of wage rates, with the result that men of similar experience and level of training were paid appreciably different amounts. In 1970 and 1971 attempts were made to tidy this up. First, the definition of face teams under the NPLA was changed so as to exclude craftsmen, though the latter were still paid NPLA rates while working on the face. Then in March 1971 the NCB and the NUM agreed on a new structure of four grades for all craftsmen. These were: colliery electro-mechanics; craftsmen I A, who had served their apprenticeship; craftsmen I B, who had gained their experience in service; and craftsmen II, who had been purpose-trained, usually in NCB establishments. Former distinctions (reflected in pay) between engineering and other craftsmen of the same grade were thus abolished and, in general, the differences between the standard shift rates of different craftsmen were reduced. But so, too, for the time being, were the differentials between craftsmen and non-craftsmen,[1] which could create among the former some feeling that their skills were not fully recognized.

Some indication of the reaction of workers to their conditions is given by levels of attendance, a good deal more by the number and character of disputes. The large amount of absence, which was a permanent feature of coalmining, showed no sign of lasting improvement, but attempts were made to distinguish more clearly its causes and components. For many years all absences were heaped together in one percentage figure, with the derogatory label, 'absenteeism'. But a distinction came to be drawn between voluntary and involuntary absence, and the former was found to be very much the minority. Of course, there is also often an element of choice in so-called 'involuntary' absence, which arises mainly from illness and injury. There are indications that the fears of management about the effect of a sick pay scheme on absence were, to some extent, justified by subsequent experience. On the other hand, it has to be borne in mind that a coal mine is a peculiarly awkward place in which to feel sick.

Total rates of absence, which had been rising since the mid-fifties, continued to climb. They passed 14 per cent in 1958 for the first time since nationalization, and by 1961 they were over 15 per cent. After three steady years they resumed their rise, to reach an average of 16.0 per cent in 1964–5 and 18.0 per cent in 1965–6. For one year there was a reduction, then for three years a resumed level of just over

[1] Ibid., 96–100.

18 per cent, and a rise to a new peak of 19.2 per cent in 1970–1. The rate came down to 17.7 per cent in 1971–2 and was similar in 1972–3.[1] These were worryingly high figures, especially as faceworkers continued to have higher absence rates than the rest and gaps in coal face teams were liable to be more disruptive to output on highly mechanized faces.[2] Absence was a contribution to the gradual reduction in the average time worked. In 1958, with the cessation of Saturday working, the average number of shifts per man per week fell sharply to 4.38, but the reduction continued thereafter even though there was no further comparable change in practices. The figure was 4.24 in 1959, fluctuated near that level, but went down to 4.17 in 1965–6.

There were also some reductions in hours. Surface workers had their working week reduced from 42½ to 41¼ hours in 1961 and to 40 hours in 1969, in all cases exclusive of mealtimes. In 1970 it was fixed as 40 hours, including mealtimes of 20 minutes per shift, which was effectively a further reduction of 1 hour 40 minutes per week.[3] It was natural, therefore, that further losses of time through increased absence should be seen as a matter for correction, if possible. But the most that could be accomplished, through enquiry and exhortation, in which unions co-operated with management, was to keep voluntary absence to fairly low levels. When the absence rate made its greatest jump, in 1965–6, it was found that almost all the increase was in involuntary absence. The voluntary absence rate was then 5.8 per cent. It came down in successive years to 5.7, 5.3, 4.7, 4.7, 4.4, and 4.3 per cent. Both the absolute level and the trend of these figures are much less disquieting, and suggest that the absence figures are not likely to have been symptoms of disaffection or growing irresponsibility. Even the higher figures of involuntary absence, which no doubt indicate both a changing attitude to indisposition and greater security during sickness, may have been affected by other factors, such as rising average age. In 1971–2 the collieries had more than twice as many men in their fifties as in their twenties, and the number aged over 60 was more than half as great again as the number under 20 years old.[4] Nevertheless, later experience (though with a younger labour force) showed that much lower rates of involuntary absence were attainable. It may be that in the later sixties there was some lack of enthusiasm about turning up when feeling below par or about getting back to work

[1] NCB *Ann. Rep.* 1971–2, II, 86–7; 1972–3, 6.

[2] Ibid. 1965–6, I, 13.

[3] Ibid. 1969–70, I, 35; 1970–1, I, 38.

[4] Ibid. 1971–2, II, 94. The absence percentages are from successive *Ann. Reps.*

quickly after illness. Any conclusion must be speculative because the causes and significance of absence cannot be stated without imprecision.

Experience with labour disputes indicates no increase of friction and, on the whole, some diminution, until the very end of the sixties. On any reckoning, miners became decreasingly strike-prone from the late fifties on. In 1957 there were 2,228 work stoppages in the coal industry, which was 78 per cent of the number recorded for all industries; in 1970 there were 221 in the coal industry, 4 per cent of the national total; and there was a fairly steady downward trend from one point to the other, interrupted in 1960 and, very slightly, in 1964. Coalmining employed a declining proportion of the nation's labour force in these years, but this cannot account for the change. Stoppages per 1,000 men employed in the coal industry fell from 3.1 in 1957 to 0.57 in 1970, and stoppages per operating pit from 2.65 to 0.50. This happened in a period when the incidence of stoppages in industry generally was rising sharply: the number in other industries rose four and a half times. In the early sixties it also happened while the numbers of disputed issues in the coal industry was still rising, but, to an even greater extent than before, they were settled through the conciliation procedures without anybody first going on unofficial strike.[1] Of course, the mere number of separate stoppages is not, for all purposes, a satisfactory index of what was happening. Most stoppages in the coal industry involved only a small number of men for a few hours, and the biggest reduction was in these. Examples of the larger and more extended strikes and other forms of non-co-operation over a period still continued, though these also diminished in number; and it was these which accounted for the greatest direct loss of output.

The amount of coal output lost through disputes therefore showed a much less regular downward trend. The peak of 7.27 million tonnes in the three years 1955 to 1957 inclusive was never remotely approached in the sixties, but the low figure of 0.99 million tonnes in 1959 was exceeded in each subsequent year until 1967–8; and 1961, when 2.13 million tonnes were lost, was (until 1971–2) worse than any other post-nationalization year except 1955 and 1956.[2] But the large reduction in the constant stream of little strikes was a great help, especially when mechanization was making continuity more and more important for productive efficiency and competitive pricing. Indirectly, it was a significant contribution to output and productivity. The annual fluctuations

[1] Handy 1981, 213–16.

[2] NCB *Ann. Rep.* 1971–2, II, 86–7.

in the losses directly attributed to disputes depended mainly on the effects of one or two particularly troublesome strikes, rather than on the general situation. In 1961, for instance, about 40 per cent of the tonnage lost was in one dispute over pieceworkers' price lists in Yorkshire, which went on in February and March and at one stage affected 61,300 men in 69 collieries.[1] In 1966–7, when there was again a sudden increase in the tonnage lost, well over half the total resulted from a ban on week-end working (apart from statutory inspections and safety duties) imposed from June to September by NACODS.[2] Even the better years were sometimes prevented from being much better still by a single incident. For instance, in 1965–6, when the loss was 1.20 million tonnes, more than 40 per cent of the total came from a fifteen-day unofficial strike by the South Wales members of NACODS over an incident at Deep Duffryn colliery. The absence of the 3,500 underofficials prevented about 31,000 men from working.[3]

The general improvement in the incidence of disputes seems most likely to be attributable mainly to two influences. One is an improvement in management. This was coming about by the gradual retirement of some who had never adapted either to the conditions of a nationalized industry or to the position of the unions in the transformed labour market of the post-war years, and their replacement by others who had been better trained and had grown up with the conditions in which they found themselves. It was also assisted by the clearer definition of managerial authority and accountability, and the stronger support given to it, since the presentation of the Fleck Report. The second, and probably greater influence, is the reaction of the workers to the more difficult market conditions in which they found themselves and which were expressed in colliery closures and the continuous reduction in the number of jobs in the coal industry. Any resulting change of attitude is likely to have been shared by local union officials. There could be no great reduction in the national total of stoppages unless there were great changes in Scotland, Yorkshire, and South Wales, because that is where most of the stoppages had always been; and great changes there were. Much the earliest and fastest decrease in the number of stoppages was in Scotland, which was worst affected by the first wave of closures. There was, however, also a large decrease in Yorkshire, where the pressure was much less. In South Wales there was a reduction,

[1] NCB *Ann. Rep.* 1961, I, 13.
[2] Ibid. 1966–7, I, 13.
[3] Ibid. 1965–6, I, 14.

but a much less definite trend before 1965; but until then South Wales was lightly treated in the matter of closing loss-making collieries. Even in the regions where stoppages had been few, they diminished where pressure on jobs was strong. The incidence of stoppages (in proportion to the number of workers) halved in eight years in Northumberland and Durham, but there was no such change in the Midlands, and the West Midlands (where the general demand for labour was intense) was peculiar in seeing a rising incidence of stoppages in coalmining.[1] So it looks as though a greater readiness to use conciliation procedures, rather than walk out, was in part a response to a perception of changed economic and employment conditions in the industry, though regional variations in the response were not altogether consistent. In particular, the downward trend in the number of stoppages appears to have been faster than might have been expected before 1965 because the relatively favourable employment situation in Yorkshire did not prevent a sharp response there to the national situation.

It had always been hoped that a more rational wages system would get rid of most of the points of impassioned argument, but it is hard to see much sign of this before the later sixties, though there were so many different district agreements affecting particular groups that it would be difficult to prove one way or the other. If there was not much influence, this is not altogether surprising, because it appears that faceworkers were more prone to strike than others and the national reform of their wage system did not come until 1966. The NPLA certainly appears to have made a difference. It was followed by a reduction in the number of issues about wages which were put forward for conciliation, and also by a further drop in the number of stoppages over wages. The tonnage lost through disputes fell by almost three-quarters in 1967–8 to only 0.45 million tonnes, and the following year went down again to the record low figures of 333,000 tonnes. The improvements were directly attributed to the NPLA.[2] But it is not at all certain that all the effects of the NPLA eased industrial relations. Though it reduced the number of strikes initially, it may well have been a factor in soon making strikes bigger. By starting a move towards greater uniformity it contributed to greater solidarity. By compressing differentials it made the men who lost most by the compression feel that they had less to gain by remaining restrained. And when the effectiveness of large strikes had been demonstrated, there began to be stoppages about fresh issues, some of which

[1] Handy 1981, 216–27.
[2] NCB *Ann. Rep.* 1967–8, I, 43.

can be associated with the NPLA and the Third Daywage Structure. The early seventies not only had bigger strikes, but after 1972 there was some increase in the frequency of strikes, without any return near to earlier levels. Many of the additional stoppages then were about such matters as work organization and refusals to accept what management maintained was reasonable alternative work. There was a problem in that the wage agreements of 1966 and 1971 had classified as equivalent large numbers of jobs previously dealt with separately; and some of these were still regarded by the men as very unequal in terms of the effort and skill they demanded.[1] When anything occurred to make workers less docile, such differences of perception were likely to create friction.

There were signs of a change of mood in 1969, when disputes caused a loss of 2.94 million tonnes. Nearly all this loss came from a single stoppage, which lasted a fortnight and was the biggest since nationalization. The occasion was the claim for surface workers' hours to be reduced to 40 a week including mealtimes. The NCB had countered with an offer of 40 hours excluding mealtimes, the NUM had reiterated the original claim, and while this renewed demand was being considered, large numbers of men supported it by an unofficial strike. At its height this involved 140 collieries, including every one in Yorkshire. Later in the year an individual ballot produced a majority for accepting the NCB offer,[2] so it might seem that the unofficial strike had been an unsuccessful attempt to coerce the majority; but it would not escape notice that the full claim which the unofficial strikers had supported was officially presented again the next year, and won.

Things became still more serious in 1970, when tonnage lost through disputes exceeded 3 million for only the second time since nationalization, with the highest rates of loss coming in Doncaster Area, South Wales, Scotland, and Kent.[3] This came as no great surprise. The annual conference of the NUM passed resolutions demanding minimum adult wages of £20 a week on the surface, £22 underground, and £30 for men paid under the NPLA, and recommending strike action if the claim (which could have cost £75 million) were not conceded. The NCB knew they could not pay higher wages without a price increase, and thought £30 million was the most they could add to prices. Even this amount was expected to cause the loss of half the industrial market over the next three or four years, which would destroy the jobs of 20,000 miners.

[1] Handy 1981, 225–6.

[2] NCB *Ann. Rep.* 1969–70, I, 35.

[3] Ibid. 1970–1, I, 36–7.

These arguments were publicized. Lord Robens reminded his colleagues that there had been no indication of the government's attitude apart from general and public statements, but he believed the government would not give way in the face of a strike. The NCB offered £2.50 a week to the lowest paid and £2.30 a week to other daywagemen, thus producing equality for the two lowest grades in the daywage structure, and £1.87½ to those conditioned to the NPLA. These proposals offered about 17 per cent to the lowest and 7 per cent to the highest paid, and averaged about 10 per cent. They would have given minima of £17.50 on the surface and £18.50 underground, and various rates under the NPLA, as national rates were not due for another year. Some members of the board thought the offer a little too low to win acceptance from the moderate majority of the NUM Executive, but most of them thought they would be accepted by a pithead ballot, though nobody had any hope of avoiding unofficial strikes, which might lose 3 to 6 million tons of output, with further adverse effects on the market. In fact, the NUM Executive rejected the offer and recommended strike action to a pithead ballot of members. The ballot produced only 55 per cent in support of a strike, which was well short of the two-thirds that NUM rules then required. So the NUM reopened negotiations and the NCB improved the offer by £0.50 a week all round. Unofficial strikes and overtime bans then began, and unofficial action was supported by NUM area delegates in Scotland and South Wales. The unofficial action lasted for almost four weeks. But the NUM Executive recommended the revised offers, which were accepted by a pithead ballot by 158,239 to 82,079.[1] It had been quite a near thing to a national strike. One important sequel at the next NUM annual conference in 1971 was the passing of a resolution which lowered to 55 per cent the majority required for a pithead ballot to authorize a strike.

All the signs then were that it would be very difficult to avoid a national dispute. The events of 1970 had done nothing to resolve the difficulties that were separating many of the interests of the NCB from those of the miners, and external pressures were making them worse. The strongest external influence was the more rapid rise in prices. This was putting up the non-labour costs of coal production and thus threatening the market position at a time when other factors were suggesting that the decline in coal consumption might be coming to an end. Rising prices also eroded the value of miners' wages faster than before. Thus one and the same influence was driving the miners to

[1] Ibid., I, 36–7.

demand bigger increases and the NCB to try to concede less. If conditions developed in such a way as to persuade most miners that they had a common interest in wage demands that outweighed local and sectional interests, and that the common interest was best pursued by national, centrally directed action, then a united opposition to any but very large wage improvements was to be expected, and that, in several ways, was how conditions were indeed developing.

Miners had experienced a long period in which their pay had risen more slowly than that of most industrial workers, and this had happened while they had, on the whole, been very co-operative in accepting measures which drastically contracted the coal industry, and while output per man had been rising faster than in most industries. It is not surprising that they should feel aggrieved. There are different averages that may be estimated in order to indicate the extent of the relative decline of coal industry wages and they give different results, but all suggest it was substantial. The NUM told the Wilberforce enquiry that over fifteen years mineworkers' earnings had deteriorated by 25 per cent relatively to industrial earnings generally.[1] This may have some element of uncertainty, but it is neither an unrealistic nor the highest possible figure. Other estimates suggest that since 1957 average hourly pay of miners had been rising about 1 per cent per annum *less* than that of all industrial workers, taken as a group.[2] Yet other calculations show (with allowance for sick pay and holiday pay) that mineworkers' weekly pay was 29 per cent above the all-industry average for adult males in 1956 and had fallen to equality with it by 1970. In the latter year mineworkers' pay was 3 per cent below the average for all *manufacturing* industries. It was a good deal further below the levels in the motor industry, ship-building, metal manufacture, printing, and transport, whereas until 1956 coal had been the best paid of all industries. If one moves to calculations which involve price movements, taxes, and social benefits, then an average mineworker with a wife and two children had a net real income which in 1957 was 22 per cent above, and in 1969 2 per cent below, that of an average manufacturing worker.[3]

This general deterioration was complicated by the large changes in differentials within the industry. The narrowing of differences between coalfields, which had been proceeding ever since the Second World

[1] *NUM Submission to Wilberforce Enquiry* (Feb. 1972), section 1, p. 5.

[2] Handy 1972, 539–40, includes this estimate in a very good summary of the main influences leading to the strike.

[3] Handy 1981, 172–4 and 230.

War, continued as before. It was sharply symbolized when, on 1 January 1972, the requirements of the NPLA were completed by the introduction of a uniform national shift-rate. All the other districts were brought up to the level of £5 a shift, which had applied to Nottinghamshire and Kent, while those two districts experienced a standstill. There was also the policy of reducing the differentials between the former pieceworkers (i.e. the faceworkers and others among the best paid men underground) and the daywagemen. This policy had been applied fairly sharply since 1966 and, even though it was agreed NUM policy, supported by the NCB, it was not going to suppress normal human reactions among people who found they were suffering for the benefit of others and would rather not suffer at all.

It was the complicating changes in the wage structure that united the interests of mineworkers in aggressive national action. The relative position of the daywagemen had deteriorated less than the average figures suggest, but they had always been among the poorest paid and were anxious for more in a time of accelerating price rises. Those who had been better paid, and reckoned to perform the most demanding jobs, had done even worse than the average, and if they worked in one of the higher paid coalfields they had done worst of all. A few of the latter had at one point had a wage reduction, and more of them had experienced some rises so small that inflation turned them into reductions in real terms.[1] The groups treated relatively worst included some (e.g. the miners of Nottinghamshire) who were both highly productive and traditionally co-operative in industrial relations, but current conditions were giving them an interest in aggressive policies which had been initiated by others. For a variety of reasons things had to be done nationally. The introduction of the Third Daywage Structure and the final move to uniform shift rates under the NPLA had ended the opportunities for any large-scale local bargaining; and, given the NUM policy on differentials, there was no way of restoring the position of the faceworkers and other skilled underground workers, relative to other industries, except to press for large increases for the lower paid and let that pull up the wages of the rest. This was, perhaps, a risky line of action, as a large increase in surface wages in the coal industry could easily stimulate emulation in manufacturing industries and lead to the removal of any superiority which other mineworkers had temporarily recovered.

There were a few other influences to make it likely that mineworkers would not only unite behind more militant policies but would do so in

[1] Ibid., 166–9 and 218.

1971–2. One was that some of the contrary pressures were reduced. Pit closures, which had been so numerous, were now few, and there was no longer a stark presentation of the possibility of choice between high wage increases and job losses. Indeed, it seemed rather more likely that, without higher wages, there might be some difficulty in keeping quite enough labour to produce all the coal that could still be sold. Another influence was the change of personnel at the top. The NUM had a new president, who was facing a new chairman of the NCB. Quite apart from any wish of new men to demonstrate that they do things differently, there was now more scope for a change of outlook in the NUM. The Scottish Area NUM, from which Lawrence Daly had come to be general secretary at the end of 1968, had never whole-heartedly followed the Paynter line of accepting small wage increases in the hope of saving more jobs. There was more enthusiasm there for giving the first priority to raising wages. It was not until 1971, with a change of president, that there was a chance to give the Scottish line more weight in national policy. There was a shift then, though not as big a one as there would have been if Michael McGahey, the Scottish Area president, had been the successful, instead of the defeated, candidate in the election for national president. The new president, Joe Gormley, though in general a supporter of Paynter, had been known to question his restraint on wages and was a more forceful character than his own predecessor, Sidney Ford. A further possible influence on the timing of action was an apparent belief that not only had the demand for coal stopped declining, but the NCB's commercial and financial position was improving, so that they could afford to pay out more in wages.[1] This was a mistaken view, but it may have added a little extra weight, though it was probably far less decisive than the other factors.

There had been no national strike in the coal industry since 1926, but in the latter half of 1971 events moved steadily towards one. At the annual conference in July the NUM agreed on claims, of which the chief were for minimum basic weekly rates of £26 on the surface, £28 underground, and £35 under the NPLA. These claims involved increases varying from 35 per cent to 47 per cent on prevailing rates, and they were put formally to the NCB in September. When the board reviewed the

[1] Cf. Handy's reference (1972, 539) to the 'knowledge' that the NCB was about to turn a deficit into a surplus. Gormley 1982, 85 suggests that the NCB in 1971 followed a traditional line of expressing optimism at the start of the year but always finding adverse factors to state at wage negotiation time. But the NCB assessment of their financial difficulties in the autumn of 1971 was quite genuine.

position they found costs rising and productivity lagging. Their best expectation was that, with unchanged prices and wages, they would break even in the current year and might keep down the net loss for the following year to £15½ million. The largest price increase that was thought commercially practicable was 5 per cent, which might turn the 1972–3 projected loss into a surplus of £20½ million. The view was that the board could not afford any increase in wages, but had to make some offer, for there was anxiety to avoid a strike, which was expected to cost at least £10 million and possibly £15 million a week. It was therefore agreed to set aside £25 million for wage increases. This would enable increases averaging 6 per cent to be offered. The board were reminded that inflation in the last twelve months had reduced the real value of wages by about 10 per cent, and both the board's industrial relations member and the one trade unionist among the part-time members expressed their views that the proposed offer had no chance of acceptance. The latter member suggested that an offer of increases averaging £2 a week, with more for the lower paid, was the least that would be considered. Nevertheless, the board stuck to their view and agreed to offer increases of £1.60 a week for all grades of daywagemen (by this time nearly everybody), with discretion for the negotiators to go up to £1.75. In fact, during negotiations they went as far as £1.80 on the minimum surface rate, while sticking to £1.75 for all other rates. These negotiations took place on 12 October.

The NUM national executive recommended the rejection of the offer, the imposition of a ban on overtime, withdrawal from consultation, and the holding of a strike ballot. All these recommendations were endorsed by a delegate conference on 21 October, and the overtime ban and withdrawal from consultation took place from 1 November. The overtime ban had been intended to prevent coal being produced in overtime but not to affect safety work or to stop activities which were necessary to enable the men to work a normal five-day week. At some collieries, however, the men did not follow the union's guidelines in these respects. The pithead ballot, held in the week beginning 22 November, produced an 88 per cent poll, with 58.8 per cent of the voters supporting a strike. On 9 December the NUM executive agreed that the strike should go ahead and gave notice that it would start on 9 January 1972.

During the month of notice there were several more meetings between the NCB and the NUM. In the course of these the offers by the NCB were somewhat increased and subjected to various slight rearrangements, but they remained far below the union demands. The last offer, on 5 January 1972, would have given increases (from 1 November 1971) of £2 a week to all adult daywagemen paid under the 1955 structure and £1.90 on other rates, five extra individual days of holiday from 1 May 1972, and possible further increases in pay from 1 January 1972 on the basis of a productivity agreement to be negotiated. All the offers were rejected. The NUM also refused to have the dispute submitted to the National Reference Tribunal. This was understandable in the circumstances of the time. Though there were no statutory limits in force, the government were speaking of restraints in public sector pay, and unions (whose suspicions of government were already roused by the Industrial Relations Act, which came into force on 1 January 1972) believed that all arbitrators were bound to be influenced by this. The NUM based their claim not just on current conditions but on the relative deterioration of the miners' position for years past, and suspected that this would not be given much weight in any conciliation or arbitration proceedings.

So the strike went ahead. The NUM paid no strike pay, though they gave £2 a day expenses to pickets who travelled away from home. The strike was very solidly supported, with a good deal of practical sympathy from some other unions, but the other unions in the coal industry remained at work. In some places and in some respects the support was rather reckless. In particular, the widespread disregard of union instructions about provision of full safety cover put prospects of a quick recovery after the strike at risk, despite the efforts of management, including members of NACODS. Representations about this, to NUM national officers, secured no improvement, nor did representations by the national officials of NACODS to the general secretary of the TUC, to whom they expressed concern not only about the security of the pits but about the violence of the picketing at some collieries. By the beginning of February the full safety cover officially enjoined by the NUM was being provided at only 31 of the 289 collieries, and at 203 no NUM member was doing anything underground; 10 per cent of all powered supports were affected by compression, and the condition of faces and roadways was deteriorating. By the beginning of February the fuel supply situation was getting difficult in some places, as the public became aware through numerous power cuts. Part of the difficulty came from the success of the NUM in picketing not just the collieries but the premises of large merchants and users, sometimes in great strength, and so hampering or preventing the movement of stocks of coal and coke. Particular prominence, as a news item, was given to the effective closure of the Saltley coke depot by a group of 'flying pickets' from Yorkshire, mobilized by Arthur Scargill, who had got wind of the operation of this depot before the union's national officers did, and acted on his own initiative.

The nature and widespread effects of the strike, and the impossibility of the NCB being able to finance any acceptable settlement from its existing resources, necessarily made it a government matter. Both sides saw the Secretary of State for Employment, Robert Carr, on the first day of the strike, but his department was not at that stage directly engaged in the search for a settlement. Both sides also saw the TUC general secretary, Victor Feather, on 19 January. He, in turn, went to see Mr Carr on his own initiative, and this led the latter to invite the NCB and NUM to see him separately on 21 January. But it was not until things got more difficult, in February, that the government became active in trying to end the strike. The NCB had had to go to the government to ask for the drafting of a bill to extend their borrowing powers and to indemnify them against exceeding their permitted cumulative deficit, which they could not avoid doing. On 4 February the BACM and NACODS jointly saw Mr Carr both to complain about the behaviour of NUM pickets and to urge the need for a settlement. The TUC had publicly said that a settlement could not be reached unless the government created a more flexible situation. On 8 February the government declared a State of Emergency, and the next day Mr Carr invited the NCB and the NUM to see him separately, and then later to negotiate together under the chairmanship of one of his officials. A higher offer emerged, but only on the basis that it would run for eighteen months instead of a year. It included increases of £3 a week for surface daywagemen, £3.50 for underground men paid under the 1955 structure, £2.75 for other men underground, and a guarantee that no adult would get less than £22 for a full week's work. The NUM rejected the offer next day and told the NCB they needed increases of about twice the size of those then offered. As the union would not go to arbitration, the Secretary of State announced that he would set up a Court of Enquiry under the terms of the Industrial Courts Act, 1919. He asked the NUM to call off the strike on the terms of the latest offer by the NCB, with the clear understanding that, if the Court of Enquiry recommended higher increases, they would be applied from the day work restarted. The NUM refused and indicated that, though they would give evidence to the Court of Enquiry, they would not undertake to be bound by its findings.[1]

[1] NCB *Ann. Rep.* 1971–2, I, 32–4 gives a succinct summary of the course of events. Gormley

The composition of the Court of Enquiry was announced on 11 February. The chairman was Lord Wilberforce, a Lord of Appeal. The other members were Mr John Garnett, Director of the Industrial Society, and Professor L. C. Hunter of the University of Glasgow. The enquiry was held against a difficult background, with many power cuts and a rising number of workers laid off, and it was conducted with great urgency. The NUM and the NCB both prepared detailed and lengthy memoranda, which were submitted as written evidence, and the Court heard oral evidence on 15 and 16 February, and had its conclusions ready on the morning of 18 February. The Confederation of British Industry gave brief written and oral evidence, mainly concentrated on warnings against the likely adverse effects a high settlement for the miners would have on other wage settlements and on industrial prices; but the main preoccupation was with the evidence of the union and the board. The NUM relied not only on their own research and industrial relations department, led by Trevor Bell, but brought in outside guidance and evidence from Professor Hugh Clegg, Michael Meacher MP, and the Trade Union Research Unit of Ruskin College, with John Hughes from there taking part in the oral sessions. The evidence was powerfully and effectively presented by the general secretary, Lawrence Daly. The NUM statement reviewed in detail the structure, absolute level, and trend of mineworkers' wages, with many external comparisons. It brought out also some special features, such as the problem that many surface workers had wages in the region of the 'poverty trap', i.e where an increase in wages reduces eligibility for some social security payments and introduces a liability for income tax, so that the effective marginal tax on the increase is very high. The union also attempted tentative calculations of the NCB's financial resources and of the relatively more favourable financial treatment of other industries by the government. Their evidence naturally concentrated on the comparisons most favourable to their case and glossed over the extent to which the reduced differentials of some of the higher paid workers had resulted from union policy. But there was a very strong case to be made, and the union made it well. The NCB also put forward a very cogent statement, reviewing the financial position (more accurately than the NUM) and concentrating also on statutory obligations, wages, negotiation arrangements, and productivity. Both parties carried their review of the evidence back to the beginning

1982, 82–109 gives an account from the union side. NCB minutes and papers from most meetings from Oct. 1971 to Feb. 1972 provide more detail from the employers' side.

of the nationalized industry and put a lot of emphasis on changes since the mid-sixties.

The Wilberforce Report recognized that the NCB were constrained by their statutory obligations and financial condition, and stated that 'the NCB's offers were, in the light of the objective seen by them, perfectly fair'. But its recommendations conceded a large part of the union's basic claims, and the report added that, if the recommendations put up costs so much that it was impossible for the NCB to meet them without colliery closures on an unacceptable scale, 'then the Government will have to provide the necessary finance, because it is unreasonable to expect miners' wages to be held down to finance uneconomic operations'. The report argued that the wage increase should have two components. One was a normal periodic increase, and for this the Court of Enquiry appears to have had in mind something around the 8 per cent which members of the government had been suggesting in public ought to be the current maximum for everyone. The other was an 'adjustment factor' to compensate the miners, who were a special case, for the falling behind which they had experienced in recent years. This appears to have been put at 11 to 13 per cent and to have amounted to an addition of £70 million to the total wages bill. The recommended increases, to run from 1 November 1971 for sixteen months, were £5 a week for surface workers, £6 for underground workers on the 1955 daywage structure, and £4.50 for those on the NPLA and the Third Daywage Structure. These increases gave minima (for non-craftsmen) of £23 on the surface; £25 below ground on the 1955 structure, £27.70, £29.42½, and £34.50 for the three grades of the Third Daywage Structure; and £34.50 for those on the NPLA. The Court also recommended that the parties should complete negotiations about the age at which adult rates became payable, and on increased holidays; that they should try to negotiate within six months an agreement on a scheme whereby productivity would be reflected in wages; and that, if the present arbitration machinery was unsatisfactory, they should revise it.

An agreement to end the strike was negotiated through a long session that began at 10 a.m. on 18 February and ended at 1 a.m. on 19 February. There was such urgency to reach a settlement and the negotiations were so wearyingly protracted that the NUM were able to push for every concession they felt there was even a remote chance of getting, and to obtain a high proportion of them. The two sides first separately met the Secretary of State for Employment and then their negotiating teams met together, with adjournments for consultation with their colleagues. At one stage the whole national board met the whole NUM executive. Later, the TUC general secretary had discussions with the NUM executive. Later still the Director-General of the CBI, the TUC general secretary, the NUM negotiating team, and the national board all had separate interviews with the Prime Minister, Edward Heath, after which negotiations were resumed with even greater urgency.

Although the NCB agreed from the outset to accept the Wilberforce recommendations, the NUM at first declined to do so and demanded £1 a week more than the recommendations for all men paid under the 1955 daywage structure. The NCB twice refused to make this concession, and the Prime Minister also told the NUM that the government (which would have to finance any settlement) would not agree to it. But the NUM had many supplementary demands, some on fairly small points of detail and some quite large, and the NCB were accommodating about nearly all these. Even when first refusing the demand for wages above the level of the Wilberforce recommendations, the NCB gave the NUM a reply on nineteen separate items and either conceded demands (as for a phased lowering of the age for adult wage rates from 21 to 18) or offered to start negotiations (as for a scheme for subsidized transport). Later in the day immediate action was offered on some items which initially it had been proposed should be examined over a period. The NUM also added some further demands. At a late stage the NUM brought forward one of their largest demands, that the payment for the bonus shift (introduced when the five-day week was negotiated in 1947) should from 1 June 1972 be incorporated in the ordinary shift rate, thus increasing the shift rate by one-fifth, with consequential effects on the calculation of overtime payments. This also was conceded on condition that the NUM executive would agree to a settlement without making further demands. This was done, though even then the excutive voted only 17 to 8 for acceptance. Besides this concession and the lowering of the age for adult wage rates, the NUM had obtained an extra five individual days holiday in the year from 1 May 1972; increases in pay from 1 November 1971 for weekly paid industrial staff, coke workers, opencast semi-industrial workers, canteen workers, and (for this occasion only) clerks; full increases (corresponding to Wilberforce recommendations) for men on personal rates; a new wage structure for lorry drivers; an extra 80p a week on top of the recommended £5 for winding enginemen, from 1 November 1971; discussions with a view to introducing a national subsidized transport scheme by 1 May 1972; an undertaking that the calculation of next summer's holiday pay would not be reduced because of the overtime ban and strike (as it would have been if the existing scheme had been strictly applied); and a twelve-month spread for the payment of rent arrears on NCB houses.

The NUM agreed to recommend acceptance of the recommendations of the report and the other arrangements that had been negotiated with the NCB, to call off picketing immediately, to ballot their members in the following week, and if the recommendation was approved, to resume normal working on Monday 28 February 1972. The ballot took place on 23 February and its result was available on 25 February. Only 7,581 voted against acceptance, and 210,039 voted in favour.[1]

Agreements with NACODS and BACM to restore the relative position of their members quickly followed, and an agreement was made with the clerical unions for a £5 a week increase in basic rates from 1 November 1971.[2] Not many problems in industrial relations arose immediately after the strike. The most serious was a disagreement with the NUM executive about concessionary coal, which the NCB had withdrawn during the strike, on the ground that it was 'payment in kind'. The NUM demanded its restitution and supported their claim by a reminder that the supply of concessionary coal had continued during previous unofficial strikes, and that in four wages districts there were agreements specifically providing that entitlement lapsed only after four weeks' continuous absence without leave. The NCB rejected the demand but offered to submit it to the National Reference Tribunal (whose chairman, Z. T. Claro, had written to both the NCB and NUM to reaffirm its independence and impartiality) or to an independent arbitrator. The NUM preferred the latter; Lord Gardiner agreed to act; and in August 1972 he ruled that mineworkers should have half the amount of concessionary coal (or half of any payments in lieu) to which they would have been entitled if there had been no strike. Members of the board thought this a very odd ruling, based on no precedent, but, having agreed to arbitration, they carried it out.

Apart from this, though some people in the industry detected signs of

[1] NCB *Ann. Rep.* 1971–2, I, 34–5 is the most concise summary of the resolution of the dispute. The recommendations are in Department of Employment, *Report of a Court of Inquiry into a Dispute between the National Coal Board and the National Union of Mineworkers*, Cmnd. 4903 (1972). For the evidence before the Court the following typewritten collections were used: Wilberforce Court of Enquiry, *Minutes of Proceedings*; *NUM Submission to Wilberforce Enquiry*; *Introductory Oral Submission by the National Coal Board*; *The National Coal Board's Submission to the Court of Enquiry–Supplementary Notes*. The course of the final negotiations is set out in Gormley 1982, 112–17. A detailed account of the issues involved in the strike and the settlement, mainly from a union standpoint, is Hughes and Moore 1972.

[2] NCB *Ann. Rep.* 1971–2, I, 36.

lingering bitterness, things went fairly smoothly, though the negotiations to establish some link between productivity and wages made little headway. Immediately after the strike, levels of attendance were higher than usual and productivity began to pick up. This helped physical recovery from the effects of the strike, which depended even more on careful management planning, during the strike as well as after. About half the coal faces had been affected by deteriorating conditions (twenty-five faces had to be abandoned completely), and on the first day of resumed work, output was only 47 per cent of the average for the previous October. Within four weeks output had recovered to about 90 per cent of normal, and sales were increased by efforts to lift stocks much faster than usual. This activity was particularly helped because, throughout the strike, most opencast sites (very few of whose workers were in the NUM) produced coal normally but did not distribute it. Restrictions on working hours at opencast sites were temporarily relaxed so that the accumulated stocks could be moved.[1]

The great strike of 1972 was one of those events which has the character of tragedy because it was a clash between two sides, both of which, from different standpoints, were largely in the right. The mineworkers, in a period when their performance had improved and they had co-operated in a drastic contraction of the industry and in the adoption of new methods, had been left behind in pay by workers in many other industries; and they had no prospect of closing the gap by customary methods of negotiation. The NCB, with the existing financial resources and market opportunities, could not continue to run the industry as they were statutorily required to do, and as hitherto they had succeeded in doing, if they paid the mineworkers what they needed to restore their lost level of earnings. There was no way out unless someone—and it could only have been the government—had, in advance of a possible conflict, changed the rules about what was required of the coal industry. But nobody in authority sought such a change, so it was only after a conflict that the rules were modified somewhat. For a number of years one contribution—by no means the only one—to the supply of national needs of coal at a commercially competitive price had been to keep the lid on mining wages. In 1972 the lid was blown off and things could never be the same again.

Yet the strike had not resulted in the establishment of a clear set of new conditions from which a fresh start could be made. In the fairly short run its results could be seen to be rather messily ambiguous. The

[1] NCB *Ann. Rep.* 1971–2, I, 5–6.

president of the NUM has said that, at the outset of the negotiations, he privately told both the chairman of the NCB and at least one minister that it would need an offer of rises of about £3.50 a week to avert a strike. The NUM had asked for far more than that, and by their strike they got more. They got more, that is, by being more militant than was implicit in that original suggestion of the president, who was himself persuaded by the failure of negotiations to push as hard as possible to obtain the maximum. The lesson of how success had been won was not likely to be forgotten. Lord Gormley, in retrospect, has expressed uncertainty about the balance of good and bad in the strike. On the bad side, he says, 'its success led to an attitude of mind . . . where people, the moment they don't get what they want, think and talk immediately of strike action.'[1]

Another, most important, aspect of the outcome was that the Wilberforce recommendations did not for long restore the relative position of the miners to the degree they had been seeking. This was probably partly due to an under-estimate, by the Court of Enquiry, of the size of the 'adjustment factor' that was needed, but it was due even more to the rapid rise of other industrial earnings in the nine months immediately after the miners' settlement. Surface daywagemen had continued to gain on the best paid underground workers, and their relative position remained not only better than it had been before the strike but better than it had been ten years earlier. For men on the NPLA and Third Daywage Structure things were different. By October 1972 the average weekly earnings in coalmining were only 5.6 per cent above average male earnings for all manufacturing industries, and the ratio of faceworkers' earnings to average male manufacturing earnings was only one percentage point more than it had been a year earlier.

The full effects of this change were delayed. The government imposed a statutory wage freeze in November 1972, so the miners could not get any immediate wage rise when the Wilberforce settlement expired at the end of February 1973. The next phase of the pay policy was to permit wage rises of £1 plus 4 per cent, a formula which tended to narrow all differentials. The NUM made a quick settlement on this basis in April 1973. The available increase was shared equally, so that all mineworkers and weekly paid industrial staff got a rise of £2.29 a week, plus another 50p to offset their contribution to the cost of improvements in the mineworkers' pension scheme. An agreement with NACODS gave officials an increase of £2.56 a week. By October 1973 average mineworkers'

[1] Gormley 1982, 86, 93–4, and 118.

earnings were only 2.3 per cent above the average for all manufacturing industries, which was about 10 per cent less than had been thought appropriate by the Wilberforce enquiry. The ratio of faceworkers' earnings to the average for all manufacturing was actually lower than it had been in October 1971.[1] So within two years, most of the gains from the strike and the Wilberforce recommendations had been eroded. The sense of unfairness, which had aroused so much militancy in the winter of 1971–2, had been allowed to return, with the chance to arouse it all over again in the winter of 1973–4.

## viii. International relations and entry to the European Coal and Steel Community

In some ways the British coal industry from the late fifties onward might seem very insular in its preoccupations. Though it was acutely conscious of the pressure from competing forms of fuel that were obtained abroad, its own direct involvement in international markets was small. Once the last remaining contracts had run their course, early in 1959, no more coal was imported to the UK for over a decade; but the conditions which closed the home market to foreigners were repeated in many other countries, even those with much smaller indigenous coal resources. Such international markets as remained could be supplied to a large extent by the low-priced coal made available from Poland and the USSR, and, provided that shipping freights were low (as they often were for appreciable periods), by cheaply produced coal from the USA and South Africa. British coal, once the mainstay of one of the world's great export trades, was unable to maintain even the greatly reduced export levels of the earlier post-war years. After 1957 total exports continued their recent steady fall and then stabilized. In nearly every year down to 1964–5 they were just under or just over 5 million tonnes. The one exception was 1963–4 when they reached 8 million tonnes. This was mainly because of a coal strike in France and restocking after the fierce European winter of 1962–3. There was still talk of rebuilding exports to 10 million tonnes but it was in vain. In fact, from 1966–7 onward the annual total was always under 4 million tonnes and in some years was little more than 2 million. Most was general purpose coal for steam raising, and there was a small surviving market for some of the special coals of South Wales. Federal Germany was the largest customer. In the early seventies it appeared that rather more could have been exported if more

[1] Handy 1981, 183–4; NCB *Ann. Rep.* 1972–3, 6.

had been produced, but in the strike year of 1971–2 things were so bad that imports were more than twice the amount of exports.[1]

International links, though weakening in trade, continued and strengthened in politics and in consultative organizations. The relationships were conducted mainly through the Council of Association between the UK and the High Authority of the ECSC, and through two other bodies. These were the *Comité d'études des producteurs de charbon d'Europe occidentale* (CEPCEO), in which British influence was informal but strong, and the Association of European Coal Producers, which consisted of the countries of CEPCEO plus Britain and Spain, and of which Britain was a full member from 1957. E. F. Schumacher played a leading role in the two latter institutions. All these bodies were concerned primarily with the exchange of information and the promotion of enquiries. But the Council of Association continued to keep Britain in close touch with the formulation of ECSC policy, and the other bodies sought to prepare suggestions for action to strengthen the coal industry and to urge them upon governments. It seems probable that few people concerned with the direction of the British coal industry around 1960 wanted anything much closer. Indeed, as the market for coal became more difficult there was suspicion that too close involvement with any multinational bodies might hamper freedom of action in dealing with the difficulties. In 1960 the UK government, for general political reasons, was preparing its ultimately unsuccessful attempt to join the European Communities. The NCB then accepted the view of their General Purposes Committee 'that, on the basis of the present ECSC Treaty and judged solely from the standpoint of the British coal industry, the disadvantages of UK membership of the Community outweigh the advantages'. The board, however, wanted to participate in the negotiations if the government decided to go ahead with an application. The NCB's suspicions were reciprocated within the ECSC, all of whose coal producing members found their coal industries in even greater difficulty than Britain's. They were alarmed not only by the larger size of the British industry but by the fact that it belonged to a single enterprise, the NCB. It was thought that this might upset competitive arrangements to the disadvantage of the original members of the ECSC.

During the sixties opinions changed. On the British side this depended much on the influence of a few individuals. D. J. Ezra (who had been NCB representative in the British delegation at the High Authority of the ECSC) and Lord Robens, who was firmly convinced of the

[1] NCB *Ann. Rep.* 1964–5, I, 16; 1969–70, I, 21; 1970–1, I, 18; 1971–2, II, 84–5.

value of a European connection,[1] were strong advocates. But the course of events also had its effect. The old fear that desirable policies would be thwarted by a supranational authority was shown to be groundless. The limitations of the power of the High Authority had, in fact, already been exposed in 1958–9, when its attempt to invoke the crisis provisions of the Treaty of Paris in order to impose production quotas was overruled by the Council of Ministers; but not a lot of weight was attached to this at the time. The freedom with which member states were able to adopt their own protective measures for their beleaguered coal industries was more convincing. The Treaty of Paris forbade subsidies (subject to certain exceptions under article 95), but members nevertheless illegally introduced them to deal with coal industry losses. The ECSC had to acquiesce and was left to draw up rules to define what subsidies were permissible. These rules began to operate in the mid-sixties but their form was not finalized until some time after the UK joined the Community. Continental producers, on longer and closer acquaintance,[2] lost any suspicion of the NCB as a potential discriminatory threat to fair competition, and in any case there was a second large organization, as most of the mining undertakings of the Ruhr were merged into a single undertaking, *Ruhrkohle AG*. The coal undertakings and trade unions of all the member states found a common interest in acting together to formulate common programmes, which they could each press on national governments, to safeguard the use of coal. One of their problems was that the ECSC, unlike the EEC, did not have a community policy on export and import trade with third parties. Each member state could pursue its own policy towards non-members, and one worry of the producing countries was that so much of the coal imported by member countries came from outside the ECSC.[3]

[1] See his contribution ('The lessons of coal and steel') to the *Daily Telegraph*'s series of articles on 'Common Market Issues' in August 1971.

[2] Much of the acquaintance came through the Council of Association. It may have been a helpful symbol of the co-operative British attitude that the annual reports of this body appeared only in French. Any of these reports indicates the areas of common interest. A typical example lists the studies promoted by the Council's Coal Committee: (1) trends in the supply of the domestic sector in the Community and the UK 1961–6; (2) legislation against atmospheric pollution (a matter of concern because of the effect on costs and markets of the wide differences in national policies); (3) technical progress in the mines (which included pit closure and concentration, and productivity); (4) social security systems and the methods of financing them; (5) the evolution of energy consumption in the Community and the UK. (Conseil d'Association, *Douzième rapport annuel*, 16 janvier 1967–31 décembre 1967 (Apr. 1969).)

[3] Ezra 1970 stresses the dominant influence of individual governments on the operation of the ECSC market.

So far as there were common policies within the ECSC, which was mainly in pricing practices and the regulation of competition in the dealings of members among themselves, the British coal industry was well acquainted with them and followed very similar practices. So there was neither fear of the unfamiliar nor any problem of adaptation.[1] When the UK government in 1970 again sought entry to the European Communities the NCB had swung round to full support. The new chairman of the board, Derek Ezra, reminded his colleagues that:

> [the Board] will still be free to select their own markets and will have the advantage of freer access to the Community markets, into which between 25 and 30 million tons of coal a year is currently being imported. If the Board's investment proposals meet with ECSC approval, they will be able to draw on Community funds at special interest rates but the Community will have no power to restrict investment by the Board of their own funds.

There was, of course, a financial contribution to be made in the other direction as the ECSC had the power to make a levy on producers, but the UK government had already undertaken to pay the initial contribution of both the NCB and the British Steel Corporation. The entry of the UK to the European Communities was predominantly motivated by political considerations. For the coal industry the prospect of much increase in exports to ECSC countries was not very strong; but access to support for investment and research was important, and the chance to play a leading part in the formation of west European energy policies was much valued. When UK entry to the Communities took place on 1 January 1973, coal was the only large British industry that was fully integrated with Community institutions and regulations immediately, without any transitional period.[2] The NCB and the unions immediately began to take a prominent part in the work of the ECSC Consultative Committee, and the NCB continued their role in CEPCEO, which chose the NCB chairman as its president. The main concern was to press on the Community authorities the need for a new and clearer energy policy with a larger role for coal, and to start using the financial facilities available within the ECSC. The first grants from the ECSC began to be spent in 1973, and other projects involving higher expenditure from ECSC funds were then being prepared.[3]

[1] Findlay 1972, 76–7.
[2] NCB *Ann. Rep.* 1972–3, 4.
[3] Ibid. 1973–4, 15 and 31.

## ix. The condition of the coal industry in the early seventies

To an even greater extent than most periods, the sixties and early seventies were a time of transition for the coal industry. These years were transitional primarily in two ways. They involved rapid contraction and concentration in order to match a shrinking of the market and a relegation of coal to a much smaller role in government plans for the future pattern of energy use; and they involved the technological transformation of production methods in such a way as to bring about a large and sustained rise in the productivity of labour. Both types of change were cost-reducing and therefore lessened the difficulty of maintaining a position for coal in a market where the competitive forces were felt most strongly through the pressures of lowered real prices for other fuels. But a transition should have a definable outcome. The anticipated goals of this particular transition included one for each aspect. The change to much more sophisticated technologies, mechanical, electrical, and electronic, meant (it was hoped) the building in of productivity advance as a permanent process in the coal industry, so that it would continue to keep up with the normal progress of the economy in efficiency, and perhaps surpass some of its own immediate competitors. But the experience of contraction and concentration was assumed to be once-for-all. It was expected to have a clear end, with the industry stabilized at a much smaller size which would be sustainable (and perhaps be a base for some re-expansion) because the permanently uneconomic and surplus capacity would have been shed and what remained would be characterized by the ever-improving technology which would keep it competitive.

In the middle sixties there seemed no reason to doubt that both goals would be attainable, though only through several years of great difficulty and effort. But by the early seventies the worrying thing was that the characteristics of the two goals seemed to be getting interchanged in the most undesirable way: the rise in productivity was becoming slower and less sustained, and the attainment of stability on the basis of commercial competitiveness kept on receding. Overall OMS rose from 24.9 cwt. (1.26 tonnes) in 1957 to 42.5 cwt. (2.16 tonnes) in 1968–9, and the rise was accelerating: the annual increase was 6.6 per cent in 1967–8 and 9 per cent in 1968–9. But then progress began to falter. The next two years saw increases of only 2.1 per cent and 1.7 per cent respectively, and the strike year of 1971–2 brought OMS below the level of 1968–9. The rapid recovery after the strike brought a 9 per cent annual increase in OMS, but most of that was a retracing of advances made before the strike. At 45.8 cwt. (2.33 tonnes), OMS in 1972–3 was only 7.75 per cent higher than it had been four years earlier.[1] It was clear that tentative projections of what results the industry might achieve, physically and financially, in the seventies would have to be reconsidered, as these had assumed in the late sixties that recent trends of productivity would be sustained. On this basis it had been thought reasonable in late 1968 to assume an overall OMS figure of at least 56 cwt. for 1972–3 (and more if it were possible to retain fewer uneconomic pits), and Lord Robens in 1969 was stating his belief that an OMS of 75 cwt. could be attained by 1975 if there was a co-operative attitude from the unions, the men, and the underofficials.

Though projections on various assumptions were made, figures of that order of magnitude were much used until they were falsified by events. But it was difficult to maintain the momentum of advance in productive efficiency. There was less to be gained by the removal of less efficient capacity, simply because the lowest productivity pits had nearly all been closed already. There was no new innovation quite so drastic in its effects as the replacement of hand by mechanized methods of getting coal, which had been almost accomplished. The hoped for transformation by methods of electronic control had had to be slowed down because so many technical problems were still unsolved. Until that check was overcome, improved technical efficiency probably depended most on the recent innovations in tunnelling methods and underground haulage, which had great potentialities, only some of which had yet been turned into achievement, and the greater co-ordination and more continuous operation of the techniques introduced in the sixties. But co-ordination and continuity require the co-operation of everybody involved, and neither the wage system currently in use nor the morale of the workers was inducing that co-operation wholeheartedly. There were still possibilities of achieving rising efficiency in the seventies, but the conditions for doing so were more difficult than in the previous decade.

A steady improvement in technical efficency was, of course, one of the major contributions needed for the move to stability and commercial viability in the coal industry; but much more was involved. Within the industry there had to be a movement away from all that belonged to the days when maximum output rather than minimum cost was the overriding aim. The requirement was summarized in 1968.

[1] NCB *Ann. Rep.* 1967–8, I, xi; 1968–9, I, xi; 1971–2, II, 4; 1972–3, 3 and 36.

> The Board are quite clear that their whole policy must be directed towards maximising the size of the industry by reducing costs and being thoroughly competitive in the 1970s. . . . The efficiency of the industry has been substantially improved year by year, but the full effects of this improved efficiency cannot be realised until the high-cost collieries are closed. They cannot be closed any more quickly than the present programme.

But the success of the policy depended also on outside influences. The NCB aimed to maximize the output of their cheapest production, but were concerned lest, having done so, they might find this coal excluded from its markets by national policy. They added that, 'above all, it is clear that the industry cannot be viable, whatever cost reductions are achieved, if coal is relegated in the electricity market to the role of a peak-load fuel.'[1] They might also have noted that the cost reductions which they achieved could be jeopardized by increases in other costs over which they had little control, to the detriment of their competitive ability.

The mixture of these various influences changed repeatedly between 1965 and 1973, and the nature of the adaptation required of the coal industry changed accordingly. On the whole, it became more protracted and more difficult. For several reasons there was a strong wish not to let the coal industry shrink below a fairly substantial minimum size. If it were too small the inescapable minimum of overheads could become a burden tending to push up unit costs, and there were conceivable levels of output so low that they would leave no employment for some regularly profitable capacity. Moreover, stabilization at too small a size would involve a contraction either so rapid or so protracted that morale could be badly damaged. No one of talent would want to join an industry going through such an abysmal experience, and many of the best already engaged in it would seek to get out if they could. Once an irreducible minimum size had been determined, the need was to remove capacity above this level, to ensure that what remained in production was the lowest cost capacity capable of providing the required quantities of the different types and qualities of coal, and to ensure that, on the average, the new level of output could be supplied at competitive prices which at least avoided a financial loss. This did not mean that there would be no loss-making collieries in operation—unavoidable short-term fluctuations in operating costs would make this almost impossible at any given moment—but it did involve an effort to keep down cross-subsidies as far as possible. Modifications to closure policy for social

[1] NCB *Ann. Rep.* 1967–8, I, 7–8.

reasons were likely to ensure that such clear lines of action could not be followed without some deviation, but it was not to be expected that the deviations would be very large. A much greater difficulty was that the indicated lines of action kept changing as average costs of production changed both for the coal industry as a whole and for particular parts of it, and as the required levels of coal output and prices changed, whether because of altered supplies and prices of other fuels, changing commercial demands, or arbitrary government decisions.

In 1965 the NCB drew up a five-year programme of colliery closures, which provided for the closure of 113 collieries plus those which would be recommended for closure by the Working Party on South Western Division,[1] then still in session. At that time it was intended to put all the closures of uneconomic collieries in the first two years, and to leave the last three for closing collieries whose reserves were becoming exhausted and for any additional closures of uneconomic collieries that might be necessary if the demand in 1970 had been over-estimated. But in 1965 the NCB still expected to sell about 180 million tons in 1970, though government plans at the time were using a figure of 174 million tons. The government had already been urging speed in the completion of the programme of closures,[2] and the NCB assumed that the end of a rapid rundown could be foreseen for around 1970. But in 1967 the government introduced a new target for a reduction of coal output to 120 million tons by 1975, and a further acceleration of the rate of closures immediately followed. Stability was again a receding prospect. After two more years of this the NCB thought the situation was threatening to become unmanageable. They urged:

> there must now be a slowing down in the number of closures so that confidence in the industry can be restored. . . . With continuing heavy redundancy and increased wastage (the consequence of low morale), the industry could not secure the necessary recruitment or transfers in the right places, and the rundown would be unmanageable and destructive of the industry's potential.[3]

Although capacity remained well above the government's proposals for the ultimate size of the industry, it had been brought far below what the NCB in 1965 had assumed as the likely size five years ahead. The rate of closures was markedly slowed down in 1969–70 and still more in 1970–1, when, by historic standards, it had become quite low.

[1] See above, p. 263.
[2] NCB *Ann. Rep.* 1965–6, I, 6.
[3] Ibid. 1968–9, I, 5.

So by 1970 it seemed as though, in some important respects, a measure of stability had, after all, been restored. In 1971 the NCB declared:

> the immediate and longer term prospects of the coal industry are now brighter than they have been since 1957.... Despite the rapid contraction of past years, the industry is now in good shape and is increasingly seen to offer a secure and promising future to recruits.[1]

The optimism of that statement was seized upon by the NUM and used to embarrass the NCB in the unhappy circumstances of the Wilberforce Enquiry less than a year later. Was it ever justified? The main contributions to optimism came from changed external conditions. 1970–1 was the third successive year in which the demand for coal exceeded output and stocks had to be reduced. In this year it seems highly probable that, if a little more coal could have been produced at home, it would have found a ready market. During the year the prices of both fuel oil and coal imports rose steeply. Home-produced coal got a general small price advantage over oil in the industrial market and was at most times and for most grades appreciably cheaper than available imported coal.

But some of the economic features of the British coal industry were less encouraging. The most serious was the trend of costs. After these had been held down most successfully for a decade, they began to rise rapidly from 1969, mainly because of the general inflationary pressure in the economy, which put up the prices of supplies and services to an extent that could not be contained by improvements elsewhere. This change of trend began while it was still possible to keep reducing the proportion of labour costs in total costs; but this was becoming more difficult because of the slowing down of the growth of labour productivity and the signs of increasing discontent with wage levels.

The rise of costs had several consequences. Although the hardening of the market for coal in 1970 enabled it to absorb a large rise in prices, and deep mining was therefore slightly more profitable in 1970–1 than in the previous year, the rising trend of costs showed that those competitive advantages that had appeared in 1970 might quickly disappear, as, indeed, they did. And as there remained a wide spread of production costs between coalfields, general rises in costs were likely not only to worsen the financial results of deep mining as a whole but to increase the number and proportion of unprofitable collieries. The elimination of the highest-cost capacity had helped deep mining to keep price rises

[1] NCB *Ann. Rep.* 1970–1, I, 5–6.

below the general rate of inflation and still at least break even, but many unprofitable collieries remained and fewer could be carried if their losses increased and the profits of the others were squeezed even a little. So a possible need to resume a larger scale of colliery closures was looming, just when it had been hoped that that traumatic experience was over. Vulnerability to the pressures of rising costs had been increasing for some time because of the growing proportion of coal that had to be sold in the lowest priced market, i.e. mainly to power stations. There was coal with superior qualities which were no longer wanted, with the result that it had to be sold as general purpose industrial coal and priced accordingly. This sort of disadvantage affected some collieries which had been financially successful—the shrinking of the market for coal to make gas and domestic coke which, *inter alia*, affected parts of Staffordshire and Yorkshire is an illustration—and so they could do less to offset unprofitability elsewhere.

A little more detail emphasizes the strength of these influences. In 1970–1, a reasonably successful year in that the NCB ended with a small surplus after meeting all their interest charges, only eight of the eighteen mining areas made an operating profit (all of them, except East Wales, in the Midlands and Yorkshire), and only the five areas in the Midlands made a profit after interest charges. Costs per ton in the worst area (Kent) were almost 94 per cent above those in the best (North Nottinghamshire) and the special qualities of Kent coal enabled only a fraction of this to be offset by higher prices. Losses per ton were, in fact, three times as high in Kent as in the next worst area, Scottish North.[1] With such variations the NCB did not look for a high degree of financial success from deep mining. They told the Wilberforce Enquiry in 1972 that they expected deep mining only to break even, and looked to opencast mining and the Coal Products Division to provide the surplus which would cover interest charges.[2] As deep mining was by far the largest activity, it might be thought that this piece of financial policy was a symptom of pessimistic realism. In fact, the other activities did not make quite enough profit to cover all the interest charges, and about £16 million was needed from the collieries to bridge the gap. This, as it happened, was a little less than the operating surplus of deep mines in 1970–1. In that year, of the 291 collieries, 144 (employing almost exactly half the workers) made operating losses, but the results of the 147 profitable collieries more than offset these.[3] Clearly, a lot of collieries risked

[1] Ibid., II, 46–55.
[2] Wilberforce Court of Enquiry, *Minutes of Proceedings*, 2nd day, 9 and 13.
[3] Ibid., 19.

moving into a hazardous commercial condition if the recent trend of costs continued or worsened.

The small number of six colliery closures in 1970–1 came from an original programme of twenty-two, which had also provided for numerous possible additions. As colliery losses remained widespread, with no early sign of cure, such restraint was bound to be questioned. Before the end of 1971 the NCB chairman was telling his colleagues that, because of rising costs and a weaker market, there must, after a two-year gap, be more colliery closures. These were already under discussion with the Department of Employment and posed serious problems because most of them would have to be in places with high unemployment. When the NCB in December 1971 completed their long term industry plan for submission to the Department of Trade and Industry, it provided for a reduction of 108 in the number of producing collieries over the next five years, with the numbers more than halved in Scotland, North East England, and the North Western Area. Even this rate of rundown was only on the basis of estimated results which assumed an average increase of 4 per cent in overall OMS, which recent experience suggested was very optimistic, although it was well below the projections made a few years earlier. The proposed closures were expected to reduce output capacity by 27 million tons. Though the collieries were those with the poorest production prospects, economic conditions had to be given appreciable weight. Only one-third of the collieries were expected to reach physical exhaustion within five years.

Every problem was exacerbated by the 1972 strike and the large increase in costs resulting from the Wilberforce settlement. The immediate estimate was that, at the new level of costs, 61 collieries would move from operating profit to loss, and this would leave the NCB with only 86 profitable collieries (out of a total of 291), and these collieries produced only 54.3 million tons (55.2 million tonnes) a year. Moreover, there was the added uncertainty because the strike had raised doubts about the security of supply to consumers. Although a couple of years of recovery could be expected to increase the size of the profitable sector a little, it was a discouraging position to have reached after all the drastic surgery and the many successes in increasing efficiency in the sixties. In 1965 it had been expected that adjustment would have been largely completed and conditions of stability achieved by the early seventies. When the early seventies had arrived it looked as though another spell of adjustment just as drastic was still needed; but adjustment on similar lines was impracticable. After the strike of 1972 it had to be taken for granted that wage trends like those which had helped the industry's results in the sixties would meet insuperable resistance if they reappeared, and that there would be much less union co-operation in any further contraction of capacity. Indeed, as the NCB started to replan their finances for the next few years, they became doubtful of earlier proposals and tentatively assumed that no collieries would be closed unless their reserves were exhausted.

Although 1972–3 was a year of recovery, with few serious interruptions and a financial outcome a good deal better than the forecasts made just after the Wilberforce settlement, the results showed how serious things were. NCB collieries produced a little less than in any year before the strike year of 1971–2, yet total coal output (including opencast) exceeded sales and the collieries made an operating loss of £88.8 million. In money terms, the lowest cost area had higher operating costs than the average for the whole industry two years earlier, and in Durham, South Wales, and Kent costs had gone even higher than in the strike year. In Kent the proceeds covered only 58 per cent of operating costs. The three very high cost areas, North Western, South Wales, and Kent produced only 14 per cent of the output but made 36 per cent of the losses. Only two areas, South Yorkshire and North Nottinghamshire, made an operating profit. The strength of the pressure to maintain sales is illustrated by the fact that North Yorkshire, which had temporarily displaced North Nottinghamshire as the lowest cost area, nevertheless returned an operating loss because its coal, mainly for power stations, was sold so cheaply.[1] There was a tendency to fix the prices from low cost areas so attractively that consumers (mainly the CEGB) could be encouraged to take rather more from areas with higher prices than they otherwise would have done. Yet, despite the losses, in the first nine months of 1973 coal prices at power stations were higher than those of oil, except in Yorkshire and the Midlands.[2] Nearly all the competitive advantages which had been appearing in 1970 had been swallowed up by the higher costs of the coal industry.

Not everything was adverse. The most significant change was probably in public policy. In the later sixties it sometimes seemed as though the government's basic attitude to the coal industry was that it was expendable and that the only reason for not getting rid of it was the intolerable social cost. That was no longer the case. It appeared likely that the demand for coal in the seventies, though much below the NCB forecasts, would remain somewhat higher than the government had

[1] NCB *Ann. Rep.* 1972–3, 3 and 36–7. [2] Ibid. 1973–4, 11.

projected. The price movements in 1970 and the subsequent political and economic discussions of the oil-producing states had indicated that assurance of cheap alternative energy supplies was not as great as had been assumed. And the experience of the 1972 coal strike had been a reminder of the disruption which could be caused by a breakdown in home production of coal. So government policy was once again prepared to provide a secure context in which the coal industry might operate and plan, as was shown by the Coal Industry Act of 1973 and various public statements. There was a price to be paid for this. Greater dependence on government policy and more frequent recourse to the government for larger sums of money could hardly fail to produce some tendency towards more detailed political interference, even if it were occasional rather than systematic. Already, before the financial losses of 1971–2, there had been moves in this direction by the Conservative government. They included indications that the NCB should shed some of their non-mining business (which helped to pay the interest charges), and the curious additions to the bill which was to become the Coal Industry Act 1971, including a clause which would have given the Minister drastically increased powers of direction to the board. These were matters which helped to persuade Lord Robens to refuse to serve for a third five-year term as NCB chairman.[1] In the end, the offending clause was withdrawn and, though the NCB disposed of most of their brickworks, the new policies on ancillaries were pressed only with modifications. The tendency remained for government to dabble occasionally in rather more matters below the level of strategic general policy.

Nevertheless, external conditions, both political and commercial, were by the end of 1973 offering more positive opportunities to the coal industry than for a good many years past. Within the industry there still were enormous difficulties to be faced, as shown particularly by the level and structure of costs and by the rapid relapse of many miners' wages to conditions which the 1972 strike and settlement had been expected to cure. But there were also some large opportunities: for the industry to spread its best practices, to make fuller use of its new technology and push the innovations further forward, and to exploit unused resources with the maximum efficiency. It was a precarious balance of risks and opportunities, handicaps and advantages; but at least the conditions for the development and operation of the coal industry were appreciably different by 1973 from those of the previous fifteen or sixteen years.

[1] Robens 1972, 314 and 320–4.

CHAPTER 7

# Competitive Opportunities and Economic Recession 1973–1982

## i. Competition and costs: the oil price rise and the 1974 strike

In 1973 there were several events which strengthened the indications, evident from 1970, that the long decline in the use of coal might be ending, with the prospect of much more secure conditions for the operation of the industry. There were others which indicated difficulties that had to be overcome if advantage were to be derived from that prospective security and, indeed, if the security itself were to be maintained.

For the first half of the financial year 1973–4 demand from nearly all markets increased. To the end of September power stations consumed 4 million tons of coal more than in the previous year and added 4½ million tons to their stocks, which reached a record of 20 million tons. Demand from coke ovens was also high and so was the domestic demand for manufactured smokeless fuels.[1] In the spring of 1973 Yorkshire and East Midlands coal was just about on level competitive price terms with oil in the power stations. The NCB recognized that there was a very wide range of possibilities about the future movement of relative prices, but reckoned that, on almost any reasonable estimate of probabilities and any basis of calculation, the national average cost of coal would be less than that of oil not later than 1980, and that plans ought to be made accordingly. If the rate of increase in the real cost of coal could be kept to the average rate of the period since 1960, or if oil prices rose at high rates that were beginning to be widely expected, a general competitive advantage would be achieved a good deal earlier.[2]

In fact, the transformation came much sooner and was more drastic than expected. The increasingly assured semi-monopoly position of OPEC had created assumptions that the governments of oil producing

[1] NCB *Ann. Rep.* 1973–4, 5 and 11.

[2] The use of the average rate of increase since 1960 was misleading as an indication of trend because increases were slight from 1960 to 1969 and much faster thereafter.

states would continue to demand and obtain higher payments from the oil companies, and that the organization would be able to control supplies sufficiently to maintain higher prices to purchasers of crude oil. The outbreak of the Yom Kippur War on 6 October brought a demonstration that the Middle Eastern producing countries were prepared also to use oil supply as a political weapon. Within days they announced that they would cut output and that the price of crude oil would rise by 70 per cent. This was only the first of a rapid series of rises. Between October 1973 and the spring of 1974 the price of crude oil increased roughly fourfold and, though the costs of distribution and refining did not rise similarly, the price of fuel oil in the UK was about doubled, or rather more. In March 1974 the delivered cost of fuel oil for electricity generation was the equivalent, on average, of about 6p per therm, which was far above the average level for coal. It was then reckoned that general purpose coal sold to power stations had an advantage of £5.50 per ton in the central coalfields and between £2 and £3 elsewhere. Some of this price difference was offset because a good deal of coal-fired generating capacity was old and of low efficiency. For this reason the NCB judged that coal would keep the competitive edge only if its price, therm for therm, were no more than four-fifths that of fuel oil. A greater margin than this had emerged generally, though there was not a lot to spare in South Wales or in the case of coal delivered by sea from north-eastern collieries to south-eastern power stations.

These changes had both immediate and more lasting consequences. Immediately, there were attempts, where equipment permitted it, to burn less oil and more coal. For the longer term there was a reconsideration of the assumptions underlying national fuel policy. Recent confidence in both the security and cheapness of oil supplies was shaken. It had become prudent to plan for a larger supply and use of alternatives to oil.

The British coal industry could hardly fail to derive some benefit from the new circumstances, which provided strong political arguments for governmental promotion of increased coal use and for financial provision for the improvement of the industry. But there were no grounds for expecting the benefit to accrue quickly, or to be very large, or necessarily to remain certain for an indefinite period ahead. For one thing, there were other competitors better poised for rapid adaptation. The gas industry had virtually completed the switch to natural gas and it was national policy to go for rapid use of North Sea reserves, encouraged by relatively low prices. Gas and coal were not interchangeable for all uses, and there was reluctance to see more than a small proportion of gas reserves consumed in non-premium uses such as the firing of industrial and power station boilers. But a lot of interchangeability was possible and, in practice, over the ensuing years gas gained much more than coal at the expense of oil. Some potential competition was also provided by foreign coal. Imports were permitted, all coal producers shared the competitive effects of the oil price revolution, some countries produced coal more cheaply than Britain did, and some producers appear to have been ready to sell abroad more cheaply than at home, whatever their costs of production.

On the other hand, there were serious limitations within the British coal industry. For years it had been so continuously cut back, and so restricted in the amount and range of its investment, that it was in no position to cater for much market expansion, and could not be until there was a renewal of physical capacity that would have only very slight effects for several years. The risk was also present that British coal might fail to keep the advantages which changes in oil supply had provided. The 1972 strike had damaged the faith in security of supply which the domestic location of coal reserves should give, though in 1973 the suspicions were beginning to lessen. And for the last few years there had been obvious difficulty in controlling the steep rise in the costs of coal production. This problem was being made no easier, for the rising rate of inflation was making some inputs more expensive and was creating much uncertainty about the level of living costs in the near future. Such uncertainty provoked demands for large rises in money wages, not only in the coal industry but throughout the economy; and a competitive scramble for higher pay produced many anomalies in wage differentials between industries. It happened that underground workers in the coal industry were coming out rather badly in these changing comparisons. It was only to be expected that the unions would seek to alter this. So the coal industry, which had hoped that its labour costs had been put on a sound and stable basis after the Wilberforce settlement, was faced with the need to work them out all over again.

It was, indeed, through wage demands that the first great challenge to the new position of the coal industry presented itself. The quick settlement at the end of March 1973, with the NUM's acceptance of the NCB's offer of the low legal maximum permitted when stage II of the government's incomes policy replaced stage I, resulted from a coalfield ballot. The national executive of the NUM had recommended the rejection of the offer and asked for authority to call a national strike, but the ballot went against the national executive by 143,006 votes to 82,631.[1] This result may have given the impression that mineworkers were in a pacific and co-operative mood, but in July the national conference of the NUM approved a claim bigger than that unsuccessfully put forward earlier in the year. The new claim was for minimum weekly basic wages of £45 for faceworkers, £40 for other underground workers, and £35 for surface workers, with appropriate differentials for craftsmen and others; a substantial increase in allowances for working afternoon or night shifts; and a reversion to the pre-1972 settlement date of 1 November. This claim amounted on average to about 31 per cent, and far exceeded anything allowed by the incomes policy. It also involved, yet again, a bigger rate of increase for surface workers than faceworkers, although it was the faceworkers whose pay, relative to that in other industries, had slid back to the position before the 1972 strike, whereas the surface workers still retained much of the relative improvement they had gained. The rate of increase claimed for workers elsewhere underground was, however, highest of all. The NUM sought to justify the claim by pointing to the relative decline of coal industry wages as a whole (ignoring the difference of trends for different grades of worker) and by the increased amount of voluntary wastage among mineworkers.

The NCB also were anxious for wage arrangements that would encourage adequate recruitment and retention of labour where it was needed, and anxious also to preserve both industrial peace and the relative pay position of mineworkers. But they were bound to work within the government's incomes policy, stage III of which was in force when formal negotiations were being conducted. The NCB offered all they thought was legally permissible: a general wage increase of 7 per cent or £2.25 per week (whichever was the greater), a 17p allowance for each hour worked between 8 p.m. and 6 p.m., an additional 1 per cent for the improvement of holiday pay, a 'threshold' agreement which would give an extra 40p per week if the cost of living rose by 7 per cent and a further 40p for each additional percentage point rise in the cost of living, and a possible productivity payments scheme if one could be negotiated within the terms of stage III. The NUM expressed willingness to sign a threshold agreement, but rejected the main part of the offer.

The government got involved directly at a very early stage. While stage III of the incomes policy was in preparation the Prime Minister, Edward Heath, accompanied by the head of the civil service, Sir William Armstrong, had held a secret meeting on 16 July 1973 with the President

[1] Gormley 1982, 123.

of the NUM, Joe Gormley. The very holding of such a meeting helped to confuse all later negotiations. The government obviously recognized that politically the most important and difficult group to satisfy within the incomes policy would be the miners. This early meeting was bound to be taken as an indication that the effective party in negotiations would be the government rather than the NCB. At the same time the impression was created that, if the government could do a deal with the NUM President, he, and he alone, could carry his union with him. This was false optimism on the part of the government. In fact, at the July meeting Mr Gormley dropped the hint that special payments for unsocial hours might be a way to give the miners both the absolute increases and the restored differentials over other industries that they sought. When provision for limited extra payments of this kind was included in the incomes policy in October, it failed to mollify the NUM because it did nothing for differentials. Yet illusions apparently persisted that some sort of deal between government and union could be reached, although every contact between these parties left at least one of them disappointed. In such conditions the negotiating role of the NCB was bound to be very difficult, all the more so because government activities were pursued in a rather dispersed manner, which added to the confusion. In particular, the Department of Employment, which might have been expected to play a major part, was at first left on the periphery.

The July meeting had been secret and personal, but the government openly appeared on the stage early in the negotiations when, on 23 October, the Prime Minister received a delegation from the NUM. This did nothing to help labour–management relations and also left little confidence in the government's willingness to deal impartially with the points at issue. The NUM got no concessions from the Prime Minister, and on 26 October received authority from a national delegate conference to call a national overtime ban. It was made plain that the ban would be absolute and no NUM member would do any work at weekends, when essential maintenance and safety work is normally done. Although it was found possible to reinterpret some of the rules of stage III so as to permit a very slight improvement in the NCB offer, this had no effect. The overtime ban came into force on 12 November, although some collieries in Leicestershire and Warwickshire started it a week earlier. The NUM immediately added several supplementary demands to their previous claim. These related to such things as sick pay, holiday pay, retirement lump sums, and the inclusion of clerical staff in any productivity scheme, and the NCB were willing to do something about most of them. One demand they agreed to was the issue of a statement of their intention to increase mineworkers' wages in real terms in future years.[1]

In the ensuing weeks and months there were various meetings involving the NCB, the NUM, the Pay Board, the Prime Minister, the Secretaries of State for Employment and for Energy, in various combinations. The discussions involved the permissibility of various additions to the pay offer. A ruling was obtained from the Pay Board about the kind of productivity agreement that might be allowed. In late December representatives of the NCB and NUM jointly had meetings with the Pay Board about the possibility of pay for waiting and bathing time at the beginning and end of the shift, and a ruling was given that if it were shown that such time was regularly required, extra pay could be approved for any of it that was outside a 40-hour week. The NCB conducted a hasty exercise at forty-two collieries which revealed large variations in the time involved, but produced an average of 40 minutes per shift, which agreed with the findings of an earlier investigation. This, however, only brought up the total week to 40 hours 25 minutes, so that the Pay Board ruling permitted only 25 minutes to be paid as overtime, an addition of 60p per week, and the NUM were not interested in anything so little.[2]

There were, in fact, few changes of much substance in the offer under negotiation, which the NUM in late November, after seeing the Prime Minister again, had refused to put to a ballot. The government pay policy was a barrier to any additions and the government, fearful of a general uprush of wages, had no intention of relaxing it for the time being. The Prime Minister on 28 November told the NUM that if they would accept the maximum offer that was available under stage III, the government would start a wide-ranging review of the coal industry and the position of the miners. William Whitelaw, who had recently become Secretary of State for Employment, told them much the same on 20 December, promising that immediately after acceptance, the government would consider with the union and the board 'the pay arrangements appropriate to a modernized industry in the longer term'. There

[1] Gormley 1982, 127–30.

[2] Ibid., 132–6 puts a lot of stress on payment for waiting and bathing time as a lost opportunity to settle the dispute. He accuses the NCB of producing figures to show that waiting and bathing took only five minutes a day, but this was not so. The five minutes a day was all that the Pay Board ruling would allow to qualify for extra pay. Gormley also accuses the Leader of the Opposition, Harold Wilson, of taking up the issue and misusing it for political purposes.

seems to have been a belief in some government circles that the current offer was sufficient to restore miners' wages to the position established by the Wilberforce settlement, and that the refusal of the NUM to accede to the government proposal must therefore be an act of political defiance. If that was, indeed, the belief, it was incorrect on both counts.

In the circumstances, the dispute was bound to be seen as having a political aspect. Government policy limited the room for negotiation, ministers had involved themselves personally very early in the dispute, and members of the public experienced increasing restrictions as a consequence of it. The government declared a state of emergency just after the start of the overtime ban and there were measures to reduce the consumption of coal and electricity. Coal stocks declined gradually from their high level. In mid-December the government announced that industry would be put on a three-day week from 1 January. At that time the Department of Industry and the NCB agreed that, at the rate of depletion since the start of the overtime ban, power station stocks would fall by the end of January to 7 million tons, the level below which it would become difficult or impossible to maintain general continuity of electricity supply. With the new restrictions, it was estimated that power station stocks would still be 8 million tons at the end of March, after which the coming of spring should make things easier.

In the early weeks of 1974 the dispute was in a state of deadlock and the public in a state of frustration. The TUC made an attempt to secure progress by offering that, if the government would permit a relaxation of the incomes policy to get a settlement of the miners' dispute, no other union would use that settlement as an argument in support of its own claims. This offer was made three times, at a meeting of the National Economic Development Council on 9 January, at a special TUC delegate conference on 16 January, and by a TUC delegation to the Prime Minister on 21 January. It was rejected by the government, no doubt on the sceptical view that it was the TUC which made the offer but the individual unions which did the negotiating, and the former could not guarantee the conduct of the latter.

Public statements by members of the government were more and more taking the line that the government was carrying out its duty to govern and was resisting coercion, and this line made any negotiating movement almost impossible because of the 'loss of face' involved. Coal stocks were holding up rather better than expected, so that pressure for a settlement was not overwhelming. The NUM therefore increased the pressure. On 24 January the national executive resolved by 16 to 10 to hold a national strike ballot, and the next day gave the NCB provisional notice of a strike from midnight on 9 February, even though legislation currently in force required 30 days' notice of a strike. The ballot was held on 31 January and 1 February and the result announced on 5 February. There was an 81 per cent poll and the strike call was supported by 188,393 to 44,222. The NUM announced that if there was no satisfactory offer, the strike would go ahead as notified. The response of the government seemed odd to large numbers at the time and to even more in retrospect: on 7 February it was announced that there would be a general election on 28 February, and the miners' strike was made the main issue of the campaign, with 'Who Governs Britain?' as the theme question. This was odd because the government had a working majority and was in no difficulty about carrying on. It was odder because it was proposing a purely political (or constitutional) attempt to answer an almost purely economic question: basically (with a few additional trimmings) what the miners were seeking was the restoration of the wage position they thought had been assured to them, with general (including government) support, by the Wilberforce settlement. It was oddest of all because the government had already set in train alternative moves towards a settlement, thus making an electoral verdict on the strike superfluous: the case of the relativities of mineworkers' pay was referred to the Pay Board for examination and recommendation as quickly as possible. The NUM were told by the Secretary of State for Employment that the government undertook to backdate to 1 March any recommendations made by the Pay Board. The NUM on 12 February agreed to give evidence to the Pay Board, but only on the understanding that they would not necessarily be bound by its findings.

So the strike and the election campaign went on side by side, neither of them very pleasant experiences, with much recrimination. For the coal industry there were a few mitigating conditions. The clerical workers in APEX decided not to strike. The Pay Board ruled that a request by NACODS (which the NCB supported) that officials should have a 10 per cent differential over the pay of chargemen in face teams was permissible and could be applied from 7 November 1973. This mollified NACODS a little, though before the NUM strike was over, NACODS decided to hold a ballot on possible industrial action to back up their main wage claim. In fact, the officials and colliery managers worked together throughout the strike, as they had during the overtime ban, to keep the pits safe and in workable condition, and co-operation from the NUM was better than in the 1972 strike.

In the middle of all this, from 18 to 22 February, the Pay Board held its enquiry. Evidence was taken from the NCB, the mining unions, the TUC, the CBI, and government departments. A strange feature was that on 21 February the Pay Board published figures which purported to show that the miners were entitled, under the existing stage III of the pay policy, to significantly more than they had been offered. The figures published were calculated on a different basis from that used in the Wilberforce Report and therefore adopted by both the NCB and the NUM in their evidence to the Pay Board. The publication provoked some press comment suggesting that it must have been deliberately intended to damage the government's election campaign. It was certainly extraordinary that the Pay Board should make such a public statement while the enquiry was still in progress; and, in view of the number of discussions which the parties involved had held with the Pay Board's officials while the overtime ban was in force, if they were wrong about what was permitted, this would seem to reflect no credit on the Pay Board. It was an episode that might have reduced the acceptability of the Pay Board's report, which was made available on 4 March. In fact, the report proved very helpful. It not only recognized the case for an improvement in miners' wages relative to those in other industries, it concentrated on the difficulties caused by the NUM's continued insistence on narrowing the differentials that underground workers had over surface workers, while the NCB had stressed in evidence that it was in underground occupations that there were serious shortages of manpower. The report stated: 'Our recommendations have dealt with the underground workers because we judge that it is their pay relative to that of all other workers which needs to be changed. We are conscious that in doing so we have altered the internal relativities between underground and surface workers.'[1] In fact, they were not able to alter them as much as they wished, for the report indicated a recommended, not an imposed, settlement, and it was followed by negotiations.

By the time the Pay Board Report was published the general election was over. No party obtained a clear majority, but Labour was the largest party and came into office when the Conservative government resigned. The new Secretary of State for Employment, Michael Foot, indicated that negotiations for a wage settlement in the coal industry could proceed without regard to the pay policy restrictions of the previous government and ought to be concluded quickly. A settlement was reached on 5 March and recommended, together with a proposal to

[1] Pay Board, *Special Report: Relative Pay of Mineworkers*, Cmnd. 5567 (1974).

resume normal working with the first shift on 11 March, by the national executive of the NUM to a proxy vote of the areas. The latter gave unanimous backing to the recommendations.

The settlement included a threshold agreement, improved holiday pay and retirement benefits, and a commitment to an extra week of paid holiday a year later. On the main pay claim the NUM refused to follow the Pay Board more than a little of the way on internal differentials. The Pay Board had proposed even higher rates of pay for faceworkers than the NUM had asked for, but the NUM settled for its original claim for faceworkers and got for other workers more than the Pay Board suggested, though less than the original claim. Surface workers still got a bigger percentage rise than faceworkers.[1] The details are in Table 7.1.

Table 7.1. *Mineworkers' pay claim and settlement, 1974*
(£ per week (basic))

| | 1973 (actual) | Claim | Settlement |
|---|---|---|---|
| Minimum surface | 25.3 | 35 | 32 |
| Minimum underground | 27.3 | 40 | 36 |
| Coal face (NPLA) | 36.8 | 45 | 45 |

*Source*: NCB *Ann. Rep.* 1973–4, 7.

The settlement restored the gains in external relativities obtained from the Wilberforce settlement and did a little more for faceworkers and a lot more for surface workers. The relativity of surface workers' wages to the average for all manufacturing industries had previously been greatest in 1956, but in the autumn of 1974 it was far higher than it had been then; but faceworkers had recovered only the relative position of 1969. If the comparison is extended further back the reduction in the advantage of faceworkers is more striking. Only in South Wales were the hourly rates of faceworkers higher in proportion to the manufacturing average of their region than they had been in 1960. For the country as a whole faceworkers' hourly rates had been 180 per cent of the average for manufacturing industries in 1960, and after the 1974 settlement they were 156 per cent; but in 1970 they had been only 137 per cent and

[1] On the whole course of the overtime ban, the strike, and the settlement see Gormley 1982, 123–45. It was a continual preoccupation of NCB meetings from early October 1973 to early April 1974. Summary references and figures are in NCB *Ann. Rep.* 1973–4, 3, 5, and 7. A fuller and very useful summary is provided by Fay and Young 1976.

this was perhaps a better remembered comparison. By contrast, the national ratio for surface workers, which had been 85 per cent in 1960 and 84 per cent in 1970, was 112 per cent after the 1974 settlement.[1]

This settlement made the mineworkers more satisfied with the fairness of their wages, but it did not put an end to the problems arising from wage levels. Quite apart from the possible inadequacy of faceworkers' differentials as a stimulus to maximum effort, it was necessary to try to avoid a repetition of what had happened after the Wilberforce settlement, when the benefits arising from a better satisfied labour force were removed by wage movements and wage limitations which arose outside the coal industry. There were immediate risks because the Labour government permitted large wage increases fairly generally. At first these arose mainly as a result of payments due under the threshold agreements encouraged by the previous government, and as the coal industry had such an agreement, wages there maintained their position fairly well, though not completely. But in July 1974 the government abandoned statutory pay restraint, wage rises generally accelerated, and between August 1974 and March 1975 the ratio of coalmining to average manufacturing wage rates fell by 6 percentage points. A 19 per cent basic wage increase in March 1975 enabled the mineworkers to remain ahead as other industries made settlements. But subsequent agreements between the government and the TUC for voluntary limitations of wage rises first to £6 per week and later to 4½ per cent had reduced the relative advantage of mineworkers' wages substantially by the middle of 1977.[2] What had become evident was the dependence of smooth working in the coal industry on high rates of increase in money wages. This was a sharp contrast with conditions in the sixties. Much of the difference simply reflected the much higher rate of general inflation, but not all. There was a risk of the destruction of the recently improved competitive position of coal, especially as the competition depended to a large extent on prices set abroad and not necessarily influenced by the same rates of inflation. So there was a very strong need to secure greater productivity of labour to offset (and improve on) the trends in wages, and to contain other costs in the coal industry as tightly as possible.

Just how much difference the 1974 wage settlement made to costs is impossible to state precisely because there are costs of the preceding disruption to take into account, and because there were continuing additions—the settlement, for instance, included a threshold agreement

[1] Handy 1981, 175–7 and 185–6.

[2] Ibid., 186–7.

which was triggered within a very short time by the rise of the cost of living index. The NCB had costed the original NUM wage claim (most of which was achieved), with consequential increases for other grades and for officials, at about £144 million for a full year, and added £52.5 million for other items, such as night shift allowance, holidays, sick pay, and threshold payments—all this at a time when colliery turnover was normally of the order of £900 million. But by the time a full year had passed, other changes had occurred. Costs in 1973–4 were increased and distorted by the effects of the overtime ban and the strike, which together directly reduced output by 20 million tons, while many items of cost were maintained and productivity reduced.[1] Nevertheless, despite the distortions, the annual changes in money costs of production give some indication of the exceptionally large effect of wage rises in 1974. Total costs per unit of output in deep mining rose by 23 per cent from 1973–4 to 1974–5, but the wages element rose by 31 per cent.[2] This is only partly explicable by the fact that the reduced output of 1973–4 had caused a disproportionate rise in non-wage items of cost in that year.

Because of external events these movements of costs were not too serious provided they were not a matter of annual repetition. An overtime ban and a national strike, interrupting fuel supplies and greatly adding to costs, for the second time in three years, were bound to leave deep suspicions about the reliability of the coal industry as a supplier of energy at a commercially acceptable price. But for the time being the suspicions were not directed only at the coal industry. It was while the overtime ban and the strike were taking place in the winter of 1973–4 that the oil suppliers were adding far more to costs than the miners, and public opinion began to worry about the possible political insecurity of oil supplies. So for the time being, it was possible for disillusioned consumers to regard coal as, at worst, the lesser evil and to be prepared to wait and see how things turned out in the next few years. Had wages gone on leaping ahead of other prices, and had the biggest increases in other wages charges coincided with those in wage rates, then disillusion would have grown and cost trends might have become commercially destructive. But, partly because the wage settlement of 1974 was something special to correct an accumulated shortfall, these things did not happen. Wages and wages charges rose to 51½ per cent of total costs of

[1] NCB *Ann. Rep.* 1973–4, 7.

[2] Calculated from MMC, *National Coal Board*, II, 25. These figures relate to wages only and exclude other wages charges.

deep-mined production in 1974–5, instead of the immediately preceding 50 per cent, and were just over 52 per cent in 1975–6, but thereafter the proportion declined.[1] Nevertheless, the absolute rise was rapid: in money terms nearly threefold in five years from 1972–3. If the rise in the ratio was arrested and reversed, that was partly because other costs were rising so fast, and the smallness of the changes in the structure of costs indicated the disappearance, at least temporarily, of the old role of labour costs (falling as a result of greater efficiency) in offsetting those movements. Most of the rapid rise in money costs came from general inflation in the economy, but not quite all of it. So there was a risk of losing some or all of whatever competitive advantages had appeared. The coal industry could not rely on the oil suppliers every year pushing up their prices in huge leaps. It needed a resumption of smooth working and rising efficiency if it was to exploit the opportunities which external conditions had presented.

## ii. Organization

In dealing with its problems the coal industry had established organizations that showed a high degree of continuity. Major changes in personnel and institutions were few and came about gradually. On the national board Sir Derek Ezra remained chairman until July 1982 (he became Lord Ezra at the end of that year), and the policy of having a mining engineer as deputy chairman, which had been resumed when he became chairman in 1971, was maintained. W. V. Sheppard, who had been appointed then, retired in 1975. In October 1973 Norman Siddall was also appointed a deputy chairman and he continued in that office until he became chairman in 1982. In the NUM, after the elections of 1968 and 1971 there was more than a decade of continuity. The president, Joe Gormley, was in office until 1982 and the secretary, Lawrence Daly, until 1984. There was an election for vice-president in 1973, when Michael McGahey (the defeated presidential candidate) was successful, and he continued alongside the other two and remained vice-president after their retirement.

The national board had become, and remained, a body whose full-time members had mostly spent their entire careers in the coal industry

[1] Calculated from MMC, *National Coal Board*, II, 25. The arrangement of the figures differs a little from those used earlier in this book, and consequently wages and wages charges appear as a very slightly higher proportion of the total, but the difference is not significant and the trends within this period are unaffected.

and, as time went on, there was a growing preponderance of mining engineers. The choice of part-time members was also somewhat more specialized than it had been. It became the regular practice to include people from the chief coal consuming industries, always someone from electricity supply and, for a few years, someone from the steel industry as well. In the late seventies and early eighties there was also regularly at least one NCB Area Director among the part-time members of the national board. The members of the board, as before, were not appointed with specific functional fields of responsibility, but while in office, every full-time member was assigned several activities for special attention and guidance, though executive responsibility remained with the appointed heads. The professional specialism of board members ensured a close and detailed understanding of the range of specific problems in the coal industry and of the practical possibilities and limitations of different approaches to them, to an extent which did not exist in the earlier years of nationalization. On the other hand, there were the potential risks of institutional inbreeding, which can have results in either direction. The coal industry has so exceptional a degree of individuality that it needs a directing board with a large stock of professional expertise of its own kind, but there are other needs. When the Monpolies and Mergers Commission enquired into the coal industry in 1982 it clearly thought that the practice of recruiting board members from within the organization had gone too far, and recommended the appointment of additional full-time members with a wider range of experience in industry and commerce. It also wanted all part-time members to come from outside the coal industry.[1]

The formal structure through which the coal industry was administered was very much what had emerged from the reorganizations of the late sixties,[2] and there was a strong wish that any further change should come through piecemeal adaptation to specific circumstances. It was the easier to keep to this evolutionary character because the detailed studies which were set up at the instigation of the 1968 working party, especially an enquiry into the organization and staffing of areas and collieries, led to the modification of the more standardized patterns that had been adopted in 1967. It was recognized that individual differ-

[1] MMC, *National Coal Board*, I, 371.

[2] For an account of the structure as it was in the mid-seventies, see D. G. Brandrick, 'Organisation of the Coal Board', *Colliery Guardian*, Dec. 1975 and Jan. 1976. See also NCB *Ann. Rep.* 1977–8, 20–5, which, besides giving descriptive information, restates and updates the principles of managerial structure set out in the board's report for 1948.

ences in the conditions and problems of areas, and still more in those of collieries, would be better accommodated with some freedom to vary staffing arrangements and responsibilities. So there was a move to prescribe formally a simplified general structure, within which some variations could be 'tailor-made' to match local needs; and as the needs changed, the structure could be further modified or brought back nearer to standard form.

In the structure of areas there were, indeed, what were nearer to two standard forms, each with local modifications. This duality resulted from the mergers which were made in order to carry further the adaptation of the scale of business units to the reduced size of the industry. A single Scottish Area and a single South Wales Area had been formed in 1972 and 1973 respectively. The process was carried further on 1 April 1974 when North Western and Staffordshire Areas were merged into a new Western Area, and a North East Area was formed by combining the former Northumberland, North Durham, and South Durham Areas. The units resulting from these four mergers were each responsible for the whole of one or more distinct coalfields and became known as 'coalfield areas'. When they were formed they took over nearly all the services which had been provided for more than one area by regional outstations.

The remaining areas, four in Yorkshire and four in the Midlands, each covered only a part of a larger coalfield, so it was practicable for them to continue to use services provided regionally for several areas in close proximity. Their area administrative structures and functions were therefore rather less comprehensive. Among these, the South Midlands Area was something of a hybrid, and became more so from 1 December 1975, when the three Kent collieries (previously under a General Manager who reported directly to national headquarters) were added to its responsibilities.[1] This area had complete coalfields within its boundaries, but the unworked extension of one field linked up with South Nottinghamshire; in terms of output and manpower it was rather smaller than any of the 'coalfield areas'; and it was geographically convenient for some services to be shared throughout the Midlands. So South Midlands was not organized as a 'coalfield area'.

The survival of some regional services was sometimes felt as an irritant by area managements, who would have liked to have them under their own control. But reconsideration from time to time suggested only modifications of detail. The mergers that produced the large 'coalfield

[1] NCB *Ann. Rep.* 1975–6, 16.

areas' provided useful savings in staffing and overheads and demonstrated that there need be no reduction in the effectiveness of control if collieries were at quite long distances from area headquarters; and regional outstations were a more economical way to provide some services than a more dispersed and smaller scale operation would have been. The lines of development indicated by these changes were carried no further after 1975, though a few years more experience and observation of them might well suggest some fresh thoughts about the optimum scale of the subordinate formations in the NCB organization.

As would be expected, even though there was no drastic reorganization, there were developments in departmental structure and procedures, both at headquarters and area, in attempts to remove known weaknesses and to find more effective ways of dealing with changed circumstances. The one which received most public attention, though not necessarily the most important, was connected with an attempt to arouse public disquiet about alleged waste and corruption in NCB purchasing activities. In the spring and summer of 1972 two NCB employees, Dr R. D. Leigh and Mr W. A. Grimshaw, made allegations about waste of money through excessive prices paid for powered supports and spares as a result of inefficient procedures and malpractice. Dr Leigh was in the Operational Research Executive; Mr Grimshaw, who had been in the coal industry since he was 14, had been an area purchasing and stores manager, first in Cumberland and then in Northumberland, until 1967 when his post disappeared in the reorganization of that year, but he was kept on the strength without a specific managerial function. A critical internal memorandum by Dr Leigh was leaked to the press and two months later Mr Grimshaw wrote to the NCB's external auditors, Thomson McLintock and Co. His allegations included the claim that the placing of orders was influenced by shareholdings which the responsible officials had in supplying firms.

The NCB had the share registers of the relevant companies examined and found no trace of any holdings by any existing members of staff. They also asked Thomson McLintock to carry out a special investigation. Their report was ready in January 1973. It rejected all the allegations but recommended some minor procedural changes.

The affair had attracted a good deal of sensation-seeking comment in the press and on the television. The NCB were anxious to counter this and anxious also to discover any practicable solutions to various difficulties of price negotiation of which they were well aware. One problem was that the NCB had a monopsonistic position in the market for many items of mining equipment and that there was a monopoly or oligopoly of the supply of most of the most expensive items; and neither buyer nor seller had much information about the effects of any particular orders on the business of the other. Another was that cost expectations at the time an order was placed could be falsified by variations in the number of repeat orders or, above all, in the demand for spares. An increase in these purchases, by spreading overheads over longer production runs, could dramatically increase suppliers' profits. This was a problem long familiar to all large undertakings, and they usually tried to deal with it by renegotiating prices for later orders or, occasionally, by negotiating retrospective refunds.[1] The existing NCB policy in dealing with monopoly suppliers of equipment and spares was to try to ensure that over a period of several years their profit did not exceed 10 per cent on turnover, and if it was necessary in order to achieve this objective, to negotiate a refund. This was not always easy because there were limits to the exactitude of information on costs available to the purchaser. But in 1973 a refund of £1.3 million in respect of powered supports was obtained. Grimshaw regarded this as showing the truth of his assertion, though such an argument was quite illogical.

To try to find an answer to the problems, the NCB commissioned an independent enquiry into purchasing procedures. The three men who conducted this were Sir William Slimmings, who (besides being a partner in Thomson McLintock) was chairman of the Review Board on Government Contracts; Sir Henry Benson (of Coopers and Lybrand), who represented contractors before that body; and Mr E. C. T. Humberstone, Executive Director (Purchasing), Hawker Siddeley Aviation Limited.[2] The report of the enquiry was completed in March 1974. It made numerous recommendations about departmental organization and procedures, many of them new, some of them (for example, defining more carefully the responsibilities of area purchasing and supply managers so that they could not inadvertently commit headquarters commercially) confirming changes already made. It proposed to deal with the monopoly problem by requiring equality of information

[1] Cf., for instance, on high profits from long production runs and on the negotiation of rebates, W. Ashworth, *Contracts and Finance (History of the Second World War: UK Civil Series)* (1953), 100–4, 126, and 170–7.

[2] The appointment of Mr Humberstone was given adverse publicity in *Private Eye*, on the ground that one of the supplying firms of mining equipment was a subsidiary of Hawker Siddeley. This objection had already been pointed out by the NCB to the Department of Energy, which had proposed his name. As the Department thought the objection immaterial the NCB agreed to the nomination.

between supplier and purchaser, so that all spares could be correctly priced in themselves and related in price to the parent equipment, and that each item should be correctly priced in relation to other equipment. Although the NCB thought it impossible to act on this recommendation for all items supplied monopolistically, because of the time, staff, and expense required to verify all the relevant information, they felt bound to act on it in relation to selected items. Before deciding on the appropriate steps they consulted British Rail, the CEGB, the Post Office, the British Gas Corporation, and the Ministry of Defence about the practices of those institutions. They found enough variety to suggest that there is no one final solution to the problems. Nevertheless a large number of detailed improvements came out of the report, so many that there was a welcome for its last words: 'if our recommendations are accepted, the Purchasing and Stores Department should be afforded an adequate period to deal with them free from interruption of further inquiries.'

In practical terms this was the improvement to which the 'Grimshaw affair' was one stimulus, though, in fact, in the later seventies there were further evolutionary changes in the Purchasing and Stores Department, which cumulated to something quite substantial, especially in response to the continuing need for greater centralization. The 'Grimshaw affair' itself rumbled on. The Select Committee on Nationalised Industries in 1973 decided to take up the allegations about the purchase of powered supports and reported in June 1974. The Select Committee concluded that 'there was very little substance in the specific charges made'. It also took a different view from Sir William Slimmings and his colleagues about the best way to deal with the problems of monopoly supply. Grimshaw also pursued his allegations in evidence to the Royal Commission on Standards of Conduct in Public Life. He was declared redundant by the NCB early in 1975 and this produced claims of victimization, which were eventually referred by the House of Commons to the Committee of Privileges in February 1976. The latter reported that it was 'satisfied that there was no indication of Mr. Grimshaw's treatment being adversely affected by his having been a witness before the Select Committee on Nationalised Industries'. A year later Grimshaw was still pursuing the subject in his own privately printed memoir, but it had long lost its news value.[1] Even in its early stages it

[1] The 'Grimshaw affair' and the enquiries and changes involving NCB purchasing and stores produced a record extending over several years. Besides the reports of the SC on Nationalised Industries and the House of Commons Committee of Privileges, there was an account by

received much more attention than it deserved, for the allegations which attracted the press were found to be without substance. The pressures of publicity hastened some changes of organization and practice but tended to divert effort in a one-sided way. Over a period of years modifications of detail were needed, and gradually took place, not only in purchasing activities but in other administrative procedures. Even in a period of basically settled administrative structure, such gradual evolution, in both narrower and broader matters, was a normal feature.

The broader changes of practice that could be observed were mainly extensions of what had been developing over the years. In particular, there had been the gradual attempt to free those responsible for the principal activity, deep mining, from the distractions involved in other activities. This had been done by creating specialized administrative units for the latter, with greater delegated authority. The establishment of the Brickworks Executive and the Coal Products Division in the sixties had pointed the way. A much smaller innovation in 1971 was the establishment of the Minestone Executive, separately from the Marketing Department, with responsibility for the commercial disposal of all the non-coal minerals in which the NCB had an interest. The chief concern was with the disposal of burnt and unburnt colliery shale, particularly as a bulk filling for motorway and other construction projects, but there was also something to be done in the extraction of surface minerals from some NCB land before tipping began on it.[1]

In 1971 the NCB carried all these rationalizing steps a stage further by groupings its activities into four sets of 'profit centres': deep mining, opencast, coal products, and ancillaries. Within each group there were self-contained management units, each given its own operational objectives.[2] To some extent this change was hastened by the clause in the Coal Industry Act 1971 which, in its pre-amendment form, had led Lord Robens to refuse reappointment for a third term. This clause gave the Minister power to restrict or discontinue any of the NCB's non-colliery activities, and there were consequential discussions between the NCB and officials of the Department of Trade and Industry. It also became government policy for the time being that no more of the money borrowed from the Exchequer should be invested in non-mining

Grimshaw himself (Grimshaw 1977). The fullest information is in NCB records. *The Times*, 15 Mar. 1983, had an obituary notice of Grimshaw which indicates the persistence, despite all the evidence, of the idea that Grimshaw was a martyr whom Parliament refused to protect.

[1] D. Turnbull, 'Minestone Executive', *Colliery Guardian*, June 1977.

[2] NCB *Ann. Rep.* 1971–2, I, 1 and 44.

activities. So although the logic of managerial needs led to the concept of the four profit centres, political pressures influenced the form in which the concepts were embodied. In order to insulate the non-mining activities from the rest in a financial sense, and to lessen any excuse for direct government interference with them, incorporation under the Companies Acts seemed appropriate. So two holding companies, NCB (Coal Products) Limited and NCB (Ancillaries) Limited, were incorporated on 15 March 1973 and began trading on 1 April 1973. The former controlled the NCB's businesses in carbonization, chemical by-products, and North Sea gas and oil, and the latter controlled all their other non-mining businesses. Each holding company was wholly owned by the NCB and each operated through a number of subsidiary companies (nearly all of them wholly owned), to which the NCB transferred the relevant assets, and through various associated companies and partnerships, shares in which were transferred from the NCB to the holding companies.[1] The new system made it possible for the NCB to be released from the potentially lethal restriction of having to get ministerial approval before making *any* change in the nature and range of their non-mining activities. There remained, however, some limitations because of agreement to consult the Minister before either holding company, or any of their subsidiaries, made changes beyond a defined scope in capitalization, directorate, or type of activity.[2]

The new system remained as a permanent feature of the NCB administrative structure. Though there were some changes of detail—for instance, carbonization activities came to be direct recipients of government money—there were no fundamental changes. An apparent modification was that the NCB began to refer publicly to the structure as consisting of three, rather than four, profit centres, with deep mining and opencast put together under the combined heading 'mining'.[3] While there was a possible case for this, on the ground that it put all coal production together, it caused some confusion and criticism when the published accounts showed only the combined figures for mining. The very different results of deep mining and opencast were thus concealed. After a few years the practice was dropped and the separate figures were again published. In the projection of output and sales, deep mining and

[1] NCB *Ann. Rep.* 1973–4, 50 and 58; Brandrick, 'Organisation of the Coal Board', *Colliery Guardian*, Dec. 1975 and Jan. 1976. The Minestone Executive was treated as part of the mining organization, as it was concerned with a direct colliery product. For more detail on the non-mining activities see Chap. 9 below.

[2] The subject is further discussed in Chap. 9 below.

[3] e.g. NCB *Ann. Rep.* 1977–8, 20.

opencast were considered together as complementary contributors, but the two activities had separate budgets and were organized and operated quite distinctly. Four, not three, groupings of activities remained the reality.[1]

Although from the early seventies the form of NCB organization was characterized by continuity and stability, there were probably some respects in which the substance changed to an extent that was not readily apparent in the form. The most significant was the increased centralization of the making of business decisions. This was almost inevitable because of the changed nature of coal's business. In the seventies there was one dominant purchaser of coal, the CEGB, and two others, much smaller customers but still substantial, the BSC and the SSEB; and by 1980 the BSC was becoming both absolutely and relatively less significant. It could be said that the sale of coal had centralized itself, and administrative practice had to follow suit. There were many other transactions, and plenty of them required different methods; but the decisions that affected much the greater part of the NCB's mining turnover had to be highly concentrated. And that requirement was all the greater because coal was being produced mainly for the use of two other nationalized industries, so that there was a blurring of the line between business and politics. Only the top layer of the organization had the weight and authority, as well as the full range of relevant information, to deal both with the large customers who provided most of the revenue and with the government who provided most of the capital. The result was that the areas were no longer the NCB's main *business* units in the sense, or to the degree, that they had once been. The areas were, rather, a level of *operational* management, with collieries as their operational units, all working within agreed budgets which should require them to meet the maximum practicable financial targets, but all working with the amount, source, and price of many of their supplies, and the destination and contractual terms of most of their output, predetermined by decisions taken at a higher level. The practical relations between headquarters and areas were therefore bound to be modified, with the accountability of the latter to the former being strictly appropriate in operational and physical terms, but with the fulfilment of financial budgets as an indirect feature.

[1] Deep mining and opencast results were combined from 1974–5 to 1977–8 inclusive in the published accounts. From 1975–6 onward there was a note which could be applied as a corrective to the figures for each deep mining area, but it did not detail the results for opencast regions. Publication of the results in a clear and distinctive form was resumed in 1978–9. Surprisingly, the NCB still referred to their *three* profit centres in evidence to the MMC in 1982.

This state of affairs was not new; it had been developing from the sixties onward. It might seem to provide less than the maximum pressure to achieve the highest financial returns, as it did to the Monopolies and Mergers Commission. This body suggested the adoption of revised accounting practices, with separate balance sheets, profit and loss accounts with allocated interest charges, and cash flow statements, for each area. The purpose of the suggestion was 'to enable the deep mine Areas and other formations to have adequate financial information and to be operated, as far as is possible, as separate business units'.[1] But this was merely to recognize a problem and to propose a solution that was obsolete by fifteen or more years. To implement it as things were would have been to impose on areas some responsibilities without corresponding power, and to change the organization so as to make the proposed practices realistic would have involved a diversion of business transactions from their most economic channels. A clearer inference was that, as the change in the nature of the business compelled changes in the way the organization was used, perhaps the gradual evolutionary development of the administrative structure was coming to the end of its usefulness, and the occasion for some wider ranging rethinking was at hand. But it needed to be more forward-looking than that of the Monopolies and Mergers Commission. In fact, in the early eighties attention was increasingly given to this question within the NCB.[2]

There were also changes in the political context within which the administration had to operate. Some of these were formal. For instance, from the beginning of 1974 there was a new Department of Energy, headed by a Secretary of State, which was very much the old Ministry of Power writ large, but with the major additional preoccupation of North Sea oil. Thus the interests of the fuel industries could again be more weightily represented to the government, instead of taking their place as just one among the many concerns of a vast Department of Trade and Industry. There was also the possibility that a more specialized department might try to exert more detailed supervision over the nationalized industries. As far as the objectives of the NCB and the major elements in progress towards their achievement were concerned, the likelihood of this was lessened by a recent agreement with the Department of Trade and Industry.

In the sixties the amount of detailed information regularly supplied by the NCB to the sponsoring ministry had proliferated, without much

[1] MMC, *National Coal Board*, I, 52–3 and 370–1.

[2] Ibid., 372.

indication that a lot of it was put to any practical use there. In 1970 the accounting firm of Cooper Brothers, in the person of one of its senior members, Sir Henry (later Lord) Benson (who had been the most influential member of the Fleck Committee), was called in to advise on the appropriate range of information to be supplied, and the form and frequency of its presentation. On the basis of his recommendations, agreement was reached in May 1971 about what was to be supplied in future. The new practices, embodied in what were referred to as 'the Benson brochures', came into use from the start of the 1972–3 financial year and continued through the ensuing decade without formal alteration, though aberrations in the detail of procedure crept in. The flow of information covered proposals annually and reports of actual progress quarterly, and was still very wide-ranging. But it was more precisely arranged than before, in relation to the central issues of policy and its performance, and therefore lent itelf, if required, to more fruitful discussion on an agreed and fairly tight timetable, though the government rather neglected the opportunity. The change in practice also made it possible to reduce the number of statistical, financial, and other returns made by the NCB to the department, and agreement on this was reached in the summer of 1972 and implemented thereafter. Out of eighty-nine returns previously made regularly, thirty-five were discontinued, three were simplified, and three were made less frequently.[1] These further changes provided useful economies in administrative and clerical time, and the whole exercise facilitated rather than weakened the ability of the department to appreciate and generally supervise policy, and to monitor its implementation.

There was still the possibility that the government might seek to exercise control more closely or differently, at least in relation to some specific questions, and this might have been done through legislation or through *ad hoc* decisions by government to get involved at particular points. The NCB were particularly exposed to this possibility because they had become financially more dependent on the government. In both 1972 and 1974 the strikes dragged the NCB into such large losses that only special government grants could save them from breaching statutory financial requirements. From 1973 onwards government

[1] It was regrettable that, at the same time as there was this improvement in the private presentation of policy information, the NCB's published annual report (which was formally a report to the Secretary of State) became much more sketchy. The NCB needed all the informed public understanding they could get and the deterioration of the annual report spoiled a chance of useful communication. Only in the later seventies did the quality of the report markedly recover.

grants towards revenue became a normal element in financing, and the arrest and reversal of the industry's decline required heavy new investment, financed mainly by borrowing from the government. So there was a necessity for frequent financial negotiations which gave the opportunity for government to call the tune, if general policy or inclination suggested it, though membership of the ECSC set a few rules as a limiting framework that had not been there before.

It had looked from the terms of the Coal Industry Act 1971 as though tighter restrictions were likely to be applied to some of the powers of the NCB in policy making, but in practice no clear trend of this kind emerged. New legislation from time to time gave the NCB a financial basis, however far from ideal, from which they could meet their commitments while retaining managerial freedom. The Labour governments of 1974–9 took away the NCB's interests in North Sea gas and oil because they had decided to establish a large state-owned undertaking to play a major role in the development of the energy resources of the North Sea, but in other respects they were willing to see some extension of the powers of the NCB. The Coal Industry Act 1977 contained various provisions of this kind, including some giving new powers to undertake business outside Great Britain. The 1976 report of the National Economic Development Office on nine nationalized industries, including coal, urged the government to become more directly involved and to limit the power of nationalized boards to make policy. It wanted the creation of a Policy Council, with functions distinct from those of executive management, for each nationalized industry. This was a proposal that roused anxiety in the NCB, where it was thought both very difficult to work and unnecessary, and the report had, indeed, recognized that coal was an industry in which there already was explicit agreement on policy between government and nationalized board. It was a proposal that did not commend itself to government as an answer to the problems of the public sector, and nothing came of it. Only when there was a fundamental change in the general economic objectives of government was there any serious move towards greater restriction of managerial discretion in the nationalized industries. For coal this was made explicit in the financial provisions of the Coal Industry Act 1980, which were so drawn as to make it almost impossible to operate the industry in the way it had been operating until then.[1]

In some ways, however, it seemed as though, perhaps unconsciously,

[1] See below, p. 415.

most nationalized boards, including that for coal, were regarded in the seventies as more subordinate politically. Ministers would provide direct answers to Parliamentary questions on detailed subjects well below policy level, which would once have elicited only the reply that this was a matter for the NCB (as, by statute, it was) and perhaps that the Minister would ask the NCB to communicate with the questioner. Likewise, there was the tendency for nationalized boards to have to negotiate on major questions with civil servants at a lower level than hitherto. Matters which in the fifties and sixties would have gone personally to permanent secretary or deputy secretary were in the seventies often worked out entirely at the level of under-secretary or assistant secretary. In these symptoms of the downgrading of the affairs of nationalized industries there was perhaps something which was a preparation, though hardly a precedent, for the greater intervention in matters of detail which occurred under the Thatcher government.[1]

But there was one new element in the arrangements for policy review in the coal industry which was deliberate and was fairly constant for some eight years. This was the role given to the Coal Industry Tripartite Group from 1974 onwards. This group represented the government, the NCB, and the unions, and it provided a forum where they could exchange views and information and seek to reach agreement about the major elements of policy and strategy, or make explicit the limits to which one of the parties would go in support of, or objection to, some particular part of the programme. In some cases the group established its own working parties to examine particular problems or possible lines of development before any proposals were put forward in a definite form. This was an additional help in ensuring that each party was quite clear about the position of the others, so that the conditions in which a particular proposal would have to be pursued, if it were pursued at all, were known in advance. Among the subjects considered in this way were, in 1974, the constituents of a possible research and development programme for coal,[2] and, in 1978, the problems and prospects of the coal industry in South Wales.

The original specified purpose of the Tripartite Group in 1974 was to

[1] By the spring of 1980 there was worry in the national board at the extent to which the junior minister immediately concerned with coal, Mr Moore, was concentrating his discussions on isolated aspects of operational detail, and at the renewed demands by the Department of Energy for financial details which it had been agreed under the Benson Formula were not required.

[2] Dept. of Energy, *Coal Technology: Future Developments in Conversion, Utilisation and Unconventional Mining in the United Kingdom* (n.d. [1978]).

carry out what was known as the Coal Industry Examination, which had as terms of reference 'To consider and advise on the contribution which coal can best make to the country's energy requirements and the steps needed to secure that contribution'. At that time it had not only a practical task but also a symbolic, almost therapeutic, purpose. The insecurities aroused by the revolution in oil prices compelled a reconsideration of energy policy which might lead to new requirements of the coal industry. But two recent national coal strikes had produced feelings of insecurity about coal also, and the suspicions and antagonisms exacerbated by any major dispute were unlikely to fade quickly without some strong antidote. A revival of the coal industry depended on complementary contributions from the government, the NCB, and the unions. An opportunity for all three to consider together the extent to which renewal of the industry was necessary and practicable, and the means to turn potential into achievement, was a helpful way to re-establish trust. The provision of the fullest information, for evaluation by all three, provided a better chance that they would all appreciate the nature and scale of what they would be involved in. Without that trust and knowledge they were less likely to make contributions that were truly complementary.

There was, of course, a risk that an attempt to get three parties to express equal commitment to a new programme might blur responsibilities when, for effective performance, these needed to be precisely located. But in fact, specific proposals both in general and detail, and the examination of the means by which they might be carried out, originated in the NCB. The other parties looked at the proposals both as a whole and in relation to their own roles, and were able to suggest emendations and the nature of their own commitment. Responsibility for making policy specific within the general requirements of the government and for managing its execution remained firmly with the body statutorily entrusted with it. The terms of reference for the Coal Industry Examination were very broad and were most easily observed by starting with a careful estimate of the prospects for demand and supply over a period long enough to permit significant changes in the adaptation of productive capacity to the market, but not so long that sensible estimates had to be replaced by hopeful guesses. The NCB were already engaged in planning several years ahead, and the contribution of the Tripartite Group is best observed by looking at the way in which an agreed longer-term programme could emerge from continuous planning exercises of this kind.

### iii. New Plans for Coal

In 1973 there were both the means and the opportunity to take a new look at the prospects for the coal industry, and to plan accordingly. The NCB had always had an obligation to settle the main lines of investment policy with the sponsoring minister. At this time they were discharging the obligation by presenting for approval each year what was called the 'long term investment review'. The form of this, in considerable detail, was one of the things agreed as a result of the enquiry by Cooper Brothers and embodied in the Benson brochures. The review was rather inappropriately named, for it covered only the next five years, in relation to a retrospect of the last five years, and five years is far from long term for major mining developments. In fact, the review was renamed the 'medium term development plan' in 1976–7. But whatever the title, the plan required detailed exercises that could easily be extended, and provided a basis on which plans for a longer period could be based. It was more practicable than it had been to attempt this because in 1973 there had been established at headquarters a small Central Planning Unit which could undertake specialized exercises and reconcile and co-ordinate the planning data and projections in which all the major administrative departments had to be involved.

The occasion for new plans was the changing state of the energy market over the last few years. By 1972 the NCB were getting increasingly concerned about the cumulative effects of recent low levels of investment on the capacity of the industry if its future were to be something other than irreversible decline. The expectation that there would be increasing pressure on international oil supplies had begun to grow since 1970. In the course of 1973 political worries about the Middle East were leading to widespread assumptions that oil prices would rise, though few people anticipated the speed and extent of the rise after it began in the autumn of that year. The NCB initiated discussions with the government in 1973, before the oil price rises started, and based them on the proposition that coal would become increasingly competitive with oil, but would need a large increase in investment. The government encouraged the development of ideas on these lines.

The long term investment review for 1973–4 to 1977–8, which was submitted at the beginning of 1973, necessarily involved assumptions about the period a little more distant, as investment estimates had to be made for the fifth year ahead. The review told the government:

> If closures were allowed to continue at a rate constrained only by social considerations, deep mined output would probably fall to 80 m tons in 1980. In the Board's view such a course would be wholly incompatible with coal's future potential in the light of developments in the energy market. On the other hand, probably the maximum level of deep mined output in 1980 would be some 120 m tons.

Even that maximum was nearly 10 million tons below the current actual deep mined output and was believed to be attainable only by ceasing to close any collieries on economic grounds alone and by proceeding rapidly with almost every physically practicable scheme, including some in high cost coalfields.[1] Right from the beginning of the attempt to plan for a secure, and eventually rather larger, place for coal in a revised energy policy there was thus an avoidance of airily optimistic assumptions about what was attainable[2] and a recognition that the physical output that seemed likely to be required was not compatible with a financial break-even on deep mining. Indeed, in February 1973 the NCB carried out a theoretical exercise (partly for the information of the Department of Trade and Industry) on the achievement of profitability (with a margin for contingencies) on deep mining as a whole by 1976–7, and found it quite impracticable unless there were what then seemed improbably large opportunities to raise prices well above the rate of inflation. If the targets were sought entirely by closures (physically the only quick method) it would involve closing 160 collieries and reducing output by 60 million tons in three years. Such a policy would have chaotic effects on the power stations and the steel industry, and hence generally, and would be socially so devastating as to be self-defeating.

One of the important things was that all concerned, including the government, went into the planning of a revived coal industry with their eyes open about the need for continued deficit financing, which was introduced in a more systematic way, but with an initial time limit, by the Coal Industry Act 1973. The justification for a new energy policy that could embrace such plans thus needed careful financial probing. Quite apart from such normal exercises as the investigation of the

[1] Though five-year investment forecasts were included in the plan, most of the other projections were on this occasion made for only three years ahead, because of financial uncertainties until the Coal Industry Bill completed its passage through Parliament.

[2] When 1980–1 arrived the actual deep mined output was the upper quartile of the figures in the projected range in these early forecasts.

soundness of calculations of the expected rate of return on investments designed to give specified additions to physical output, there were other questions, such as the amount (if any) it was prudent to pay, above the short-term competitive price, for hidden costs and benefits, including, above all, long-term security of energy supply from domestic sources. It was also necessary to consider the competitive position of home produced coal in terms not just of market prices but of market prices plus the unit cost of government grants, or at least those grants used directly to offset some element of mining costs, including interest. The last point might have led on to the further question of how far grants might be reduced by loading a bigger proportion of total costs on to market prices, but this question was usually left in the background, partly because it is much more complicated than it appears at first sight.

What the NCB did in the spring and summer of 1973 was to work out a new longer term strategy for coal. The scale of what was proposed was based on estimates of the relative trends of fuel oil and coal costs, and of the practical possibilities of future production at each existing colliery (data on which were kept constantly updated) and at entirely new sites of which the resources were at least partly known. At the same time the Department of Trade and Industry requested NCB views on energy policy, and the same exercise provided the means of answering this request. These activities led to the formulation of what came to be known as *Plan for Coal 1974*. It was considered in detail by the national board in autumn 1973, ready for submission to the Secretary of State in November. The oil price revolution was proceeding apace while the plan was being considered. A revised version was therefore prepared. This was approved on 1 February 1974 for submission to the Department of Energy a fortnight later, despite preoccupation with the debilitating strike then taking place. But the calling of a general election and the subsequent change of government delayed the submission. The NCB used the interval for further revisions and the plan was submitted to the Secretary of State for Energy on 5 April. The *Plan for Coal* then became the basis for the Tripartite Discussions (known as the Coal Industry Examination) which began later that month.

The interim report on the Coal Industry Examination, in June 1974, accepted *Plan for Coal* as a broad strategy for the industry, subject to annual review; and the government gave an assurance that short-term fluctuations in the price and availability of competing fuels would not be allowed to interfere with progress with the Plan, while it accepted that coal had to remain competitive in the light of long-term trends.[1] The final report, in October, reiterated support for the Plan, including the proposed exploitation of the new Selby coalfield and the expansion of opencast production. On the vital question of finance, the government endorsed the strategy of creating 42 million tons of new capacity, while requiring each individual major project to pass appropriate investment appraisal tests, and undertook to consider dealing with any special problems of financing which would arise from the long interval between expenditure and the consequential creation of revenue. On this theme the report concluded:

> We welcome the establishment of a financial framework for the industry which will give it the objective of long term competitiveness while covering its costs of production and contributing towards financing the new investment programme, but at the same time recognising the special burdens of the past, the need to provide safeguards against short term fluctuations in the price of competing fuels, and the need to take appropriate action if other public policies prevent commercial pricing or impose exceptional burdens on the Board.[2]

*Plan for Coal* was an examination, on various stated assumptions, of the prospects of the energy market, in terms of prices and physical demand, down to 1985, and the ways in which the demand might be met. The contribution that coal might make rested on an appraisal of the potential output capacity of each colliery, the investment required in order to maintain or increase capacity, and the number of men likely to be needed and obtainable, and their cost. The case for investing in a revival of the coal industry was made principally in terms of the cost advantage of coal over oil for electricity generation. In 1973 the average delivered cost of power station coal was 3.5p per therm, or 3.1p per therm for coal from the Midlands and Yorkshire. Even if costs increased in real terms by 4 per cent per year they would still be less in 1985 than the delivered costs per therm of fuel oil at the prices introduced early in 1974. This advantage could not be made fully effective, however, unless the generating boards had enough efficient coal-fired capacity, and so there was repeated emphasis on the need to build additional up-to-date coal-fired power stations as a complement to the Plan.

It should be stressed that, though there were hopes of expansion in the longer future, *Plan for Coal* was not proposing great increases in coal output within its period of application. The physical capacity of the coal

[1] Dept. of Energy, *Coal Industry Examination: Interim Report* (1974), 12–13 and 16.

[2] Ibid. *Final Report* (1974), 16 and 23.

industry had been so reduced that only over a long period could output rise substantially again. The Plan assumed further large increases in the use of natural gas and nuclear power, and estimated the probable market for coal (including exports) at between 120 and 150 million tons in 1985, with the mid-point of 135 million tons used as a general guide. Losses of capacity were expected to average 3 to 4 million tons a year, with a risk of acceleration of the rate of loss unless investment increased, and the maximum practicable creation of new capacity by 1985 was put at 42 million tons. Of this new capacity 9 million tons would come from making access to new reserves at collieries which would otherwise close, 13 million tons would come from major schemes at existing collieries, and 20 million tons from new collieries, including up to 10 million tons from the Selby coalfield. Thus new capacity would roughly equal the loss of old. The existing deep mining capacity in 1974 was calculated as 120 million tons, though output was running at an annual rate of only 113 million tons while the Coal Industry Examination was in progress. The only way to meet a possible demand for 135 million tons appeared to be to increase opencast output to 15 million tons a year. This was considered feasible and such an expansion was part of the Plan. The Coal Industry Examination expressed the belief that an extra 8 or 9 million tons a year might be got from existing capacity if every constraint on optimum performance were overcome; but though some specific contributions, such as an incentive wage scheme and the more general adoption of current 'best practice', were mentioned, there was also some rather vague aspiration for 'a new attitude and spirit in the industry'. Essentially, the Plan had to be, as it was described by the NCB, 'for overall stabilisation of the industry in the years ahead'.[1]

The capital cost of the Plan to 1985 was tentatively estimated at £600 million at March 1974 prices. This was in addition to the £70 to £80 million per year needed for ordinary purposes outside the Plan. It was clear that this was much more than could be generated from internal sources unless coal prices could be raised enough for the NCB to retain a significant revenue surplus throughout the reconstruction period. The resultant fnancial problem was obvious and was particularly serious because of the long time lag before major investment schemes could be completed and produce revenue. The NCB would have preferred that any finance by government loan should be interest-free until the completion of the Plan. As already noted, the government

[1] Ibid. *Interim Report* (1974), 10–12.

expressed readiness to give special consideration to problems of this kind, but did not commit itself to any particular type of solution.

The uncertainties in planning were known to be great, especially in proposals for new capacity where many details still remained to be investigated, and the tentative nature of estimates of quantity and cost was stressed from the outset. Some of the influences taken into account turned out to be more variable than was foreseen, and there were new uncertainties from other factors. One difficulty was that *Plan for Coal* was launched just after the breakdown of the international monetary system had led the world to adopt floating exchange rates. The effect was to cause substantial fluctuations (over which producers had little control) in the relative prices of fuels from different sources. Moreover, the oil price revolution helped to make international financial disequilibrium worse than it otherwise would have been, and helped to stimulate higher rates of inflation. The latter, in turn, made it more difficult to forecast and control the rise in the cost of inputs, helped to put up the cost of new borrowing, and also prompted restrictive policies that slowed down the rate of economic growth. Rising fuel costs caused other complications. It was premature for coal to rejoice if oil prices rose even faster than its own costs, for general rises in the prices of fuel had some direct adverse effect on the rate of economic growth, and the rise in fuel costs prompted a search for methods of energy saving, for which there was ample scope. These effects tended to reduce the demand for coal, but it was hard to foresee by how much and how quickly. The uncertainties of planning in the chief coal-using industries also had secondary repercussions. All the proposals about coking coal in *Plan for Coal* were based on the government's and the steel industry's own estimates of steel output and sales, but in the uncertain international economic conditions of the later seventies, these were vulnerable to increasing error. In the electricity generating industry, delays in the completion of some power stations, and in decisions about the type and starting date of others, increased uncertainty about the likely future position of many of the coal-fired power stations in the merit order, and therefore about the likely amount of coal to be burned.

There were also some general difficulties in implementing plans as completely and as quickly as had been hoped. These had not been fully appreciated in advance. One was the effect of the low level of investment since the mid-sixties. Much more than had been expected had to be done urgently in order to stop the industry sliding backwards in both the quantity and quality of its productive capacity. That meant that fewer resources were available to carry through new projects. This was a serious matter because some specialized resources were scarce. The NCB commented in 1974:

Most of the tasks of planning and implementing major projects must be undertaken by engineers and surveyors trained and experienced in the coal industry. These staff normally gain their experience among the engineering and surveying posts (to meet the demands of continuing production from existing collieries). Their numbers have declined since the implementation of the Board's investment programmes of the 1950s. Some increase as resources allow and some redeployment . . . are now necessary. . . . The use of consultants and contractors for certain aspects of major projects work will need to continue.[1]

There were other difficulties as well. The great contraction of the coal-mining industry in the sixties had been a severe blow to the firms manufacturing mining equipment and those offering specialized engineering services on contract, and they, too, had suffered a loss of staff who were trained to launch new projects. It took time to overcome the shortage, and when schemes under *Plan for coal* were launched there was not enough suitable experience to go round. In all this there were unhappy echoes of the decade following nationalization. Then, too, short-term schemes to remedy the defects of previous under-investment had to be given disproportionate attention. Then, too, major long-term projects suffered delays from the shortage of planning engineers, for whom, in the immediate past, jobs and prospects had been inadequate.

In the seventies and eighties there was also one new external constraint on the speed of progress with new projects. This was the time taken to get planning authorization. The delays resulting from public objections were most numerous in relation to opencast projects, but in these cases the difficulties were usually resolved without serious hold-ups in reaching production. But deep mining schemes could be subject to more complex and protracted frustration. This was most fully illustrated by the second largest new project, that for mining the North East Leicestershire coalfield from three proposed new collieries. The NCB applied for planning permission in 1978; the Secretary of State for the Environment called in the planning application and held a highly detailed and expensive public enquiry which lasted from October 1979 to May 1980; the inspector conducting the enquiry recommended that, subject to the elimination of tipping at two colliery sites, planning permission should be granted; the Secretary of State reversed the inspector's

[1] Dept. of Energy, *Coal Industry Examination: Final Report*, 13.

verdict, but took until March 1982 to reach his decision, and even then indicated that a revised and reduced planning application could be submitted.[1] But it was more than a year before there was a decision on this, even though it took the NCB only three months to prepare and submit a revised planning application. So five years went on legal processes and political argument, involving substantial cost, before anything practical could be started.

It was a requirement of the Department of Energy that progress with *Plan for Coal* should be reviewed annually, and this was one of the purposes which the medium term development plan served. The effects of the difficulties that have just been reviewed, and of others in the current operation of the industry (especially the failure of labour productivity to rise as anticipated), soon showed themselves and cumulated severely. Consequently, there were repeated revisions of forecasts, all in the direction of reduction and delay for the immediate future, which was seen to be full of constraints, even though the longer term prospect still seemed unimpaired. The plan for 1978–9 to 1982–3 inclusive, which was submitted at the end of 1977, illustrates what was happening. In this plan the 1980 projections were 107 million tons for deep mined output, 122 million tons for total disposals, and 47½ cwt. for overall OMS. In the plan made a year earlier those 1980 projections had been 110 million tons, 126 million tons, and 49½ cwt. respectively; and two years earlier they had been 118 million tons, 132 million tons, and 52½ cwt. Yet even the reduced projections made in 1977 were thought to be attainable only if everything went as well as possible. At the same time, high inflation rates and high interest rates were making the financial burdens much larger than had been anticipated. The internal view of the plans as they were revised in late 1977 was summarized thus:

> What emerges . . . is a further postponement of the move towards the attainment of Plan for Coal objectives and increasing calls on Government finance. . . . This emphasises the Board's major problem of credibility. Nevertheless, given (a) the continuing strong case for the long-term investment programme, (b) the absence of an effective programme for closing uneconomic capacity, there is little alternative to the prospectus put forward.

Belief in the strength of the long-term case, despite early discouragements, remained the dominant influence on planning policy. A major review of *Plan for Coal* took place in 1976. The estimated capital cost of the projects in the Plan had risen, because of inflation, by 74 per cent in

[1] NCB *Ann. Rep.* 1982–3, 6–7.

only two years from the figure of £600 million at March 1974 prices. In addition, it was found that at least half the output from new collieries would be unachievable until after 1985, that more reconstruction projects (including some to reduce costs without adding to output) would be needed, and that the ordinary routine investment outside *Plan for Coal* would have to be increased. All these revisions (and others in greater detail) called for a 29 per cent increase in real terms in estimated total capital expenditure for 1974–85, on top of the effects of inflation. So, at March 1976 prices, estimated costs had risen to £1,500 million for *Plan for Coal* plus £1,650 million for ordinary capital expenditure. Nevertheless the Tripartite Group agreed that it was right to press on with the policy,[1] and this was done. Capital projects representing more than half the *Plan for Coal* programme had been approved by September 1976, and the medium term development programme at the end of 1977 showed an intention to commit investment expenditure for almost all the remainder by 1982–3.

This persistence was backed up by attempts to look at a rather more distant prospect. The NCB put before the Tripartite Group, in October 1976, outline proposals for extending *Plan for Coal* from 1985 to 2000. These proposals envisaged a range from 135 to 200 million tons as the demand for coal in 2000, and indicated that an output near the mid-point of this range, with 150 million tons from deep mines and 20 million from opencast, was feasible. But even if *Plan for Coal* were fully implemented by 1985 (which was very improbable), its extension required about 60 million tons of additional new capacity, roughly half of it in replacement of exhaustions, i.e. the commissioning of 4 million tons of capacity a year from 1985 to the end of the century. Most of the additional capacity would have to be in new mines. In the light of current and recent past experience, this looks to have been a wildly optimistic target. The estimates of demand were based on the assumption of average annual rates of economic growth of not less than 2 per cent nor more than 3 per cent, and on some possible switches to coal as a result of adverse supply and price conditions for fuel oil, limits to the nuclear power programme, and some possible need for substitute natural gas, made from coal, as output from the North Sea dwindled. It was also assumed that steel output would rise by about half.[2]

These proposals, named *Plan 2000*, were encouragingly received and

[1] Dept. of Energy, *Coal for the Future: Progress with "Plan for Coal" and prospects to the year 2000* (n.d. [1977]), 10–11 and 22.

[2] Ibid., 15–17.

taken into account by the government as a new energy policy was framed. A Green paper in 1978[1] used figures that were the mid-points of the ranges in *Plan 2000* and wrote approvingly of the need to create 4 million tons a year of new capacity. This Plan, however, was never formally endorsed by the government, though the Tripartite Group in 1979 again approved in general terms the need to continue investment at a high level. After that, conditions changed markedly, both economically and politically, with the result that plans for the future had to be reconsidered in a more pessimistic frame of mind. *Plan 2000* was never of much value as a framework for action. Its quantitative range of possibilities was too wide to offer practical guidance. For those who did not take a euphoric view of the speed with which new productive capacity would be created, it was a helpful indicator that a market might be available for as much new economic capacity as it was physically and administratively practicable to put in hand, and this contributed to a sensible case for investing after 1985 much as before. But that was about all.

*Plan 2000* soon ceased to be used as a reference point. Official estimates of economic growth and of primary energy demand were being greatly reduced. Long-term expectations for coal had to be modified also, even though in 1979 and 1980 the competitive advantage of coal over fuel oil greatly increased: by March 1981 the dollar price of crude oil had risen 120 per cent in two years,[2] but this helped coal consumption to contract less than oil, rather than move on to an upward trend. After 1979 there was no medium term development plan accepted by the government, as the latter sought to establish a different kind of financial framework. A Coal Policy Review, submitted by the NCB in January 1981 to cover the next five years on lines taking account of government financial thinking, was made obsolete by an almost immediate change of government proposals. But there was a new longer term development plan to 1990. This was submitted by the NCB to the Department of Energy in July 1981 in order to show the order of magnitude of investment that would be needed in order to maintain continuity with *Plan for Coal*. The new plan was based on the assumption that the level of energy consumption in 1979–80 by final users would not be reached again in the twentieth century, though a case was made for suggesting that the share of coal might increase somewhat because of increased direct use in industry, mainly after 1990. This last point was

[1] *Energy Policy: A Consultative Document*, Cmnd. 7101.

[2] NCB *Ann. Rep.* 1980–1, 4.

one on which the Monopolies and Mergers Commission expressed doubt and urged caution.[1] The method of presenting a plan now, however, was in terms of a reference case, based on a set of explicit assumptions, and measures of the sensitivity of the case to specific variations in the assumptions. Thus it was possible to show what modifications would be needed in plans and their execution as the unfolding of events revealed their departure from particular forecasts.

It might seem as though the value of the whole exercise in planning was lost or greatly reduced because of the repeated revisions and especially the drastic change in the economic and political climate after 1979. There was, indeed, something of an aberration in the presentation of the fantastic upper bounds of *Plan 2000*, which encouraged some vain hopes and beliefs inside and outside the coal industry. But *Plan for Coal*, and the annual exercises which took account of its basic propositions, provided a useful frame within which to work. In some important respects, forecasts proved very accurate (even if aided by the balance of compensating errors) until the deep depression of 1980. The demand for coal in 1980–1 was almost exactly what had been forecast in 1974, except from the coke ovens, which contracted with the iron and steel industry. Total energy use was well below forecast, but coal's share was higher because of the continued delays in getting nuclear power stations completed and fully operational, and the renewed great rise in oil prices. Total deep mined coal output came neatly in the middle of the range projected in 1974. New capacity was created more slowly and productivity rose less than projected, but exhaustion of old capacity also proceeded more slowly. In comparison with other forecasts the NCB had quite a good record on market demand and their methods became more sophisticated and more continuously adaptable.[2]

If the planning methods and presentation were open to more criticism it was probably in the apparently small adaptation to the worsening financial outlook as capital charges soared, though, in practice, this objection should have been met, at least in part, by the appraisal of each capital project separately in accordance with detailed procedures and the requirement to meet approved financial criteria.[3] In view of the long period it takes to introduce new and better quality productive capacity, the coal industry needs a plan looking ten or more years ahead, quantified as carefully as possible, but devised in such a way that sensitivity

[1] MMC, *National Coal Board*, I, 78.

[2] Ibid., I, 61–70.

[3] Ibid., I, 80–93 describes the NCB plans and planning methods; 99–100 makes a number of criticisms.

tests could be applied at the outset and continually thereafter, so as to permit rapid adjustments as circumstances change. It was not only the accuracies, but also the things that were delayed or falsified, that confirmed the value of the 1974 *Plan for Coal*.[1] The basic strategy underlying it seemed even more necessary in some respects when the implementation of parts of it was being unexpectedly frustrated. The centres of difficulty looked more prominent in the context of the strategic plan. The thing to do was to use their easy identification as the start of attempts to remove them. Then the continuity, that was so essential to the strategy, would be less under threat.

### iv. Productivity and incentives

One of the frustrating things that had not been expected was the long delay in achieving any further sustained rise in the productivity of labour. There had been a serious slowing down at the end of the sixties, but the improvement after the 1972 strike raised hopes that, in settled conditions, a steady improvement would be resumed. But this did not happen, as may be seen from Table 7.2. Recovery in the immediate aftermath of the strike was quicker in 1974 than 1972, but after the first few weeks the pace of improvement slackened. For several years productivity fluctuated around a slightly declining level and it did not start to rise again until 1978, as appears from the figures of overall OMS. For various reasons the average number of shifts worked by each man in a year was also declining until the late seventies, so annual output per man was affected rather more.

These disappointing trends hampered the execution of the plans that had been approved. They not only caused projections of output to be lowered, but they increased revenue costs and cast doubt on the likely achievement of the projected results from some approved capital schemes. Some of the latter doubts could be suppressed for the time being, mainly because recent investment had been so low that there was little current evidence that would falsify projections of capital productivity. In addition, there was less evidence of a relevant kind available for challenge, because nobody had been able to devise a satisfactory *general* measure of the productivity of capital,[2] and the most that could be done

[1] MMC, *National Coal Board*, I, 83 notes that the NCB, though by 1982 they had greatly changed the detail of their planning assumptions, still based strategy on principles in accord with those in *Plan for Coal*.

[2] On the NCB's lack of capital productivity measures, the difficulties of defining and calculating them, and suggestions of possible approaches to the problem, see MMC, *National Coal Board*, I, 120–3 and II, 88–91 and 93–4.

Table 7.2. *Labour productivity, 1972-3 to 1982-3*
(tonnes)

| | 1972–3 | 1973–4 | 1974–5 | 1975–6 | 1976–7 | 1977–8 |
|---|---|---|---|---|---|---|
| Output per man-year | 480 | 390 | 474 | 462 | 448 | 441 |
| Overall OMS | 2.33 | 2.15 | 2.29 | 2.28 | 2.21 | 2.19 |
| | **1978–9** | **1979–80** | **1980–1** | **1981–2** | **1982–3** | |
| Output per man-year | 448 | 470 | 479 | 497 | 504 | |
| Overall OMS | 2.24 | 2.31 | 2.32 | 2.40 | 2.44 | |

*Note*: The statistics are affected by changes of definition and practice adopted in 1979, including changes in the attribution of shifts and output to revenue and capital activities respectively. The figures used here are little affected, though some other measures of productivity have a significant discontinuity at this point. For comments on the measures and definitions see MMC, *National Coal Board*, Cmnd. 8920 (1983), I, 106–9 and II, 91–2. The NCB adopted metric units on 1 Apr. 1978 and figures for earlier years have been converted to aid comparison.
*Source*: NCB *Ann. Rep.* 1982–3, 20–1.

was to try to measure the return on individual major projects, of which there were hardly any recently completed examples. Partial measures of capital-related productivity, such as output per machine shift and daily output per face, caused less alarm than the trends of OMS, but were not very encouraging in the mid-seventies. Daily output per face in a normal week of March rose from 625 tonnes in 1973 to 670 in 1975, but fell back to 608 in 1976 and 638 in 1977 before following a sustained upward trend (interrupted to 1981) which carried it to 754 in 1982 and 773 in 1983.[1] But OMS showed the productivity of the largest single input and was therefore the nearest (though still rather remote) proxy for total factor productivity, and its persistently adverse trend could not be ignored.

Naturally, there were attempts to explain it and to do something about it as soon as possible. It was generally admitted that there were physical causes in the mines, partly the result of under-investment, but it seems probable that too little weight was given to these. The greatest stress was put on labour attitudes. Some of these might be irrational, but it was believed that the daywage system and the existing differentials between grades failed to encourage maximum effort and maximum use of the highly productive equipment which was already installed, even if

[1] NCB *Ann. Rep.* 1982–3, 6. Because of the aftermath of the strike there was no 'normal' week in March 1974.

it was not being added to as fast as it had recently been. This lack of incentive was believed to apply particularly strongly where it could make most direct difference to output, i.e. at the coal face and in underground development work. In late 1974, when the evidence of stagnation was not yet so prolonged but some sort of improved incentive was being anxiously sought, attention was directed to shortage of faceroom as well as poor labour performance, but it was pointed out that the inadequate provision of faceroom was not solely the fault of management but was partly the result of delays caused by lack of effort. Impromptu discussion then suggested that the causes of sub-standard OMS varied from place to place, but lack of effort by the men was picked out as clearly evident in Scotland, South Wales, Kent, and a group of pits in the North East. By contrast the main cause was believed to be geological difficulties in South Yorkshire and lack of faceroom (aggravated by the objections of the NUM to the employment of outside contractors on development work) in North Yorkshire. In other areas, where OMS was more or less in line with budget, it was nevertheless believed that higher output could be achieved if conditions were changed. A general problem was the reduction in total output at long-life collieries, which it was impracticable to match with an equivalent reduction in underground labour away from the face. A tendency for the productivity of such labour to decline was the obvious arithmetical result.

More sophisticated attempts were made to explain changes in productivity. One statistical enquiry indicated that throughout the third quarter of the twentieth century a relationship involving only power loading, the use of powered supports, and labour morale could be made to give a very good fit with the observed movement of overall OMS,[1] though it seemed likely that there was something fortuitous in this result from an over-simplified exercise. The latter suspicion was borne out by later detailed studies of the variance of face productivity, in which a colliery productivity model, based on regression analysis, incorporated thirty-five parameters.

In the light of such studies it seems probable that some of the anxiety in the mid-seventies resulted from too high expectations. The particular technological changes that had done so much in the sixties had been

[1] The relationship was expressed as $OMS = 23.6 + 0.11M + 0.13P - 0.12D$, where OMS is measured in cwt; M is the output from mechanized faces as a percentage of the total; P is the percentage of major longwall faces with powered supports; and D is the tonnage lost in disputes (in millions of tons), used as an indicator of morale.

carried nearly as far then as they could be, and nothing else with demonstrably comparable strong influence was in rapid progress, though increases in face length, with associated improvements in design, were helpful. The benefits of a regional redistribution of output were available to a much lesser extent after the cessation of the huge programme of colliery closures, and the same change tended to diminish the opportunities to gain the small, but positive, productivity returns to scale. On the other hand the distance between shafts and coal faces was still tending to increase, with an adverse effect on OMS; and the additional improvements in manriding and the installation of very long trunk conveyors and their extension along new drifts to the surface, which might offset this effect, had not become very common by the early seventies. So it was not altogether surprising if further large rises in productivity were not practicable without further restructuring of the industry and a revival of investment, some of it devoted to the adoption of more advanced technology.[1] All this was bound to take several years, but there was still concern that labour relations prevented the optimum use of existing resources and could, if not changed, prevent the realization of maximum benefits from any large new programme of capital schemes. This was a subject that seemed appropriate for immediate treatment and capable of producing quicker improvements; so it received constant attention.

There were three aspects of the wage system on which management was particularly anxious to make progress. One was to try to lessen the risk of disruption through frequent conflicts in the course of wage negotiations. The way to this was seen to be through long-term agreements, which obviously would have no appeal unless they provided for cost-of-living increases at frequent and regular intervals and for other increases at stated times on a basis agreed in advance. A second was to continue reducing the number of potential points of disagreement by further simplifying the grading and wage structures along the lines that had been adopted in the three major agreements of 1955, 1966, and 1971. Indeed, the ideal was to try to combine into one the three structures thus created, and to incorporate also the various additional groups, such as the craftsmen whose grading structure was comprehensively revised in March 1971, and the motor vehicle drivers for whom a new structure was introduced in August 1972. A third was the introduction of some kind of incentive scheme, and the form of this was intended to be a contribution to the rewidening of differentials in favour of underground

[1] Chap. 3 above for the innovations of the seventies in various branches of technology.

workers. On all these aspects a great deal of work was done in the NCB in 1972 and 1973, with an effort to combine all aspects into a single comprehensive wages strategy. All this work was coming together at the time of the overtime ban in the winter of 1973–4, when specific proposals for the longer term were being prepared.

On the union side the idea of long-term agreements never had much attraction. The NCB were ready to sweeten the prospect with a commitment to high wages. But long-term agreements are a curb on freedom of action when pressures change, and the high rates of inflation in the seventies made it seem risky to be tied down for any lengthy period. So nothing came of this approach. Further revision of the wages structure was, however, very welcome to the NUM. At the beginning of 1974 they expressed a wish for a complete review, and the NCB had proposals for a single wages structure complete and ready for discussion. This harmony of outlook made it possible to introduce the Comprehensive Daywage Structure in 1975. All mineworkers were then grouped in a single structure. For non-craftsmen there were seven underground and six surface grades, and for craftsmen three underground and four (later reduced to three) surface grades. There were different levels of pay for craftsmen underground when they were deployed to faces from those when they were deployed elsewhere.[1] Union attitudes to incentive schemes were more mixed. There was a lot of lip service over a period of several years, but whenever action was proposed, all sorts of antagonism emerged. There was, indeed, a contradiction between the idea of incentive schemes, which gave extra reward varying with performance, and the principle of national uniformity of pay for the same job, which the NUM had proclaimed from the beginning. So movement towards incentive schemes had repeated setbacks.

The Wilberforce Report in 1972 had urged the adoption of an incentive scheme and this had led to discussions between the NCB and NUM, which produced at least the agreed (and minuted) conclusion that no effective scheme could be devised on a national basis, i.e. with payments based on the national performance and made at uniform rates. The obvious inference might seem to be that there should be an incentive scheme on some more localized basis, but for the time being the union seemed to lose interest in having any scheme, though there was no formal break in negotiations. The Pay Board Report at the time of the 1974 strike again urged the negotiation of an incentive scheme, and

[1] MMC, *National Coal Board*, I, 265–7 for grades and differentials in basic rates as they had become in 1981–2.

later in the same year the reports of the Coal Industry Examination, which the unions committed themselves to support, referred to the need for an 'effective scheme of incentives' to help stimulate production. The NCB therefore decided in July 1974 to start fresh negotiations with the NUM.

These negotiations proved extremely difficult, mainly because of very sharp divisions within the NUM. A draft incentive agreement, under which there would be nationally uniform percentage rates of bonus but they would be applied to production norms measured locally, by face or by pit, was drawn up and considered by the NUM national executive and then by a delegate conference in September. There was fierce opposition, led by the Yorkshire Area President, Arthur Scargill (who walked out of the conference with his delegation), and in October the NUM told the NCB that they would accept only a scheme which was based on national OMS and gave the same bonus to everybody.[1] This was what the NUM in 1972 had agreed could not be an effective incentive scheme.

The NCB refused to agree to such a proposal, but offered amendments to the previous draft, including the possibility that the production bonus for all except faceworkers could be averaged over a wider span than the individual colliery. On this basis the NUM national executive gave the NCB a written statement accepting the general lines of the proposal but asking for three small amendments and for the rate of bonus to other workers to be raised above the 50 per cent of the average incentive payment to production workers. All these demands were met, whereupon the NUM national executive declared that it would put the improved proposals to a national ballot, with a recommendation for rejection. Not unnaturally the NCB took the recommendation as evidence that some members of the national executive had been negotiating in bad faith. The NUM President subsequently described the recommendation as 'one of the daftest decisions the Executive of our Union has ever made. ... But the "antis", the wreckers I should call them in this case, had had their way.'[2] One of the problems of the NCB was to ensure that the details of the scheme were fully known to the mineworkers, especially in Yorkshire, where it was stated that Mr Scargill had threatened to prevent the distribution of a leaflet setting them out. Wide publicity was, in fact, secured, but there was a good deal of hostile campaigning and in the ballot the NUM recommendation for rejection received 61.5 per cent of the votes.

[1] Gormley 1982, 148–51.

[2] Ibid., 152.

That was the end of hopes of an incentive scheme for the time being. In the following year the NCB, in what looks like unwise desperation, agreed to pay national bonuses in one quarter at rates based on the excess of output over an agreed national base tonnage in the previous quarter. The agreement had no noticeable incentive effect, and only in the first quarter of its operation was there any eligibility for bonus.[1] The scheme was thus the total failure always predicted.

But in 1977 the prospects of an incentive scheme revived as the government sought to keep down earnings rises to 10 per cent for the year, but excluded self-financing productivity deals from the limit. At the NUM annual conference in July the idea of an incentive scheme was only narrowly defeated, and in some NUM areas there was talk of seeking local incentive schemes. In August the NCB decided to take the initiative in offering an incentive scheme, but first sought government approval and got it after long discussion. Negotiations went on through September and October, the NUM national executive agreed to a scheme very similar to that rejected in 1975 and this time put it to a national ballot with a recommendation to accept. But the ballot was held only after the Kent Area had taken the NUM to court (and to the appeal court) in a vain attempt to get the ballot declared unlawful, and after an intense campaign, including press advertising, by the Yorkshire Area to have the scheme rejected. And at the end of October the ballot did reject it, with almost 56 per cent of the votes going against. But fifteen of the twenty-two NUM Areas voted in favour of the scheme, ten of them with 65 per cent or more in favour. Moreover, the government's pay policy was having the usual effect on coal industry wages and miners were starting to slip down the earnings league. An incentive scheme was the one ready way to give them a chance to earn appreciably more, and it was a device which might take other industries further ahead if the miners passed it by.

So within a few weeks several NUM Areas approached the NCB with requests for local incentive schemes, and the South Derbyshire Area sought authority from the NUM national executive to negotiate such a scheme. On 8 December the national executive gave such authority to any area and there was a rapid conclusion of agreements, all on the basis of the scheme rejected in the national ballot, by one area after another. By the beginning of March 1978, though there was still no agreement in Kent, 84 per cent of all faces and 68 per cent of all drivages were the subject of agreements, and payments had already started at 177 collieries.

[1] NCB *Ann. Rep.* 1975–6, 13.

Two months later the scheme was operating throughout almost the entire industry.[1]

The new scheme was, strictly, a combination of direct incentive payments to those with a directly measurable output and related bonus payments to all other workers. Incentive payments went to men working on installation agreements, and, for this purpose, an installation was a coal face or a development drivage with a planned advance of at least 30 yards or a comparable underground workplace where work was exactly measurable and repetitive. For each installation a standard manning level was agreed and a standard task laid down, in terms of area extracted (for faces) or linear distance advanced (for drivages). The standards were normally established by conventional methods of work study, but there were fallback procedures for any cases where agreement could not be reached on this basis. Incentive payments were made, at a rate agreed nationally from time to time, for each percentage point of performance achieved above the 'basic' task, which was defined as 75 per cent of the standard task. There was no upper limit to the incentive payment. Payments were related to performance on each drivage separately, but in each colliery the faceworkers could opt to have payments related to performance on each face separately or on all the faces collectively in the one colliery, and in practice the choice was fairly evenly divided.

Bonus was paid to all other workers as a percentage of the incentive pay earned by installation workers. For colliery workers the incentive pay used as the relevant basis could be the colliery average or the NCB area average or the NUM area average. Everywhere except North Derbyshire and Kent the workers opted for the colliery average. The rate of bonus (as a percentage of the incentive pay) was 100 per cent for underground officials and for craftsmen deployed to faces, 75 per cent for men completing their training at faces, 65 per cent for other workers while they were deployed to faces, 50 per cent for other underground workers, and 40 per cent for surface workers. Thus the scheme did something to widen differentials in favour of direct production workers. Mainly in order to avoid further erosion of differentials, bonus was also paid to all clerical staff and to managerial staff other than those in the most senior grades. These payments were mostly at the 40 per cent rate, though colliery line management and engineers got 100 per cent and certain others with underground responsibilities got 50 per cent. Nearly all bonus payments to non-industrial staff were related to the incentive average for the NCB area.

[1] Gormley 1982, 152–3; Handy 1981, 278–81.

When the scheme began, the level of incentive pay was fixed so as to produce £23.50 a week for the achievement of standard performance. This went up to £26.50 from March 1979 to January 1981, and to £30.00 from then until November 1982. The amount earned in incentive pay varied quite widely between areas, but for the industry as a whole it kept appreciably above the amount due for standard performance. For mineworkers as a whole, in a normal spring week, incentive and bonus pay contributed 12 per cent of total earnings in 1978 (when the scheme was just starting) and 17 per cent or 18 per cent in each of the next three years.[1]

The acceptance of the incentive and bonus scheme was a significant contribution to the restoration of coal industry earnings relative to those of other industries at a time when government restrictions on pay increases were making this difficult. The settlement after the 1974 strike, and the effect of wage changes in all occupations (including coal) in the next few months, made coal the best paid of all the twenty-seven industries in the standard industrial classification used in the New Earnings Survey of the Department of Employment. After 1975 the position was changing. In 1977 coal had dropped to third place in the 'earnings league' and average gross weekly earnings of industrial workers in coalmining had fallen to 108 per cent of the average for all manufacturing industries, after being 125 per cent in 1975. The incentive scheme put the relative position back above 120 per cent in 1978, and subsequent changes in the rate of incentive pay helped the annually negotiated wage settlements to keep it there. Miners were again top of the earnings league in 1978, 1979, and 1980, and second in 1981 and 1982.[2] Average weekly earnings (including the value of allowances in kind), which had gone up from £40.09 in 1973–4 to £60.53 in 1974–5, were £90.12 in 1977–8 and rose to £108.30 in 1978–9. They reached £150.08 in 1980–1 and £167.21 in 1981–2.[3] The hope was that the combination of relatively high total earnings and an incentive element would induce in workers an attitude to the job that would lead to greater productivity.

[1] MMC, *National Coal Board*, I, 266 and 268–71; Handy 1981, 277–8 and 281–5.

[2] MMC, *National Coal Board*, I, 274. The relative figures are for weekly earnings unaffected by absence or overtime. The percentage relativities could be calculated in other ways and with slightly different data. For instance, the New Earnings Survey figures (which are compiled on a sample basis) are not quite the same for coalmining as the more comprehensive figures produced by the NCB. But the relative position from year to year is reasonably presented by the figures cited, which are prepared on as nearly uniform a basis as practicable.

[3] NCB *Ann. Rep.* 1982–3, 20–1. There was a change in the method of calculating the figures in 1976 but comparisons over the period are not seriously upset.

It is impossible to say precisely how fully the hope was met. One possibility was that workers would not only be more effective while they were on the job but would also be on the job more regularly and thus make the management of operations smoother and easier. In this respect the results were conflicting. Past experience had suggested that the more numerous the complications of the wage system the greater the number of disputes, and the incentive scheme reversed the previous trend towards drastic simplification by adding one new complexity. Sure enough, in 1978–9 the number of manshifts lost through disputes almost doubled the previous year's total, and well over half the lost time arose from disagreements on incentive payments and bonuses. The proportion of coal lost through disputes, 1.39 per cent, was higher in 1978–9 than in any other year without an official strike since 1970–1. But time lost through disputes arising from the incentive scheme fell steadily thereafter, and was more than halved within three years: in 1981–2 it amounted to 1 shift a year for every 5 workers.[1] This was probably not too high a price to pay, especially when it is borne in mind that by late 1977 dissatisfaction with the general level of pay was building up and, in the absence of an incentive scheme, would have led to greater losses through disputes. The other factor affecting time at work was voluntary absence, and in this respect the incentive scheme appears to have been wholly beneficial. Once there was the regular possibility of earning extra for a full day's effort, fewer men were content to miss a day if they could avoid doing so. The absence percentage rose from 16.7 in 1975–6 to 17.3 in 1976–7 and 17.6 in 1977–8. After the introduction of the incentive scheme it started to come down, slowly at first, then more rapidly. It was 17.1 in 1978–9, 14.8 in 1979–80, 12.4 in 1980–1, 11.4 in 1981–2 and 10.4 in 1982–3, the lowest level since nationalization.[2]

The widening of differentials in favour of those most directly engaged in production was accomplished to a significant degree, though not quite as much as had been intended, because the terms for which the NUM pressed when negotiating on other elements in wages tended to go somewhat in the opposite direction. Over the first eighteen months the differential of the highest graded underground men over surface workers was about 12 percentage points more than it had been before incentive payments started, and their differential over lower graded underground workers was 7 or 8 percentage points more.[3] The signs of

[1] MMC, *National Coal Board*, II, 189, Handy 1981, 289–90.

[2] NCB *Ann. Reps.* for years stated.

[3] Handy 1981, 282–5.

response were in the figures of productivity, with the change of trend coinciding with the start of the incentive scheme. The largest changes came where incentive pay, rather than bonus, was due. The number of manshifts worked at the face per machine shift immediately began to fall sharply, and by 1980–1 was about 12 per cent less than before the incentive scheme. Production OMS probably rose about 22 per cent in the four years to 1981–2, though the precision of the figure is lessened by changes of definition. And the excess of performance achieved over standard task was greater, on the average, on development drivages than at the coal face. Nevertheless, the level of overall OMS still did not rise as much as might have been hoped. If the projections made in *Plan for Coal* had been accomplished, it would have been 26 per cent higher in 1981–2 than it actually was. By 1978 the achievement of those projections was no longer expected, but something a little nearer might not have seemed out of reach. And indeed, in the first year at least, and perhaps for longer, it was doubtful whether the incentive scheme had broken even financially—again, it is partly a matter of definition—as the increase in productivity was less in proportion to the total of incentive and bonus payments than had been anticipated.[1]

There were two obvious limitations. One was that while productivity increased on the surface as well as at the point of production, it did not do so among workers elsewhere below ground, whose OMS in 1981–2 was less than it had been ten years earlier and showed no significant change of trend when bonus payments were introduced. By this period about half of all mineworkers were in this category, for the proportion of workers at the face had fallen to about 28 per cent as a result of greater mechanization.[2] So the low productivity of workers elsewhere below ground was a very serious drag on the overall figure. It seems unlikely that these workers had collectively a different response from others to bonus payments. The observed trend of their productivity indicates that the incentive scheme was far from being a cure-all and that other explanations have to be sought. These are likely to be found in the way labour was deployed. This may have involved some overmanning in order to be able to cope with a rate of output that was in progress for only part of each shift, but it must also have reflected the retention of various jobs that were discontinuous and only modestly aided by mechanical and electrical equipment. Such jobs were a proportionately heavier burden for an industry with a lower total output.

[1] MMC, *National Coal Board*, I, 109, 281–3; II, 73–83.

[2] Ibid., I, 124.

The other limitation was shown by the wide dispersion between areas, and to some extent between collieries, in the amount of incentive and bonus payments earned.[1] This could have arisen from inconsistencies in the measuring of standard tasks. But the interesting thing was that incentive earnings tended to be highest where levels of OMS were already highest before the new scheme. This suggests that low productivity may encourage habits and practices that tend to prolong it even when opportunities for change arise. Both limitations indicate that there was scope for raising the rate of increase of productivity faster than had been accomplished in the late seventies. They suggest also that an important influence on that rate of increase was the extent to which output was concentrated at collieries where the most advanced technology was most widely applied and in locations where the pursuit of high productivity was familiar enough to be taken for granted.

## v. The modernization of capacity

Although there was hope that some re-expansion of coal output might eventually be both demanded and practicable, the basic strategy attempted for the dozen years from 1974 was to keep output approximately level and to produce it more efficiently. This was indicated by *Plan for Coal*, which postulated losses of old capacity more or less equal in amount to the maximum new capacity that could be created in the available time. But a quantitative balancing of capacity changes, which was the best that could be hoped for, was to be accompanied by qualitative improvement. One of the explicit premisses of *Plan for Coal* was an average increase of 4 per cent per year in OMS. This was to be achieved not only by more efficient working at existing collieries, but by bringing the new capacity into use in good time.[2]

There were at least two significant implications. One was that the more efficient industry was likely, in one sense, to be somewhat smaller. An annual 4 per cent rise in labour productivity, with a more or less constant output, implies by inescapable arithmetic a gradually declining labour force. This was not stressed at the time, partly because in 1974 labour shortage seemed a more likely prospect than labour surplus. The NCB were conscious that, whatever the total labour force, the level of new recruitment would need to be higher than the average for the previous decade, and there was doubt whether this increase would be

[1] Ibid., I, 270–1; II, 198–9; Gormley 1982, 153.

[2] Dept. of Energy, *Coal Industry Examination: Interim Report*, 11.

easy to attain. A second implication was that the capacity that would be lost would be much inferior to the new.

The forecasts of capacity loss were based mainly on prospective exhaustion,[1] but it was highly relevant to the achievement of greater efficiency that collieries approaching exhaustion included many of those with the lowest productivity and the highest costs. The rate of improvement in efficiency could be affected by the time at which such collieries were deemed to be exhausted. 'Exhaustion' is to some extent a matter of definition. An otherwise exhausted colliery may sometimes be revived by investment to reach previously inaccessible coal. On the other hand, a colliery may be visibly dying long before the last piece of coal is extracted. In its last years the amount of coal remaining is often so small in relation to the size of the workings and the number of men and machines needed that operating costs go on rising steeply, and the physical problems of operation may also become more difficult. In such conditions the term often used in the sixties as a reason for closure, 'exhaustion of realistic reserves', seems appropriate. But the view may be pressed that this is not total exhaustion and that the colliery should continue to operate, although this must work against the objective of increasing the efficiency of the industry.

The unstated but clear implications and the problems of definition eventually made it difficult to proceed with as rapid a modernization of the industry as had been hoped and intended. All the members of the Tripartite Group, including the trade unions in the coal industry, collectively declared in 1974 'our conviction that an efficient competitive coal industry has an assured long-term future'; and they immediately added, 'The NCB's Plan for Coal has our full support as a general strategy.'[2] The combination of new developments and closures of some old capacity was central to the strategy. It determined the quantitative aspects of supply and was essential to the achievement of increasing competitive efficiency, which was an explicit major objective. In practice, however, the commitment to new developments proved stronger than the commitment to closures. The strength of the latter commitment tended to diminish when competition increased in the labour market, and the possibility of alternative definitions of what might seem to be agreed criteria for closures gave rise to disagreement about the nature of the commitment. Some proposed closures were criticized and resisted, especially by the NUM, whose attitude appeared

[1] Dept. of Energy, *Coal Industry Examination: Interim Report*, 10.

[2] Ibid. *Final Report*, 23.

to have political approval in the later years of the Labour government. The general scale of anticipated closures of capacity was, however, clearly stated as part of the *Plan for Coal* for which all the parties concerned publicly expressed their support. If it were not applied in practice the agreed objective of rising efficiency was unlikely to be fully attained.

By the early eighties there was little sign of an approach to a state of affairs in which worn-out or obsolescent capacity had been replaced by new. The rate of closures assumed by the Coal Industry examination would have meant the disappearance of around 28 million tonnes of old capacity by 1982. But closures to that date were only of collieries which had produced 8.74 million tonnes in their final year and which, before they entered terminal decline, probably had total capacity of about 14 million tonnes; and even this reduction included a disproportionate amount in the first year, which related to proposals which preceded agreement on *Plan for Coal*. This was a major (though not the only) reason for the continued existence of a large number of high cost collieries. Of all the collieries operating at the end of March 1982 almost a third had made an operating loss in every one of the five financial years to 1980–1, and just over a third lost more than £10 a tonne in 1981–2, though these latter produced less than one-fifth of the total deep mined output in that year.[1]

It may seem surprising that the outcome differed so much from what had been intended, especially when slightly earlier practice is compared, so it is desirable to review the course of events. The decision of the NCB to respond to rising costs by seeking to resume colliery closures had already been adopted late in 1971 and the procedures set in train, before the outcome of the 1972 strike made it seem impolitic to press for any more closures that could be avoided.[2] But it takes time to complete all the procedures already begun, and most of the closures already agreed did not take place until 1973. In the year 1973–4 eighteen collieries closed and four were merged with others. A few of the closures, including that of Gresford, in North Wales, which employed more than 1,000 men, took place on economic grounds, though exhaustion was the commonest reason. The programme spilled over into 1974–5, when ten more collieries were closed and three merged, but, subject to all the problems of definition, closures that would be regarded as predominantly economic had ceased by then. After that, colliery closures

[1] MMC, *National Coal Board*, I, 168, 171, and 174.

[2] Chap. 6 above, pp. 326–7.

became very few. There were five in 1975–6, three (plus one merger) in 1976–7, five (plus two mergers) in 1977–8, nine in 1978–9, and five in 1979–80.[1] None of the collieries affected were very large and none were closed on mainly 'economic' grounds, except that, under a special agreement, one or two in Scotland closed in anticipation of exhaustion. In the course of 1980–1 the NCB, under increased financial pressure from the government, sought to increase the rate of closures and to stop restricting them to cases of physical exhaustion of reserves, a change which was successfully resisted by the NUM.[2] The nine closures carried out in that year resulted from decisions taken earlier and were in the same category as those of recent years. Altogether, in the first eight years of *Plan for Coal* (i.e. to the end of 1981–2), there were fifty-six closures (excluding mergers), but the ten in the first year related to earlier decisions. The forty-six accomplished in seven years represented rather more than half the number expected when the plan was adopted, but less than half the amount of capacity. One feature that was in accordance with the plan was the geographical distribution of closures: thirty-eight of the fifty-six were in the high cost, peripheral areas of Scotland, the North East, and South Wales. There were also ten in Yorkshire, which was one of the chief regions for low-cost development, but although much of the Yorkshire coalfield has large modern collieries, it also included pits that had continued from an earlier period than any elsewhere, apart from Durham.[3]

The strongest reason for the small contribution of closures to modernization was the opposition of the NUM. This union adopted a national policy of opposition to closures other than those caused by total exhaustion of reserves of coal or safety reasons. The fact that this policy conflicted with the union's pledge to support *Plan for Coal* made no difference in practice. The other unions also did not accept financial non-viability as the principal ground for closures (and *Plan for Coal* had never expected it to be the *main* ground), but they were more flexible than the NUM, at least to the extent of recognizing impractical mining conditions as a justification of closure,[4] though NACODS took a view of closure policy generally similar to that of the NUM.

There were, however, some arguable, though not conclusive, grounds

[1] MMC, *National Coal Board*, I, 173–5 gives summary figures, not distinguishing mergers from closures.

[2] See below, section vii of this chapter.

[3] An unavoidable side effect was the further contraction of that part of the Yorkshire coalfield that has a long tradition of very good labour relations.

[4] MMC, *National Coal Board*, I, 174.

for going slowly with closures for a few years. One was that from 1974 no reduction of capacity was envisaged, and the larger projects to create new capacity, even those (the great majority) at existing collieries, took a few years to plan and complete. Moreover, many of these schemes improved the quality of capacity rather than increased its quantity. Smaller schemes could be carried out more quickly but they probably did little more (if anything) than offset the loss of capacity which is the normal accompaniment of the operation of continuing collieries. So there was a rational case for delaying closures until the fruits of increased investment were available, and much of this new capacity came rather late. This was partly because nearly half the new capacity was planned to come from new mines which would need longer than any other projects, and what proved practicable was both less in quantity and later in date than expected. It was also partly because, although there was fair progress with many projects at existing collieries, they took longer to complete, on the average, than had been hoped. By 1977, too, eyes were looking further ahead to a future in which coal sales and output might greatly increase, so perhaps the longer preservation of capacity that had appeared ready for removal looked, for the moment, less undesirable; but such an attitude would be less rational.

In fact, by this time the economic case for delaying closures was weak. The medium term development plan prepared at the end of 1977 noted that the coal industry still had a potential deep mining capacity of 120 million tons, which was under-utilized to the extent of about 15 million tons. It was expected that this under-utilization would continue for the next few years, if only because the introduction of a voluntary early retirement scheme might create some shortages of skilled manpower. The difficulty of getting full output from existing capacity could be used as an argument for keeping some capacity in hand. But the opposite argument for shedding a little old capacity to bring the total nearer to the figure of actual output looks the stronger, first, because there was doubt about the ability to achieve the projected level of sales, and second, because some investment schemes were being completed and so adding a little new capacity of better quality.

Something was being attempted in specially difficult cases. Despite the drastic contraction from the late fifties onwards, deep mining in Scotland was still making heavy losses which the NCB wanted to reduce. Provided that the existing Scottish market of 11 to 12 million tons (about three-quarters of it for the power stations) could be secured, it seemed better to try to supply it with Scottish coal at an improved level of costs than to incur the transport costs of supplying more of it from England, thereby perhaps putting up prices enough to lose some of the market. The only solution to the problem seemed to be to increase the proportion of opencast coal in total Scottish output and therefore to close some of the most uneconomic deep mines, with this perhaps making it possible to end the operating losses on the mining of steam coal. But none of this could be done without some agreement with the SSEB on coal sales and with the unions on pit closures. So in April 1976, under the chairmanship of the Minister of State, Scottish Office, joint discussions were held with representation from the NCB, the mining unions, the SSEB, the NSHEB, and the electricity industry unions. These discussions paved the way for a five-year agreement from October 1977. This involved a guaranteed minimum purchase of coal by the Scottish power stations, in return for concessions on price. These concessions were made possible partly by some earmarked financial assistance from the government, and partly by the closure of certain heavily losing collieries, to which the mining unions agreed in return for NCB undertakings to invest in major projects at other Scottish collieries. The unions accepted the objective of making the Scottish coalfield financially viable for steam coal production by the end of the five-year period. Seven Scottish collieries were, in fact, closed in the five years starting in 1977–8, and in one or two cases the unions accepted the closure of collieries before their reserves were physically exhausted, although they were approaching exhaustion. It was a braver try than most, but the financial objective proved elusive. In the fifth year (1981–2) every one of the twelve collieries in Scotland made an operating loss, and six of them made such a loss in every year of the five.[1]

The situation in South Wales was comparable, but financially even worse. The NCB in 1978 discussed with the government two heavily losing collieries, Abernant and Mardy, which could not be closed without causing more than 2,000 redundancies, to which there was no chance of getting union agreement when the collieries were reviewed. Alternative closures that would cause no redundancies were also considered. Discussion with the Welsh Office produced only a referral to the Department of Energy, but the Secretary of State for Energy (Mr Benn) reconvened the Tripartite Group to examine the whole problem of the coal industry in South Wales, and the Group set up a working party in December 1978. All this suggests government reluctance to take any difficult action. It was symptomatic of prevailing views that, at the first

[1] MMC, *National Coal Board*, II, 30 and 42.

meeting, the NUM tabled a paper implying that viability must be achieved without pit closures. However, it proved possible in April 1979 to produce an agreed report which outlined the measures needed if the coalfield's loss were to be reduced, over five years, to £2 million a year. They included joint efforts of management and unions to improve performance at loss-making pits, and closures when necessary, though there was a warning that closures would be difficult in places where no replacement capacity was in prospect; new NCB projects on the basis of normal investment criteria; and additional interim assistance from the government.

The weakness of all this (which doubtless made agreement easier) was that it was merely descriptive and committed nobody to anything. It was believed in the NCB that the exercise had had some educative influence and had, for instance, caused some moderation in the level of wage demands proposed from South Wales. But the government accepted no additional financial obligation; there were few capital projects which could meet normal financial criteria for investment; pit closures did not accelerate and in one case, Deep Duffryn, there were delays and additional expense in order to convince the union of the unsoundness of their claim that there were still practical possibilities of working additional coal. And the condition of the South Wales coalfield as a productive business did not improve. Nor did a further enquiry into the development of employment opportunities in South Wales, conducted by the House of Commons Committee on Welsh Affairs in 1980, produce anything more effective.

It might well seem that, not just in South Wales but generally, progress in modernization by the removal of obsolescent capacity was excessively slow. To any suggestion of this kind the reply of the NCB was 'that the necessity of maintaining good industrial relations precluded faster progress'.[1] There were, indeed, both particular and general ways in which relations with the unions had to be taken into account. The particular concerned the joint management-union procedures at area level which were instituted in 1972 for the review of the progress of each colliery every quarter, with the object of agreeing on measures to deal with any problems that were identified. Any proposals for closure could come only as part of this procedure. The effect was that most closures were carried out by agreement made locally. If there were no agreement there was provision for the matter to be referred to the national board and, if the disagreement persisted, for the unions to make

[1] Ibid., I, 175.

a case on appeal to the national board. If, after all that, the NCB still decided on closure they were bound to give four months' notice.[1] This was a procedure that probably had a bias towards limiting the number of closures: in the eight years after the adoption of *Plan for Coal* only five collieries were closed after an unsuccessful appeal.[2] The more general aspect was that closures were only one sensitive point among several. It did not always appear that the gains from pressing hard on closures would offset the loss of goodwill on other matters, and after the upheavals of the early seventies, there was much anxiety not to have a repetition. Even so, that still leaves doubt whether all the possibilities of trading off other benefits in return for some additional closures were fully explored.

The complementary, and much more readily appealing, element in modernization was investment in new capacity. In *Plan for Coal* the original proposals would have averaged about £60 million a year for major projects plus £70 to £80 million a year for smaller, routine investments, at 1974 prices. The time taken in planning inevitably meant that large projects could not be approved and started immediately, but capital expenditure on major projects reached the original proposed level, in real terms, in 1976–7 and thereafter regularly surpassed it, and the proposed investment level was increased from 1976. Investment in the smaller schemes exceeded the planned level right from the start. The contrast with the level immediately before *Plan for Coal* was very great, as is evident from Table 7.3.

There was thus no holding back on the commitment to the positive side of modernization. By September 1978 approval had been given to about 100 major projects at seventy-four long-life pits and to four projects for new mines. The investment at existing collieries was designed to add 20 million tonnes of new capacity, and that at new mines a further 11½ million tonnes. This was in addition to schemes intended to improve revenue and operating efficiency rather than add to capacity. The latter type of project included improved coal preparation facilities, replacement of steam by electric winders, improved surface buildings and equipment, and rapid loading systems for railway wagons. The bringing of additional capacity into use took longer. By March 1978, 38 of the major schemes (costing over £500,000) under *Plan for Coal* had been completed but only about half these were for additional capacity. If

[1] MMC, *National Coal Board*, II, 191–2 gives details of the procedure.
[2] Ibid., I, 174–5.

Table 7.3. *Trends in mining capital investment, 1973-4 to 1981-2* (£ million at outturn prices)

| Year | Projects costing over £500.000 on completion | Total capital expenditure in each year |
|---|---|---|
| 1973–4 | 7 | 68 |
| 1974–5 | 21 | 112 |
| 1975–6 | 53 | 211 |
| 1976–7 | 89 | 266 |
| 1977–8 | 130 | 334 |
| 1978–9 | 182 | 454 |
| 1979–80 | 252 | 617 |
| 1980–1 | 333 | 736 |
| 1981–2 | 381 | 722 |

*Source*: NCB *Ann. Rep.* 1981–2, 12.

one takes the rather higher category of schemes costing over £1 million, then 35 of these were completed before March 1979, and the total had risen to 43 one year later and 69 two years later; by March 1982 this total had reached 97.[1] These figures meant that very significant improvements of capacity had been achieved, and by 1980 they had become substantial in quantity. There was then an ample supply of productive capacity, even without the use of all the very high cost collieries that were still being operated.

Though things changed rather more slowly than had been hoped, the provision of extra capacity at existing collieries proceeded reasonably close to the scale intended. The main deficiency was in the establishment of new collieries. This had always been the most speculative side of the programme. In 1973–4 the opportunities for locating new collieries were known, in most cases, only in very general terms. Further investigation was necessary before detailed planning and construction were possible, and this investigation indicated that some of the prospects were less promising than had been expected. Only four new collieries were constructed, though planning went a long way on several others. The new collieries fell into two categories. The first were drift mines of only moderate size in districts that had already been heavily

[1] NCB *Ann. Rep.* 1977–8, 15; 1978–9, 16; 1979–80, 19; 1980–1, 20; 1981–2, 12.

worked but where substantial reserves were found to remain at not very great depths, though too far below the surface for opencasting. The other projects were for larger and deeper collieries.

Three new drift mines were completed in the seventies, Royston and Kinsley in Yorkshire, and Betws in Dyfed. The possibilities at Betws had been known for several years, but the long and rapid contraction of the industry had ruled out any chance of turning them into reality. These drift mines, though highly mechanized and laid out for high rates of production and the greatest possible integration of operations, were relatively simple in basic design and could be constructed quickly. At Royston the nature and extent of the reserves were still being investigated by boring in 1973, but the completed mine began producing coal in September 1976: the drifts were driven in only 25 weeks. Betws was a bigger undertaking, going to much greater depths with drifts 3,246 metres long (about five times the length of Royston's) and was designed from the outset to operate two faces, whereas Royston originally had only one face and was adapted in 1978 to operate with two. Betws therefore took four years to complete, with two and a half years spent on the main physical construction. It was inaugurated in March 1978. Kinsley was the quickest project of all and was producing coal less than two years after work began. It came into operation in August 1979, about a year ahead of schedule, as it was able to work a shallower seam than the one with which it was originally intended to begin. These were, on the whole, very successful projects. Though the new mines were not very large physically, the three together produced about $1\frac{1}{2}$ million tonnes a year, and they were able to do this because they were laid out for high productivity. Their overall OMS was about twice the national average (rather more for the two Yorkshire mines), and their face output was very high; Royston several times raised the world record for the greatest length of face cut in one shift. Despite some early difficulties, Betws by 1980–1 became that near miracle, an anthracite mine with an operating profit, and the other two mines, though producing coal at the bottom end of the market, were, in general, usefully profitable.[1]

The major activity in constructing new collieries was for the exploitation of the Selby coalfield. The extent and character of what was possible there could not be fully appreciated until, with the change in the

[1] MMC, *National Coal Board*, II, 34, 41, 46, and 53. The early figures for Kinsley are distorted because they relate to revenue production while the construction of the mine was still being completed.

market, it seemed worth while in 1972–3 to follow up the reconnaissance of the early sixties with much more detailed exploration. The abundance and cleanness of the coal (which could be used without going through a coal preparation plant) created a desire to go ahead on a large scale, although it was recognized from the outset that Selby was a difficult mining proposition. One great problem was that the whole district is flat and low-lying and receives a great deal of water, and any disturbance of natural drainage by subsidence needs to be minimized. That requirement in turn reduces the proportion of the coal which it is safe to extract. The colliery workings also needed to be widely dispersed around the coalfield in order to help to reduce the subsidence in any one locality; and the natural drainage presented problems in the sinking of the mines. Despite these inhibiting influences, proposals to work the coalfield soon went ahead. It had the advantage of location close to existing power stations and to rail transport. At first it was cautiously suggested in public that it could produce 5 million tons a year, but this was soon at least doubled, and the figure of 13 million found its way into an early draft of *Plan for Coal*, but in the summer of 1974 it was decided that 10 million tons a year was the fastest rate of exploitation that was manageable. A plan was prepared for the development of the coalfield by a highly integrated, large-scale mining complex, using the latest technology. There were to be five collieries, at Wistow, Riccall, Stillingfleet, Whitemoor, and the rather imprecisely named North Selby. These would all be linked underground, and two drifts would carry the output of all of them to the surface at Gascoigne Wood, near Monk Fryston. It happened that it was not only geologically convenient to bring the drifts to the surface there, but there was at this point a large area of little used railway siding land, which could be equipped for the rapid loading and dispatch of the coal.

There was a lot of negotiation with the local authorities and other public bodies, and an effort to keep the public informed about what was proposed and what its effects were likely to be. The NCB's planning application was made the subject of a public enquiry in the spring of 1975. On the whole, though there were disagreements, there was little public hostility, despite the novelty and extent of the likely economic and social impact on the district, but there was public pressure for many specific safeguards. Most of these were voluntarily accepted by the NCB and were endorsed and slightly strengthened in the conditions attached to the planning permission which was granted in April 1976. Partly to meet the need to minimize subsidence, the plans were based on extraction only from the thick Barnsley seam, with nothing taken from the five lower seams which had been proved, one of them with thicknesses up to six feet. There was agreement to leave large unworked pillars of coal for the support of much of Selby, including the Abbey and the River Ouse, and another pillar under Cawood to prevent flooding. The planning decision also made mining near to the River Derwent conditional on county council approval. But the planning enquiry supported the NCB's resistance to demands to refrain from mining beneath the York to Selby railway line. Instead it was decided to build a new stretch of main railway line to the west of the existing one.[1]

Besides the fairly ready grant of planning permission, the way was smoothed by the removal of one legal obstacle. It was found that some people retained copyhold rights in the coal, so that although the NCB owned the freehold they were not free to work the coal. Nothing had been done about this problem before because nobody was aware of it. The government agreed to the inclusion of a clause in the Coal Industry Act 1975 to give the NCB full rights of operation in such cases.[2]

So the construction of the Selby complex of mines went ahead. It was the largest single capital project ever undertaken in the British coal industry and was planned as the largest deep-mining complex in the world. It was expected to employ between 3,000 and 4,000 men and to produce 10.2 million tonnes a year when in full production. That full production was scheduled for 1987–8, but the first revenue production, with one colliery in partial operation, was originally projected for some time in 1981–2 and then altered to 1982–3. In fact, the beginning of regular production, at the Wistow mine, was not until 4 July 1983. It was hoped to time the build-up of production to match the rundown of some West Yorkshire collieries which would be exhausted, and some of the early recruitment of labour was from this source. When the full-scale commitment to carry the project through was made in February 1978 it was estimated to cost £552 million at August 1977 prices. With the effects of inflation and some modifications in the detail of the project

[1] Various Authors (Vielvoye and others) for NCB 1976, 97–128 describes the Selby project in two articles by Michael Pollard and Jeremy Bugler respectively. NCB *Ann. Rep.* 1974–5, 21–5 has a briefer description.

[2] Dept. of Energy, *Coal Industry Examination: Final Report*, 23; NCB *Ann. Rep.* 1975–6, 21. The discovery of retained copyholds was more or less the legal equivalent of finding a coelacanth. Probably most experts believed that copyhold rights had never extended to sub-surface minerals. The nineteenth-century agreed procedures and early twentieth-century legislation, which were intended to abolish copyhold, the nationalization of coal royalties in 1938 and of the coal industry in 1946, had all failed to bring any relevant example to light.

this had gone up to £1,005 million by March 1982 at the prices then current. These figures include the interest payable in the period up to completion of the scheme. This was originally estimated at £213 million, and in March 1982 at £346.5 million.[1] Some £452 million had been spent by March 1982. It was still too early to say how well the scheme would live up to its hopes but, despite the delays in starting production at Wistow, which soon ran into serious difficulties with the inflow of water, it was still believed that the whole project could be completed in 1987, as planned.

Selby was the only large new mine project on which practical progress was made in the seventies, but there were several others on which a good deal of preliminary planning was done. Much the largest came from the combination of two prospects originally considered separately. The first was in the Vale of Belvoir where exploration proved so encouraging that in 1975 consultants were appointed to make feasibility studies. In the summer of 1976 they suggested the sinking of a mine at Hose, to produce 3 million tons a year from the lower two of the four seams that had been proved. By this time total recoverable reserves had been assessed at 360 million tons. It now began to seem sensible to work the two upper seams simultaneously and produce from them 2 million tons a year, with the construction of another shaft, about five miles from Hose, for manwinding and services. The Vale of Belvoir coal had been thought to terminate at a fault to the south, though it was known that there was unworked coal still further south, in a field known as East Leicestershire. But intensive exploration of the latter had established by mid-1976 that this was a continuous extension of the Vale of Belvoir reserves, and a search began for another suitable site in this southern section, where a mine to produce another 2 million tons a year might be located. Right from the start, it was recognized that there would be environmental problems, chiefly associated with dirt disposal, at any mine site in the Vale of Belvoir, and one reason for selecting Hose was the belief that it would be thought the least objectionable. In fact, there was very strong opposition to the whole proposal, led by the principal landowner, the Duke of Rutland, who also happened to be the chairman of Leicestershire County Council, which at first showed a more neutral attitude.

Nevertheless, the preparation of plans went ahead. The coalfield was conveniently situated in relation to the Midlands power stations and its development was supported by the CEGB. Capacity there could, it was

[1] MMC, *National Coal Board*, I, 224–30.

hoped, be phased in as replacement for collieries that would decline in the existing Leicestershire field and elsewhere in the South Midlands. Apart from the tipping question, there were no expected severe technical problems, and it appeared that a financially satisfactory investment could be made. There were estimated to be 520 million tonnes of recoverable reserves.

By the summer of 1977 a scheme had been prepared for the development of the combined fields, now jointly renamed the North East Leicestershire Prospect, with proposals for mines at Hose and Saltby in the north and at Asfordby in the southern extension, with a projected labour force of 3,800 and a total output of 7 million tons a year on completion, which was expected to take about twelve years. Application for planning permission was made in August 1978 and the application was formally called in by the Secretary of State for the Environment. But although, after an immense public enquiry, the inspector recommended in favour of the application, subject to restrictions on tipping at the colliery sites, his recommendation was overturned. As has been noted, it was not until 1983 that planning approval was given for development in the southern extension only, with one colliery at Asfordby. So there had been years of partly wasted effort and much abortive expenditure.[1]

The largest of the other prospects that were pursued were a possible new colliery at Park, just north of Stafford, and the reopening (or, in effect, replacement) of Thorne colliery, to the north-east of Doncaster. The Park prospect was under detailed consideration from 1975 onward. There were some 130 million tons of coal reserves, and a scheme was drawn up for a colliery that would produce a little over 2 million tons a year. But the scheme was unattractive to the CEGB because of the location and nature of the coal, which had a high chlorine content of 0.6 per cent. Chlorine can be damaging to boiler tubes. The Park coal would therefore need to be blended with low chlorine coal from other sources, and there were discussions about the possibility of doing this at a particular power station or at Park or at some other supplying colliery. Whatever arrangement might be adopted would add to costs and increase the fnancial risks of the project. There also seems to have been some worry in the CEGB that the development of Park might give rise to pressure for the building of another coal-fired power station west of the Pennines, which was not wanted. With the encouragement of the Secretary of State for Energy, planning permission for the Park project was applied for in September 1978; but the continued unwillingness of

[1] There is a brief summary of the project in NCB *Ann. Rep.* 1978–9, 16.

the CEGB to accept this coal, or to support the project at any public enquiry, caused the NCB to cease pursuing the proposal.[1]

Because it was a reopening that did not need planning permission, there were fewer problems at Thorne. Reserves of 50 to 75 million tonnes of good steam and house coal had been established, though new operations there could not be as profitable as those at Selby, because the coal needed to be washed before sale. The problems of flooding and the resultant damage, which had closed the colliery in 1956, had been remedied by 1966, but there had never been market conditions or investible funds to permit resumed operations. But in January 1979 it was decided to go ahead, refurbishing parts of the old colliery for further use but sinking a new shaft and opening new reserves, from which annual output of 2 million tonnes was planned for the mid-eighties. A major worry was that Thorne colliery had been notorious for labour indiscipline, so not only were assurances sought from the unions about future manning, but everything (except, curiously, that nobody seems to have thought of a new name) was done to treat the project as a new venture, not a revival of an old one. But by 1982–3 the market had so deteriorated that output from the project was thought to be dispensable, and it had become very difficult to provide enough capital within existing financial limits. Although a good deal of work had been done, it had not gone so far as to make completion the least uneconomic course, and the project was suspended.[2]

No other scheme got nearly so far. For a couple of years from late 1977 the possibility was being considered of a new colliery at Margam, in South Wales, to mine reserves of 51 million tonnes of prime coking coal. It was thought it might be a useful item when efforts were made to get the unions to agree to a complete reorganization of the South Wales coalfield, but that was not something to rely on. Mining conditions were such that it was thought unlikely the colliery could produce more than 800,000 tonnes a year, costs per tonne were projected to be more than two and a half times as high as at Selby, and the steel industry's demand for coking coal was contracting. So an economic return could not be expected and nothing was done for the time being.

Another possibility of developing new reserves as part of a programme involving the closure of high cost capacity was in Scotland, as a sequel to the 1977 agreement on power station coal burn. This proposal was for the working of 51 million tonnes of reserves beneath the Firth of Forth

[1] Ibid. 1978–9, 16; 1981–2, 12–13.

[2] Ibid. 1978–9, 17.

at Musselburgh. The scheme was not for a new colliery but for the extension of the existing Monktonhall colliery with an additional shaft and surface buildings at Musselburgh. Reckoned as part of the financial return was the saving on the closure of Dalkeith and Lady Victoria collieries before their reserves were physically exhausted, a procedure to which the Scottish NUM would agree if, but only if, the scheme were approved and jobs thereby created at Monktonhall colliery for the displaced men. This was somewhat stretching the normal criteria for financial appraisal and the scheme was also not without physical problems, including those of a cramped surface site. In the end, though a lot of preliminary work was done, the scheme was left in abeyance and other reserves within Monktonhall colliery were opened up instead. Dalkeith colliery closed promptly in 1978 but Lady Victoria colliery stayed open until 1981.[1]

All in all, the contribution of new collieries to the modernization of capacity in the seventies and early eighties was rather disappointing in amount, though the long delay before the contribution could be large was recognized when *Plan for Coal* was first drawn up. The greater disappointment was not the small amount of production coming from new collieries in the early eighties, but the clear indication then that the amount in the later eighties would be a good deal less than had been planned. Yet in total quantity, capacity had been kept well up to market requirements, both by the many completed projects at existing collieries and by the retention of collieries beyond their expected life. The doubts were about quality and its relation to costs and profitability. The restructuring of the industry's productive capacity had been a good deal less comprehensive than had been expected. Some of the new capacity was not commercially successful. In 1981–2, of fifty-nine collieries which had had seventy-seven major schemes, twenty-two (which had absorbed just over a third of the investment) had a problematical financial prospect. This was not because the schemes were physically unsatisfactory. In some cases the forecast costs had been exceeded, and most often the market had turned sour to an extent not foreseen: eleven of the twenty-two collieries produced coking coal.[2] But, in general, new investment brought better quality into productive capacity, and even when the financial results turned out unsatisfactory they were less unsatisfactory than those of the old, high cost capacity which was retained.

The Monopolies and Mergers Commission concluded that the NCB

[1] NCB *Ann. Rep.* 1981–2, 13.

[2] MMC, *National Coal Board*, I, 210–11.

'should henceforth adopt a much more cautious approach to investment intended to increase or maintain production capacity, at least until such time as it is fully satisfied that it will be able to eliminate high cost excess capacity'.[1] It was an understandable suggestion in the light of the excess of supply over demand at the time of the report. Yet it was also near to a counsel of despair. Additional capacity usually incorporates higher standards of quality, and a restriction of investment to schemes aiming at qualitative benefits only is usually too piecemeal to achieve more than a patchy and small improvement in operating efficiency. Without additions the coal industry was likely to find the struggle to achieve competitive prices and to control its costs impossibly difficult. But that was equally true if it went without subtractions. In the late seventies and early eighties the subtractions were too few and the additions, though numerous enough, were completed rather too slowly.

### vi. Prices, sales, and the financial outturn

By 1973 the marketing and political situation of the coal industry had begun a continuing course of change that made rather different pricing strategies appropriate. No longer was most coal at a cost disadvantage for most uses, so it was no longer necessary to try to preserve a market by having appreciably lower list prices for the most cheaply produced output, as had been done in the sixties. Prices could be fixed with an eye on the ability of competitors, who often had higher costs for much of their production, to match them. But the margin between British coal costs and those of competitors was constantly changing. It was also more difficult than ever before to predict what the margin would be a year ahead; yet it was desirable, in the interests of both customer goodwill and the maintenance of strict budgeting within the coal industry itself, to keep prices stable for up to a year at least. The difficulty arose not only in predicting what competitors might decide, but from the unprecedented and varying rates of general inflation and from the existence of floating exchange rates that led to large fluctuations in the value of sterling. The latter had serious effects on the price of oil (which was charged in American dollars, even for British production) and the price of imported coal. In deciding what prices to charge for coal, attention was given mainly to competition from fuel oil, though by the end of the seventies more regard was having to be paid to the price of potential coal imports, especially of coking coal. Gas was a major competitor, but

[1] Ibid., I, 100.

its prices had less influence, simply because it was impracticable to match them as long as the policy adopted from the outset for the depletion of North Sea gasfields was maintained. All the coal industry could do was recognize the large parts of the industrial and domestic fuel markets which were irrecoverable in the immediate future, and plan and budget accordingly.

In these conditions it was desirable to have a price structure which would maximize revenue while clearing the market. Even if the prices of competitors were matched or beaten, there was a risk that, if fuel prices as a whole rose too far or too quickly, the response would be some reduction of energy consumption, and the objectives of the price structure would not be attained. But there were a lot of pressures towards large and frequent price increases. The one thing that might seem to work in the opposite direction was the provision of larger government grants after the passage of the Coal Industry Act 1973. The financial objective of the NCB became the achievement of break-even after payment of interest and receipt of government grants. But government grants were subject to variation or possibly withdrawal after a few years, and they were for defined purposes. To rely on them as a general cushion would have been both bad management and foolish policy. Revenue from sales had to be the chief source of cash and it had to meet rapidly increasing outgoings. The real cost of coal production was expected to rise throughout the seventies, even with the most favourable scenario, and in practice, with the continued lag in labour productivity, achievement fell short of the best hopes. Prices had to rise with real costs. There were also extra charges to meet because of the resumption of investment on a large scale. Most of this was financed by loans, so that ever-mounting sums of interest were due. The rest of the finance had to be generated internally, and though hopes of providing about half in this way were not fulfilled, there was still a need for extra revenue for this purpose. There was also a temptation to play safe when guessing at the immediate prospective effects of inflation and not be caught with prices of which the revenue yield failed to match the inescapable growth of the cost of inputs.

The pressures for price rises were particularly strong in 1974. The overtime ban and subsequent national strike of the winter of 1973–4 had caused losses which needed to be at least partly recouped; the wage settlement increased costs, and the policy of keeping miners at or near the top of the wages 'league' gave a commitment to continued increases in outgoings; and the huge rise of oil prices gave coal a hitherto unknown room for manoeuvre on prices. The result was that coal prices soared. In July 1973 they had gone up by 7½ per cent. On 1 April 1974 industrial and coking coals went up by varying amounts, all high, the highest being 48 per cent on those produced in the East Midlands and Yorkshire. They went up again on 1 October by an average of 32 per cent for coking coals and ranges between 20 per cent and 27½ per cent for industrial coals. On 1 March 1975 there were further increases averaging just over 32 per cent for coking coals and just over 30 per cent for industrial coals. There were increases also in the prices of domestic coals, from slightly differing dates, though these increases in 1974 were less than for other coal, an imbalance which later increases sought to redress.

Within one year coal had thus completely changed its prices relative to those of other commodities. It was, of course, only following the example of oil and had not thrown away all its competitive advantage, though it had reduced it. But in making this leap it was not seizing a market opportunity to make high profits. It was only starting to cover a lot of the enormous increase in costs experienced in the last five years. Not surprisingly, the immediate sequel to the revolution in energy prices was a small drop in energy consumption in 1975–6. The use of electricity fell for the first time in any year since the Second World War (apart from the effects of the three-day week in 1973–4). Even though the main impact was on oil, the consumption of coal fell by 5 million tons.

It was a warning. After March 1975 increases in coal prices were made at yearly intervals, and for four successive years were held within the general rate of inflation. In 1979 market conditions were again seriously disturbed by very high rates of inflation and by another series of great rises in oil prices. Coal prices were held for only four months and went up again on 1 July by 7½ per cent for coking and 13 per cent for industrial coals. This was followed at the usual time of 1 March 1980 by increases of 15 per cent and 20 per cent respectively for the two classes. Domestic coal prices also increased with similar frequency, but were differently phased and dated. After 1980 price arrangements changed in ways which make comparable general figures inapplicable.[1] A rough idea of the change in levels is that average price per tonne of coal produced was £7.12 in 1973–4 and £35.93 in 1981–2,[2] but this understates the price rise, because of the increasing proportion of lower-priced coals in the total.

The assessment of changing market conditions led to substantial changes of relativities within the price structure. The earliest was the

[1] NCB *Ann. Rep.* 1973–4, 12; 1974–5, 13–14; 1975–6, 6–7; 1978–9, 7; 1979–80, 11; 1980–1, 10.

[2] Ibid. 1974–5, 47 (with conversion from long tons to tonnes) and 1981–2, 65.

raising of prices in the central coalfields to the level of most of the rest. Quality for quality, in 1973 central coalfield prices were about 35 per cent below those in the peripheral coalfields, but the point had been reached where, for both steam raising and coking coal, demand for central coalfield output was threatening to outrun supply, while difficulty still remained in disposing of all the output from the rest. An additional argument was the need for a near-standardized national price to match the practices of the oil industry. Another was that the heart of the industry must be shown to be financially sound so as to give maximum justification for the increased investment needed there.

That the Board make substantial losses in Scotland and South Wales is something that causes no surprise by now to either informed commentators or the lay public, but our failure to break even let alone make a viable profit in the high productivity low cost central coalfields is inexplicable and exposes us to very damaging criticism, especially when we are trying to raise substantial additional capital.

So the policy of using central coalfield output, as in effect, a loss leader was abandoned, and only a few of the peripheral coalfields were still allowed to charge extra to try to recoup a little of their higher costs. That was why the price increase on East Midlands and Yorkshire coals in March 1974 was nearly double the other increases, and why further steps towards equalizing prices from nearly all coalfields were taken on the occasion of subsequent general price increases. By March 1976 pithead prices for industrial coal were the same for all coalfields except for a slightly higher level for Lancashire, a still higher one for South Wales and Kent, and the highest (about 14 per cent above the general level) for Scotland. For coking coals there was a general common price list, but the North East had a very slightly higher level, Cumberland, North Wales, and Lancashire a rather greater excess, and Scotland the highest prices.

Other structural changes concerned domestic coals. Here, too, relatively more was charged for what was most attractive. When prices were raised an above-average increase was put on the best quality. This was done again in April 1977 when the number of different grades was reduced from five to three, a simplification which was probably justified by the shrunken size of the market. A further change of practice was the abandonment of the system of zone delivered prices from 1 May 1979. From then on, domestic, like all other coal, was sold at pithead prices and merchants made their own financial arrangements for transport.

This, too, was a response to changed marketing practices. The system of zone delivered prices was related to arrangements involving railway rates fixed by a single carrier. It was designed when nearly all coal was carried by rail and it had allowed for fixed, slightly higher, prices for the minority of coal collected and delivered in other ways. But by the late seventies less than half of the supply of domestic coal was railborne, and it was impracticable to accommodate the varied charges of many competing carriers within a standard set of delivered prices.[1]

Another adaptation to a contracting market affected coking coals, demand for which fell off sharply in the late seventies because of the difficulties of the steel industry. The largest influence differentiating the price of one coking coal from another was the 'coking factor', which was based on the average quality of coke produced from different types of coal. Coking coals were grouped into six categories, each with a different coking factor. The method of calculating the coking factor was modified from 1 July 1978, and from 1 March 1980 the types of coal with the lowest coking quality were reclassified to be priced as industrial coal. From 1 January 1981 only the very best coking coals (mainly those which had formed the two top categories of the old six-category grouping) were still priced as such. The point was that in a reduced market the others could no longer command any premium if priced as coking coals, and it was therefore better to price them on the same basis as other industrial coals.

The changes in pricing practices, and especially the changes in price levels, did not proceed without external argument. The NCB were freed by membership of the ECSC from the possibility of statutory price control, but they regularly informed and consulted the government about prices and they had to take account of the reactions of their principal customers. As these were other nationalized industries, it was sometimes useful to resolve disagreements in discussions which involved representatives of government. Every major increase in coal prices brought a letter of protest from the chairman of the CEGB. His argument in 1975 that fuel costs were threatening electricity sales received some support from the drop in electricity consumption in the following twelve months, but in general there was the effective answer that the power stations found it worth while to burn more and more coal. When, in February 1977, the announcement of a forthcoming coal price increase at no more than the forecast rate of inflation attracted the usual letter of protest, this letter included a complaint about the practice

[1] NCB *Ann. Rep.* 1976–7, 5; 1979–80, 11.

of raising the price of Yorkshire and East Midlands coal as much as the rest. It was alleged that this invalidated the assumptions, based on statements by the NCB, which had guided the location of new power stations in the sixties. The NCB chairman was able to reply:

> the price your power stations have had to pay for their coal during the past 12 months has been far below the equivalent price for fuel oil. The major new stations on the central coalfields must have been paying, on average, £5 or £6 a ton below what would have been the case had that capacity been built oil fired on the Coast as has been the case in many other European countries and indeed, of course, in your own last phase of power station ordering. So I really do not think that CEGB has any need to regret their decision to build so much coal fired capacity in the Midlands and Yorkshire.

The CEGB returned to the same point two years later, with the implication that they were being charged far above cost for Yorkshire and East Midlands coal which, for those areas as a whole, was not so. The true contrast was that they paid so much less than cost for coal from some other areas. In 1977 it was reckoned that if Yorkshire and East Midlands coal were competitively priced it would raise another £200 million.[1] Despite all the complaints, power station consumption of coal rose from 70 million tonnes in 1972–3 to nearly 89 million in 1979–80, after which the economic recession caused a slight reduction, and very little of this was imported. The NCB sold 85.6 million tonnes to power stations in 1979–80, and coal generally at this time supplied well over 70 per cent of the fuel used for electricity generation.[2]

The NCB had, however, to pay a lot of attention to the needs and influence of their largest customer, and there were various arrangements they wanted and might not get without persuading the CEGB to support them. The competitive strength of coal was not determined by its own prices and qualities alone; it depended also on the number, location, and quality of coal-fired power stations. The NCB wanted to see the building of an additional large coal-fired power station and also the refurbishing of some of the older coal-fired stations, so that they would keep or improve their position in the merit order of operation. Even by 1976 the CEGB's use of oil-fired plant had dropped near to the technical minimum, so the feeling was that oil could not lose much more from

[1] It was considered impracticable to raise prices in this way because with the traditional system of a uniform price for each quality, some coal from elsewhere would be unsaleable, and the CEGB would not accept a change of pricing system which simply raised the cost of its most economic supplies.

[2] NCB *Ann. Rep.* 1972–3, 11; 1979–80, 11–13.

competitive changes, whereas coal could, if the swings went the wrong way. This was particularly so in relation to power stations used only for peak load. These mostly combined high delivered cost of coal with low generating efficiency, and the competitive advantage of coal in such stations was, at best, fairly small.

In fact, coal maintained its position very well. The increase in fuel oil tax in 1977 may have helped a little, though much of its effect was soon offset by the appreciation of sterling against the dollar. The building of a new coal-fired station, Drax 'B', was approved in 1977, the first new coal-fired station approved since 1966. The recurrence of large price rises for oil in 1979 improved the advantages of coal, so the CEGB gave attention to the refurbishing of 500/660 megawatt coal-fired units and thereby appreciably extended their design life.[1]

But there were also risks from fluctuations in current purchases of power station coal and from the possibility that some of the needs of the power stations might be met by imports. Some smoothing of the fluctuations was achieved by an agreement, continued over a series of years, that the generating boards would take extra coal, from time to time, for stocking on their own sites, in return for provisions about deferred payment for their surplus stocks. Various measures were taken to promote the use of home-produced coal. The agreement running from 1977 to 1982 has already been discussed for its part in the attempt to restructure the Scottish coal industry. Nearly all the coal used in Scottish power stations was mined in Scotland and was therefore higher priced. Under the agreement the SSEB and the NSHEB got their coal at Yorkshire/East Midland prices plus 5 per cent, and in return agreed to minimize the use of oil. The government made available up to £7 million a year to cover part of the cost of the price reduction, and the agreement also applied to any surplus Scottish coal which was taken by the CEGB or the Northern Ireland Electricity Service. In the same year the government agreed for a couple of years to provide up to £2 million a year to assist the extra use of Welsh coal in some power stations as compensation for the loss of sales for Aberthaw B power station, which was designed to burn a particular type of Welsh coal and was kept non-operational for long periods because of technical problems. Of wider application but shorter duration was an agreement, for the winter of 1978–9 only, whereby the government made available a grant of up to £19 million to enable the NCB to reduce the price of some coal to the CEGB, and the latter agreed to burn an extra $2\frac{1}{2}$ million tonnes of coal

[1] Ibid. 1976–7, 6; 1977–8, 8; 1981–2, 16.

and to reduce coal imports by 750,000 tonnes. This arrangement was supplemented by a government grant of £1½ million to enable the SSEB to continue to burn coal at the Kincardine power station where coal had, for the time being, lost its competitive advantage.

The most far-reaching arrangement came in 1979. In the summer of that year, after the NCB had announced their second price increase in four months, the CEGB told them that they would increase their imports of coal. They did, indeed, place several short-term contracts, the effect of which was that imports were higher in 1980–1. But in October 1979 the NCB and CEGB reached a Joint Understanding, to operate from 1 March 1980. This provided that for five years the CEGB would 'use their best endeavours' to take at least 75 million tonnes of coal a year from the NCB, and would adjust their coal import plans so as not to jeopardize their ability to do so. This undertaking was conditional on NCB coal prices moving in line with the retail price index, using February 1979 as the base for comparison. It was an agreement that gave the NCB a better basis for the planning of operations and sales.[1] It was even more important as the start of a fundamental and continuing change in the relationship of principal supplier and principal consumer.

Difficulties with the sale of coking coal were greater because the steel industry was in a prolonged depression which eventually would be identified as a severe and lasting contraction. Steel production was 23.3 million tonnes in 1974–5, already well below its peak, and never reached that level again. After annual fluctuations which left it at 20.3 million tonnes in 1978–9, it went rapidly down to 14.2 million in 1980–1 before recovering slightly to 15.8 million in 1981–2. With such a trend the steel producers were bound to try to reduce the cost of inputs as much as they could, and this had very adverse effects on coal sales. A smaller steel industry needed smaller coal stocks and, for several years, could meet its needs partly from stocks rather than new supply. The search for economies led to the closure of the least efficient plant, so that coal consumption fell faster than steel output. And as the depression of the steel industry was worldwide, a surplus of coking coal appeared on international markets, with consequential price cutting in the search for sales. The NCB had the one bargaining counter of being among the British Steel Corporation's largest customers, but, in such adverse commercial conditions, this could do no more than slightly modify the concessions that were unavoidable. The coal price rises of 1974 and 1975 drew protests from the BSC and led to a short-term arrangement

[1] NCB *Ann. Rep.* 1976–7, 6; 1977–8, 7–8; 1978–9, 7–8; 1980–1, 10–12.

whereby the BSC deferred payment for some coal until it was lifted from stock, and the NCB had a similar concession for some purchases of steel. But sales of coking coal declined and the BSC soon began to increase coal imports. This was particularly serious because in the mid-seventies coal supplied for the manufacture of metallurgical coke was probably the only marketing category of coal which, as a whole, consistently yielded a significant profit.[1]

This state of affairs did not last much longer. The BSC before the end of 1976 had already got contracts and options for nearly 2 million tonnes of imported coal and (partly on the basis of forecasts of steel production which proved to be too high) was already arranging for a progressive annual increase in imports. Only the highest grades of coking coal (which were becoming scarce in Britain) were being sought, particularly from Poland and Australia; and the BSC and the NCB disagreed about the proportion of such coal which was needed, in preference to medium quality coking coals (which were more abundant and cheaper), in blends suitable for metallurgical coke.[2] This argument about the appropriate quality of blends was to have other serious effects, for it led the BSC to change the specification for coal requirements for the new steelworks at Redcar, and reject some British coal in favour of imports, after the NCB had undertaken new investment in Durham coastal collieries in order to supply the new plant. As the BSC developed additional port facilities, and entered the eighties with the means of directly unloading imported coal at Port Talbot, Redcar, and Hunterston, close to three major steelmaking plants, the NCB's competitive and bargaining position was further weakened. Things were specially bad in 1979 and 1980 when the large appreciation of sterling put the NCB at a huge disadvantage relative to foreign coal (which was cheapening in dollar terms because of general depression), at the very time when the BSC was experiencing appalling financial difficulties.

The outcome was the negotiation of special price agreements between the BSC and the NCB. The first of these was made in July 1979, and was followed by a further agreement early in 1980 with the object of limiting BSC's coal imports to 4 million tonnes in 1980, an amount which, in the event, the BSC could not easily get and which exceeded

[1] There were all the problems of products jointly supplied to different markets and these made it difficult to be precise about the profitability of different categories.

[2] The arguments about the quality of coals needed in the blends seem to have become less fierce after NCB coal was supplied at lower prices, though the BSC used arguments about the quality of particular British coals to try to build a case for further price reductions.

needs. Concessions on price made to the BSC had to be given also to the other producers of metallurgical coke (of which the chief was the NCB's own subsidiary, National Smokeless Fuels), and for 1980–1 the NCB's budget was framed on the assumption that coking coals (which had a fairly wide price range) would on the average have to be sold at 16.1 per cent below list price. In October and November 1980 a longer term agreement was negotiated, under which the NCB agreed to align to world levels the prices charged to the BSC, the latter agreed to try to take at least 4 million tonnes a year from the NCB, and some concessions were made on the prices of steel products sold to the NCB. This new agreement, which the NCB had the option of extending to the end of 1983, involved for the whole of 1981 a still greater reduction below the list prices of coking coal. The detailed application of the agreement had to be worked out afresh each summer in relation to the following calendar year. Though a general principle had been agreed, a lot of scope was left for particular disagreements, for instance about what quality of British coal was being compared with what imported coal, and what expected movement of what currencies should be used in establishing the reference level of world prices. For the NCB the agreement ensured that revenue would be squeezed, but offered some hope that the output of some hard-hit pits could be maintained and that total turnover might be very slightly greater than it would otherwise have been.[1]

The agreements with the CEGB and the BSC effectively ended the policy of selling nearly all coal at list prices, which had been pursued throughout the existence of the NCB. Instead, most coal was now to be sold in accordance with the terms of individual contracts with major customers. Such a practice might once have caused worries about its compatibility with the legal requirement of 'avoidance of any undue or unreasonable preference or advantage'.[2] But it could be argued that there was nothing 'undue or unreasonable' in what was being done, and the new practice was not unacceptable to the ECSC. There was no great increase in the number of purchasers for whom individual prices were arranged, and there was not significant discrimination in coal prices charged to firms competing against each other. The discrimination offered was for buying British coal rather than a competing fuel. For 1980–1 price rebates were estimated at 6 per cent of turnover, and there had been earlier years when they amounted to 5 per cent of turnover. The essential change was that all three of the biggest customers (CEGB, SSEB, and BSC) were now getting regular rebates for the first time.

[1] NCB *Ann. Rep.* 1980–1, 12; 1981–2, 16–17.

[2] CINA s. 1 (1)(*c*).

The NCB had, in fact, started to make some agreements with individual firms in the late fifties in order to match the tactics of the oil companies; but, in the course of the seventies, higher oil prices enabled the NCB to reduce in real terms the rebates to the industrial market. Some of the agreements did little more than provide for a period of extended notice before increases in list prices were applied to sales to the individual customer. The NCB argued that there were many advantages in following the practices of most large manufacturing firms and gaining the flexibility of having high list prices and making sales at various discounts. Nevertheless, in 1980–1 only about 6 per cent of coal production was being sold at list prices, whereas until then 85 per cent had been, and the change resulted from increased pressure in the market-place and the readiness of a new government to let that pressure be exerted. In 1973 hopes were raised because competitive conditions appeared to be changing to coal's advantage. So they were in one particularly important way, but there were always influences operating in opposing directions. The movement in the relative strength of these influences was a major feature of the business environment in which the coal industry had to operate; and the industry's ability to take advantage of the favourable, and successfully to resist the unfavourable, movements was an indicator of its state of health.

These conditions are reflected in the level of production and sales which the NCB were able to achieve and the related financial results. Some of the most illuminating figures are summarized in Table 7.4. The precise degree of commercial realism in these figures is subject to argument because of the existence of government grants. Some people like to class every penny of government grant as a 'subsidy' to the coal industry, but such a classification is inaccurate and misleading. It is, however, sometimes illuminating to total everything and measure the scale against the industry's turnover. It may also be revealing to consider how far total government grants and the payment of interest (mainly to the government) on borrowed capital merely offset each other. Government grants fall into three categories. Social grants are not usually classed as subsidies, though they may have some slight subsidy aspect. They include grants towards the extra cost of pensions for mineworkers made redundant before reaching normal retirement age, and grants to assist the redeployment of NCB employees in the course of the elimination of uneconomic capacity. Operating and deficit grants are direct aids to production (including carbonization as well as coalmining) and can hardly be regarded as anything other than subsidy. Regional development grants

Table 7.4. *NCB commercial and financial results, 1972-3 to 1981-2*

| | 1972–3 | 1973–4 | 1974–5 | 1975–6 | 1976–7 | 1977–8 | 1978–9 | 1979–80 | 1980–1 | 1981–2 |
|---|---|---|---|---|---|---|---|---|---|---|
| NCB coal output (m. tonnes)[a] | 139.2 | 107.7 | 126.1 | 124.9 | 119.9 | 119.9 | 119.0 | 122.3 | 125.6 | 123.2 |
| UK coal stocks at year end (m. tonnes) | 31.0 | 19.1 | 21.6 | 29.9 | 28.1 | 29.8 | 28.8 | 27.7 | 38.4 | 43.5 |
| NCB financial record (£m.): | | | | | | | | | | |
| Turnover | 1,033.1 | 913.6 | 1,589.6 | 1,982.2 | 2,426.6 | 2,733.1 | 2,989.4 | 3,740.4 | 4,186.5 | 4,727.5 |
| Profit/(loss) on trading:[b] | | | | | | | | | | |
| Mining activities only | (35.8) | (116.1) | 13.3 | 35.7 | 101.1 | 92.9 | 77.7 | (6.2) | 61.9 | (48.2) |
| All activities | (39.4) | (98.7) | 40.4 | 52.2 | 109.8 | 108.7 | 121.1 | 27.6 | 69.5 | (84.5) |
| Interest | (43.9) | (32.2) | (36.2) | (50.0) | (78.0) | (87.0) | (138.0) | (184.7) | (256.2) | (341.0) |
| Other items[c] | (0.4) | 0.2 | (4.2) | 3.1 | (4.6) | (0.8) | (2.5) | (2.2) | (20.1) | (2.8) |
| Deficit grants | | | | | | | | 159.3 | 149.0 | 428.3 |
| Surplus/(deficit) | (83.7) | (130.7) | — | 5.3 | 27.2 | 20.9 | (19.4) | — | (57.8) | — |
| Profit on trading restated without direct government aids to production | (39.4) | (132.8) | 32.1 | 52.2 | 98.7 | 84.7 | 53.4 | (2.2) | 43.6 | (111.3) |

*Notes*: [a] Licensed mines added about 1 million tonnes each year to these figures.

[b] Up to 1978–9 all government grants were included in the figure of trading profit. An alternative statement of trading profit excluding these operating grants is given in the bottom line of the table.

[c] Mainly taxes, profits and losses on foreign exchange fluctuations, and extraordinary items.

*Source*: NCB *Ann. Rep.* 1981–2 and MMC, *National Coal Board*, II, 22–3.

were available generally under the Industry Act 1972 in order to promote the economic needs of assisted areas. The NCB were also specifically made eligible for them from 1973–4 to 1981–2 inclusive, under provisions of the Coal Industry Acts 1973 and 1977,but this eligibility was removed by the Coal Industry Act 1980. It is a matter of definition and argument how far the element of subsidy goes in regional grants. What is clear is that if there is an element of financial privilege, this is one respect in which coal has been less privileged than many other industries, including privately owned industries.[1] Payments of the various classes of government grant are summarized in Table 7.5, under headings intended to make clear the main distinctions.

Tables 7.4 and 7.5 indicate that 1973–4 was a particularly significant year of change. In terms of production and finance it was disastrous for the NCB, who needed a great deal of government assistance to put them back on a sound business footing. But it was also the year when market conditions moved so strongly in coal's favour that the NCB were able to set prices much nearer to the coal industry's needs, and, within a couple of years, financial health was greatly improved. For the rest of the seventies the NCB had rather better financial results than were called for, in the sense that, after receipt of grants and payment of interest, there remained a small surplus. Government grants were at a very modest level until 1978–9 and were not much devoted to directly aiding production. In fact, a trading profit was regularly achieved even without the inclusion of direct government aids to production. At the end of the seventies the results began to decline. Nevertheless, over the ten years ending with 1981–2, despite the problems at the beginning and the end, the operating profit almost equalled direct government aids to production; i.e. at the trading level break-even was achieved without government grants.[2] The trading profit was, however, not big enough (mainly because of interest charges) to render government aid to production superfluous. Only in 1975–6 and 1976–7 did the final surplus exceed government grants for production, and the achievement of break-even over all was, as intended, after the receipt of government grants. Yet in five of the ten years the NCB paid out more in interest than they received in government grants of every kind.

[1] MMC, *National Coal Board*, I, 23–6 and II, 24 describes the different classes of grants and their purpose.

[2] On a formal accounting basis the result was better than break-even, but the special government grant to wipe out the 1973–4 deficit (less that part of it which offset interest) ought to be reckoned in the balance, although it was outside the accounts. Its inclusion makes the trading result over ten years slightly worse than break-even.

Table 7.5. *Government grants to the NCB, 1972-3 to 1981-2*
(£ million)

| | 1972–3 | 1973–4 | 1974–5 | 1975–6 | 1976–7 | 1977–8 | 1978–9 | 1979–80 | 1980–1 | 1981–2 |
|---|---|---|---|---|---|---|---|---|---|---|
| Social | 15.4 | 21.0 | 22.1 | 32.4 | 43.4 | 51.0 | 54.3 | 62.0 | 79.0 | 119.7 |
| Regional (included in trading profit) | | 75.0 | 37.8 | | | | 50.0 | | | |
| Direct aids to production: | | | | | | | | | | |
| (a) included in trading profit | | 34.1 | 8.3 | | 11.1 | 24.0 | 67.7 | 29.8 | 25.9 | 26.8 |
| (b) excluded from trading profit | | | | | | | | 159.3 | 149.0 | 428.3 |
| TOTAL | 15.4 | 130.1 | 68.2 | 32.4 | 54.5 | 75.0 | 172.0 | 251.1 | 253.9 | 574.8 |
| Total grants *less* interest | −28.5 | 97.9 | 32.0 | −17.6 | −23.5 | −12.0 | 34.0 | 66.4 | −2.3 | 233.8 |
| Total grants as proportion of turnover (%) | 1.5 | 14.2 | 4.3 | 1.6 | 2.2 | 2.7 | 5.8 | 6.7 | 6.1 | 12.2 |
| Government payments outside NCB accounts[a] | 6.4 | 141.8 | 12.4 | 13.5 | 15.7 | 17.6 | 17.8 | 15.7 | 15.6 | 48.6 |

*Note*: [a] This item consists entirely (except in 1973–4) of expenditure borne directly by the government on the Redundant Mineworkers Payments Scheme, provided for by the Coal Industry Act 1967, as subsequently amended. In 1973–4 there was also a special government grant of £130.7 million to extinguish the deficit for the year, which arose mainly from the national overtime ban and strike.

*Source*: MMC, National Coal Board, II, 22–3 and Table 7.4.

This modestly favourable financial performance owed a good deal to the financial reconstruction of 1973,[1] which greatly reduced interest and depreciation charges, and the effects of this could not last very long unless the financial reconstruction were the basis for a continuation of the business on more appropriate financial lines than before. This is something that did not happen, and it was soon evident that new financial difficulties were being stored up. Productive capacity having been compulsorily run down in the sixties, it had to be partly replaced in quantity and wholly restored in quality, and the method of financing the necessary investment was on the same lines as had created problems before, but in more adverse conditions. The NCB could not generate a volume of sales at a sufficiently profitable level to enable most of the investment to be financed internally, and the external finance was not in the form of equity but entirely in fixed interest loans, raised at a time of exceptionally high interest rates, and these had to be serviced for years before the schemes they were financing were ready to produce anything. Suggestions in relation to *Plan for Coal* that there might be loans with deferred interest met with no response before 1980, and no useful response then. The government's undertaking to look at the problems of financing long-term schemes was carried out to the extent of ensuring that adequate capital was supplied, but no further. So the burden of interest charges began to increase quickly and its weight was a principal element in the deterioration of financial results in the early eighties.

In March 1973 the outstanding capital liabilities of the NCB had been consolidated into a single 20-year loan of £412.7 million with interest at 5.519 per cent. Three years later the total outstanding loans were still only £692.8 million (including £29.3 million of temporary advances), but those raised from the British government since March 1973 carried an average interest rate of 10.580 per cent, those from other sources 9.130 per cent.[2] By March 1982, however, outstanding loans under the Coal Industry Acts totalled £3,428.3 million. Of this sum, £1,633.1 million constituted funded loans raised from the government since 1973 at interest rates averaging 13.820 per cent, and £798.3 million was borrowed from other sources at an average of 9.665 per cent. The rest consisted mainly of temporary advances, and there were also the outstanding principal of the March 1973 consolidated loan and £109 million in loans made direct to wholly owned subsidiaries of the NCB.[3] Interest paid in

[1] Chap. 6, pp. 284–5 above.

[2] NCB *Ann. Rep.* 1975–6, 45.

[3] Ibid. 1981–2, 60. The loans from non-government sources were guaranteed by the Treasury,

1981–2 amounted to practically 10 per cent on the entire capital of the undertaking at the end of the year.[1] To have a very large business, which is not required to do more than break even, pay nearly 14 per cent on almost half its capital, while devoting a substantial part of that capital to schemes which will need four to ten years before they become revenue producing, looks like finance with a flair for the bizarre. Even if all the investment eventually produced results as good as those estimated in advance after elaborate appraisal, the rapidity of the growth of interest payments seemed almost certain to bring trouble before that happy issue had a chance to arrive.

There are a few other features of the results that might cause concern. One was the increasing difficulty of selling all the coal that was produced, with a consequent need to carry high and increasing stocks. Output never returned to the level before the disruption of 1973–4, but in 1975 consumers' reactions to the huge increases in coal prices sent sales down and stocks up. In the next few years there was little inroad into stocks despite a 4 per cent reduction of coal output. In 1979 the revival of productivity and output was initially accompanied by a somewhat greater growth of sales, but it did not last, and the early eighties saw the stockpiling of enormous quantities of unsold coal. The stocks would have been even higher but for a doubling of exports to 9.4 million tonnes in 1981–2, a result achieved by selling abroad a good deal more cheaply than at home, simply in order to increase cash flow and save some of the cost of financing stocks.[2]

A further worry was that, while deep mining was by far the biggest activity, it was never the main contributor to profits. The trading profit on mining activities, achieved in most years, came from opencast, not from deep, mining. And the effort to break even after meeting interest charges relied on a contribution from non-mining activities which was much larger in relation to their size than that from mining. If they began to perform worse, as some of them (especially carbonization) did at the end of the seventies, there was less to offset weaknesses in the financial results of deep mining. In fact, from 1977 the main positive contribution from non-mining activities was not from ordinary business at all, but from the sale of colliery houses, which realized a surplus over the written-down book value.

which had to be paid for this. Inclusion of this fee means that the cost of servicing these loans was equivalent to an effective average interest rate of 12.0 per cent (ibid., 54).

[1] The balance sheet value of total assets exceeded outstanding loans by only £82.2 million.

[2] MMC, *National Coal Board*, I, 76–7.

Questions would be asked about the financial results in any case, simply because they fell appreciably from their best, even though for a decade they came up to what was required, on the average. The questions acquire some extra sharpness when particular features of performance are emphasized. They almost pose themselves. Could revenue and profitability have been increased either by charging more for what was sold or by selling more as a result of charging less? Was it possible to have lower costs so as to permit either lower prices or higher margins? Were the trends in financial results—marked improvement, a few years on a plateau, and then marked deterioration—attributable mainly to the state of the coal industry or to the commercial and political environment? They are, of course, all questions which relate to the mining activities, the proper place for emphasis.

For putting up prices generally more than was done there was little scope save for very short periods. Even the huge price increases of 1974 and 1975 did not quite go to the limits permitted by oil prices, and 1979 probably offered some brief opportunity for higher prices of industrial coal. But greater increases would only have caused reductions in business in both the near and longer future. As appeared in 1975–6, very high price increases led to decreased consumption. Still greater increases would have reduced the prospects of businesses installing new and additional coalburning equipment, and would have caused attempts to import more coal. Although British coal retained a cost advantage over oil almost anywhere in the home market that suitable combustion equipment was available, other aspects of competition imposed constraints. Power stations were the main consumers of coal, fuel costs were the largest single item in the cost of generating electricity, and electricity competed with gas, for which until the eighties there was a national policy of low prices. British coal also faced limited competition from imports. Port and transport facilities, as well as concern with continuity of supply at acceptable prices, ensured that in the shorter term coal imports could not provide more than a small proportion of total fuel supply; but for users with plants close to suitable ports, such as the steel industry and some waterside power stations and industrial establishments, imports or potential imports could either cut the market for British coal by several million tonnes or bring British prices closer to those of available foreign supplies. Prime coking coal was available from abroad below British prices, and the British coal industry lost sales and had to cut prices. There were also times when imported coal was cheaper than British on delivery at some southern (especially Thamesside) power stations, though not for the greater part of the electricity generating industry.[1] By 1979 this competitive situation was a source of pressure on the NCB to limit price increases, for rising unsold stocks meant that they needed to secure for themselves every bit of the coal market that they could. When sales of marginal tonnage were so important, it was even an irritant that supplies of home-produced non-vested coal (cheaply obtained from old tips and sold mainly to power stations) began to increase again: the amount, which had dropped to $1\frac{1}{2}$ million tonnes or less in the mid-seventies, rose to 3 million tonnes in 1981–2. Further market losses to imports needed to be resisted as strongly as possible.

Price reductions and mitigations of price rises thus had a role, but it was to try to prevent reductions in sales rather than to increase them. There was little opportunity for increased sales for the time being. In 1981–2 UK consumption of energy was 22 m.t.c.e. less than in 1972–3,[2] a year in which North Sea gas had not quite built up to its full production and, as much the lowest priced and most convenient substitute, was ready to take over any net market share lost by oil.

So it appears that external constraints increasingly limited the scope for the British coal industry to enlarge its sales volume or its revenue by a change of price policy, at least until it could adjust by shedding its highest cost output. The interrelated problems may be summarized in this way. There was a mounting glut of energy on world markets, especially from 1980 onward, and this drove down prices in real terms, particularly for internationally traded coal. For the British coal industry to reduce output to match the contraction of the market would have involved (as long as the closure of high cost capacity was successfully resisted) producing below capacity across the whole industry, including the sacrifice of the lowest cost marginal output at long-life pits. This could only worsen the financial results. So the NCB felt constrained to keep up the level of output and sales as far as possible, by aligning prices with those of imports wherever there was a serious threat, as there was in both the electricity and steel industries (the two principal users), and by trying to increase exports at low prices. The effect was to make it more difficult to cover costs by prices, yet any different policy that was practicable in the short term would have made the adverse gap even greater.

In all this argument there is the important proviso about the difficulty

[1] MMC, *National Coal Board*, I, 75.

[2] NCB *Ann. Rep.* 1973–4, 11; 1981–2, 15.

of getting rid of high cost capacity. This is a central matter in relation to the control of both costs and levels of output, and it is considered elsewhere in some detail.[1] But it is also appropriate to look more generally at costs of production, and to consider whether these might in some ways have been kept lower, with beneficial consequences for financial results.

The steep rise in costs in the seventies was a serious problem for the coal industry, which needed the headroom created by the oil price revolution. Many of the increases came from outside and were almost impossible to control, but this was not true of all of them. The MMC carried out a somewhat crude exercise of comparing actual deep mining costs since 1971–2 with what the costs would have been if the input costs of fuel, materials, and manpower had followed the national trends for each of those items. It estimated that by 1980–1 deep mining costs had risen 16 per cent more than they would have done if they had kept to those national trends.[2] Its general conclusion, relating specifically to the last five years of the period, was that 'the largest contribution to cost increases was from wages and associated costs, and the largest contribution to cost improvement resulted from changes in the balance of production between Areas'.[3] Some of this is rather misleading. There had to be some allowance for 'other costs', many of which are specific to individual industries and for which there is no relevant national trend for comparison, and a shift of base year for the comparison could make a slight difference. It also has to be borne in mind that wages and associated costs may make the largest contribution to cost increases merely because they are the largest single component of costs, not because they are increasing disproportionately fast. Another way of looking at charges is simply to measure the rate of increase of each component, as in Table 7.6.[4]

This exercise indicates that in the first half of the decade the items rising disproportionately fast were materials and repairs, and power, heat, and light (a little in both cases), and wages charges (a lot), though wages and wages charges together were not seriously out of line.[5] The spurt in wages charges was caused partly by national insurance costs

[1] pp. 378–84 above and 437–42 below. [2] MMC, *National Coal Board*, I, 102–6.

[3] Ibid., I, 123–4.

[4] Ibid., I, 31–2 calculates the movement in total costs from 1976–7 to 1981–2, but excludes social costs and so gets an increase of 104 per cent rather than the 107 per cent in Table 7.6. This is compared with a rise of 84 per cent in the wholesale and 85 per cent in the retail price index. On this basis deep mining costs rose by an average of 3 per cent per annum in real terms over the five years.

[5] The combined percentage increases for wages and wages charges were 406 for 1972–3 to 1981–2, and 96 for 1976–7 to 1981–2.

Table 7.6. *Percentage increases in unit operating costs at NCB deep mines, 1972–3 to 1981–2*
(current prices)

| | 1972–3 to 1981–2 | 1972–3 to 1977–8 | 1976–7 to 1981–2 |
|---|---|---|---|
| Wages | 345 | 150 | 98 |
| Wages charges | 576 | 307 | 91 |
| Materials and repairs | 387 | 226 | 81 |
| Power, heat, and light | 458 | 219 | 106 |
| Salaries and related expenses | 357 | 150 | 106 |
| Other colliery expenses | 760 | 204 | 240 |
| Overheads and services | 374 | 180 | 100 |
| Depreciation | 382 | 93 | 193 |
| TOTAL EXPENSES | 423 | 192 | 107 |

*Source*: Calculated from MMC, *National Coal Board*, II, 25.

and partly by the concessions made after the 1972 strike. In the second five years the items of rapidly accelerating costs were depreciation and 'other expenses'. The latter were 7.0 per cent of total expenses in both 1972–3 and 1976–7, but were 11.5 per cent in 1981–2. Increased depreciation charges were an obvious consequence of the large investment programme undertaken under *Plan for Coal*. 'Other expenses' is a heading that might rouse suspicions that here is a ragbag hiding a multitude of wastes. In fact, the main items responsible for the high figure under this heading were three. One was the Voluntary Early Retirement Scheme introduced in 1977, enabling mineworkers, if they wished, to retire at ages progessively lowered to 60 and receive a lump sum followed by weekly payments until they qualified for pension at 65. It was not a funded scheme and was charged against revenue, and in these years the cost was bound to increase progressively. This item might, without disregard of function, be alternatively classified as a wages charge. The other two items were both concerned with improvement of the physical environment. Larger sums than before were spent on restoration of surface damage and compensation for it, and there was a special programme, begun in 1976–7, to improve the safety, as well as the appearance, of disused pit spoilheaps and mineshafts.[1] It might be

[1] See the notes on 'operating results' and 'deferred liabilities' in 'Notes to the accounts' in NCB *Ann. Reps.* 1976–7 onwards.

queried whether all this could be afforded, but it was expenditure required by the standards of society which had, for the most part, been given legal force.

The possibilities of achieving a general slowing down of the rise of costs were very restricted. It is difficult to say that the most rapidly rising items were both unjustifiable and avoidable. Something could have been saved on depreciation if there had been less new investment (which would also have made savings by reducing the need to borrow), and this would have been helpful if the unsuccessful schemes alone had been eliminated; but not otherwise. The other place to seek large improvements was the biggest single item, wages and wages costs. Here the disappointment was not that labour costs rose out of line with the rest, but that, despite the move to more capital-intensive technology, they did not significantly fall as a proportion of the total. The smallness of the rise in the figures of productivity indicates one of the main limitations in the containment of costs. The other is the insufficiency of the shift of the balance of production between pits and between areas.[1] If these two factors had gone as far as was hoped and planned, then the rise of costs would have been significantly less, and the financial outturn would have been better.

These considerations lead towards an answer to the question whether the course of the financial results depended more on influences within or outside the coal industry. They suggest that the absolute level of the results in any year could have been better if a number of things had been done differently (in quite feasible ways) by both labour and management in the coal industry. But the pattern leading from improvement to deterioration was imposed almost entirely by outside influences. There was nothing to indicate that the industry was worse run when the financial outturn was getting worse. If anything, the opposite seems true. The rise of costs was less controlled, the failure to improve productivity more complete, in the mid-seventies, when the financial results were best, than in the following years. Indeed, it could well be that the achievement of an operating profit, without taking account of direct aids to production, in the severe business conditions of 1980–1 was the best managerial achievement for ten years. It was the level of interest charges and general economic recession that made the early eighties look so much worse.

[1] This point is discussed more fully in section ix of this chapter, below.

### vii. Economic recession and government restraint

For the coal industry the effects of the economic recession which set in by the end of 1979 were complicated in several ways. One was initial uncertainty, gradually moving to pessimistic conviction, about the longer trends on which the recession was imposed; and associated with this was the uneven impact of adverse conditions on different types of economic activity. It appeared that in both the very short and the longer term the hardest hit included the most energy-intensive activities. This had already been appearing for several years in the state of the steel industry and its declining demand for coal and coke. Now similar, though not quite so drastic, experiences befell many other manufacturing industries. The contrasting movement of different economic indicators illustrates the point. If 1980 is taken as 100, then gross domestic product, on average estimates, fell from 102.6 in 1979 to 98.5 in 1981 and recovered to 100.4 in 1982; but manufacturing output fell from 109.4 in 1979 to 93.6 in 1981 and 93.7 in 1982, and even at the end of 1983 was probably still 13 per cent below the average level of 1979. No wonder the national consumption of energy fell sharply: 348.0 m.t.c.e. in 1979–80, 322.7 m.t.c.e. in 1980–1, 318.6 m.t.c.e. in 1981–2, and 309.6 m.t.c.e. in 1982–3. Inland consumption of coal, which had been slowly rising and was higher in 1979–80 than in any year since 1972–3, fell accordingly, in fact rather more steeply: it was 128.4 million tonnes in 1979–80, 120.3 million in 1980–1, 117.0 million in 1981–2, and 110.4 million in 1982–3.[1] Some uncertainty about the precise degree of seriousness of these figures was created by the loss of coal sales resulting from the long steel strike in the early months of 1980, and the industrial disruption on the railways in January and February 1982, which reduced coal deliveries to power stations; but their generally gloomy implications could not be mistaken.

The figures had an obvious significance for coal industry policy. It had always seemed sensible that investment in the modernization of capacity should be maintained without pausing for each short-term economic fluctuation. But a protracted downward trend was a different matter if the time of its reversal could not be foreseen. It seemed certain that in the mid-eighties, when the programme of *Plan for Coal* ought to have been completed, the consumption of energy, and the market for coal, would be appreciably less than had been forecast. So there was a need to look again at the productive capacity of the coal industry. This

[1] Energy and coal figures from NCB *Ann. Reps.* for years stated.

was where another of the problems came in. The general recession was accompanied by rising unemployment, which reached levels unknown for half a century. The effect of this was to alter the social climate for any redeployment of labour and any attempts to remove superfluous capacity.

The other complication for the coal industry was the election in 1979 of a new Conservative government with strong ideas about reducing the demands of nationalized industries on public expenditure and about the reduction of public sector borrowing as an instrument of anti-inflation policy. All loans raised by nationalized industries—even from sources other than the UK government—were classed as part of the public sector borrowing requirement, unless it could be proved that a loan could not possibly compete with the private sector demand for capital; and such a proof was practically impossible. In applying these ideas the government initially showed little awareness, or even willingness to become aware, of the empirical evidence about the conditions of each of the nationalized industries. The general line appeared to be to keep them tight for money, on the assumption that this would compel them to become more efficient, and in the belief that this was necessary in order to fit in with a presumed uniquely effective type of monetary policy. An essential element of control was therefore to tighten the imposition of cash limits, which the previous government had applied since 1976–7. For each nationalized industry the controlling restriction was an external financing limit, i.e. a maximum amount of new borrowing plus government grants, fixed for each financial year. This was a clumsy device as it failed to distinguish clearly between revenue and capital, or even between subsidy and investment. It was, however, supplemented by some proposals for periods longer than a year, and it was doubtless hoped that these would render such distinctions unnecessary.

For the NCB the government worked out, in its early months, a set of requirements which were embodied in the Coal Industry Act 1980. Social grants were continued, but most of the operating grants for specific purposes were reduced or discontinued; access to regional grants was withdrawn, despite heavy involvement in some of the relevant areas, such as South Wales; and there was, for the time being, to be one main grant, known as a 'deficit grant'. But the total of grants was to be reduced rapidly in real terms, year by year, and by 1983–4 all grants, except social grants and some relief of interest payments on some loans, were to cease. This was an almost impossible result to achieve in the circumstances of the time unless the relief of interest went to lengths which were not being contemplated. It represented a very large change in a very short time from the objective which governments in the recent past had required. It had to be sought in recessionary conditions when, however miraculous an improvement of management was accomplished, it was almost impossible to increase sales and sales revenue, and when industrialists were clamouring for a reduction of energy prices, on the alleged ground that they paid more than their foreign competitors. It had to be sought also while the NCB were still expected to continue to increase the future security of energy supplies by maintaining and modernizing the coal industry. Something had to give. The NCB kept within the external financing limit of £832 million for 1979–80, but the figure for 1980–1, £882 million, was lower in real terms and caused great difficulty. There was dispute about whether it had or had not been exceeded. The NCB adopted various expedients, including the transfer to a consortium of banks of the financing of loans to employees, which they thought could be treated as separate items. They believed that this device enabled them to stay within the limit.[1]

Large savings of various kinds had to be sought and some of these could be damaging to efficiency and prospects in the not very distant future. There were, for instance, drastic reductions in the ordering of new equipment, a measure which had serious repercussions on the activity of the mining equipment firms. But such economies were not enough. Things came to a head in February 1981. The NCB sought a meeting with the unions (held on 10 February) and put to them a plan with four related objectives: (1) to maximize sales by keeping prices competitive, (2) to increase efficiency and reduce unit costs, (3) to maintain a high level of investment in new and replacement capacity, (4) to bring supply and demand into better balance by maximizing sales, expanding output at pits with viable reserves, and diminishing capacity where realistic reserves were exhausted or where, for geological or other reasons, there could be no long-term financial contribution from a pit. The last point was spelled out and roused the immediate opposition of the NUM. The unions were told that about 10 million tonnes of capacity would have to be shed as quickly as possible, and, though each area director would discuss with the unions within a week the proposals for particular pits in the area, the search for speedy action meant a departure from the normal colliery review procedure. The NUM told the NCB that, if this proposal were retained, they would ballot the

[1] NCB *Ann. Rep.* 1980–1, 3 and 6.

members with a recommendation for strike action, and they asked for a meeting of the Tripartite Group, which was then arranged for 23 February. In South Wales, where the closure of five pits in 1981 was proposed, sporadic strikes began at once and the whole coalfield was on strike by 17 February. In Kent some miners were on strike and the rest banning overtime, and Derbyshire and Scotland had voted at area meetings to strike the next week.

The meeting of the Tripartite Group was brought forward to 18 February. The next day the Secretary of State for Energy, David Howell, made a statement to the House of Commons about what had happened at that meeting. After describing the proposals made earlier by the NCB to the unions and commenting that 'fears and anxiety among the work force arose through rumoured and distorted impressions of what was being proposed', he went on to say that, at the previous day's meeting,

> three main points were raised—closures, financial constraints and coal imports. I said that the Government were prepared to discuss the financial constraints with an open mind and also with a view to movement. The chairman of the National Coal Board said that in the light of this the Board would withdraw its closure proposals and re-examine the position in consultation with the unions. I accordingly invited the industry to come forward with new proposals consistent with *Plan for Coal*. As regards imports, I pointed out that these would, in any case, fall this year from their 1980 levels. The industry representatives said that they wished to see this figure brought down to its irreducible minimum. I said that the Government would be prepared to look, with a view to movement, at what could be done to go in this direction.[1]

The bland wording of this statement might suggest that the government was making a reasoned and conciliatory response to arguments put forward in the discussion. In fact, the essence of the statement represented a unilateral political decision, taken by the government before the meeting but unknown to any of the other parties until the meeting was assembling. The invitations of the Secretary of State were, of course, accepted. The NUM managed to get the local strikes stopped within a few days. In the last week of February there were meetings between the NCB and the unions, and then between all these parties and the Secretary of State, at which the NUM pressed for particularly large increases in financial provision by the government. The latter abandoned the policy expressed in the Coal Industry Act 1980 and agreed to much bigger grants, and consequently a much higher external financing

[1] The statement is reproduced in MMC, *National Coal Board*, II, 2.

limit, for 1981–2. The requirement of break-even with none but social grants in 1983–4 was dropped. Where amended legal provisions were needed they were subsequently included in the Coal Industry Act 1982.[1] The government also put pressure on other bodies to fall in with the amended policy. In particular, the CEGB had almost to phase out coal imports, even though there were continuing contracts for some imports from Australia. Financial compensation was paid to the CEGB by the government and coal delivered under these contracts was stocked at continental European ports instead of being brought into the country. Such measures created resentment, which had to be taken into account in the NCB's negotiations with their customers.

Nobody emerges from the events of February 1981 with much credit. Ministers imposed a policy without paying attention to its practicability. When a consequential crisis arose they ran away and then started to patch, and inevitably they left speculation that some day they might again act similarly. The NUM had again shown reluctance to translate the commitment to *Plan for Coal* into acceptance of the loss of old capacity as counterpart to the creation of new. This time the more difficult employment situation and the limitation of normal procedures justified some sympathy, but the attitude expressed was not new. The NCB were pressed into urgent economies, but adopted a procedure that was certain to run into opposition. The most notable thing about pit closures in the seventies, apart from their fewness, was the high proportion which were agreed the second time round. Most of the pits which were the subject of appeals against local proposals for closure were, in fact, kept open after appeal; and longer experience and patient argument usually showed that to go on working in them was not a very attractive and secure prospect, so closure was eventually agreed on the spot. It might well appear that this patient approach should have been tried at more pits in the seventies and was the right method for February 1981. But at the latter date the government's financial pressure on the industry made this impracticable. Previous experience was, in fact, repeated. By October 1981 agreement had been reached locally for the closure of nine of the twenty-three collieries listed in February for accelerated closure, and already, of those nine, five were wholly and two partially closed. Most of the twenty-three were closed within two years, though there were a few, still very unprofitable, where closure was successfully resisted. The

[1] MMC, *National Coal Board*, I, 12 and 82. There is an account of this episode from a personal and NUM standpoint in Gormley 1982, 173–82.

number of operating collieries fell by twenty in two years after March 1981, the largest two-year drop since 1973–4.[1]

So perhaps the events of February 1981 stimulated a little push in an unpopular but useful direction. And they had the merit of making the Cabinet aware of some of the conditions in the coal industry and the interaction of their own policies with them. But neither the measures agreed in the wake of the crisis nor the closure of twenty collieries were sufficient to solve the industry's problems. There remained the tasks of determining what was likely to be required of the coal industry in the medium and longer term, what in the existing industry needed to be preserved and what needed to be improved and added in order to meet that requirement, and how the industry should be assisted and reshaped in the current difficult economic climate. These were questions to which the answers must suggest actions, some of which would be irreversible, but the balance of which must be susceptible of continuous adjustment.

It was in these circumstances that the NCB went on to draw up their new and more cautious development plan to 1990, and also that the government, late in 1981, announced the intention of referring to the Monopolies and Mergers Commission the question of the efficiency and costs of the NCB in the development, production, and supply of coal. This reference was formally made in March 1982 and the Commission was given six months in which to report, a period subsequently extended to nine months.[2] These exercises took the questions some way forward, but scarcely within sight of a solution, for the economic context for planning evolved in rather confusing ways, and the recommendations of the MMC, though some of them were both perceptive and important, were reached and expressed in a somewhat piecemeal fashion, which caused them to cumulate rather than cohere.

## viii. International activities

The most obvious change in the international position of the British coal industry from 1973 was the membership of the ECSC. Britain was the largest coal producer in the Community and was able to take a leading role in any attempts to formulate and implement a policy to strengthen the coal industries of the Community and increase the consumption of their output within the Community. This role was

[1] NCB *Ann. Rep.* 1982–3, 21.

[2] MMC, *National Coal Board*, I, 1. Partly because of problems about the inclusion of commercially confidential information, it was 15 months before the report was published.

performed both in relations directly between the NCB and the Commission of the European Communities, and through CEPCEO (the association of coal producers of the European Community), of which Sir Derek Ezra was president in 1974 and 1975. Pressure was also exerted through the ECSC Consultative Committee, which represented Community steel and coal producers, trade unions, consumers, and traders. Sir Derek Ezra was president of this body in the year 1978–9.

Yet the effect on the fortunes of the British (or any other West European) coal industry, though favourable rather than unfavourable, was disappointingly small. In December 1974 the Council of Ministers agreed on the objective of stabilizing Community coal output to 1985 at about 250 million tonnes or rather more, a figure likely to require some small expansion in Britain and West Germany to offset contraction elsewhere. But there were no significant new steps by the Community to secure the achievement of the objective, which in practice steadily receded. There were three important influences. The first was the failure of the energy market to grow as predicted, with a particularly sharp contraction in the steel industry, which provided a bigger proportion of the coal market for the Community as a whole than for Britain alone. This market change led to surpluses of oil and gas, and of coal from third parties. Consumers were led to believe that there was no risk in not relying on Community coal supplies, and even the scare caused by the resumption of oil price rises in 1979 was damped down by the re-emergence of energy surpluses in the depressed economies of the next few years. The second influence was that most countries in the Community had little or no interest in coal production. Only Britain and West Germany were large producers. Belgium and France continued to keep their coal industries in existence at greatly reduced levels. Very little other Community coal was produced,[1] and even France took the line that, as the Community was never likely to be able to supply its own long-term needs of coal, it should establish long-term sources of imports, including overseas mining projects undertaken by the Community's own coal producers. The third influence was the absence of any treaty requirement for a common Community policy on coal trade with third parties, each member state being free to make whatever arrangements it chose.

The combined result of these influences was a large and rapid increase

[1] Coal production ceased in the Netherlands after 1974. There was a continuing annual production of about 50,000 tonnes in Eire. Greece produced a little lignite, but West Germany was the only large producer of lignite.

in imports of coal from outside the Community, and a decline in coal production within the Community, which was reversed after 1978, though output remained well below that of 1973. Imports from outside the Community rose from 29.8 million tonnes in 1973 to 43.7 million in 1976, 59.3 million in 1979, and 74.5 million in 1980 (of which only 0.5 million was attributable to the inclusion of Greece in the Community in that year). The main sources in 1973 were Poland and the USA. Poland maintained its level of supplies until the political upheavals there cut them down in 1981, but there were large increases in imports from the USA and South Africa, which became the two principal suppliers, and smaller but significant increases in imports from Australia.[1]

In these conditions it was difficult to get much response to representations by the coal producers, and they were probably not helped by the unrealistically large estimates of the growth in Community coal consumption by the end of the century (generally about 100 per cent) which they put forward at the end of the seventies. The argument that the Community should reduce its dependence on imported oil—about half of total energy supply around 1980—had some effect, and by 1982 the Commission was proposing to the Energy Council of Ministers a policy of development of the coal market in their countries, with more efficient production achieved through concentration of investment in new and improved mines. But it remained hard to secure new actions from the Council of Ministers, and even the Commission proposed no new short-term measures of assistance, such as additional aid for stocks and for intra-Community trade in coal.[2]

From all this it is evident that membership of the ECSC did not give the British coal industry much opportunity to increase production and sales. There were, however, some direct practical benefits. One was access to loans for capital purposes at rather lower rates of interest than were charged by the UK government. For instance, it was reckoned that loans from the ECSC resulted in a saving on interest charges of £1.8 million in 1976–7, £2.1 million in 1977–8, £3.7 million in 1978–9, £5.7 million in 1979–80, £6.5 million in 1980–1, and £6.3 million in 1981–2. Another benefit was the availability of grants from the ECSC and the EEC for specific pieces of research and some smaller grants to assist with the readaptation of the industry. These various grants amounted to £4.0 million in 1976–7, £6.7 million in 1977–8, £3.6 million in 1978–9,

[1] Ezra, *European Community Coal Policy*, 1978, summarizes the main points. There is a section on Europe or on international affairs in each NCB Annual Report.

[2] NCB *Ann. Rep.* 1981–2, 29; 1982–3, 18.

£7.9 million in 1979–80, £5.0 million in 1980–1, and £4.2 million in 1981–2. These benefits, in cash terms, were generally somewhat larger (though not greatly so) than the annual levy on the coal industry, payable to the ECSC. This rose from £2.0 million in 1973–4 to £6.4 million in 1977–8 and £9.3 million in 1981–2.[1]

Perhaps just as significant, though not measurable, were the closer relations with foreign coal producers. In western Europe these relations had been close ever since the creation of the Council of Association between the ECSC and the UK, and membership of the ECSC did not in itself make them much closer. However, the unification of much of the West German coal industry into *Ruhrkohle Aktiengesellschaft* (RAG), a body more comparable to the NCB in scale than any other West European producer, increased both the opportunities for reciprocal exchanges of views and their potential benefits. In 1974 and 1975 a Performance Improvement Team of the NCB, with the co-operation of RAG, made a detailed comparative study of the British and West German coal industries. It learned a good deal, particularly about the technical advantages of heavy duty equipment and of shield supports, and about the benefits from the combination of mines into very large units. There was clear evidence of the contribution which these features made to the working of coal faces with a much larger area and the achievement of high daily outputs per face. These comparative studies undoubtedly influenced British thinking and practice in the immediately following years, though German methods were not copied wholly, and not at all without a good deal of questioning and trial. This was partly because the proportions of dirt raised with the coal and the costs of production in the Ruhr, despite the advantages, were higher than in Britain, and partly because conditions were different in many respects. In particular, the British industry could not discard its legacy of small collieries. Even in 1981 the largest British colliery, Kellingley, was below the *average* size of the collieries in West Germany, and the latter average was five times as great as that for the British coal industry as a whole. Two of the conclusions drawn by the NCB team in 1975 gave the empirical support of foreign experience to what seemed obvious policy. These were that 'the worst of the loss-making collieries should be closed so that more resources can be devoted to improving the industry where the potential is high', and that 'collieries should be

[1] NCB *Ann. Reps.* for years stated, 'Notes to the Accounts'. In the three years before 1976–7 a total of £5.1 million was received in ECSC grants, but there was no significant saving on interest until that year.

combined or extended where appropriate'. But as has been seen, there were serious obstacles to the following of these examples from Germany.

Despite the differences in national circumstances, there were advantages in getting a closer acquaintance with foreign practices and exploring their possible relationship with differences in performance. Not only was there concern to continue building on the good relations established with RAG, but the NCB made agreements with several other state and private bodies, mostly outside the ECSC, for the exchange of information on a wide range of topics concerning mining techniques, safety, and coal utilization. Such agreements were made with the US Department of the Interior in 1974 and the Polish Minister of Mines in 1975. There was also an agreement with Canada for the exchange of research information, and one with the USSR. An agreement was made in 1976 with the second largest West German undertaking, *Saarbergwerke AG*.[1] Some of these agreements led to a certain amount of collaborative research.

There were some examples of still wider international collaboration. The most prominent was the International Energy Agency, set up by agreement between governments late in 1974, at the suggestion of the USA. It comprised members of OECD with a common interest in minimizing their dependence on imported oil. Its main purpose was to promote research into new energy sources and methods of conversion, and to develop them. The UK government was invited to take the lead in its Coal Research and Development Group and, at the request of the Department of Energy, the NCB agreed to take responsibility. Leslie Grainger, the board member for science, became chairman. In 1975 five initial schemes were agreed and the NCB set up a wholly owned subsidiary, NCB (IEA Services) Ltd., to manage the work to be carried out in the UK. Four of the schemes were services, mainly concerned with the collection and provision of different classes of information, but the fifth, and most important, was a project for pressurized fluidized-bed combustion, to be carried out at Grimethorpe over a five-year period. This project was later transferred to a separate company, NCB (IEA Grimethorpe) Ltd., and the plant was commissioned in 1980, with a four-year programme of experiments. The purpose of the project, which was funded equally by the governments of the UK, USA, and West Germany, was to assist in the design of

[1] Ibid. 1974–5, 16; 1975–6, 21.

large commercial plant to operate at high combustion efficiencies and in an environmentally acceptable manner.[1]

These various activities suggest that there was in the seventies a wider international outlook than before, however difficult the current international commercial conditions might be. It was, indeed, impossible to forget that prices and the level of sales for British coal depended on the international energy market, and that it was not sensible merely to remain passive about this. In the early summer of 1975 the NCB began to consider new possibilities of coal production and trade overseas. At that time, since coal imports by the major British customers were no longer avoidable, the NCB favoured an attempt to become the buying agents for UK consumers and to act as importer merchants, either jointly or independently. For example, they hoped they might make an agreement to supply all the coal needed by the BSC, whether from home production or imports. They also considered the possibility of mining overseas, about which views were divided, though no one favoured it except as a collaborative undertaking. For the time being they limited themselves to reviewing the approaches that had been made to them to participate in mining ventures overseas. One of the problems was that they could engage only in activities conducive to the discharge of their basic duties laid down in CINA, and, under the terms of the Coal Industry Act 1949, any activity overseas needed a relevant Order by the Secretary of State. No existing Order covered the possibility of mining overseas. Both the secretary of the board and the head of the legal department urged that, if the NCB really wanted to extend their oversea activities, they should seek corresponding statutory powers and not have to ask for a separate Order every time they wanted to start a new venture. This advice was followed. The Coal Industry Act 1977 gave the NCB (subject to the written consent of the Secretary of State for Energy) power to do anything outside Great Britain which they judged requisite, advantageous, or convenient and which they were required or authorized to do in Great Britain, whether or not it was related to the working and getting of coal in Great Britain.[2]

Nothing came of the idea of becoming an importing merchant or agent. The NCB's major customers were big enough and experienced enough to make their own coal contracts without needing a middleman, and the inescapable importing activities of the NCB's first decade did not provide a happy financial precedent, though market conditions had

[1] NCB *Ann. Rep.* 1975–6, 23; 1980–1, 28–9; 1981–2, 27; 1982–3, 17.

[2] Ibid. 1977–8, 5–6.

changed since then and offered better prospects. But from 1977 there were two developments that showed determination to seek more opportunities abroad. One was the establishment, under NCB leadership, of British Coal International Ltd., which embraced the exporters of coal and coke, mining equipment, and mining consultancy services. The object was to engage in combined promotional activities abroad and to permit the arrangement of packages of supplies and services for foreign customers, and in these ways to increase export earnings. The other development was the creation by the NCB of a Directorate of Overseas Mining and the start of a collaborative venture in Queensland.[1]

The Queensland project began with an authority to prospect for coking coal at German Creek, 450 miles north-west of Brisbane. There was competition from several international firms for the authority, which was granted by the Queensland government to Capricorn Coal Developments Pty Ltd. (CapCoal), a wholly owned subsidiary of the UK registered firm, Overseas Coal Developments (Queensland) Ltd. (OCDQ). The latter company was owned 30 per cent each by Commercial Union Assurance Co. Ltd., Intercontinental Fuels Ltd., and Austen & Butta Ltd. (an Australian coal firm), and 10 per cent by the NCB. The financing was further complicated because Intercontinental Fuels Ltd. was owned 20 per cent by Commercial Union and 50 per cent by the British Fuel Company; and the latter was a 50–50 partnership between the NCB and AAH Ltd. (the successor of the pre-nationalization coal-mining firm, Amalgamated Anthracite Ltd.). The NCB also agreed to supply services in return for an allotment of A$100,000 of loan stock in CapCoal. If they supplied services above that value, the NCB had the option of being paid for them in cash or taking more loan stock. When total expenditure passed A$1 million, the NCB had the option of either putting up more money or reducing their proportionate contribution. If and when a mining lease was granted, which it was hoped would be in about two years, the NCB had the option to increase their total interest in the project to 20 per cent by taking an allocation of shares and loan stock from the other shareholders in OCDQ. There was also provision at a later date for possible financial participation by RAG which, along with the main steel producers of Belgium, the Netherlands, Sweden, and the UK, and the French governmental coal-importing agency, gave written support to the original application by CapCoal to the Queensland government. In fact, RAG did take a financial interest from September 1977.

[1] Ibid., 30, 43, and 50.

The initial commitment of the NCB was small and there was no compulsion to increase it if progress looked unpromising. In fact, exploration (which started in May 1977) was speeded up, partly to keep pace with the activities of the holders of the two adjoining leases, and the NCB put up extra money to maintain their 10 per cent share, as the early drillings indicated a quantity and quality of coal in line with expectations. The project went quite well. By 1979 a mining lease had been obtained and an additional one-mile strip had been added to the lease, reserves established amounted to 60 million tonnes workable by opencast and 1,100 million by deep mining, and the quality was very high and suitable for the best blends in modern coking ovens. Ownership had changed somewhat with the entry of Shell Company of Australia, but the NCB expected to retain either 16⅔ or 18 per cent of the venture, and in August 1979 agreed, subject to consent by the Secretary of State for Energy and the Treasury, to invest up to £40 million. That consent was quickly forthcoming. Opencast output was expected to begin in 1982, but no deep-mined output before 1985–6. By mid-1980 negotiations were in progress for a sales contract with Japanese interests, and physical progress was so good that a few trial shipments were planned for 1981. In fact, it was March 1982 when the first shipment was delivered—to Egypt.

In 1981 most of the investors except the NCB left OCDQ and operated directly through CapCoal. OCDQ (which was renamed Coal Developments (Queensland) Ltd.) therefore became an 89 per cent owned subsidiary of the NCB, who thus indirectly owned 18 per cent of CapCoal. In 1982, however, the Australian Foreign Investment Review Board ruled that 50 per cent of the equity in this venture must be Australian-owned. Coal Developments (Queensland) Ltd. therefore had to sell a third of its holding in CapCoal, which was done at a price well above book value. From then on the NCB had only a 12 per cent interest in the joint venture.[1] It was not a large investment but so far it had gone satisfactorily. It was not likely to have any impact on the British coal market, but it promised the NCB a small entry into other international markets in which they could otherwise not compete.

British Coal International, the other innovation of 1977, was, of course, quite different. Results were expected much more quickly, but they were bound to be less tangible and not strictly measurable. The company steadily increased its promotional activities, with publications, exhibitions, 'British Coal Days' in major business centres (in the

[1] NCB *Ann. Rep.* 1981–2, 29, 52, 57, and 59; 1982–3, 45 and 47.

form of presentations of information and seminars for decision-makers in government, industry, and finance), and delegations to major economic institutions. Some success was achieved in increasing the export sales of members. In 1976–7 these sales were £126 million and in 1977–8, when British Coal International was just beginning its activities, they were £160 million. Table 7.7 shows how they moved after that. The exceptionally high figure for mining machinery in 1979–80 was due to a large once-for-all order from China. The leap in coal exports in 1981–2 resulted from an attempt to increase cash flow and reduce stocking costs by seizing the opportunity to replace supplies which Poland was unable to produce because of internal political difficulties. Much of the increase in consultancy earnings from 1980 was caused by the entry of Horizon Exploration Ltd., a joint undertaking of the NCB and English China Clays Ltd., which had special expertise in seismic techniques that were relevant to exploration for oil as well as coal. The drop in 1982–3 illustrates the effects of the widespread economic recession.

Table 7.7. *Export sales of members of British Coal International 1978-9 to 1982-3*
(£ million)

| Year | NCB sales of coal, coke, and coal products | Mining machinery and equipment | Consultancy services of NCB and associates | Total |
|---|---|---|---|---|
| 1978–9 | 78 | 91 | 2 | 171 |
| 1979–80 | 100 | 182 | 5 | 287 |
| 1980–1 | 184 | 129 | 9 | 322 |
| 1981–2 | 348 | 133 | 21 | 502 |
| 1982–3 | 273 | 169 | 17 | 459 |

*Source*: NCB *Ann. Reps.*

Yet, while the international activities were substantial in themselves, they may seem very limited in comparison with the total operations. The obvious test is provided by international trade, which was small in relation to both the production and consumption of British coal. Throughout the seventies exports of coal were generally less than 2 per cent of output, and there were only two years in the decade when exports exceeded imports and one when they were equal. In the early eighties the level of exports was rapidly raised to reach 9.1 million tonnes in the calendar year 1981 and 9.4 million in the financial year 1981–2, which was near the maximum which British ports could then handle. (This was itself a comment on the decline in the assumed export capability of the British coal industry.) But as has been noted, the rise was the result of special circumstances, which were not symptomatic of economic health. The increased exports mostly went to power stations within the European Economic Community and, though the delivered price kept within Community rules and was not much different from the price to the CEGB, the revenue accruing to the NCB averaged about £12.50 a tonne less. This export effort was undertaken to avoid greater losses on output which could find no immediate market and would otherwise have had to be added to stock.[1]

Such an experience raises the question whether British coal found itself uncompetitive when exposed to international market forces. There is no single, precise answer to such an over-simplified question, but some approximate indications can be attempted. Competitiveness in the home market has already been considered and it appears that, on the average, British steam raising coal was cheaper than imports at most times and in most locations. But by the late seventies the best coking coal could not complete with imports unless it was sold at an appreciable loss, though it is not clear that the prices of available imports always covered their long-term costs.

Experience after 1973 made it fairly plain how British coal stood in the European Community. The growing scarcity and relatively high cost of British coking coal was clearly a weakness, for coking coal was a large part of the Community's international market and Britain could not compete on either quantity or price. As far as price was concerned, however, the position relative to other Community producers depended more on subsidies than costs. There were two available subsidies in the ECSC. There was a Community-financed sales aid with a fund jointly provided by the ECSC, the iron and steel industry, and governments, for up to 14 million tonnes of intra-Community trade a year; but the UK government never joined this fund, which had been started before Britain entered the Community. So British coal did not benefit. There was also a permitted maximum level of government subsidy for the production and sale of coking coal, but the UK government never paid more than a small proportion of the permitted amount and paid much less subsidy than the other governments. In general, though there can be some disagreements about what is and what is not an 'aid to

[1] MMC, *National Coal Board*, I, 76–7 and 79.

production', it is clear that British coal had much less financial assistance than that of other Community countries, though the discrepancy was lessening in the early eighties. Direct aid to production in 1977 was calculated at £19.8 per tonne in Belgium, £13.3 in France, £2.6 in West Germany and £0.5 in the UK. By 1979 these figures had gone to £28.5, £15.2, £12.6, and £1.5 respectively, and in 1982 to £16.47, £17.72, £8.65, and £3.19.[1]

Some of the variations in price and cost comparisons arise from fluctuations in the sterling exchange rate, but the differences bear some relation to real differences in the costs of production. Differences in accounting practices make any figures less precise than they appear, but those available indicate both British superiority and its declining margin. Average costs per tonne (including interest and depreciation) in 1978 (1978–9 for the UK) were put at £56 in Belgium, £42 in France, £40 in West Germany, and £24 in the UK. In 1980 these had gone to £61, £45, £44, and £35 respectively, and in 1981, when the continental countries showed little change, UK costs had risen to just over £40.[2] Here again, exchange fluctuations are part of the explanation, but there are some indications that performance trends were, in a few respects, less good in British mines, and it is also probable that in some respects the comparisons are slightly too favourable to the British. The comparative study of RAG, for instance, found that some items which the NCB capitalized were charged against revenue in West Germany; and British performance benefited from having a higher proportion of opencast than any other Community country. The most notable adverse feature of British performance was that, from 1977, output per manhour underground was growing only about half as fast as in France, where as a result of greater concentration into large units, output per colliery day was also growing faster. On the other hand, output per manhour underground remained in France absolutely lower than in Britain; and in West Germany, where this ratio was much higher, it was improving only half as fast as in Britain.

The MMC had hoped to test the efficiency of the NCB against that of

[1] NCB *Ann. Rep.* 1977–8, 19; 1980–1, 8; 1982–3, 4. The figures for 1979 take account of retrospective amendments.

[2] Ibid. 1978–9, 12; 1980–1, 6; 1981–2, 7. These comparative figures are derived from ECSC sources. Those for the UK, though not quite identical, are very similar to those given in table A3 except for 1981–2, when there is a substantial discrepancy. Part of the discrepancy in that year appears to arise from a different treatment of gross social costs, and the figure may also be affected by the ECSC choice of exchange rates when making inter-country comparisons. But there may also be an element of unexplained error.

mining abroad, but found that valid comparisons between the mining efficiencies of whole countries, or even whole mines, could not be made because of irreconcilable differences in the size and layout of mines, their working practices and facilities and geological conditions. It did, however, commission detailed enquiries into some British and West German faces where conditions were not very dissimilar, and found nothing to choose between the two countries in efficiency, though there was scope for improvement in both. The NCB interpreted the available statistics to indicate that performance in the larger British mines in the early eighties was significantly better than in the West German.[1]

It seems fairly clear that, in terms of costs, British coal, on the average, was the most cheaply produced in western Europe, although the British industry carried a longer tail of small, high-cost collieries. But its advantage was not big enough for it to be competitive, in terms of delivered price (without financial loss), in all parts of the EEC, and differences in subsidy levels tended to reduce the advantage further. Against the larger non-European producers, notably the USA, South Africa, and Australia, it was in most foreign markets not able to compete effectively most of the time, though the position varied when there were large fluctuations in exchange rates or shipping freight rates.[2] This was not necessarily because of inferior efficiency, and was much affected by differences in physical conditions. There were countries with thicker, more extensive, and more uniform coal seams at shallower depths than in Britain, and some of them were laxer about damage to the environment. In the world as a whole a much bigger proportion of coal could be worked by the cheaper opencast method than in Britain. And the practice of differential pricing by exporters could make international competition more fierce. The effects of such differences in conditions might, with effort and ingenuity, be lessened but could not be avoided; and for the British coal industry, these were among the unwelcome facts of international life.

## ix. The state of the coal industry in the early nineteen-eighties

The current position and immediate prospects of the coal industry in the early nineteen-eighties could hardly fail to produce feelings of disappointment. Since the great contraction set in at the end of the fifties

[1] MMC, *National Coal Board*, I, 148–50 and II, 98–124.

[2] In October 1980, when sterling was appreciating, it was noted that, in some locations, imported domestic coal, notably from the USA, had a price advantage of as much as £10 per tonne over home produced.

there had been at least two periods of what appeared to be well-grounded hope. The first came during the last and most intense phase of contraction in the later sixties. Then it seemed that nearly all the high cost and highly unprofitable collieries were being pruned away and that a continually profitable industry would remain, strengthened by the indefinite extension of the recent rising trend of productivity and by the reduction in the number of stoppages, and able to hold its own competitively with no further contraction. That hope perished with the rapid rise in costs from 1969, the pressures of inflation, the great reduction in productivity gains, and the rebellion against the prolonged relative deterioration of miners' wages, which culminated in the national strikes of 1972 and 1974.

Yet the second of those strikes ended with hope reviving, because reconstruction had removed some of the worst financial burdens of the past, and because, despite the large increases of the last five years in the cost of producing coal, the oil price revolution had made coal more, rather than less, competitive and had suggested that it might be worth while in future to rely more on coal as a fuel likely to be more secure in supply and cheaper and steadier in price than oil. The new hopes were embodied, in a cautious and rational way, in *Plan for Coal*, approved in 1974. They were apparently justified by the results of the next few years. In both 1975–6 and 1976–7 the NCB were able to earn, even without direct grants to aid production, a trading surplus more than sufficient to cover interest charges, and they almost did this again, with higher interest charges, in 1977–8. These were the optimistic days that produced the less cautious and less rational *Plan 2000*, which looked forward to a renewed rapid expansion of the coal industry from the late eighties onward. But it is doubtful whether the grounds for optimism were ever very strong, as productivity, output, and sales remained stagnant or slowly declining and costs continued to rise in real terms. In such conditions customers were unlikely to buy coal in steadily increasing quantities unless they had nowhere else to turn; and they had other options. Energy conservation, plant closure and reduced employment in energy-intensive industries, gas, imported coal—these were all possible, and not unlikely, contributors to the reduction or elimination of the projected market expansion for British coal. And indeed, within a few years (especially after 1980–1) it was evident that even the sensible expectations of *Plan for Coal* could not be realized, despite substantial progress with the modernization of capacity, and the NCB found themselves in great difficulty with sales and finance.

The basic differences between what had been expected and what had been achieved can be quickly summarized. Costs were higher in real terms, much higher than would have been thought possible back in 1967 or 1968, somewhat higher than had been foreseen in 1973 and 1974, though by then future projections included the assumption of a gradual sustained rise in real costs. Productivity was lower than had been planned. Even though a rising trend had been resumed, levels in the early eighties were much below those which, a decade and a half earlier, had been thought to be attainable by the mid-seventies. The market was smaller than had been forecast and had become more and more biased towards the lowest-priced sector, especially after the great shrinkage in the demand for coking coals. Capital needs were greater than expected, both in real and financial terms, because of the enforced neglect of investment in the sixties and the high inflation and interest rates of the seventies. The details of the results which were shaped by these features have already been set out and analysed.[1] It remains to be considered how the general structure and prospects of the coal industry were affected.

One of the accompaniments of the recession from 1979 onward was a chorus of complaint from British manufacturing industry about the high price of energy. It was alleged that competitors, especially in western Europe, paid less for energy, with the result that British industry generally was put at a disadvantage in international markets. The implication was that all the British energy industries were either inefficient, high cost producers or were adopting unreasonable pricing policies. In either case they were damaging the UK's international trade and, incidentally, reducing their own sales. Eventually the National Economic Development Council was goaded by the complaints into setting up a task force to examine them. It reported in March 1981.[2]

The NCB had always expected that this enquiry would reveal little that reflected adversely on the performance of the coal industry or was likely to make much difference to it. The expectation was largely borne out both by the discussions (in which industrial consumers made no complaint about coal prices charged to them) and by the report. The findings of the task force indicated that energy prices generally were not higher in the UK than in the rest of the EEC, though some adverse movements were currently in progress, not because of bad performance by the British energy industries but because of movements in the sterling exchange rate. The main difference was in the structure of prices

[1] See especially section vi of this chapter.

[2] The report was summarized in the press and its publication gave rise to a good deal of careful press comment. See, e.g. *The Times*, 5 Mar. 1981, and the *Sunday Telegraph*, 8 Mar. 1981.

as it affected different classes of consumer. Though the average prices might be similar, very large users got bigger reductions in the price of electricity and gas in other EEC countries; and the UK put more of its petroleum taxes on fuel oil and less on petrol than the rest of the EEC. For coal the main relevant findings were that steam raising coal was generally as cheap as anywhere else in the EEC, despite much lower production subsidies, but foundry coke was 30 per cent dearer than in the rest of the EEC and 50 per cent dearer than in France, a difference entirely explained by subsidies. This was well known and the NCB had already agreed to match import prices when selling coal to coke ovens. The coal industry could also be affected by the attention drawn to the £8 per tonne duty on fuel oil in the UK, but coal was still generally competitive with oil even without that duty. There was, too, a risk that a demand for cheaper electricity for very large users might lead to requests for cheaper power station coal rather than a redistribution of electricity charges over all consumers.

Perhaps just as significant was the comment of the task force that the big fuel-burning industries in the UK were more energy-intensive than those in continental countries, and they had a lot of scope for savings from energy conservation. If they acted on that comment they would, of course, tend to reduce the market for the primary fuel industries. This was something the coal industry had to heed. Other, later, surveys confirmed that British industry, in general, was not relatively badly served by the energy suppliers.[1] But it remained relevant that higher energy costs tend to discourage the expansion of economic activity and to encourage technical measures to economize in fuel. The coal industry might not have a specially bad performance record relative to most of its competitors, but it retained a powerful market incentive to seek every practicable limitation of costs.

It has already been suggested that most of the elements of cost increase were not reducible to any great extent,[2] but the MMC concentrated attention on two, which need further comment. The suggestion was that labour costs added heavily to the rise in total costs (as was likely, since they were the largest single component), and that shifts of production between areas had been the most important source of cost

[1] e.g. although the CBI and some individual industries continued to complain about electricity charges, the 1984 survey of twelve countries' energy costs by National Utility Services (an American organization) indicated that energy (including electricity) charges were advancing more slowly in the UK than in most of the other countries (*The Times*, 11 May 1984).

[2] pp. 411–13 above.

savings. Regional differences in costs had always been appreciable and were likely to remain so. In the late sixties it was hoped and expected that their significance would diminish, because there was such a wholesale closure of unprofitable collieries that, for a time, it seemed likely that even the highest cost pits still in production would not be much of a burden on the finances of the industry. All this was changed by the large rise in costs in the seventies. If collieries were arranged in rank order of profitability, the line between the profitable and the unprofitable, instead of coming near the bottom of the list, as hoped, was thrust higher, so that in 1981–2 only 29 per cent of collieries, producing 42 per cent of output, made an operating profit.[1] Though there were often large differences in costs between collieries in the same area, there was a long experience indicating that results improved if more output was taken from the central, and less from the peripheral, coalfields. The emphasis on Scotland, South Wales, and the North East as the location of most of the small number of colliery closures after 1974 was thus a probable contribution to better financial results. The figures, however, as given in Table 7.8, suggest that the shift in the location of output, though in the usually advantageous direction, was not taking place very rapidly, especially after 1977. That is what might have been expected if most collieries had covered their costs most of the time. Closures through exhaustion would still have produced a slight shift away from the peripheral coalfields, but no more than that. When so many collieries had become unprofitable, the smallness of the change looks unbusinesslike.

If the central coalfields were losing their advantage in relative costs and profits, the smallness of the redistribution after 1977 would not matter and might be economically advantageous. Some care is needed when looking at regional relativities because, as a result of new rules for the allocation of rebates on coal sold below list prices, there were changes in internal accounting practice in 1980–1 which were specially unfavourable for the profit and loss figures of South Yorkshire and Western Areas and specially favourable for those of South Wales. In 1981–2 further modifications were made which had the effect of easing the burdens which had been placed on the South Yorkshire and Western figures and lessening the advantages to the figures of South Wales (and also of Durham and Kent). The accounting practices in 1981–2 were not the same as they were down to 1979–80, but broad general comparisons are not seriously disturbed.[2] It is clear that, though there

[1] MMC, *National Coal Board*, I, 168.

[2] In 1980–1 the lower cost coking coal collieries of South Yorkshire and Staffordshire were, in

Table 7.8. *Regional shares of deep mined output 1972-3 to 1982-3* (percentages)

| | 1972–3 | 1977–8 | 1981–2 | 1982–3 |
|---|---|---|---|---|
| Yorkshire | 28.4 | 29.0 | 28.7 | 29.3 |
| East Midlands | 23.6 | 25.8 | 27.1 | 27.6 |
| South Midlands[a] | 6.8 | 8.0 | 7.9 | 7.9 |
| TOTAL: central coalfields | 58.8 | 62.8 | 63.7 | 64.8 |
| Scottish | 8.7 | 7.9 | 6.7 | 6.3 |
| South Wales | 8.5 | 7.1 | 7.0 | 6.6 |
| Others | 24.0 | 22.2 | 22.6 | 22.3 |
| TOTAL | 100.0 | 100.0 | 100.0 | 100.0 |

*Note*: [a] Kent, which is obviously not a central coalfield, was absorbed into South Midlands Area in 1975 and has been included in the South Midlands total throughout. The effect on the figures is negligible. Staffordshire (part of West Midlands in previous chapters) was later transferred to Western Area and is here grouped with 'Others'.

*Source*: Calculated from NCB *Ann. Reps.*

were some worrying aberrations in parts of Yorkshire, the central coalfields were not losing their superiority. If one lists for each of the eleven years from 1972–3 to 1982–3 the three areas with the lowest costs per tonne, the list consists mainly of the three areas in the East Midlands. North Nottinghamshire appears in ten years, North Derbyshire in nine, and South Nottinghamshire in eight. The only other areas to appear are North Yorkshire and South Midlands, three times each. A similar list of the three highest cost areas includes South Wales every year, Kent in every year while it was still separately administered, and the North East in eight years. Scottish Area was included in each of the final three years. North Nottinghamshire had the lowest costs in every year from 1977–8 onwards, and South Wales the highest costs in every year from 1976–7 and, indeed, in every one of the eleven years if Kent is not counted as an area.

Similar lists based on operating profits and losses per tonne rather than on costs are not identical, but point in the same direction. North Nottinghamshire was the one consistently profitable area, along with South Yorkshire down to 1979–80. North Derbyshire was usually profitable from the mid-seventies on, and South Midlands achieved a

effect, bearing in their accounts part of the larger rebates granted by the higher cost coking coal collieries of South Wales, Kent, and Durham.

profit in a few years. So did South Nottinghamshire occasionally, after grants are taken into account. South Wales was one of the three heaviest losers in every year from 1975–6 on and had the highest rate of loss in all but two of them. Scottish Area (Scottish North in 1972–3) appeared in the list of heaviest losers in seven of the eleven years, including all the last three. The one serious discrepancy was that there were two Yorkshire areas among the three heaviest losers in 1978–9 and 1979–80, and one in 1980–1. It was alarming that Doncaster appeared in this position twice in three years, for Doncaster was the one Yorkshire area without an appreciable sector of very old, rather small pits, though it had, in general, been developed earlier than the Nottinghamshire areas immediately to the south. The North East, though a high cost area, appeared only once in the list of heaviest losers until the last two years, after the collapse of prices for prime coking coal. The contrast in area performance had been slightly reduced but remained wide. To take the usual extremes: South Wales costs were 183 per cent of North Nottinghamshire's in 1972–3, 202 per cent in 1977–8, 177 per cent in both 1981–2 and 1982–3. (In 1972–3 the extremes were different, and Kent costs, the highest, were 201 per cent of the lowest, North Yorkshire.)[1]

These were contrasts related to even wider differences between collieries, also locationally related. In 1981–2, when average costs were £37.81 per tonne, three collieries in South Wales and one in Kent had costs over £100 per tonne, whereas three in North Nottinghamshire and one in South Midlands had costs below £25 per tonne. Those same four highest cost collieries all had losses of over £60 per tonne (and one of them had lost over £20 per tonne in each of the last six years), whereas two in North Nottinghamshire and one each in South Midlands, Western, and South Wales Areas made over £10 per tonne profit. Altogether there were three collieries in South Wales and one in Kent with losses over £10 per tonne in each of the last six years, and two in North Nottinghamshire and one each in South Yorkshire and South Midlands with profits over £5 per tonne in each of those six years.[2] Whatever indicators are used, the regional contrast remains sharp and the problems of cross-subsidy are evident.

The financial consequences of the survival of this contrast were

[1] All the foregoing comparisons are based on figures published with the annual accounts in NCB Annual Reports. South Yorkshire was not a low cost area but its costs were much lower than those of other areas producing a high proportion of coking coal, which commanded premium prices until the end of the seventies, and this explains much of its superior performance as a profit earner.

[2] MMC, *National Coal Board*, II, 30–53.

tending to become more serious. In many of the higher cost districts a large proportion of the coal produced had special qualities—notably either natural smokelessness or very good coking properties—which traditionally had commanded higher prices. These had gone part way (though seldom the whole way) towards offsetting the high cost. By 1980 the demand for special qualities had decreased, and so had the size of the premium paid for good coking characteristics. Some of the output of special quality coals had either to remain unsold or go for steam raising (mostly at power stations) just like any cheaper general purpose coal. Revenue, and the prospect of future revenue, suffered accordingly.

The smallness of the decline in the share of the higher cost regions in total output also affected the problem of labour costs, which critics stressed. In 1982–3, overall OMS in South Wales was less than half the level which North Nottinghamshire, North Derbyshire, and North Yorkshire had all achieved, and in Scotland it was less than two-thirds of that level. This was a slightly greater difference than had existed in 1972–3, though in that earlier year the relative productivity advantage of South Nottinghamshire and South Midlands over South Wales and Scotland was somewhat greater than in 1982–3.[1] The continued operation of collieries with low productivity also absorbed investment merely to provide them with working capacity as faces came to the end of their life and new ones had to be opened; but such investment did little for productivity other than maintain it. All the evidence was that OMS was higher in new or reconstructed collieries. So the more capital had to be applied to routine uses in old, low productivity collieries rather than to more fundamental reconstruction, the less the contribution to higher efficiency was likely to be.

When the MMC investigated the NCB in 1982 it supported, with particular emphasis in its report, the NCB's view of the adverse effects of the large surviving number of high cost collieries. It stressed equally that the market for coal had contracted so much that the coal industry had a permanent surplus of productive capacity. Without specifying the detailed manner of carrying out the implied policy, it concluded that, in order both to adjust to the level of demand and to reduce costs and losses, the NCB needed to close about 10 per cent of capacity, and it pointed out the probable financial benefits if the closures could be concentrated on the pits with the highest losses per tonne, viz. an improvement of about £300 million a year in the profit and loss account. It

[1] NCB *Ann. Rep.* 1972–3, 36; 1982–3, 52.

indicated that this was a state of affairs that had been developing for several years and that there had been only limited and inadequate adaptation to it.[1]

This diagnosis and prescription have so much resemblance to those of the sixties that the question inevitably arises whether the call was for a repetition of something that had failed to work the first time round. That would not be wholly fair, for some necessary permanent reshaping of the industry was achieved in the sixties, and the activities of the eighties included a good supply of new investment, in contrast with the paucity in the sixties. Nevertheless, the manner of the contraction in the sixties directly contributed to some of the weaknesses of the seventies, and the MMC was counselling greater caution in new investment unless and until more closures were achieved. So there remained some grounds for unease. There were wide questions to be faced. What kind and size of coal industry was needed, not just immediately but for the long-term future? What immediate adaptations were both desirable and feasible, and would not be incompatible with the answer to the first question? The MMC showed sensitivity about influences limiting the feasibility of particular forms of adaptation, but much less awareness of the need to make them compatible with long-term objectives, perhaps on the assumption that, if severe short-term problems are left unsolved, the long term will never be reached.

Coal industry policy in the seventies was framed with a view to a future always extending beyond the next few years. Temporary short periods of recession were expected, and were expected to be overridden in the execution of policy for the modernization of capacity. The trouble was that by 1980 three sorts of influence on policy for output and capacity had changed simultaneously, and there was uncertainty about the relative strength of each. In the short term there was a business recession that was deeper and longer than the experience of recent decades had led anyone to anticipate. Secondly, there were structural changes in the economy—some of them associated with the technological developments that enabled the electronic performance of functions previously requiring mechanical equipment—which appeared to be permanent and which reduced the demand for fuel. Energy-intensive industries, such as steel, contributed a smaller share of national product, and there was a reduction in the energy input associated with a given increment of national product. Thirdly, even though nothing had yet happened to invalidate assumptions about the eventual rising share

[1] MMC, *National Coal Board*, I, 171–9 and 366–9.

of coal in the total supply of energy, it was clear that the timescale for such a development had to be extended: what had been expected to happen within twenty years might well take thirty or forty, and even then the increase might be slower. There were several reasons for this. Economies in energy use meant that all non-renewable supplies would last longer. The higher prices for energy meant that more oil and natural gas, including UK reserves, had become economically recoverable. And there was the usual tendency for caution to have been exercised in the public appraisal of fuel reserves, just as the NCB had shown in the modesty of the early statements of coal discoveries in the Selby field. Thus it was not very surprising that by 1984 UK reserves of oil and gas were publicly recognized to be appreciably greater than had been admitted in earlier years.[1] For the coal industry the long term of abundant and rising demand had moved further away, and the pace of getting there could slow down.

In some important ways these changes made things more difficult. There was still a requirement for a newer, and possibly somewhat larger, coal industry that must grow out of the existing industry; but currently and for a good many years ahead the market for British coal was rather smaller than it had been, and it was difficult to obtain, for the current level of sales, average prices that would regularly cover current average costs. Policy had to be directed so as to retain the ability to move from the current to the later condition as opportunity served. But this could not be done simply by shutting down part of the industry for the time being and then restarting with something better when economic conditions improved. Once mining has been well established in a region, a fairly high degree of continuity of operation is essential to efficiency and safety, to the preservation of the quality and accessibility of coal reserves, and to the retention of a willing labour force, which experience has shown is very difficult to create from scratch. Capacity thus has to be retained in use in some places as a bridge to the future, even though arguments from current revenue would suggest that it is superfluous. In a sense, though not in conformity with any normal accounting principles, this is an indirect form of investment, necessary in order to make possible the opening of better quality replacement capacity in the vicinity. This adds to the financial problem of shaping the industry to meet future needs, but on the other hand, when circumstances have extended the time available for achieving any increase in capacity, there is an offsetting reduction in the annual investment requirement.

[1] *The Times*, 9 May 1984.

In practice the basic financial difficulties around 1980 arose from slightly different causes. One was the need (hard to achieve in a static or declining market) to get higher utilization of all new or reconstructed capacity, in order to obtain better revenue results, because too little cash was being generated to keep the industry viable and contribute to necessary investment. The other was the financial structure, the disadvantages of which became more acute in the seventies than ever before. Sensible investment plans, especially for schemes which might take ten or more years before they produced any revenue, were very difficult to appraise and implement when they had to be financed entirely by fixed interest loans and carried out with costs and interest rates distorted by high levels of inflation, but with the expectation of less inflationary conditions when revenue began to flow. An institution with the size and tasks of the NCB needed a large equity capital and the means to maintain it. The NCB, in fact, proposed to the government in 1979 that half their capital should be in the form of public dividend capital, but the idea was rejected in 1981.[1] Whatever the merits or weaknesses of the particular type of scheme put forward, the principle involved was not demolished and remained urgently relevant. The MMC gave only cursory attention to the subject, but did, at least, remark that 'it might be easier to evaluate the financial performance of the industry in future if it had funding arrangements closer to those of a commercial organisation.'[2]

The need for continuity of mining operations in order to retain the opportunity for more efficient future development of a coalfield is not an argument for keeping open every heavily losing pit with some unworked coal. It is one very important constraint, among several others, which will apply in some cases and which, at any particular time, is almost certain to be relevant somewhere. But some collieries reach a persistent level of loss that no conceivable future benefit could counterbalance. In some districts the operation of one or two out of several neighbouring collieries may enable all the requirements of continuity to be met. In other districts there may be no perceptible opportunities for new developments and therefore no need to have regard to continuity.

[1] The arguments for rejection are given in section VI of the HC Treasury and Civil Service Committee Report, *Financing of Nationalised Industries* (12 Aug. 1981). The Coal Industry Act 1980 provided for the possibility of loans, on terms (which might include deferred interest) to be decided by the Secretary of State, in addition to the normal loans made from the National Loans Fund. But discussions between the NCB and the Department of Energy showed that the terms for deferment of interest would be such as to provide no advantage for the NCB.

[2] MMC, *National Coal Board*, I, 54.

The choice of capacity to close, to retain, to develop, or to create is a complicated matter. Whether the right choices were made in the seventies and early eighties is an important element in an assessment of the performance and prospects of the industry.

Various influences reduced the advantages and increased the difficulties of closing large numbers of collieries. In the first place, it was not all saving because a closed colliery left things that had to be done and paid for, with no revenue coming in. In 1981 it was estimated that, on the average, costs continued at 20 per cent of pre-closure level in the first year after closure, 7 per cent in the second year, and 3 per cent in the third.[1] At some collieries certain costs went on indefinitely (e.g. for pumping and the maintenance of barriers to prevent the diversion of underground water) in order to protect continuing pits. If expenditure were avoided, it could happen that unworked coal seams could be destroyed or damaged, sometimes miles away. Coal which would have been appreciating in value while it was undisturbed could become waterlogged and valueless. Secondly, closures could create unemployment and social damage and arouse opposition in mineworkers, which was liable to spill over into reduced co-operation in other matters of industrial relations. The NCB were, in fact, still very successful in avoiding redundancies at closed collieries, but this was dependent on reducing new recruitment at continuing collieries to a trickle, or, for extended periods, to nothing at all, and this meant a loss of job opportunities in some communities in which coalmining was prominent. Problems of this kind were exacerbated because, as the evidence on costs and losses indicates, most of the candidates for closure on economic grounds were concentrated in parts of the peripheral coalfields. Most of the districts affected had above average rates of unemployment, which by the early eighties had become absolutely very high.[2] The general state of the labour market was an inescapable influence, for the higher the level of unemployment the greater the feelings of insecurity aroused by proposals for pit closures. Already by the mid-seventies things were more difficult than in the sixties, but they were less difficult than they became after 1979.

These considerations indicate that there were strong pressures for coalmining to remain to a large degree where it was, whereas all the financial data suggested that it ought to have switched much more and much faster than it did to the lower cost central coalfields, and to large modernized collieries, which showed better results and which alone had

[1] Ibid., I, 170. [2] Ibid., I. 175–9 examines some of these problems.

the scale to justify the use of the latest technology, with heavy duty equipment and ever more varied applications of electronics. The strength of the forces operating against the economic ideal was so great that, in practice, some compromise was unavoidable. A programme like that of the sixties could not be repeated without modification because the circumstances were different.

One element in the compromise was that, though new investment had to be unevenly spread, it needed to be used to preserve and create jobs in or near to the existing centres of coalmining. This is what was done, and the peripheral coalfields did not get such a tiny share as had been expected in 1974. Most of the investment under *Plan for Coal* went to improve and reconstruct existing collieries. The new drift mines were all in locations that had already been intensively worked but where fresh opportunities were perceived. Of the large schemes, which were on fresh ground, Selby was not far from the reserves of the existing large Kellingley colliery and was an outlet for labour from the exhausted collieries of West Yorkshire; North East Leicestershire was the best located offset to the prospective declines of output and employment in Leicestershire and South Nottinghamshire. Exploration revealed many other interesting prospects on newer ground. But, for the time being, it was better business sense to leave the coal undisturbed and let its value appreciate until it was needed at some later date.

On the removal of old capacity, the record is less appropriate. Some that was apparently uneconomic was better retained, for one or more of the reasons already discussed, and some that was not strictly necessary probably had to be carried to secure goodwill in labour relations; but not the number of very uneconomic collieries that still remained. In the mid- and late seventies the NCB, faced with repeated opposition, made too few attempts to keep colliery closures in line with *Plan for Coal*. In this respect both the NUM and the Labour government did too little to support commitment to the Plan with deeds. Far too big a burden was carried into the eighties, when it was more difficult to shed.

The state of the coal industry in the early eighties was thus somewhat fragile, its immediate prospects difficult, its very long-term prospects highly uncertain but probably fairly bright for fewer people in a less labour-intensive activity. It was neither practicable nor necessary for the NCB to operate without some collieries which, at any particular time, were making a loss. But they could not carry an appreciable number of collieries that never made anything but large losses year after year, and still conduct commercial business. There was a lot of reshaping to be done, and if this did not cut away the worst drains on resources, it was not going to be long before someone suggested that there might be a bigger return on increased port facilities for importing coal than on increased investment in British mines.

The difficulties of the market were such that there could be no guarantee that sales would be maintained in the short run, though there was a good chance of doing this. Existing customers had to be retained and additional customers attracted, without reliance on the oil industry to frighten them towards coal, which had not happened as expected in the seventies, though the increased oil prices of 1979 were having that effect on the decisions of a few large industrial users. For that attraction, price (based on cost and not permanently on grants), quality, and security of supply were what mattered. The price record in the seventies had not been good, but looked tolerable because it was better than that of oil (especially in 1979), whereas in the early eighties it threatened to become worse than oil's. Hence another argument for a cost-reducing reshaping. Security of supply had been greatly disturbed by the strikes of 1972 and 1974, which were still vividly remembered. It probably needed more than ten years of freedom from serious interruption before many more large customers were willing to commit themselves again to the combustion systems that would make them overwhelmingly dependent on coal, and they still had enough options to be able to wait.

These were questions that could be dealt with only by positive action and not by leaving things to take their course. They needed efficient organizations and they needed continuous co-operation of government, NCB, and unions, which was likely to be more difficult to achieve than before unless there were fairly drastic changes in both institutions and attitudes. And a change of attitudes meant new thinking, not a reversion to old slogans that were the natural ammunition of conflict. There were whole coalfields where the attainment of business viability had been postponed so often and delayed so long that faith in its possibility had almost disappeared. Their people and their resources could not be abandoned, but they could not go on with the same activities that had been crumbling away so unrewardingly for so long. The questions they posed could not nearly be solved in terms of the coal industry alone, but every possible answer had important demands on it and important consequences for it. There were many signs after 1979 that new trends had been established in both the energy and labour markets, and in politics. Such a change in the environment of operations and the market meant the start of a new phase in the history of the coal industry.

PART C

# OTHER ACTIVITIES

CHAPTER 8

# Opencast Mining

Opencast mining has a curious place in the history of the British coal industry. Though there are signs of the digging of outcrop coal with hand tools many centuries ago, the practice seems to have almost ceased after medieval times, even though there has been an increasing amount of opencast mining for other minerals, including iron ore, and despite the features which opencast mining has in common with quarrying, which has a long, continuous history. Even in an exposed coalfield much of the outcrop of coal seams has a thin covering of some other material, which has to be dug out if the coal is approached from the surface, and primitive excavations, with no permanent drainage systems, have often been stopped by rising water levels when attempts were made to extend them. So it has usually been found preferable to go underground to get at the coal, even when it is only a little way below the surface.

In some other countries, notably the USA, with some very thick coal deposits almost on the surface and covering a vast area, it has been worth while to have very large continuous opencast projects, with much more elaborate engineering arrangements, and opencast became established as an important part of a modern coalmining industry. But in Britain, opencast commended itself only as a small scale desperate expedient in circumstances of wartime shortage, when any extra fuel, however costly and unattractive, was welcome. The expedient was meant to be strictly temporary and its imminent termination was several times proclaimed. It was, indeed, assumed that the amount of unmined coal in Britain that was accessible to opencast methods could not be great enough to keep the expedient going for more than a couple of decades or so, at most. But the temporary expedient became a permanent part of the coal industry, fluctuating more, but in the long run contracting proportionately less, than deep mining. In Britain, as elsewhere, it became established as the cheaper way of getting coal, and instead of being a desperate last resort, it became probably the best business proposition in coalmining. The drawback remained, however, that only a small proportion of British coal could be reached by opencasting; but there was enough to give opencasting a life many times longer than had been thought possible.

It is as well to be clear what sort of activity is comprehended in the term 'opencast'. Several terms are familiar, including 'strip mining' and 'open pit' and others, as well as 'opencast'. Sometimes they are used loosely in similar senses, sometimes they are separately distinguished. 'Open pit' would certainly be commonly used in a restricted rather than a general way, as it is only in a minority of cases that mining from the surface requires the excavation of the sort of wide arena which the word 'pit' suggests in this context. One definition offered for opencast coal mining is 'the method of winning coal by open excavation . . . where strata or overburden overlying the coal seam is cast back into an earlier excavation from which the coal has been removed'.[1]

Opencast mining is a civil engineering operation, with many variations of detail, but with a sequence of activities that usually runs something like this.[2] The topsoil of the operational site is first scraped off and stored in mounds round the periphery of the site. These mounds act as baffle banks to cut down the amount of noise going out from the site, and the topsoil is kept separate and free from traffic, so that it is all uncompacted and available for replacement after coaling operations have finished. Secondly, this activity is repeated by scraping off the subsoil to a depth of about 0.6 metre and storing this separately around the site. A cut is then made downwards into the site. Within the cut the overburden is removed until all the coal which it is desired to take from that cut is exposed. The surface position of the cut, and its direction, depend on a variety of factors, particularly on the location and inclination of the coal seams and the need to retain stable, horizontal positions for the mechanical equipment. The overburden is removed by blasting and mechanical excavators, with the use of draglines which shift the overburden to suitable storage locations. The last bit of overburden, immediately above the coal, is usually removed by smaller equipment so as to ensure the precise separation of the coal from the overlying strata without wasting any coal. The coal is then dug out from the seam by mechanical shovels and transferred to heavy dumper trucks (normally larger than those used on public roads), which carry it to a disposal point.

It may be that, at some stage, the first cut will have to be deepened, so that coal can be taken from lower seams, but a second cut is presently

[1] Arguile 1975.

[2] Various articles include a description of the operations, e.g. Brent-Jones 1971.

made nearby and the whole process repeated, and so it goes on until all the accessible coal has been dug from the site. As soon as all the coal has been taken from one cut, the overburden removed in making a later cut is used directly to fill in the void left in the earlier cut, instead of being stored on the surface. In this way only a part of the site is left void at any one time, the area needed for storage of overburden is reduced, problems of drainage and stability are simplified, the time needed to occupy the site is shortened, and costs are reduced. When coaling in the final cut has finished, the last remaining stock of overburden can be used to fill in the void. Then the original subsoil is spread over the whole site in two separate layers, each about 0.3 metre thick. After this the original topsoil is spread and the site is ready for the detailed work of restoration.

Sometimes things are done rather differently. A very small site may be best worked from a single cut. Sometimes the sequence is not possible because too small a proportion of the coal is winnable from any one cut. This may be because the seams are steeply inclined or irregularly formed or occur at awkward intervals, or because there are geological faults. In such cases a large open pit may have to be dug before it is practicable to take out coal. This is a more elaborate and protracted, and therefore more costly, method. It is likely to be worth while only on extensive sites containing unusually large quantities of coal. It is noteworthy that the largest opencast site ever worked in Britain, at Westfield in Fife, is also a particularly impressive example of the open pit. Work on this site began in the late nineteen-fifties and it was still producing coal in the eighties.[1] But operations in this manner and on this scale are quite untypical of British opencasting.

A narrative of the main events in the history of opencast mining is fairly simple.[2] The alarming wartime drop in coal output was particularly sharp in 1941 and there was an inclination to clutch at anything that might make it less bad. A useful bit of pressure was exerted by Major A. N. (later Sir Albert) Braithwaite MP, who belonged to the civil engineering industry and who suggested that it ought to be possible for civil engineers to exploit remaining outcrops of coal. The Mines Department of the Board of Trade responded by setting up a small organization which included civil engineering firms with plant available. Some preliminary work was done late in 1941, the first coal was produced in 1942, and a surprisingly quick increase of opencast output was achieved.

[1] See Costain Mining Ltd. n.d. [? 1975] for details.

[2] Arguile 1975 is the best summary for most of the period.

The work was carried out by civil engineering firms on behalf of government departments, acting on the authority of wartime Defence Regulations. Formal responsibility, which had shifted between departments, returned in April 1945 to the Ministry of Fuel and Power, whose Department of Opencast Production arranged and supervised all the work.

Opencast mining was physically messy and was expensive and it was sending out coal in very poor condition. It was therefore unpopular and had never been intended to go on for long, but it had done something to alleviate a fuel scarcity. So it was surprising that in the autumn of 1946, when the fuel shortage was approaching its worst, the Minister, Emanuel Shinwell, stated publicly that opencast mining would cease by 1948, with the result that some of those in the industry sought and found other jobs. But without opencasting there was not enough coal. Shinwell changed his mind within weeks and asked the NCB to consider taking over opencast by about May 1947.[1] The NCB jibbed at accepting a lossmaker with a short expectation of life, and, though discussions continued, the Ministry of Fuel and Power retained responsibility until 1 April 1952. At one stage it had seemed likely that a transfer to the NCB might be agreed for 1949,[2] and the Ministry had suggested that, in that event, the NCB should run down opencasting with a view to its cessation in four or five years. There was certainly a need to get opencasting out of its civil service environment: one of the problems in sorting out the details of the eventual transfer to the NCB was that the Ministry never kept separate income and expenditure accounts for opencast, though it was claimed that its numerous statistical and accounting records enabled production to be costed fairly accurately. But uncertainty about present finance and future commitment always stood in the way, until discussions in late 1951 led to agreement in January 1952.[3]

Important changes had taken place. The techniques and available equipment had improved, and on the Ministry's figures, opencast production was just above break-even. Enough reserves had been proved to make possible continued production at existing levels for several years, and the government was prepared to face some hostile criticism and announce that production would go on. Government support and

[1] NCB 28th meeting, min. 283, 18.10.46.

[2] NCB 209th meeting, min. 743, 21.5.48.

[3] NCB 340th meeting, min. 14, 18.1.52; 343rd meeting, min. 49, 21.3.52. PRO, POWE 40/4 deals with discussions of possible transfer to NCB from late 1946 to early 1952.

direct co-operation were essential because the NCB had no direct authority to enter opencast sites unless they had been purchased. The legal powers were still those taken by the government under the Defence Regulations and this was the position until the Opencast Coal Act was passed in 1958.

The NCB established an Opencast Executive to deal with everything concerning opencast and provided it with an annual budget.[1] The actual operations continued to be done on contract by civil engineering firms. For a short time things did not go quite as well as had been hoped. In 1952 the NCB made a loss on opencast, and there were long delays and arguments in reaching a financial settlement for the transfer from the Ministry of Fuel and Power. But from 1953 the operations became regularly, and on the whole increasingly, profitable; and, after the Treasury had agreed to the use of H. A. Benson, of Cooper Brothers, as a sort of unofficial arbitrator, the financial terms of the transfer were finally settled early in 1955, much closer to the NCB's claims than the Ministry's.[2] Opencast operations made good progress, both physically and financially, until 1958, when output was greater than in any previous year. But the market had turned down in 1957, and though opencast was more profitable, its contraction caused fewer job losses than a comparable reduction of deep mining. So from 1959 it became government policy not to allow the opening of new opencast sites, except for the production of scarce special coals, such as anthracite in South Wales, which did not threaten employment in deep mining. So, as existing contracts were completed, opencast output rapidly diminished.

But there were three or four large sites on long contracts which had to be kept going; total demand for coal in the early sixties did not fall with the catastrophic speed of the late fifties; and even the small remnants of the opencast programme for general purpose coal remained profitable. So by 1964 opencast production began to reverse its decline. But the 1967 White Paper on Fuel Policy argued against opencasting on grounds of amenity and job creation, and announced a more severe return to the policy of 1959: no new opencast sites except for special coals, and pressure on the NCB to take account of this restrictive policy in the way existing sites were operated. In practice these measures were

[1] NCB *Ann. Rep.* 1952, 69–70.

[2] PRO, POWE 43/13. In Dec. 1953 the NCB estimates of assets and liabilities were such that the Ministry owed them £5.7 million, whereas the Ministry's version was that the NCB owed the Ministry £1.0 million. After Benson's investigation, settlement was agreed on the basis that the Ministry paid £4.0 million to the NCB.

less effective than those earlier, because by 1970 the deep mines had been so drastically reduced that they were having difficulty in meeting current demand. The need for opencast output became still more prominent in the coal strike of 1972. The industrial workers of the civil engineering contractors were mostly in the Transport and General Workers Union, and the opencast sites produced coal throughout the strike. Though it was all stored on site, instead of being distributed as usual, its availability immediately the strike was over was very helpful in the quick resumption of supplies of coal to power stations and industry. The NUM gained some members on opencast sites after this and there was no opencast production during the strike of 1974, though even then, normal production could be restored more readily on opencast sites than in deep mines.

But by 1974 it appeared that the decline in demand for coal had ceased, and the stability of output envisaged to 1985 could not be achieved without an increase from opencasting. *Plan for Coal* projected a gradual increase of opencast output to 15 million tons a year, a higher figure than the previous maximum achieved in 1958.[1] This target was reached in 1980–1, though there was a slight falling-off in the next two years. Each year from 1977–8 onwards more than 10 per cent of British coal output was obtained by opencasting, and nothing like this had ever happened before. Moreover, the movement of costs was such that opencasting could benefit far more than deep mining from the higher prices which the oil price revolution made possible. In the late seventies and early eighties the rising profits of opencasting contrasted sharply with the generally declining financial results from deep mining.

In the economic recession after 1979 there were suggestions that, as in 1959 and 1967, opencasting should be used as an output regulator to help protect the rest of the coal industry. These suggestions were strongly resisted by the NCB.[2] It was pointed out that it had become more difficult to reduce opencast output quickly because there had been a move to a higher proportion of relatively large sites which involved longer contracts. The average coaling life of an opencast site was between four and five years, whereas it used to be just under three. Moreover, to slow down and extend the duration of work on an approved site, or to alter the date of starting, would greatly damage relations with local planning authorities and their public, on a subject which had become increasingly sensitive. A few people took the opposite view, that opencast should be

[1] Dept. of Energy, *Coal Industry Examination: Interim Report* (1974), 12.

[2] MMC, *National Coal Board*, I, 252.

expanded and deep mining contracted, in order to improve the financial results. This, too, was thought impolitic, if not impracticable. Quite apart from the resistance it would have provoked, on employment grounds, from the mining unions, there was a limit (which was getting lower) to the environmental impact of opencast mining which public opinion would accept and which would not have seriously adverse effects on social, economic, and physical conditions. In the late seventies and early eighties the Opencast Executive took the view that this limit was somewhere near to 15 million tonnes a year, and this judgement was supported by the Royal Commission on Energy and the Environment. Nevertheless, the profitability achieved by opencast mining was attractive, and from 1979 there was growing pressure from private operators to be given more opencast licences, on rather better terms, by the NCB, and the Department of Energy added a little weight to this pressure.[1] But licensed operators were currently producing only about 5 per cent of total opencast output, so a slight yielding to this pressure made only a marginal difference for the time being. The main contribution of opencast mining is indicated in Table 8.1.

The survival and growing financial success of an activity that began as a despised and unprofitable stopgap is remarkable and obviously demands explanation. The answers are to be found mainly in technology and in managerial organization, though the rapid growth of profits after 1974 was much helped by the market changes arising from the great increase in the costs of the main competitors, oil and deep mined coal. Most of the improvement in technology did not come from the application of fundamentally new principles or methods of operation. It was attributable more to great increases in the productive capacity and operational efficiency of items of equipment that worked on principles long familiar. Great improvements were achieved in excavation, exploration, and the quality of the product.

Opencast mining depended for a long time on the supply of American equipment. It began in wartime Britain with a large number of rather small items of obsolescent equipment, supplied under lend-lease. In 1945 the Ministry of Fuel and Power managed to buy some more modern equipment in the USA, and when the NCB took over in 1952 they sent a special purchasing mission to the USA.[2] The result was that the industry in Britain, from that time on, began to operate with much

[1] Ibid., I, 254.

[2] Caseley 1959, 5–8; NCB 377th meeting, min. 154, 18.12.53; Ministry of Works, *Report of the UK Opencast Coal Mission to the USA, Dec. 1944* (1945).

Table 8.1. *Output and profits of opencast mining, 1942 to 1982–3*

| Year ended 31 Dec. | Opencast output (m. tonnes) | Opencast share of total output (%) | Profit/(loss) per tonne £ | Year ended 31 Mar. | Opencast output (m. tonnes) | Opencast share of total output (%) | Profit/(loss) per tonne £ |
|---|---|---|---|---|---|---|---|
| 1942 | 1.3 | 0.6 | NA | 1963 | 6.6 | 3.3 | NA |
| 1943 | 4.5 | 2.2 | NA | 1964[c] | 5.6 | 2.8 | 0.90 |
| 1944 | 8.8 | 4.5 | NA | 1965 | 6.5 | 3.3 | 0.88½ |
| 1945[a] | 8.3 | 4.5 | (0.86½) | 1966 | 6.9 | 3.7 | 0.70½ |
| 1946[a] | 9.0 | 4.7 | (0.47) | 1967 | 6.8 | 3.9 | 1.09 |
| 1947[a] | 10.4 | 5.2 | (0.15) | 1968 | 6.8 | 3.9 | 0.91 |
| 1948[a] | 11.9 | 5.6 | (0.07) | 1969 | 6.4 | 3.9 | 0.78½ |
| 1949[a] | 12.6 | 5.8 | — | 1970 | 6.3 | 4.2 | 1.16 |
| 1950[a] | 12.4 | 5.6 | 0.04 | 1971 | 8.1 | 5.5 | 2.03 |
| 1951[a] | 11.2 | 5.0 | 0.05 | 1972 | 10.1 | 8.3 | 1.43 |
| 1952[b] | 12.3 | 5.4 | (0.10) | 1973 | 10.1 | 7.2 | 1.92 |

| | | | | | | | |
|---|---|---|---|---|---|---|---|
| 1953 | 11.9 | 5.2 | 0.19 | 1974 | 9.0 | 8.3 | 1.74 |
| 1954 | 10.3 | 4.5 | 0.30½ | 1975 | 9.2 | 7.2 | 5.08 |
| 1955 | 11.6 | 5.2 | 0.41½ | 1976 | 10.4 | 8.3 | 6.07 |
| 1956 | 12.3 | 5.5 | 0.72½ | 1977 | 11.4 | 9.4 | 5.74 |
| 1957 | 13.8 | 6.1 | 0.68½ | 1978 | 13.6 | 11.2 | 6.48 |
| 1958 | 14.5 | 6.5 | 0.58½ | 1979 | 13.5 | 11.3 | 7.07 |
| 1959 | 11.0 | 5.3 | 0.37 | 1980 | 13.0 | 10.5 | 8.43 |
| 1960 | 7.7 | 3.9 | 0.31 | 1981 | 15.3 | 12.1 | 10.27 |
| 1961 | 8.6 | 4.4 | 0.78½ | 1982 | 14.3 | 11.5 | 10.93 |
| 1962 | 7.3 | 3.6 | 0.75 | 1983 | 14.7 | 12.2 | 13.08 |

*Notes*: [a] The profit and loss column for these years relates to the financial years 1945–6 to 1951–2 inclusive.

[b] The profit and loss figure relates only to the period of NCB operation, 1.4.52 to 31.12.52.

[c] Because of a change in the financial year the profit and loss figure relates to a period of 15 months, starting 1.1.63.

NA = not available.

Halfpence are ignored in all profit figures over £1 per tonne.

The opencast total and percentage are very slightly understated by the exclusion of licensed opencast output, which was recorded as part of the output of licensed mines, nearly all of which were deep mines. The size of the error is insignificant.

*Source*: NCB. The figures for 1972–3 to 1981–2 are those cited in MMC, *National Coal Board*, I, 234, as they incorporate slight adjustments for a few years. Otherwise they are from Annual Reports, converted where necessary from tons to tonnes.

bigger and more powerful equipment. An important change in the late fifties and early sixties was the replacement of much diesel by electrically powered equipment, though, at a later date, there were some improved designs of diesel operated equipment, which restored its advantage for some purposes. In the Second World War the capacity of the best equipment in use was just over 3 cubic yards for dragline excavators, $\frac{7}{8}$ cubic yard for drag shovels which dug out the coal, and 15 or 18 cubic yards for scrapers, which were then not self-propelled but had to be mounted on tractors. Most of the equipment was much smaller. Twenty years later, except for scrapers, even the best that had been available in wartime would have been thought too small for most uses, as well as being under-powered. Fewer, bigger, and more powerful items were wanted in the later periods.

In 1944 there were available, for the removal of topsoil and overburden, 745 dragline excavators and 568 scrapers; in 1973 (when output was rather more) only 276 dragline excavators and 104 scrapers. But at the earlier date just over half the excavators had a capacity of no more than $\frac{5}{8}$ cubic yard; at the later date only 5 were so small. In 1944 only 25 could shift more than $3\frac{1}{4}$ cubic yards, but 172 could do so in 1973. The three biggest each had a capacity of 30 cubic yards or more, including the biggest of all (familiarly named 'Big Geordie'), which had a capacity of 65 cubic yards and was acquired in 1969 by the firm of Derek Crouch (Contractors) Ltd. Of the scrapers, 524 in 1944 could move no more than 12 cubic yards and none could move more than 18 cubic yards, but in 1973 all but 3 had more than 12 cubic yards capacity, and 36 could move more than 24 cubic yards. In the early seventies the commonest sizes of equipment on opencast sites were the 7 cubic yard dragline excavator, the 6 cubic yard shovel for coaling, and the 50 and 70/85 short ton dump trucks.[1] By this time the biggest items in use had probably reached the maximum size suitable for British sites, though much bigger equipment was available in the USA, where some opencast sites were many times the size of the biggest in Britain. Indeed, in the late seventies, the tendency was to use smaller, hydraulically operated shovels for coaling, because they separated the coal more cleanly.[2]

Bigger and more powerful equipment, which was supplemented by improvements in explosives and blasting techniques, economized in

[1] Arguile 1975.

[2] NCB, *Land Use Aspects of Mining–Opencast*, Part III, sections 3–4 (written evidence for Commission on Energy and the Environment, June 1979).

manpower and greatly reduced unit costs. It was even more important that it became technically possible and economically worth while to get by opencasting coal which previously could not have been reached in this way. In the Second World War, opencast coal seldom came from more than 40 feet (12.19 metres) below the surface, and contracts then did not normally refer to the possibility of any work below 50 feet (15.24 metres). But by the seventies depths of 250 feet (76.2 metres) were not uncommon and the deepest site, Westfield, was planned to go down to about 700 feet (214 metres). In 1979 the average depth of working was about 50 metres. Greater depth was achieved because lower unit costs justified the removal of a far bigger overburden in proportion to the quantity of coal available. The ratio (by volume) of overburden to coal was usually not more than 5:1 in the early years of opencasting. The average of the ratios provided for in contracts in the later seventies was about 15:1, and the maximum 35:1. Such a change brought a greater number of coal seams within reach and some of these could never have been worked by deep mining. From an opencast cut it was possible to recover cleanly the coal in seams as thin as 13 cm (5 inches), whereas nothing less than 46 cm (18 inches) would normally be worked in a deep mine, and in the seventies it was rare to work seams as thin as that.[1]

As it became practicable to reach more coal by opencasting, the case was strengthened for more intensive exploration of the shallow coal-fields in search of additional sites, and there were large increases in exploration both in the fifties and the seventies. Progress was assisted by improvements in both drilling bits and drilling rigs, especially from the late fifties onwards. The programme, equipment, and methods were separate from those used in support of deep mining. Drilling, which was mainly of open boreholes (though 5 per cent of the distance was drilled as core samples), was to much shallower depths and therefore cheaper. So it was economically possible to drill boreholes at smaller intervals. This was necessary because the characteristics of fairly small sites had to be identified with great precision. It was the continuity and success of the exploration programmes, guided by past geological surveying and the records of previous mining, that ensured that there was always a reserve in hand of sites known to be suitable for opencast mining. By the seventies it was established that opportunities could still be provided into the twenty-first century,[2] whereas the pioneers of the nineteen-forties had assumed that they could contribute only to a brief and transient episode.

[1] Ibid. [2] Ibid., Part I.

The third aspect of technological improvement concerned the quality of the coal. In the forties opencast coal got a rubbishy reputation which was partly deserved. In the early years it was often sold with an admixture of stones and shale and with the large coal badly degraded. This was because of hurried and careless mining and loading—large coal was often dropped from a height of as much as 10 feet into a wagon for loading—and the almost complete absence of any facilities for screening and preparing coal at opencast sites. When it became clear that opencast mining would continue for an appreciable period, a great improvement took place. Efficient screening equipment and boom loaders were supplied. In districts where a succession of sites was likely to be worked, or where there were bigger and longer lasting sites, permanent disposal points were built, some of them equipped for coal preparation, and the coal was transported there. Improved mining methods helped greatly. Coal can be won in the open in much less cramped conditions than underground, and it was less difficult to design and operate equipment to cut the coal with very accurate control. Where seams include a dirt parting, a coalcutting machine underground is almost bound to cut the dirt with the coal. But it became possible at opencast sites to treat any seam with a dirt parting more than 1 inch thick as two seams and work each one separately, recovering the coal while excluding the dirt band.[1]

In general, therefore, opencast coal came to be less contaminated with extraneous material, whether dirt or water, than deep mined coal was. Less needed to be spent on coal preparation, and the advantages of quality made it increasingly common for opencast coal to be added to deep mined to provide a blend to meet the user's specifications. By the seventies, these advantages were increasing because nearly all shallow coal is very low in chlorine, whereas the general tendency to mine at greater average depths was accompanied by a steady average rise in chlorine content. Opencast also provided substantial additions to the supply of scarce special coals, which was important at least until the market changes of the late seventies, and it supplied them more cheaply than the deep mines. In the seventies and early eighties about half of all the supply of domestic anthracite came from opencast sites and, though even opencasting was a good deal more costly in South Wales than elsewhere, it was regularly possible to make a profit on naturally smokeless fuel if it was obtained by opencasting.[2]

Some of the technological advances in opencast mining were

[1] Round 1975.

[2] MMC, *National Coal Board*, I, 236–7.

5. SHAFT SINKING OPERATIONS. Wistow Colliery, North Yorkshire Area. Shaft sinking in the Selby Coalfield commenced in 1976 and production began at Wistow in June 1983. The collar of the shaft is seen above, and its segmental lining below.

6. SURFACE DRIFT WITH BELT CONVEYOR. Silverdale Colliery, Western Area, Newcastle under Lyme. The Silverdale drift was completed in 1975.

7. POWERED ROOF SUPPORTS. Barrow Colliery, Barnsley Area, October 1960. Dobson Walking Chocks being inspected by deputies.

8. ROAD HEADING MACHINE. Gedling Colliery, South Nottinghamshire Area. Dosco Road Heading Machine.

9. PIT VILLAGE. Ffaldau Colliery, South Wales Area, Pontycymmer near Bridgend, Mid Glamorgan. Colliery with typical housing on adjacent hillside in the 1960s.

10. OPENCAST DRAGLINE. 'Big Geordie' (*left*) was the largest dragline operating in Europe when it began work at Radar North Opencast Site, Northumberland in August 1969. It has a bucket capacity of 65 cu. yds. and can lift more than 90 tonnes at one time. Its bucket is shown (*right*) holding a smaller dragline.

11. PIT PONIES. Easington Colliery, North East Area, March 1971. A last visit from the farrier for two of the few remaining pit ponies at the colliery. The ponies were withdrawn soon afterwards on completion of underground mechanization.

12. MANRIDING TRAIN. Silverwood Colliery, South Yorkshire Area, May 1979. Miners coming off shift on a paddy train.

common to all types of large-scale earth moving operation, and it was because it was so similar to much other civil engineering work that there was potentially ample productive capacity for opencasting. It was essential to the development of the industry that administrative and financial arrangements should be devised that would attract and retain civil engineering capacity. The responsible government departments engaged civil engineering firms as contractors, for they had not the resources to do the job by direct labour, and the NCB continued the same practice. Although opencast mining became permanently established, it was more advantageous to do this than to set up a specialized operating organization within the NCB. The flow of new work was somewhat erratic and could be better adjusted among a number of firms, each of which could balance it with other civil engineering commitments. The firms had the skills and the operating teams already in existence. The use of some of their equipment could be shared between opencasting and other work, with a favourable influence on overhead costs. The firms and their employees were accustomed to moving their operations from one site to another every few years or oftener, and they were thus organized in a way suited to opencast needs.

It was also helpful to the NCB that, at any one time, there was a fairly large number of firms providing potential for competition. Some firms undertook opencasting for a period and then moved out of it, but others came in. Though open tendering was not sought, there was usually an extended list from whom tenders were invited. When constructional activity was on the slack side, opencasting could benefit from very keen competitive prices; in other conditions the effects of competition were less, and in the early eighties single tenders were invited for a third of the sites. Down to 1974 forty different firms had acted as main contractors on substantial opencast schemes and thirty-five others had taken at least one small contract; but a high proportion of these firms left opencasting after the Second World War, and the number of firms regularly taking a large amount of opencast work was much smaller. From 1912 to 1974 38.7 per cent of output came from four contractors: George Wimpey & Co. Ltd., Sir Lindsay Parkinson & Co. Ltd., Costain Mining Ltd., and Derek Crouch (Contractors) Ltd. Altogether the ten largest contractors produced 59.4 per cent of the total, and no other firm produced as much as 2 per cent. In 1974 eighteen different firms had contracts, and in 1981–2 the total number of approved firms regularly invited to tender was nineteen.[1]

[1] Ibid., 240–1, 244–5; Arguile 1975.

Among the attractions for civil engineering firms were the simplicity of contracts, prompt and regular payments, and clear limitation of their responsibilities to defined operational activities. The early opencast contracts were very elaborate, with a detailed schedule of quantities and a separate price for each item.[1] The NCB adopted and retained a much simpler system. For each site the Opencast Executive estimated in advance the amount of coal recoverable and laid down detailed rules of operation. Within those rules the tenderer had only to quote a rate per tonne for the whole estimated output, with a specified quantity delivered each week, a rate per tonne for haulage to the disposal point, and a sum for plant costs; then he had to add a lump sum for restoration of the site to the point where, after topsoiling, he handed it back for rehabilitation. Very often it was found that there was rather more coal than in the original estimate and, in such cases, a supplementary contract was negotiated, usually at a lower price to take account of the lower overheads which the contractor needed to cover. So simple were the contracts that, out of more than 5,000 placed by the Opencast Executive in more than 20 years, only one produced any legal dispute. It was also a welcome practice to contractors that the sums due to them from the NCB were paid on a fixed day every month. Most of the equipment was provided by the contractors, but when additional items were needed for a particular site, the NCB sometimes made loans to finance their purchase. The NCB also bought or leased some equipment of their own (usually only large or highly specialized items) and sub-leased it to contractors for use at specified sites.[2]

The contractors were spared the hassle and expense that are associated with property negotiations and with the legal submissions and political dealings involved in securing authority to operate in particular ways at particular places. All this was dealt with by the Opencast Executive, which operated with a regional organization, the number and boundaries of the regions varying from time to time, mainly because of changes in the level of output required. Thus in 1980 the number of regions was increased from five to six. This degree of decentralization made it easier to maintain close relations with local planning authorities and local landowners, and to keep up with opportunities and changes in the land market. There was a clear division of function. The Opencast

[1] Ward Walters 1948 gives (10–11) an example of the early type of contract schedule, and (13–14) particulars of a simpler type, adopted by the Ministry of Fuel and Power in 1945, which came nearer to what was subsequently used by the NCB.

[2] MMC, *National Coal Board*, I, 241–6; Arguile 1975.

Executive was responsible, within its budget and the general policy of the NCB, for devising an ongoing programme of exploration and production, placing contracts for its execution, and exercising general supervision over contractors' operations. It was responsible also for acquisition of rights of access to sites or for ownership of them, and for obtaining authority to carry out opencast coalmining on them.[1] Until the late fifties it was usual to seek rights of entry and occupation for the duration of operations, but thereafter the normal practice was to buy all the land likely to be needed and re-sell it after it had been worked and rehabilitated. Some land was always bought in advance of operational need, particularly if it happened to come openly on the market. Securing authority for operations involved the Opencast Executive in a lot of work in compliance with statutory procedures, including the presentation of a case to public authorities and representation at any public enquiries.[2] A good deal of public relations activity was a necessary supplement. The work of the Opencast Executive left the contractors free to concentrate on production without much distraction.

The nature of much of the work of the Opencast Executive shows that the establishment of opencast coalmining as a permanent and profitable activity was achieved only within many constraints. It is an activity which always interrupts the normal course of land use and causes disamenity of various kinds. It is therefore bound to be subject to some kind of governmental regulation, the nature of which may well vary in response to political pressure. The fact that opencast coalmining was for so long carried on under the authority of the wartime Defence Regulations 50, 51, and 51A was unlikely to increase its popularity or to convey an expectation of its permanence.[3] At the time of the transfer to the NCB, the government had a committee considering the replacement of wartime emergency measures by permanent legislation if they were still required, for there was a political desire to get rid of as many wartime regulations as possible. But it was found impossible to draft a suitable bill for opencast mining, so there was only the cosmetic exercise of abolishing Defence Regulation 16, which had authorized the temporary closure or diversion of highways and footpaths, while amending Regulation 51A to enable this still to be done in relation to opencast coalmining sites.[4] As long as an acute shortage of coal persisted there were not many political constraints on governmental autocracy on behalf of the coal industry (though there were a few), but by 1957 the

[1] MMC, *National Coal Board*, I, 234–5.
[2] Ibid., I, 239–40.
[3] Arguile 1975.
[4] PRO, POWE 37/275.

shortage was over. From 1958 opencast operations ceased to depend on Defence Regulations and were governed mainly by the Opencast Coal Act of that year. This statute still made it possible, for ten years, to seek compulsory rights of entry to land for opencast purposes.[1] When this power was nearing expiry the NCB considered seeking its extension but decided it was not wise to do so, lest a complete reconsideration of the legislation might be provoked. But in a few exceptional cases the lack of such a power was found to be a handicap and it was restored by the Coal Industry Act 1975, which made various other minor amendments to the Opencast Coal Act.[2] In general, however, it was desirable to use persuasion rather than compulsion, and statutory regulation was not oppressive.

The legal position of opencast mining after 1958 remained distinctive. Unlike deep mining it was not subject to general town and country planning legislation, though many similar procedures applied. But the decision on applications for opencast working by the NCB lay with the responsible fuel minister (latterly the Secretary of State for Energy), not the local planning authority or the planning minister (latterly the Secretary of State for the Environment). The preliminary activities before any application to work an opencast site have usually involved consultation with about thirty bodies, the provision of much information, and the issue of public notices. If any County or District Planning Authority, or any owner or occupier of land within the site, made statutory objection, the Minister was obliged to hold a public enquiry. If authority to work the site was granted, six weeks had to be allowed for the possibility of an appeal against it to the High Court. In practice most problems were resolved informally during the preliminary consultations, but the statutory procedures became increasingly influential in the seventies because of the growth of the conservation movement and its somewhat litigious disposition. To the end of 1973 only nineteen public enquiries had resulted from 215 applications, but from 1974 to 1981 inclusive 144 applications produced thirty-one public enquiries, five of which ended with rejection of the application.[3] It was taking longer, and requiring greater expense, to reach the point of operation after suitable land had been acquired, and there was greater hostility from part of the public.

One attempt to lessen the friction was the production of a written

[1] NCB *Ann. Rep.* 1958, I, 14.

[2] Ibid. 1975–6, 21.

[3] NCB, *Land Use Aspects of Mining–Opencast*, Part II; MMC, *National Coal Board*, I, 239–40.

code of practice (about the supply of information, consultation, and procedures) which was agreed in August 1980 between the Opencast Executive and the three local authority associations. Even this document stated that it was prepared on the understanding that there is a justifiable national need for opencasting, 'but in no way commits the Associations to departing from current policies which in the case of the Association of Metropolitan Authorities questions this need and the target production figures in a continuing dialogue with central government'.[1]

Behind this growing, but still mainly unsuccessful, resistance to opencast mining was a set of beliefs that it imposed many hidden costs which ought to have been placed against its benefits and were not. In the earlier years the arguments related mainly to agriculture, in the later years to amenity. Most opencast mining was necessarily on agricultural land and it was argued that the effect was to deprive the country of urgently needed home-grown food. In 1952 the Ministry of Agriculture proposed that its Agricultural Land Service Officers should start to oppose applications not only to work but even to prospect land for opencasting, if they considered the land to be of high agricultural value.[2] This sort of opposition was mounting at that time, and in the same year one case was referred to the Cabinet, who vetoed the proposed opencasting. This example consisted of two adjacent sites at Penston, on class I agricultural land in East Lothian. The effect was virtually to give the Department of Agriculture for Scotland a power of veto over opencast applications in respect of good agricultural land in Scotland, a power which was soon exerted to prevent some proposed prospecting in the East Neuk of Fife.[3] Not unnaturally there were demands for exemption for some sites in England and Wales. But the Penston decision was a rather special case. It was in a Parliamentary constituency won by the Conservatives but not regarded as particularly safe, and this was impressed on the Conservative government, which had been in office less than a year. There was often political anxiety not to arouse Scottish national feeling unnecessarily, and the land involved was of very high quality in a country with a large proportion of poor agricultural land. This combination of circumstances was not repeated elsewhere, and the government wanted an increase in coal output. The only other place

[1] Local Authority Associations, *Code of Practice: Opencast Coal Mining* (Aug. 1980). The quotation is from 2.

[2] PRO, POWE 37/278.

[3] 'Class I' refers to the classification by the Land Utilisation Survey.

where class I land was affected by opencasting was West Lancashire, and there were serious delays in getting authority to work one or two sites there.[1]

But in general the agricultural argument did not prevail and was not continually pressed so strongly, though the Cabinet again turned down a request for the East Lothian sites in 1956. In that same year the Ministry of Agriculture, the Ministry of Power, and the NCB agreed that, on the very best land, the value of crops lost while a site was occupied for opencasting was only one-twentieth of the value of coal recovered. This was almost certainly a conventional figure, diplomatically acceptable rather than scientifically demonstrated. The NCB argued that in other cases the value of the coal was up to 140 times the value of lost agricultural output. Price movements suggest that the margin would later have widened in favour of coal, but ratios on the subject seem to have been forgotten, though with mixed results. In 1977 an inspector conducting a public enquiry into an application to opencast 2 million tonnes of coking coal at Oughterside in Cumberland reported that he could not assess the relative value of alternative uses to the nation, and that the claims of agriculture and the environment should prevail; but the Secretary of State for Energy overruled him.[2] And even in East Lothian things changed in favour of coal. In 1978 operations started at Blindwells, near Tranent, the fifth largest of all British opencast sites.[3]

The acceptance of opencast mining owed something to the relatively short time that most sites were in use, and to the apparent absence of permanent damage once a site had been rehabilitated. Indeed, in a minority of cases it was clear that a permanent improvement in the quality of the physical environment was achieved as a result of opencasting. There were, of course, important factors qualifying these favourable conditions and there were some exceptions to them. Although the average site, until the later seventies, was not needed for coaling for longer than three years, and some restoration of part of the site could be achieved within that period if backfilling were practised, full agricultural rehabilitation took much longer. After the coaling contractor had cleared the site and restored the subsoil and topsoil, the Ministry of Agriculture, as agent of the NCB, carried out a five-year programme of land management to rehabilitate the productive agricultural quality of

[1] PRO, POWE 37/278.
[2] *Opencast News*, no. 11 (Winter 1977).
[3] NCB Opencast Executive, *Opencast Coal Mining in Britain* (1979).

the land. Wherever possible the owners or occupiers were employed to carry out the work. In the same period the NCB employed specialist contractors to restore the fixed farm equipment, such as walls, fences, hedges, ditches, and trees. If it seemed advantageous, restoration and rehabilitation could allow for slight changes in the contours of the land and an alteration in the field pattern and in the location of woodland. In such matters, wherever there was no restriction under planning law, the preferences of the owner were followed as far as possible, even though this might detract somewhat from appearance. Many farmers, for instance, wanted their new fields divided only by wooden fences, so that future changes would be easier and cheaper, although before opencasting there had been hedges, which the NCB would have been happy to replant.[1]

Sometimes the benefits of rehabilitation could not be sought and obtained in full for a long time. The biggest sites, such as Westfield in Fife and Maesgwyn Cap, near Hirwaun in Glamorgan, were repeatedly extended and so were occupied for decades rather than years. The improved techniques, bringing the ability to reach greater depths from the surface, caused a few sites to be reoccupied and excavated a second time. And the proximity of sites in the same district often meant that restoration of one little bit was counterbalanced by new workings nearby. A region such as the lower Aire valley, between Leeds and Castleford, where opencasting went on continually, first on one site and then another, for forty years, could not hope to look like a shining example of the achievements of restoration.[2]

In such cases a belief in hidden costs could be strengthened simply because disamenity seemed to go on for such a long time, even if the productive benefits also went on flowing for an exceptional duration. Although the argument between coal and agriculture could be settled with a fair degree of rationality, on a basis involving markets and measurement, the opponents of opencast mining tended to fall back on disamenity as their basic objection, and to stress it whatever the length of time a site was in use. The accompaniments of opencasting that were found most objectionable were probably blasting, noise (including that of blasting but not restricted to it), dust, water pollution, the diversion

[1] Brent-Jones 1971 and NCB, *Land Use Aspects of Mining–Opencast*, Part IV describe the methods of land restoration.

[2] Too rosy a view was sometimes published by those in the industry. *Opencast News*, no. 6 (Winter 1975) wrote of the lower Aire Valley: 'Many of the sites have long since been restored and are now occupied by a sports stadium, a Leeds city car park, a golf course and other developments.'

of roads and paths (which sometimes added appreciably to local journeys), and sometimes, where disposal points were at a distance from sites currently in operation, heavy traffic which congested local roads and made them dirtier.[1] It was a defect that no value was placed on these things when the benefits of opencast were enumerated, and those in the industry in its early years recognized that losses would have been greater if serious efforts had been made to minimize nuisance and improve restoration. But in the fifties there were substantial changes, and better standards applied from then on. Much more trouble was taken in advance to discuss with local people what would be involved and to make adjustments to lessen particular causes of anxiety or inconvenience. There were continuous efforts to reduce nuisance, though there was no way of getting completely rid of it.

Standards of care in this respect, as in land restoration, were much higher in Britain than in some other countries with a greater proportion of opencast coalmining, notably Australia and North America. It is revealing, and almost comic, to note the astonishment with which in 1974 a Canadian consultant commented on British restrictions on drilling and blasting, which he found added significantly to costs. British blasting regulations, he said, were ultra-conservative and based on vibration levels 'close to the human perception level and far from the true annoyance level'. The biggest problem for British contractors, he lamented, was not damage to property but 'the human response'. Humanity seems to have been getting a better deal in Britain. The incorporation of contract clauses for the reduction of nuisance in specific ways was always a condition for the grant of authority to work opencast sites. In the seventies the standard production contract had sixty-five clauses, of which twenty-eight related to the preservation of amenities and the avoidance of nuisance.[2]

The standards of land restoration and rehabilitation received a similar increase in care. Although something was attempted from the beginning, there is no doubt that in the wartime and early post-war years, many opencast sites were left in poor condition, and this contributed to the low esteem of opencast coalmining. There was a tendency then to rely on fairly generous monetary compensation to stifle the more serious criticisms made by farmers. The turning-point came in 1951 when the Ministry of Agriculture adopted a new system of practice and

[1] Davidson in *Colliery Guardian*, May 1975 considers methods of reducing nuisance and disamenity.

[2] NCB, *Land Use Aspects of Mining–Opencast*, Part III.

more money went on rehabilitation and less on compensation. It was from that time that the five-year rehabilitation programme became standard and the aim became to return land to farming with results as least as good as they had been before opencasting.[1] British restoration practice came to be regarded as providing useful guidance for countries, such as the USA, which had lagged behind.

In the sixties and seventies, though not to any significant extent earlier, there were also projects involving opencast mining on derelict land, often with the dereliction attributable partly to a long history of deep mining which had nevertheless not removed every bit of accessible coal. In such conditions it was often planned from the outset that, as soon as opencasting had been completed, the land would be reclaimed for entirely new uses. Probably the outstanding example was on the site of part of Telford New Town in the seventies. The land there had been mined for at least 200 years for coal, iron, and clay, and the opencast operation involved the removal of 200 disused pit shafts. Opencast mining produced over 1.3 million tonnes of coal, and sites totalling 520 acres (210 hectares) were then reclaimed and made fit for building. They passed to the Development Corporation, which otherwise would have had to pay about £10,000 an acre to clear them. Another much publicized example of reclamation was at Shipley in Derbyshire, where 1,000 acres (405 hectares) had been progressively opencasted since 1944 and the dereliction of 250 years of mining was removed. The site was turned into a country park, opened in May 1976.[2] Reclamation was, however, the exception. To the end of 1975 only 3 or 4 per cent of the restored land had been dealt with in this way.

The main test of the claims of opencast mining to be only a highly productive temporary interruption of normal land usage had to be the results of the standard practices of agricultural rehabilitation. A lot of effort and expense went into this: in 1978, at current prices, a five-year rehabilitation was costing an average of £2,125 per hectare, exclusive of the rehabilitation allowance paid in cash to each occupier, which varied with local conditions and was determined by the District Valuer.[3] In view of the relatively low cost of mining by opencast, such expenditure, if it achieved its objective, was well worth while. In the mid-seventies

[1] Ibid., Part IV.

[2] *Opencast News*, no. 2 (Winter 1973–4), no. 6 (Winter 1975), no. 8 (Summer/Autumn 1976). For fuller accounts of the creation of Shipley Lake and the reclamation of other sites in the Midlands, see Arguile 1971, 150–6, and Arguile 1978, 159–65.

[3] NCB, *Land Use Aspects of Mining–Opencast*, Part IV.

the whole sequence of restoration from filling in the void to the completion of agricultural management for rehabilitation was taking about 20 per cent of total opencast costs, with wide variations for individual sites, the extremes being at least 7 percentage points above and below the average, perhaps more.[1]

There was inevitably a little uncertainty about the completeness of the success of land restoration. It probably needs at least twenty years after the end of the rehabilitation programme for observations to be adequate, and most sites had not had so long. Moreover, detailed research on the subject, which was appreciable in the late fifties, was stopped a few years later when opencasting was being run down, and the research was not resumed for some fifteen years. So there was a deficiency of scientific inference from continuous data over a long period. The changes in plant and methods from the late fifties could also have made some difference. Heavier equipment could have done greater lasting damage to sites, but the better designed mountings and quicker, smoother movement over the ground were likely to have had the opposite effect. Practice in soil treatment during restoration was also not constant, but the changes were in response to new knowledge that was believed to point to possible improvements. The conclusions of the research around 1960, that, at least for grassland, yields were as high from restored land as from similar land that had never been excavated, related to conditions that were not identical with those of the early eighties.[2] But nothing in more recent experience and observation suggested that opencast mining had had any long-term detrimental effect on agriculture. Longer observation was still needed, but it appeared probable that, on a national basis, any uncounted social costs were small in relation to the net output of opencast coalmining.

A small proportion of opencast output was produced by private operators under different administrative and legal conditions. By early 1947 there were several requests for permission to act in this way, mostly from industrial firms with an outcrop of coal on their own land and a wish to extract it and use it themselves. This led to some delays which were embarrassing at a time of coal shortage. It appeared that the Ministry of Fuel and Power still had sole responsibility for opencast

[1] Economic Commission for Europe, Coal Committee, *Symposium on Environmental Problems*, Katowice 18–22 Oct. 1976: paper by R. T. Arguile on 'Surface Mining Restoration—the British (NCB) Story.'

[1] NCB, *Land Use Aspects of Mining–Opencast*, Part IV.

operation, but that under CINA only the NCB had power to grant licences to private operators. After many weeks of regarding this as an impasse the Ministry decided that the NCB's licensing authority might be used in straightforward cases where the licensee would get the coal from the surface. This practice was followed thenceforward, but there was not much licensed opencasting, though a few small fly-by-night entrepreneurs operated without any legal authority and then disappeared, leaving an unrestored site. There was no explicit statutory reference to opencast licenses until the Opencast Coal Act 1958, which empowered the NCB to license an opencast operator who proposed to work a site which the NCB judged was not likely to contain coal greatly in excess of 25,000 tons. In practice the NCB interpreted this limit as 35,000 tons. They also declined, in general, to accede to requests for a second licence for the same site after a first 35,000 tons had been won, as this was clearly contrary to the intentions of the act. There were a few rare exceptions, where the original estimates of the amount of coal in the site turned out to have been much too low.[1]

A licensed operator was responsible for getting all the necessary legal authority. He could not have a licence until he had authority, from the owner of the site and of any other minerals, to enter and work the site, nor until he had planning permission. Unlike the NCB, a licensed opencast operator was subject to general town and country planning legislation. His planning permission would specify conditions for working and restoring the site, but the legislation included no power to impose conditions which might be improvements. Planning permission normally required the reseeding of the replaced topsoil and some drainage works. Only with the passage of the Town and Country Planning (Minerals) Act 1981 was there the possibility of requiring the sort of five-year rehabilitation schemes which the NCB had practised since the early fifties. Even then, the improvements which the NCB sometimes carried out to meet the requirements of particular authorizations under the Opencast Coal Act could not be required of licensed operators.

In the seventies, when higher coal prices were making private opencasting more attractive, and more opposition was being displayed to applications for authority to work sites by opencast, the NCB were somewhat worried by these differences in conditions. There was a fear that private operators might arouse opposition that would rebound on the NCB and hamper the achievement of their expanding opencast programme, especially as, from late 1976 onward, it was becoming a

[1] MMC, *National Coal Board*, I, 253.

common line of attack at public enquiries that opencast coal was not needed. There was also concern that some private operators bid up the price of land in districts where the NCB were negotiating for other sites.[1] Nevertheless, in 1981 the NCB had to yield to pressure from the Department of Energy to agree to consider the grant of a second licence which would permit the working of contiguous reserves up to 50,000 tonnes in cases where the original licence had had a limit of 35,000 tonnes. It appeared that the Department (which was simultaneously urging the NCB to be flexible about granting licences for deep mines that were bigger than the statutory limit for licensing) came very near to urging the NCB to break a law which the government was not willing to alter. The NCB, however, still refused to give an operator a series of licences that would enable him to take over 50,000 tonnes from one opencast site.

Most licences until 1981 required the operator to deliver his coal to an NCB disposal point and sell it to the NCB at a specified price that enabled the NCB to make a profit on their services. In some cases earlier, and in most from then on, the operator was allowed to market the coal himself and pay a royalty per tonne to the NCB as owners of the unworked coal. The licensees complained that the rates of royalty were higher than the profit per tonne made by the NCB on their own operations, but they were reminded by the Department of Energy that they were not comparing like with like. The licensees were working selected small parcels of coal for a limited period. They were not maintaining, with all the attendant overheads, a continuing national programme which involved keeping a regular supply of varied types of site always ready for operation as they were needed. In any case, relevant comparison needed to be with the profits of the NCB plus those of their contractors. The NCB undertook to set royalties at levels which would permit efficiently managed operators to develop their business profitably, and operators continued to seek licences for quite as much output as the market was likely to take.[2] So private opencast mining continued to a larger absolute extent than before the mid-seventies, but as a fairly

[1] The NCB could, and did, as a condition of licensing, require restoration to standards approved by the Department of Agriculture, but they could not insist on any improvements or a supervised programme. The licensed operators claimed, through their associations, that they adopted as high standards as the NCB for restoration; but the programmes do not appear to have been so protracted or to have had so much expert official direction. The NCB were, however, particularly ready to grant licences for sites which involved the reclamation of damaged land with the support of the local authority, and for sites which were about to be used for building development.

[2] MMC, *National Coal Board*, I, 253–6.

steady proportion of the total, and on terms apparently somewhat preferable to the operators. In 1981–2 licensed opencast mining brought the NCB about £12 million in revenue.

Opencast mining had an exceptional place in the nationalized coal industry. Although private mining under licence remained a very small proportion, the activity as a whole involved an unusual combination of public and private enterprise, since the entire output for the NCB was got by private firms working on contract. The relationship between members of the two sectors was based on their respective possession of different kinds of resources and expertise that were complementary. It was made effective by devising suitable administrative practices, and was found financially advantageous by both parties. Opencast mining was, indeed, unusual in its long record of increasing profits, which well illustrated how the financial prospects of an industry could be transformed by the prompt and careful application of cumulative, yet non-revolutionary, advances in technology. These were results which gave opencast mining a greater importance than its relative size: at the end of 1981–2 only 1,600 (mostly non-industrial staff) of the NCB's 282,000 employees were in the Opencast Executive,[1] and the contractors' labour force in opencasting, though fluctuating a good deal, was usually under 10,000. The relatively small labour input was a major reason for the higher profits, and, even before the large rise of profits in the seventies, it was to opencasting and ancillary activities that the NCB looked for a surplus to cover interest charges, which the principal activity, deep mining, could not meet. In the pre-Wilberforce period, i.e. to the end of 1971–2, deep mining produced only about half the NCB's accumulated operating profits of £450 million, whereas £138 million came from the far smaller opencast. From 1972 the contrast in profits in favour of opencast was enormous. There must have been times when it seemed that the chief missing element from the British coal industry's competitive strength was the opportunity to win as much by opencast as producers in more favoured continents: it was estimated that in 1973 27 per cent of the world's coal output was obtained by opencast mining, or 42 per cent if lignite is included, as only a small proportion of lignite is deep mined.

It may still seem surprising—and it would have been incredible to the pioneers of 1942—that opencast should ever have contributed more than 10 per cent of any year's British coal output, that in the course of 1982

[1] Ibid., I, 235; NCB *Ann. Rep.* 1981–2, 55.

the total opencast output since its inception passed 400 million tonnes and, if the market wanted it, could continue without contraction for at least another quarter-century. A country which had been mined for so long, and which had remained the world's biggest coal producer through the great expansion of the nineteenth century, might have been expected to have left little for opencast working in the twentieth century. But early mining was often careless, and some operations rendered impossible the safe access underground to reserves that were known, or thought likely, to be present. Some coal was too difficult or too costly to extract from earlier mines. Some coal had to be left for safety's sake: late twentieth-century opencast operations were able to extract the pillars from abandoned bord and pillar workings that were not very deep. And a good deal of shallow coal that was not visible from the surface was missed and was left to be proved by more scientific and more intensive exploration later.

Because Britain was mined on a large scale earlier than other countries, and because highly mechanized, high powered opencast techniques came fairly late in mining history, much of Britain's shallower coal reserves were recovered with costs and physical toil that would have been unnecessary if they had been left until later. But the value of the exposed coalfields had not been exhausted by the activities of earlier periods. One great advantage of opencast mining is that the recovery of coal can be more comprehensive and complete than in deep mines. For that reason alone a thorough re-examination of the exposed coalfields was well worth trying, and the results surpassed the best hopes. Necessity inspired a search for new opportunities. Technology enabled both the grasping of the opportunities when they were found and the enlargement of the field for further opportunity. Out of these conditions came the continued progress of opencast mining, which was probably the NCB's best success story.

CHAPTER 9

# Non-mining Activities of the National Coal Board

## i. Origins, purpose, and structure

Long before nationalization, colliery companies, even if they were not composite firms engaged also (sometimes principally) in the iron and steel industry, often did other things with coal besides mine it. They processed it in various ways, of which making coke was the most important; and sometimes they used the by-products of those processes for further refining or manufacture. They transported and sold the coal and its products, not just to wholesale merchants but sometimes retail to the public. Some of them regularly tried to make profitable use of other products and assets of their collieries, whether it was clay and shale, which they used as raw material in their own brickworks, or a reserve of land (not yet needed for colliery buildings, or retained as a protection against claims for subsidence damage), which was farmed or (more usually) leased to others to farm. Many collieries also found it necessary, or preferable, to provide houses and other buildings for the workforce, especially if the location were rather isolated.

All these activities were pursued in order to make the coalmining business easier and more profitable to run, and to derive some extra revenue from assets that were necessarily part of the coal industry and could be made more productive if other assets were added. As the colliery companies found it both directly and indirectly helpful to engage in these other activities, it was likely that the nationalized industry would have been at a disadvantage if it had not been able to do likewise. The legal and financial terms of nationalization made fairly complete provision for this.[1] It is notable that, except where their purpose was to supply the colliery owners' iron and steel business, colliery coke ovens, coal products distillation plants, and manufactured fuel plants were treated not as supplementary items but as integral parts of the coal industry which must be transferred to the NCB. Colliery merchanting

[1] Chap. 1, section ii above.

property was also compulsorily transferred.[1] Houses, farms, and other agricultural assets owned by collieries were among the items which had to be transferred if either the existing owners or the NCB requested this. But manufactured fuel plants not owned by collieries, coal merchanting property belonging to companies in which two or more colliery concerns had an interest, and brickworks could be transferred to the NCB only by agreement or, failing agreement, as a result of arbitration.[2]

The nationalization act made it quite clear that the NCB were not only permitted but expected to conduct (so far as was helpful to the coal industry) just as wide a range of activities (other than iron and steel production) as any of the colliery companies, or the coal industry as a whole, had previously done. The functions of the NCB which were specifically laid down, in addition to coalmining and coal exploration, were:

treating, rendering saleable, supplying and selling coal;
producing, manufacturing, treating, rendering saleable, supplying and selling products of coal;
producing or manufacturing any goods or utilities which are of a kind required by the Board for or in connection with the working and getting of coal or any other of their activities, or which can advantageously be produced or manufactured by the Board by reason of their having materials or facilities for the production or manufacture thereof in connection with the working and getting of coal or any other of their activities, and supplying and selling goods or utilities so produced or manufactured;
any activities which can advantageously be carried on by the Board with a view to making the best use of any of the assets vested in them by this Act.[3]

The response of the NCB to the opportunities created by the legislation was to try to make use of practically all of them. Most of the physical assets of the colliery companies were transferred to the NCB, even those whose transfer had to be by agreement. In a few cases—for instance, some small manufactured fuel plants which did not belong to collieries—assets that might have been transferred by agreement[4] were bought by the NCB, and the long drawn out task of determining compensation under the Act was circumvented as far as they were concerned. The NCB, from the start of their existence, thus came into

[1] CINA, 1st schedule, part I.
[2] CINA, 1st schedule, parts II and III.
[3] CINA, s. 1 (2)(*b*)–(*e*).
[4] Under the provisions of CINA, 1st schedule, part III, para. 15.

possession of a large quantity and great variety of assets[1] in addition to the collieries, and had the task of promoting their productive and profitable use. The main non-mining activities fell into four groups.

The first concerned coal products, i.e. coke and its by-products (together with substances made from the by-products) and manufactured fuels, including briquettes and ovoids. This was essentially a group of activities treating the major product, coal, in specific ways, for particular uses for which there was a large market. The importance of this group was that, for most of the NCB's history, it was far larger than all the other non-mining activities, though it fell on hard times in the late seventies, and by the early eighties its turnover was well below that of the rest. It had the further significance of adding a high value to the material from which the product was made, so that, at various times and places, it was worth while to mine coking coal at a loss because there was an offsetting profit from the coke made from it. In the long run (mainly from the late sixties) the coal products activities also posed a significant question of business diversification. Once the practice has been established of refining the by-products of carbonization, how much further should the processing of the resulting materials be carried? In other words, how far can the coal industry knowledgeably and profitably plunge into the chemical industry, or at least that large part of it which ultimately is based mainly on hydrocarbons?

The other activities were smaller and, on the whole, less complex. The second group was concerned with merchanting. In the discussions leading up to nationalization, the possibility had been raised of making the divisional coal boards exclusively responsible for large-scale distribution and storage, but this had been rejected on the grounds that it would be complex and controversial and would give the NCB a very difficult problem of organization when they had little time available to attend to it.[2] So they were deliberately left free simply to continue the selling activities of the former colliery companies, and this was done. Many collieries had a selling depot which came under the colliery manager and

[1] For details see Chap. 1 section ii above. The magnitude of the ancillary assets is roughly indicated by the compensation paid for them which, when payment for stocks of products and stores is excluded, came to a little over £90 million, i.e. more than half as much as was paid for the coal industry assets, though the ancillary assets were probably more generously valued than the coal industry assets and were therefore relatively not quite as important as the figures suggest. The colliery coke ovens and manufactured fuel plants, though grouped in CINA with the collieries for compulsory transfer, were treated for compensation purposes as ancillary, not coal industry, assets (Chester 1975, 93–4).

[2] Chester 1975, 91–2.

was run as part of the colliery's mining business, but there were some wholesale and retail depots elsewhere, operated as a separate ancillary undertaking. These were improved and concentrated in the sixties in the course of attempts to increase sales in difficult market conditions, and the activities were further extended through associations with private firms of coal distributors.

The third group dealt with the manufacture of goods from materials produced along with coal, and the manufacture of goods used in the coal industry. The brickworks, pipeworks, and tileworks were the main constituent and qualified in both respects. They were never, in terms of assets, turnover, or profit, among the major activities of the NCB; but they were sufficient to make the NCB one of the larger constituents of the brickmaking industry, and because the number of separate establishments was initially large and the industry was different in techniques and markets from any of those directly based on coal, there were distinctive managerial problems to be tackled. In the sixties, as the coal industry contracted, so did its output of brickmaking materials and its consumption of bricks, and the natural strength of the relationship between the coal and brickmaking industries became less apparent. So in the seventies the NCB withdrew from most of their brickmaking activities. Participation in the manufacture of items used in coalmining, though it happened on a small scale for a few engineering components, was never appreciable. This was work left to the specialist producers, though with some important guidance in innovation from the NCB research establishments. What the NCB did for themselves was a great deal of repair work, but this was something central to the operation of the coal industry rather than a related ancillary activity. In later years, however, there were some specialized services rather than goods, which the British coal industry needed for itself and which could also find a market elsewhere. Computing, specialized mineral exploration methods, and mining consultancy services are examples. The NCB became involved in developing these and selling them in the market as well as using them internally.

The fourth group was, in general, less directly commercial. Its concern was land and property, principally houses. The estates needed to be run as remuneratively as practicable, for whatever purposes they best served, until they were needed for the main business of the coal industry; and there were good arguments for not retaining any lands which, it had become clear, were never likely to be needed for that business. The amount of land owned did, in fact, decline. But houses were the major item and locked up a lot of capital. The provision of houses, however, was never a business venture in its own right and was never intended to be so. It was undertaken as an aid to securing and retaining a big enough labour force for the coal industry and locating it where it was most needed. The provision of additional housing was undertaken mainly by housing associations owned by the NCB. The allocation and management of the NCB's housing stock was directly incidental to the operation of the main business of coalmining and, to a small extent, coal products. None of the housing activity can be regarded, like the other associated activities, as a distinct commercial operation. When, with the shrinking of the coal industry, the difficulties of recruitment were less severe, the provision of housing became much less necessary, and the NCB began to extricate themselves from it. Housing was thus mainly a service to the workforce and should not be treated as though it were on a par with most other ancillary activities.[1] Indeed, some of its costs, including all rent subsidies, were always charged as colliery costs, and in later years, for accounting purposes, the current costs and revenue of housing were classed as a mining item, though capital profits from house sales were treated separately.[2]

These four groups were obviously by no means homogeneous in either scale or character, nor do they include quite all the non-mining activities of the NCB. From the outset the Minister of Fuel and Power required the NCB to show operating profits and losses for ten separate classes of non-mining activity. These were (1) coal selling depots, (2) coke ovens (whose activities included the output of primary by-products, i.e. mainly coke oven gas, crude tar, and crude benzole), (3) secondary by-product plants, (4) manufactured fuel plants, (5) briquetting plants, (6) brickworks and tileworks, (7) railway wagons, (8) wagon repair shops, (9) houses, and (10) estates and farms.[3] There was also a separate accounting category for what were known as 'other activities', each of which was individually small, at least to begin with, though this miscellaneous category was in total bigger than some of the items separately specified. It included road transport, electricity distribution, shipping staithes, waterworks, pyrites plants, pipeworks, and limeworks.[4] Some of these activities had no lasting significance. The main

[1] Housing provision is discussed along with other welfare items in Chap. 10 below, rather than in this chapter.

[2] They appear in the consolidated profit and loss account under the heading 'Profit on realisation of fixed assets'.

[3] NCB *Ann. Rep.* 1947, 205–7.

[4] The notes to the accounts in each *Ann. Rep.* usually give some information about the activities listed in this miscellaneous category, e.g. NCB *Ann. Rep.* 1949, 174.

line railway wagons, which were rented by the Ministry of Fuel and Power from the NCB, passed to the British Transport Commission from the beginning of 1948. The NCB wagon repair shops continued to repair internally used wagons and undertook external work, but it was a steadily dwindling business and was relegated to 'other activities' after 1959. On the other hand, road transport, which was left in that category, underwent long and substantial growth. By 1970 the NCB had a larger fleet than all but a very few of British business undertakings. Manufactured fuel plants (of which for many years the NCB had only one) and briquetting plants were soon made into a combined category for the purposes of the published accounts.

It is also worth noting that there were some establishments whose services were charged to the main activity (most often collieries) for which they were performed, and which thus did not publish profit and loss accounts, but which nevertheless were treated as non-mining activities when classified in relation to the deployment of manpower.[1] Much the largest of these were the central workshops. This is a reminder that the non-mining activities were of two business types. There were those that were exclusively required to perform essential services for the main activities of the NCB and which could therefore never be properly regarded as actual or potential 'profit centres'. There were those that might sell some of their output (in a few cases a substantial proportion) within the NCB organizaton, but were principally concerned to produce goods or services for external sale. To some extent this important distinction was confused by accounting practice. There were some cases where what was essentially an internal service was required to produce separate profit and loss figures. The largest and most obvious example was housing. It was absurd that down to 1972–3, the collective contribution of 'ancillaries' to the financial results of the NCB should have had to incorporate (in some years with very distorting effect) the declared profit or loss on housing, which was calculated by highly artificial methods using many notional figures. Another possible case was road transport, the future organizational and accounting treatment of which became a matter of disagreement in 1972–3, when a proposal to put it into a subsidiary company was rejected and separate profit statements were no longer required.

The main focus of interest must be on the activities which were pursued in the business environment set by the outside world, for it was

[1] Cf. the table of industrial workers employed, which was included in the tables appended to the *Ann. Rep.* each year from 1953 to 1971–2 inclusive (vol. II from 1955 onwards).

these which were genuinely supplementary or complementary to the coal industry, rather than being only its subordinate services. One of their significant characteristics was that they all had to compete in their own fields, though to differing degrees. The NCB's monopoly in the coal industry was not repeated for the ancillaries. This was true even for carbonization, the largest of them and the one closest to coal production. So many of the coke ovens, even a few of those on colliery sites, had belonged before nationalization to firms engaged in iron and steel production, that the care shown in CINA to keep from the NCB any assets that might properly pertain to the iron and steel industry was bound to prevent any coking monopoly. NCB coke ovens never produced more than about 40 per cent of the national output of hard coke, and the average quality of coke ovens that remained with the iron and steel producers (especially those, the majority, located at steelworks) was better than that of the NCB's inheritance. For domestic and industrial coke the NCB production was in competition with that of the private and municipal gasworks and, from 1948, the Area Gas Boards. As a producer of manufactured fuel the NCB was in a market with two substantial private producers, Coalite Ltd. and the National Carbonising Company.

Initially, competition was not a constraint on coke producers because demand was high in relation to national productive capacity and was rising. But in the long run it was potentially serious and was eventually experienced in a very severe way. The other activities had to adapt to competition from the beginning. The NCB may for a time have been the country's second largest brickworks undertaking, but even when the NCB brickworks were in their strongest position they produced only about 8 per cent of national output, except for a few years in the mid-fifties; and many of the plants inherited from the colliery companies were small and of poor quality. The other ancillary undertakings were small fry in their respective industries and markets. If they were to make a living and not be a drag on the NCB, these various activities had to rely on efficiency and the finding of some locational or specialist niche in the market which they could fill.

Success in such circumstances required specialist knowledge in those directing the activities, and administrative arrangements that gave their skills adequate scope. The ancillary activities were not novelties for colliery concerns, and the NCB inherited and continued to employ people who were well versed in running them. But that was mostly at the level of the individual establishment. Top administrative responsibility, whether at national headquarters, division, or area, was at first almost entirely left to those who were primarily concerned with coalmining. Ancillaries were just a little extra load they had to carry, and they were hardly likely to be able to give the extras their strongest backing. Even for the large activity of carbonization, though there were some fairly senior appointments, they were not at the topmost level. At national headquarters a specialist branch was created but its Director was subordinate to the heads of the Production and Marketing Departments; and, though he had Officers in the divisions, it was the Area General Managers who had direct responsibility for the coke ovens, just as for the collieries, in their areas. For brickworks there was no coherent administrative structure. There was a Brickworks Superintendent in each area in Scotland, but not elsewhere, and only some of the divisions appointed General Managers for brickworks. Everywhere the brickworks were the administrative responsibility of the Area General Manager, but not everywhere did he have the relief of specialists with much seniority.[1] Only gradually was it recognized that if ancillaries of any size were to be worth while they had to have more weight and scope in the administrative structure. After 1950 a series of steps at irregular intervals showed a gradually increasing response to this recognition.

A start came in 1952 after the previous year's lengthy discussion of the distraction to coal industry management which the ancillaries caused.[2] A new Carbonisation Department was set up, headed by a Director-General; the four divisions with nearly all the coke ovens were each given a Carbonisation General Manager (roughly equivalent in status to an Area General Manager) to run them; and the Area General Managers were relieved of this responsibility. For brickworks it was decided that there must be an effort to increase efficiency and reduce costs, but it was nevertheless concluded that there was no need for a brickworks expert on the staff of national headquarters. But perhaps this view was affected by the other conclusion that it did not matter if there was no increase in NCB brick output.[3] In fact, in the following years, some technical supervision was provided from headquarters.

The next stages came in the sixties when it was decided that more specialized top management was desirable in order to get the best return

[1] NCB *Ann. Rep.* 1947, 77 and 80; R. J. Morley, 'Coal Carbonisation and Other Treatments', in *Colliery Guardian*, *National Coal Board*, 1957, 79.

[2] Chap. 5 above, p. 172.

[3] NCB *Ann. Rep.* 1952, 41; Morley, loc. cit.; NCB 340th meeting, min. 12, 18.1.52; 351st meeting, min. 113(a), 25.7.52.

from the assets in the larger ancillary activities. In 1961 it was agreed to take the brickworks out of the divisional structure and set up a Brickworks Executive, responsible directly to national headquarters. A few months later came the larger decision to establish an additional division which would take charge of all the carbonization, secondary by-products, and manufactured fuel activities, instead of the geographically based divisions. The Brickworks Executive, with headquarters in the Midlands, assumed full responsibility on 1 April 1962, and Coal Products Division did so in stages from 1 January 1963. The latter was organized in three regions, Northern, Midlands, and South Western and, besides being responsible for all production, it undertook the marketing of the industrial products, i.e. metallurgical coke, primary and secondary by-products, coke oven gas, and methane. A further small move in the same managerial direction in 1963 was the establishment of a separate Housing and Surface Estates Department, though the seemingly obvious step, on functional grounds, of keeping housing responsibility distinct was not taken.[1]

By the mid-sixties the NCB had extended their non-mining activities in an increasing number of small ways by association with other interests in the same fields, and this gave them a growing experience, as substantial shareholders, of limited liability companies as the appropriate financial and organizational forms for most of their lesser types of activity, though they still had only one wholly owned subsidiary company (Thomas Ness Ltd.) of the normal profit-making type.[2] The result was that they began readily to turn to the establishment of new companies or the acquisition of old ones when they wanted to extend their activities, whether it was to develop some more specialized products in the fields in which they were already engaged or to take up something novel. For example, in 1966, in order to extend their range of types of brick, the NCB bought Whittlesea Central Brick Company Ltd. In the same year they joined with steel companies to set up new companies to manufacture particular products from the secondary by-products of carbonization. With Stanton and Staveley Ltd. (a subsidiary of Stewarts and Lloyds) they formed Staveley Chemicals Ltd., and with a subsidiary of the United Steel Companies Ltd. they established P. A.

[1] NCB *Ann. Rep.* 1961, I, 26; 1962, I, 21 and 25; Chap. 6 above, pp 267–8.

[2] See e.g. the descriptions of investments and wholly owned subsidiaries in NCB *Ann. Rep.* 1964–5, II, 27–9 and 31–2. Most of the wholly owned subsidiaries of the NCB were housing associations and small non-profitmaking concerns for restricted special purposes. For the acquisition of Thomas Ness Ltd. see below (p. 490).

Chemicals Ltd., the name of which was changed in 1967 to Phthalic Anhydride Chemicals Ltd. In 1967 also, as a way of expanding distributive business, they implemented a proposal to share some of the activities of the British Anthracite Company Ltd. (part of the Amalgamated Anthracite Holdings group), with whom they created a partnership which traded as the British Fuel Company.[1] There were several other small ventures of various kinds in the same period, all using corporate formation. And all this came soon after the acquisition in 1965 of a majority holding in J. H. Sankey and Son Ltd. as a way of enabling the NCB to sell solid fuel heating appliances.[2] So by 1967 the NCB had an array of subsidiary and associated companies which had not existed three years earlier.

Perhaps the most striking adoption of the corporate form came when the NCB had the opportunity to enter the new and difficult activity of offshore exploration for natural gas, with the subsequent possibility of developing any discoveries and supplying gas from them. This was not something they could do by themselves. They had to act in association with experienced oil companies who could provide operational expertise. They did not wish to appear merely as clients who advanced money to others, but as participants with their own organization. The NCB could not take on this new function until they had explicit statutory authority. As soon as this was received, by the passage of the National Coal Board (Additional Powers) Act in December 1966, they formed a new wholly owned subsidiary company to exercise all their interests in searching for and producing natural gas and other petroleum products. This was National Coal Board (Exploration) Ltd., incorporated on 12 January 1967 with an original issued capital of 50,000 £1 ordinary shares fully paid and all subscribed by the NCB, who made loans to the company for the purpose of acquiring interests in North Sea concessions.[3] The new company was made an additional responsibility of Coal Products Division, which represented its interests and needs in discussion with the national board.

The proliferation of small companies, some of them only partly controlled by the NCB, was getting somewhat untidy and carried the risk that administrative control over one or two of them might be weaker and less informed than was desirable. Some change was needed on business grounds, but the catalyst was the change of attitude brought

[1] NCB *Ann. Rep.* 1965–6, I, 27 and 29, II, 24 and 26; 1966–7, I, 26 and II, 27–31.

[2] Chap. 6 above, p. 247.

[3] NCB *Ann. Rep.* 1966–7, I, 26 and II, 27–8.

by the Heath government. There was a strong body of political opinion that the NCB should be sticking to mining, and an indication that the government might take powers that would make it possible to forbid the continuation of some of the ancillary activities. There was also informal pressure for the NCB to dispose of some of them.[1] Though the government did not take the mandatory power that had been proposed, the Coal Industry Act 1971 empowered the Secretary of State for Trade and Industry to direct the NCB to review and report to him on their ancillary activities and associated companies, which he immediately did.[2] For nearly two years the NCB had not only to work out a scheme of organization and finance, which was not too difficult, but adapt it to satisfy a government that remained suspicious of ancillary activities and was prone to changes of mind. The main requirement to which the government kept consistently was that any non-mining activities should be conducted at arm's length from the NCB and should cease to receive money from the National Loans Fund. The NCB objected to any scheme to introduce private capital on terms that would reduce their own interest in the ventures, but they were keen to regroup their non-mining activities under a wholly owned holding company, if they could in this way gain greater commercial freedom (which might involve them in going to the market, by their own choice, to raise capital for some schemes) without loss of control.

A scheme along these lines was worked out with the Department of Trade and Industry in 1972. It raised a number of difficulties, which involved some modifications and a lot of careful explanation, particularly for the sake of industrial relations, as there were 9,000 NUM members whose employer would no longer be the NCB, and there were members of the BACM whose superannuation rights needed to be safeguarded despite the change of employer. At quite a late stage the Secretary of State introduced the suggestion that the NCB should accompany the announcement of the scheme with a declaration of their intent to dispose of all or part of their interest in the new holding companies. This would not only have reduced the whole scheme to a transitional expedient, but might well have provoked an immediate strike, and it was successfully resisted by the NCB. After an investigation by an outside consultant of the best way to organize the ancillaries other than those in Coal Products Division, a detailed scheme was drawn up and apparently agreed. Even then the Minister wrote to the

[1] Robens 1972, 320–3.
[2] NCB *Ann. Rep.* 1971–2, I, 19.

chairman of the NCB on 15 January 1973 that his colleagues accepted it 'on the clear understanding that the Board intended progressively to reduce or dispose of their interests in ancillary activities'. This was directly contrary to the purpose of the negotiations from the beginning and was also contrary to what had already been said, without ministerial objection, to the workers and their unions. The NCB refused to agree. The Minister backed down and the scheme went through with the relevant statutory changes incorporated in the Coal Industry Act 1973.

The new organization, which began trading on 1 April 1973, was based on two new holding companies, each wholly owned by the NCB. One, known as NCB (Coal Products) Ltd., was responsible for all the activities formerly within Coal Products Division. The other, named NCB (Ancillaries) Ltd.,[1] controlled all the other non-mining activities which had significant external business.[2] Each holding company operated through a number of subsidiaries and associated companies. Some of the subsidiaries were new creations, some continued as before, some continued with fairly large alterations.

Under NCB (Coal Products) Ltd. there were originally three wholly owned subsidiaries. One of these, National Smokeless Fuels Ltd., was an entirely new company to operate the coking and briquetting business. All the secondary by-product activities which the NCB had run without associates were transferred to the existing subsidiary company, Thomas Ness Ltd. All North Sea activities, production as well as exploration, were placed in NCB (Exploration) Ltd.[3] There were also four associated companies, in three of which NCB (Coal Products) Ltd. had a 45 per cent interest and in one of which it owned 50 per cent.

Under NCB (Ancillaries) Ltd., an existing wholly owned company, National Fuel Distributors Ltd. took over the retail sales service, the coal handling depots, the packaged fuel plants, and the Ayr shipping services. This was one of eight active subsidiaries. All the others, except

[1] The changing usage of the word 'ancillaries' can be a little confusing. Previously it had often been used as though it were equivalent to 'non-mining', and down to 1971–2, there regularly appeared in the annual accounts, under the heading 'ancillaries', a figure of the annual surpluses and deficits on non-mining activities as a whole. But from 1973 'ancillaries' was always used with reference only to non-mining activities other than carbonization and its derivatives.

[2] Activities which had formerly been classed as 'ancillaries' but which provided a service exclusively for the NCB (or nearly so) were not brought under either of the new holding companies, but remained directly under the NCB. Housing was the main example.

[3] The development of the North Sea gas interest was financed partly by leasing plant and equipment from LS Leasing Ltd., a firm in which the NCB held all the ordinary shares and others held all the preference shares. This firm became a wholly owned subsidiary of NCB (Exploration) Ltd. in Mar. 1974.

Whittlesea Central Brick Co. Ltd., were new limited liability companies, though three of them were direct replacements for unincorporated organizations. Of the eight subsidiaries, three operated brickworks, two were concerned with fuel distribution, and one each with engineering manufacturing workshops, computer services, and non-operational land and estates. There were also six associated companies and partnerships, two of which it was decided to treat as subsidiaries, though they were not wholly owned. These were J. H. Sankey and Son Ltd., distributors of heating appliances and building supplies (76 per cent owned in 1972–3 and 60 per cent in 1973–4) and Scottish Brick Company Ltd. (50 per cent owned).[1] The former had a substantial number of subsidiaries, as well as a couple of associated companies, and National Fuel Distributors Ltd. also had associated partnerships.

None of the companies in either group was publicly quoted. The two holding companies each had a capital structure of which half was in ordinary shares, a quarter was in non-cumulative redeemable 10½ per cent preference shares, and the remaining quarter was in 10 per cent unsecured loan stock repayable by 1997. NCB (Ancillaries) Ltd., however, was able to repay the whole of its loan stock in the first year of operation. The issued share capital (ordinary and preference) was £67.5 million for NCB (Coal Products) Ltd. and £20.25 million for NCB (Ancillaries) Ltd. After the passage of the Coal Industry Act 1973 the NCB could not make loans to the new companies without ministry consent, which was also required for any other loans raised, or guarantees given, by a wholly owned company or a 75 per cent subsidiary; and all borrowings by the wholly owned companies counted against the borrowing limits set for the NCB. Before the Act was passed the NCB arranged to transfer to each holding company adequate cash resources for working capital. The NCB reached a formal agreement with the Minister. This required them to consult him about the acquisition or commencement by either holding company, or by a subsidiary, of a new business which was either a new activity or significant in size in relation to the operation of the existing business of the acquiring company, 'significant' being defined as the smaller of 25 per cent of such existing business or £500,000. The NCB still exercised close control over the operation of the holding companies. They were held to account by means of annual financial budgets (including cash flow estimates), five year forecasts, and the comparison of actual results (required quarterly) with budget. If results were significantly profitable, benefit would come

[1] Scottish Brick Co. Ltd. later reverted to the status of associated company.

to the NCB by declaration of dividends and by group taxation relief.[1] It was expected from the outset that NCB (Ancillaries) Ltd. would be able to generate a surplus more than sufficient to finance its own future capital needs, but there was anxiety about the ability of NCB (Coal Products) Ltd. to meet this requirement.[2]

These expectations were largely fulfilled. NCB (Ancillaries) Ltd. always had the higher rate of profit and, despite its much smaller capitalization, in all but two years its profits were absolutely larger than those of NCB (Coal Products) Ltd., and from 1977–8 its turnover regularly exceeded the latter's. For new capital needs NCB (Coal Products) Ltd. had to borrow from the banks under NCB guarantee, whereas NCB (Ancillaries) Ltd. remained self-financing and regularly paid dividends.[3] Nevertheless, the structure devised in 1973 was maintained without fundamental alteration. There were changes in the range of activities, notably the enforced relinquishment of the offshore gas and oil interests of NCB (Coal Products) Ltd. and the commencement of one or two new service activities under NCB (Ancillaries) Ltd. So there were a few removals from and additions to the list of subsidiary and associated companies. But these changes could be easily accommodated in the existing framework. The relations of the two holding companies, with both the NCB and their own subsidiaries and associates, continued on the lines already laid down.[4]

## ii. Coal products and other hydrocarbons

The NCB started with a somewhat awkward assortment of productive assets, markets, and responsibilities in relation to carbonization and other manufacturing treatments of coal. Coke was the principal product. There were originally fifty-three NCB coke ovens in operation, of varying ages, sizes, types, and degrees of decrepitude. In 1947 they produced just under 6 million tons of coke, but that was less than half the national total of 13.8 million tons. There were three main markets: foundries, blast furnaces, and domestic and industrial users. Each

[1] Because of the existence of group relief the NCB and subsidiaries paid no corporation tax except for small sums in respect of their share in profitable associated companies.

[2] For details of the new organization and its relation to the old see NCB *Ann. Rep.* 1971–2, II, 33–7; 1973–4, 50, 55, 58–9, and 64.

[3] Ibid. 1981–2, 87, 88, 100, and 101.

[4] There were a few additional non-mining subsidiaries which were placed directly under the NCB, but these were non-profit-making or specially funded companies, such as British Coal International Ltd. and NCB (IEA Services) Ltd. See the list in ibid.1981–2, 57.

needed a different kind of coke. Foundries needed very strong and pure coke. Foundry coke was carbonized for longer than other cokes and was made from the very best coking coal, found in quantity only in South Wales and Durham, where supplies were dwindling. Blast furnaces also needed hard, non-friable coke, but it could be less good than foundry coke. It was possible to make suitable blast furnace coke from strongly caking coals, such as were abundant in Yorkshire, or from blends of coal in which they were a principal constituent. In times of shortage it could be produced by blending prime coking coal with less strongly caking coals which were particularly plentiful in the East Midlands. For domestic and industrial heating softer cokes were suitable. The NCB had an interest in making these from the cheaper, less strongly caking coals, and did so as far as possible. The gas undertakings wanted the best gas-making coals they could get, and they sold for domestic and industrial purposes all the coke that resulted.

In 1947 the NCB supplied 82 per cent of the UK consumption of foundry coke and 63 per cent of the domestic and industrial, but only 30 per cent of the blast furnace coke. But foundry coke, in which their market position was so strong, was only 14 per cent of their total output, against 42 per cent blast furnace and 44 per cent domestic and industrial coke.[1] Most of the NCB coke ovens (just over three-quarters of capacity) were in Durham and Yorkshire. The location was good in relation to coal supplies, but the market situation was not altogether comfortable. Well over half of all coke went to blast furnaces, the steel companies made most of it, and their ovens were not only on the spot, but on the whole were more modern in design and in better condition than those of the NCB. Moreover, it seems that for metallurgical coke generally, the NCB were some way behind the iron and steel producers in the techniques of blending.[2]

Nevertheless, there were helpful opportunities—the NCB coke ovens made operating profits of £926,000 in their first year and £537,000 in their second[3]—and there was a lot of pressure to maintain and expand carbonization activities. The government both wanted and expected a large expansion of steel output, and therefore sought to ensure an ample coke supply. It was one of the subjects which Gaitskell, as Minister of Fuel and Power, regarded with some satisfaction, though he felt that still more needed to be done.[4] In 1948 the NCB gave him an assurance that

[1] Morley, in *Colliery Guardian*, *National Coal Board*, 1957, 73–4.
[2] NCB 329th meeting, min. 155, 21.9.51.
[3] NCB *Ann. Rep.* 1948, 190.
[4] Williams 1979, 191–2.

they would take steps to maintain the level of carbonization activities, although they estimated that this would mean planning to invest £20 to £25 million in the renovation and replacement of capacity. Not everyone was altogether happy about this. The finance member of the board, Lionel Lowe, thought the iron and steel industry ought first to give an undertaking not to discriminate against NCB coke if demand slackened, but he was overruled for the time being.[1] There were problems, too, from relations with other newly nationalized industries. When gas was nationalized it became illegal for the NCB to sell their principal by-product, coke oven gas, to anyone except an Area Gas Board, and in the absence of any alternative outlet, there was difficulty in settling a price which both parties regarded as fair. This was not a negligible matter for the NCB who, by the mid-fifties, produced 8 per cent of all the town gas sold by Area Gas Boards.[2] There was also a requirement to consult with, first, the gas boards, and then, the iron and steel industry in order to co-ordinate plans for carbonization activities.

The NCB in 1951 were again reminded, by their Director-General of Finance, Joseph Latham, of their vulnerability. A minority of the national board would then have been willing to transfer practically all the carbonization activities to a joint board with the nationalized gas and steel industries. Those who took this view recognized that too much NCB output—all the blast furnace coke and coke oven gas—met only the fluctuating marginal element in a market served by bigger producers who were the NCB's only customers. But the majority opted for independence and expansion, especially because they expected a prolonged switch of demand in favour of smokeless fuels.[3] After this the renewal and gradual expansion of the carbonization plants went on without being seriously questioned again.

In fact, some safeguards had been secured. In 1949, when the government wanted enlarged coke oven capacity and was preparing to nationalize the steel industry, Gaitskell sent formal letters urging that the NCB should try to agree with the Iron and Steel Corporation how any future over-capacity should be dealt with. The result was the establishment of what were known as Sharing the Burden Schemes, which were slightly adjusted from time to time, mainly in respect of prices, but also to take account of denationalization of the steel industry. The

[1] NCB 225th meeting, 17.8.48.

[2] NCB *Ann. Rep.* 1956, I, 27.

[3] NCB 316th meeting, min. 38, 16.2.51; 317th meeting, min. 49, 2.3.51; 329th meeting, min. 155, 21.9.51.

schemes provided that composite undertakings (i.e. steelmakers with their own coke ovens) must not sell coke to other ironmakers except their own associated companies; and that independent ironmakers (i.e. those without coke ovens) must buy all their coke from the NCB or one or more of the six independent coke producers, at prices recommended in the schemes. Composite firms could buy additional coke but, because they went on increasing their coke oven capacity, they did so to a less and less extent. Because of expanding demand the agreements were not strongly tested for several years, but in 1958, when demand had turned down, the British Iron and Steel Federation sought to end the agreement. The NCB refused to agree except on terms which offered them equal benefit, and these were not forthcoming. But even with the schemes still operating, the lower level of demand and the increased proportion of it which composite firms could meet from their own ovens meant that the old fears of the NCB (and the private independent coke producers) came true. They were selling little blast furnace coke except to even out peaks in demand on a week-to-week basis. Even the protection of this small, erratic market was lost when, in 1964, the Sharing the Burden Schemes were dropped because of a proposal to refer them to the Restrictive Trade Practices Court.

In the period of expansion, total NCB production of coke rose to 7.2 million tons (7.3 million tonnes) in 1956, but whereas the share of the foundry coke market rose to 96 per cent, that for blast furnace coke fell to 22 per cent, and there was only a slight increase in that for domestic and industrial coke. Disposals of coke oven gas rose from 35 to 52 thousand million cubic feet between 1947 and 1956. The number of coke ovens had been reduced to 44, and 30 per cent of the output in 1956 came from plant installed since nationalization. Some £30 million had been invested in carbonization activities, most of it for the coke ovens. Production continued to expand in 1957, when 7.5 million tons (7.6 million tonnes) of coke were made, but thereafter the trend turned down, and the coke ovens made an operating loss for the first time in 1958.[1]

The by-products provided the basis for a gradually expanding industry with somewhat less vulnerability to market difficulties than coke. The NCB were the largest British producers of crude benzole and crude tar, as well as producing about 20 per cent of the national output of sulphate of ammonia. Output of crude benzole, 24 million gallons in 1947, tended to fluctuate at a slightly higher level than that, and was

[1] Morley, in *Colliery Guardian*, *National Coal Board*, 1957, 77–8; NCB *Ann. Rep.* 1956, I, 25–7; 1958, II, 34–5; NCB, *Investing in Coal: Progress and Prospects under the Plan for Coal* (1956), 11.

28 million gallons in 1956. Output of crude tar increased more persistently, from 320,000 tons (325,000 tonnes) in 1947 to 403,000 tons (409,000 tonnes) in 1956. Refined products were rather more profitable and it was policy to increase the proportion refined, but the ability to do this depended on the provision of new and improved capacity. Of the twenty-eight benzole refineries taken over, twelve were closed within ten years, but three large central refineries were built alongside new or rebuilt coke ovens at Nantgarw in South Wales, Avenue in Derbyshire, and Manvers in Yorkshire, and the quantity rectified rose from 17 million gallons in 1947 to 31 million in 1956. As market opportunity offered, a much greater proportion was turned into pure benzole, whereas earlier most of it was made into motor spirit, marketed through the National Benzole Company. Of the twelve tar distilleries taken over, most were old-fashioned, and most of the crude tar went to private firms for distillation. But the best of the original NCB plants, at Caerphilly, was doubled in capacity in 1955, a new plant was opened at Avenue in the same year, and in 1957 the NCB bought, for £55,310, Thomas Ness Ltd., one of the firms which regularly distilled part of the Durham output of crude tar. So there was a large increase in efficient capacity. The quantity of crude tar distilled increased by about 50 per cent in the first ten years, and was 164,000 tons (167,000 tonnes) in 1956. Besides the normal range of tar derivatives, some of it was made into proprietary products, of which the best known, with a long continued demand, was 'Synthaprufe', a liquid waterproofer and adhesive.[1]

The other main activity was to produce manufactured fuels. The most important of these was 'Phurnacite', a high grade, low ash, smokeless briquette. It was made from the smalls of Welsh dry steam coal, bonded with pitch, and then carbonized, and was designed for domestic closed stoves. It was an excellent alternative to anthracite, was in continuously rising demand, and, in contrast to anthracite, was consistently profitable. 'Phurnacite' was developed in the Second World War by Powell Duffryn, who built the plant to produce it at Aberaman. At vesting date the annual productive capacity was 133,000 tons (135,000 tonnes). The NCB responded to demand by repeatedly adding to capacity, and in 1957, the first year of operation of two additional batteries, the output was 654,000 tons (664,000 tonnes). An intention to build a second plant at Betteshanger was abandoned because it was found there was no way it could be profitable with Kent coal.

[1] Morley, in *Colliery Guardian*, *National Coal Board*, 1957, 78; NCB *Ann. Rep.* 1947, 79–80; 1956, I, 27–9; 1957, I, 17 and II, 16–17 and 46.

The 'conventional' manufactured fuels were briquettes and ovoids made from various small coals bonded with pitch, though there were experiments to form briquettes under pressure without a bonding agent. They were made in various sizes and qualities, some for domestic and some for locomotive use, and were intended to make up for the shortage of large coal. In the desperate shortages after the war some ovoids were made from slurry and anthracite duff, which would otherwise have been wasted, but the quality of such emergency fuel was so poor that, after a year or two, the ovoids were very difficult to sell. Other briquettes were much better but the demand fluctuated inconsistently. Briquette making was not very remunerative and tended to become less so. In most early years there was an operating profit, but scarcely enough to cover interest, and by the mid-fifties there was seldom even an operating profit. But additional imports of large coal would have involved bigger losses, and the NCB felt obliged to meet the demand. Production in 1947 was 704,000 tons (715,000 tonnes) of briquettes and 852,000 tons (865,000 tonnes) of ovoids. By 1952, as plants were closed down, ovoid production had declined to 250,000 tons (254,000 tonnes), but there were 980,000 tons (996,000 tonnes) of briquettes, after which there was not much change for several years. In 1957 the output was 378,000 tons (384,000 tonnes) of ovoids and 851,000 tons (864,000 tonnes) of briquettes, but a lot of the ovoids were being made under contract by a private firm at a heavy loss to the NCB.[1]

From the late fifties onwards the market conditions for the coal products of the NCB changed greatly, and though there were some favourable influences, the balance was adverse. In the iron and steel industry, though the trend of output was upward, it was erratic. Improved techniques reduced the amount of coke required for a given output of steel, and there were changes in methods which led to a smaller use of foundry products, though the effects of this were felt more in the seventies than the sixties. The steel producers had become not far short of self-sufficient in blast furnace coke, so the market open to the NCB was constricted. In the domestic market the government's introduction of smoke control policies, which local authorities applied to ever larger areas, created new opportunities, but they had to be sought in competition with other fuels. There was little chance of producing greater quantities of naturally smokeless coals, so solid fuel appliances, if they were retained, needed proportionately larger supplies of coke and manufactured smokeless fuels. There were both problems

[1] Morley, loc. cit. 79; NCB *Ann. Rep.* 1956, I, 28; 1957, I, 16–17.

and possibilities here. As the gas boards turned more to the use of oil they made less coke, so many consumers had to buy from a different fuel supplier. The NCB had the capacity to switch some hard coke production to the domestic market to fill any gap, and this they did by marketing a good quality coke under the brand-name 'Sunbrite'. But such changes in the sources of supply meant that some consumers turned away from coke altogether. If they wanted a more specialized manufactured fuel, they found that the level of supplies could not be increased so quickly. The NCB were well aware of market needs but, mainly because of technical deficiencies in production, their response came later than intended and too late for good business.

All the time other fuels were competing keenly in terms of price and convenience. More people turned to oil-fired central heating, and there was a phase when new building offered the all electric home; after which gas began to take more and more of the market for space heating. There was a growing tendency for new buildings to be put up without much, or sometimes any, equipment to burn solid fuel; and when equipment had to be replaced, the installation often committed the user to something other than solid fuel for years ahead. One problem was that in industrial, commercial, and service premises there was so much old-fashioned equipment, costly to run, that the owners were readily attracted to other fuels, with the result that by the early seventies the NCB had decided that it was best to withdraw progressively from the institutional space heating market. One other consequence of changes in competing fuels was that cheaper and larger supplies of gas, especially after the switch to natural gas at the end of the sixties, reduced the market outlets for coke oven gas, and sales fell even though direct sales to industrial users became permissible.

The sort of influences affecting the British market were at work elsewhere, so there was not much prospect of compensating for any losses at home by building up exports. There had not, for a long time, been a very strong international market for coke. The main opportunities were in meeting marginal fluctuations in demand, and there were occasional years in which these provided a helpful outlet for surplus supplies at home.

Adjustment to the new conditions was a little belated. The NCB sales of every type of coke reached their maximum in 1955, but output went on rising for another two years and stocks for rather longer. The coke ovens in 1958 entered a six-year period of continuous operating losses. There had to be a period of drastic reductions of capacity, including the removal of whatever was old and costly to operate, and some transfer of capacity from blast furnace to domestic production. In 1955 the NCB sold 7,268,000 tons (7,384,000 tonnes) of coke, but after 1957 the level never again reached 5 million, and in 1970–1 it fell below 4 million. Most of the decline was in sales of blast furnace coke, which were 2,795,000 tons (2,840,000 tonnes) in 1955, fell in every year until 1963, when they were 685,000 tons (696,000 tonnes), and then, after one year of slight recovery, resumed their fall to reach 156,000 tons (158,000 tons) in 1970–1 and 81,000 tons (82,000 tonnes) in 1971–2. A very active steel industry produced a revival of demand, but sales were still only 640,000 tons (650,000 tonnes) in 1973–4. The market for foundry coke held up better, though it shrank a little and fluctuated a lot. Sales, having been 1,149,000 tons (1,167,000 tonnes) in 1955, still reached a million tons three times between 1964–5 and 1970–1. Until 1970 the domestic market was best of all. Sales fell rapidly from 2,812,000 tons (2,857,000 tonnes) in 1955 to 1,970,000 tons (2,002,000 tonnes) in 1959, but then recovered and were over 2.4 million tons in every year from 1962 to 1970–1, after which a further large drop was experienced, doubtless attributable in part to the insecurity of supply in the national overtime bans and strike.

The relative stability of this market was encouraging to the development of manufactured smokeless fuels to serve as wide a range of domestic uses as possible, closed stoves and open fires as well as central heating and hot water boilers. Demand for the old conventional briquettes and ovoids fell sharply, and production was run down and plants closed. Only 341,000 tons (346,000 tonnes) were produced in 1964–5, when only three plants in Scotland and one in Kent remained operative, and production was soon down to very little: only 25,000 tons in 1970–1 and 16,000 in 1971–2. Resources were put into smokeless fuels instead. 'Phurnacite' remained the most successful. Sales usually exceeded 700,000 tons a year, and reached 800,000 in 1972–3. But there was not enough suitable coal available to expand the output much more, so a plant was built at Cardiff to produce an alternative high grade boiler fuel, which was named 'Multiheat'. It was made from small anthracite, which was subjected to a mild heat treatment. It was first marketed in significant quantities in 1965–6, but never sold much above 200,000 tons a year, which was too little for profitable operation.[1]

[1] NCB *Ann. Rep.* 1963–4, I, 22; 1964–5, I, 22 and 24–5; 1971–2, I, 14; 1972–3, 48. Most of the figures are from unpublished sources, as, indeed, is a large proportion of the material in this chapter.

Great hopes were placed on a new process of char briquetting, which involved treating finely ground high-volatile coal in a fluidized bed carbonizer, which removed the smoky constituents, and then subjecting it to pressure at carbonizing temperature in order to form briquettes. The purpose was to produce a smokeless fuel for burning on domestic open fires. Laboratory work was carried out from 1955 to 1960, a pilot plant was set up in 1959, a full scale plant to manufacture the fuel (named 'Homefire') was started at Coventry in 1963 and commissioned in 1966. But it became clear that there were major faults in both design and engineering, and five years of effort to attain an economic level of output and a tolerable yield of fuel had only limited success. 'Homefire' got a bad reputation for variable quality, intermittent supply, and poor handling and stocking characteristics. By 1971 the closure of the plant was under consideration. But in that year it was established that the characteristics of the Warwickshire coals that were used had changed since the period of laboratory work, coal from the North Midlands was brought in instead, and things began to improve. Another fuel, called 'Roomheat', was also developed by adapting the char briquetting process and using a double roll press. This fuel was designed mainly for use in domestic roomheaters, but it could also be used on an open fire. What was described as a pilot plant was built at Markham Main Colliery, near Doncaster, and it was intended to produce 'Roomheat' at several plants in different parts of the country. But this fuel was just as dogged by technical difficulties as 'Homefire'. The Markham Main plant was the only one ever built. 'Roomheat' did not get on the market until 1969–70, which was far too late, and, even in its third year, sales were only 72,000 tons (73,000 tonnes). On that basis it could not have a chance of being remunerative.[1]

Economies were obviously required in the early seventies. Both the NCB and other producers had done their best with well tried products, and the NCB had suffered from attempting full scale production of new fuels without having solved all the main engineering and chemical problems of new technology. Though prices varied from place to place with delivery costs, it was found in 1972 that 'Homefire' and 'Roomheat' were both generally more expensive than the two principal old-established privately manufactured fuels, 'Coalite' and 'Rexco', not only in cost per ton but in cost per useful therm of heat produced. On the latter basis 'Homefire' (the dearest) cost some 10 per cent more than large 'Coalite', though 'Roomheat' and 'Rexco' were much closer

[1] NCB *Ann. Rep.* 1963–4, I, 24–5; 1964–5, 24.

together. Moreover, the new NCB fuels came on the market just as the development of natural gas was increasing the competition. Thus the national capacity for producing domestic solid fuel (including naturally smokeless coal) had outrun any likely demand. In the spring of 1973 it appeared that total capacity was about 1.25 million tons more than the highest possible sales, and with the addition of likely imports, nearly 2 million tons more; and the private producers, whose productive capacity and output of manufactured fuels (i.e. excluding coke) exceeded that of the NCB, had already left a bigger proportion of capacity out of use in order to match their market.[1]

The NCB realized by 1971 that they would have to cease production of one or more of their manufactured fuels, though they proceeded very gingerly in the expectation of resistance from the workforce at a time when industrial relations were difficult throughout the coal industry. At first it was the 'Homefire' plant at Coventry that was the likeliest candidate for closure, though it was given a period in which it was watched for improvement. In the end, though there were divided views, the more reliable performance and better sales after spring 1971 saved it. Sales and output of 'Multiheat' fell so low that the plant could not be kept going, and continued losses on 'Roomheat' caused that to be closed also. The two plants, at Cardiff and Markham respectively, were both closed at the beginning of September 1973. The only NCB manufactured fuel plants remaining in use after that were the 'Phurnacite' and 'Homefire' works, and one small conventional briquetting plant in Scotland.[2]

The period around 1970 was thus in various ways not very happy for the coal products business. By-products were less plentiful and less profitable. More coke oven gas had to be allowed to go to waste: commercial output, which had fallen slowly to 46 million cubic feet in 1963–4 and 43 million in 1966–7, was down to 24 million in 1971–2.[3] Though the secondary by-product plants handled a bigger proportion of the primary by-products than they had done earlier, there was rather less quantity for them to work on and little development of products of higher value. There was scarcely time for the new ventures in associated companies in the late sixties to add much, though there was time for one of them to go wrong. The venture with the United Steel Companies in

[1] The NCB had mistakenly planned on the assumption that there could be 'a critical shortage of solid smokeless fuel by 1970–1', though one member of the national board gave the reminder that the advantage of solid smokeless fuel was for the interim period 'before gas became cheaper'.

[2] NCB *Ann. Rep.* 1973–4, 48.

[3] Ibid. 1963–4, I, 23; 1966–7, I, 25; 1971–2, I, 16.

Phthalic Anhydride Chemicals Ltd. was a failure. There were difficulties from the start in commissioning the plant, which soon had to be closed, and though the company did a little business for a few years (until late 1974) as an intermediary selling phthalic anhydride, it involved the NCB in losses, which had already passed £300,000 by the end of 1969.[1] Profits on secondary by-products had generally been running lower since the early sixties, but it was this which caused these activities to record an operating loss for the first time in 1969–70, a state of affairs which lasted three years. With 'Phurnacite' having financially to carry the rest of the manufactured fuel activities (as, indeed, it had done regularly since 1952), and all the coal products activities suffering financially from the strike of early 1972, there was a need for a turn of the tide.

In fact, thanks particularly to the perceptiveness and efforts of Lord Robens, there was by this time one novel venture which was promising well. This was the exploration of the continental shelf for gas and oil, where the first drilling licences were granted in 1964. During 1965 Lord Robens succeeded in interesting Gulf Oil in the possibility of joint operations, and the NCB received an unsolicited approach from another interested American company, Allied Chemicals. In July 1965 the Minister of Power, Fred Lee, announced that in awarding further production licences, he would take account of any proposals to facilitate the participation of public enterprise. Lord Robens urged the Minister to introduce legislation to permit the NCB to become involved. The Minister first counselled delay, but was told of the options provisionally available to the NCB, and time was found for the necessary legislation in 1966. By that time Lord Robens had also attracted Continental Oil Company (Conoco) to a potential association. National Coal Board (Exploration) Ltd. was set up to carry out the NCB share in all offshore activities. The original offer from Gulf Oil and the later one from Conoco for the NCB to share in their licences were taken up in March 1967. The offer from Allied Chemicals was left open for the time being.[2] Most of the NCB's interests were held jointly with Gulf or Conoco or the two together. The association with Conoco was particularly successful, and Conoco normally acted as the operator in production contracts.

[1] NCB *Ann. Rep.* 1969–70, II, 38–9; 1970–1, II, 38. Phthalic Anhydride Chemicals Ltd. was eventually 'sold' (in effect transferred) to Staveley Chemicals Ltd. in 1979 (*Ann. Rep.* 1979–80, 56 and 72). Phthalic anhydride ($C_6H_4(CO)_2O$) is used mainly in the production of dyestuffs. The NCB became interested because phthalic anhydride could be made from the naphthalene produced from tar by Thomas Ness Ltd.

[2] Ibid. 1966–7, I, 26; Robens 1972, 315–17.

These were not very eccentric policies, though there were some initial fears that they might provoke suspicion among mineworkers that the NCB were joining the hostile competitors. In fact, this did not prove to be a serious problem. The NCB had more experience in offshore drilling than any other British undertaking, though it was only within a few miles of the coast and had only a limited relation to the tasks of exploration for oil and gas. They were also taking the only available route towards recovery of a share in the benefits of the gas market, which they were losing to new technology. And, if the government wish for public participation were to be met, the NCB and the Gas Council (which acted similarly) were the only relevant institutions already in existence.

Apart from the original associations with Conoco and Gulf, the NCB obtained smaller interests in the licences of seven other groups. By early 1971 they had an interest in 10 per cent of the 55,000 square miles of the continental shelf that had been covered by production licences, and were becoming anxious to spread their risks by having a more diluted interest over a more dispersed set of blocks. But it was the 50 per cent interest in two Conoco production licences in the southern North Sea that proved most fruitful. Wells drilled in 1968 and 1969 confirmed the existence of Viking, one of the larger gasfields in the British sector of the southern North Sea. In 1969 the NCB also exercised their option to take a 25 per cent interest in a licence of the Allied group (which also included Shell and Esso, with Shell as operator) after a discovery of gas in commercial quantities. This was the Indefatigable South East field, which was related to the larger Indefatigable field on neighbouring blocks. A contract for the delivery of gas from the Viking field was negotiated in 1971, and work began on production facilities and the construction of a pipeline to the new terminal at Theddlethorpe St Helens in Lincolnshire. Deliveries of gas began late in 1972, and by 1974 built up to 400 million cubic feet a day, and 500 million from July. Negotiations were also conducted for the delivery of gas from Indefatigable South East, but no conclusion was reached.[1]

In 1971 interest was turning more to northern waters, where the main prospect was oil rather than gas. That August the NCB agreed to bid for a number of blocks in partnership with Conoco and Gulf, although, with a change of government, there was no longer any preference given to public bodies. The partners obtained several blocks on a 'work obligation' basis and two, one in the northern oil area and one in the southern gas area, for cash. The exploration mainly concerning the NCB was now

[1] NCB *Ann. Rep.* 1968–9, I, 25–6; 1969–70, I, 28; 1971–2, I, 17; 1972–3, 12; 1973–4, 49.

north-east of Shetland, close to the boundary between the British and Norwegian sectors. Here in September 1972 the Signal Oil and Gas Company, which was subsequently acquired by the Burmah Oil Company, discovered the Thistle field. Conoco then drilled a well which proved the extension of this field into the adjoining block, licensed in equal shares to Conoco, Gulf, and the NCB. In July 1974, when the NCB agreed to contribute their proportion to the costs of developing the field for production, it was estimated that they would have a share of $8\frac{1}{3}$ per cent in the whole field. In the summer of 1973 the Dunlin field was established immediately to the south in a block licensed to Shell and Esso, and this, too, was found to extend eastward into a Conoco/Gulf/NCB block. Plans were prepared for production from this field also. Later in 1973 the three partners discovered the smaller Hutton field, but production plans were left until later. There was also an interest in some oil that formed part of the Statfjord field, which straddled the Norwegian and British sectors.[1]

It was expected that NCB (Coal Products) would start to receive revenue from deliveries of oil in 1977, but before then the government had taken away all the interests in gas and oil. It had been decided to set up a nationalized oil undertaking, the British National Oil Corporation, and this was provided for in the Petroleum and Submarine Pipelines Act 1975. From 1 January 1976 National Coal Board (Exploration) Ltd. and its subsidiary LS Leasing Ltd. were transferred to BNOC. It might be said, indeed, that the gas and oil activities of the NCB provided the original nucleus of BNOC, for the Gas Council were permitted to retain their offshore interests. The NCB (who in May 1975 had sought £127 million in compensation) thought that their finances were somewhat damaged by this transaction, as the assets were transferred at book value, £25 million, and there was no compensation for loss of future profits. NCB (Coal Products) were allowed to keep 75 per cent of the 1975–6 profits, which was more or less equivalent to a year's profits, as annual interest, which fell due in the final quarter, did not have to be paid; but they had to wait more than a year before even that sum was handed over.[2]

It had, in fact, been rather a good investment. Exploration, when

[1] NCB *Ann. Rep.* 1972–3, 12; 1973–4, 49; 1974–5, 52. The Hutton field was brought into production in 1984.

[2] Ibid. 1975–6, 56–7; 1976–7, 52. In 1975 NCB (Exploration) Ltd. had assets with a written-down book value of over £55 million, but more than half was financed by loans which remained with the company when it was transferred to BNOC.

shared with various associates, was not particularly expensive, and even to share the cost of establishing production facilities for a major gasfield compared favourably with the cost of sinking and equipping a large new colliery. Of course, most exploratory drillings find nothing worth while, but even one success makes up for many failures. The NCB were fortunate that their greatest gas discovery was under a licence in which their share was larger than in many others. At the time in 1971 when they settled the contract for gas deliveries from the Viking field, they had spent only £6,272,000 on offshore exploration. Even two and a half years later NCB (Exploration) Ltd. was capitalized at only £16 million. Yet once gas production began in 1972, it made operating profits of £3.1 million in the first partial year and £7.6 million in the first full year, 1973–4; £8.7 million in 1974–5; and £6.0 million in the final nine months. Bringing a group of northern oilfields into production would have been far more costly. At the end of 1973 it was estimated that £200 million would be needed as additional capital over the next five years, and it was being considered whether this might be sought from the National Loan Fund or borrowed externally under Treasury or NCB guarantee.[1] The enforced departure from offshore activities meant that this difficult financing operation did not have to be faced. Nevertheless, what was also lost was the prospect, after perhaps four lean years, of a commensurately large revenue, which might, however, have turned the NCB into an institution of significantly different character.

The loss of the offshore oil interests had an important influence on another prospective venture. In 1974, when it had become quite clear that some of the NCB's oil licences were going to be commercially productive, some of those involved thought they could do better than just take the profits from selling crude oil. They had discussions about the possibility of Conoco refining the NCB share of the oil on a toll basis, and then the NCB going into association with Conoco and some other interested party, perhaps the BSC, to produce petrochemicals from the NCB's refined oil. It was claimed that, if the NCB wished to be involved in such activity, this was the only way, as no British chemical company showed interest in any kind of relationship except the possibility of buying the NCB's entire output of crude oil. The national board at first

[1] Ibid. 1972–3, 32; 1973–4, 52; 1975–6, 60. The profit figures consolidate those of the wholly owned subsidiary, LS Leasing Ltd. The estimate for additional investment to develop the oilfields relates only to the share of NCB (Exploration) Ltd.; the whole cost, to be shared with Conoco and Gulf, was much larger.

showed extreme caution, but allowed discussions to continue. The Department of Energy was kept informed and the Secretary of State, Eric Varley, was encouraging as BNOC was unlikely to be involved in petrochemicals for some years.

During 1975 some members of the national board built up an increasing momentum of enthusiasm for the project, even though it had become clear that the NCB would not become an oil producer, but would have to buy crude oil from BNOC. The argument was pushed that coal would eventually replace oil as a feedstock and this was the right way to get the necessary experience. The Department of Industry, to which Mr Varley had moved, and the Department of Energy both expressed support, subject to consideration of detailed proposals, and the BSC were keen to set up a new joint company with the NCB and perhaps use it for the rationalization of benzole and tar interests as well as for petrochemicals. The NCB sought to modify their role somewhat by taking a financial interest in an ethylene cracker operated by Shell and a vinyl chloride monomer plant operated by ICI instead of building their own, though in 1976 there was a return to the idea that the NCB and BSC might jointly have their own plant to manufacture vinyl chloride monomer. In April 1975 NCB (Coal Products) Ltd. registered a new wholly owned subsidiary, NCB (Hydrocarbons) Ltd., with the intention that it should operate the petrochemical activities.

The planning went on through 1976. In the autumn of that year the Secretary of State for Energy (Mr Benn) promised legislation to make explicit the NCB's powers to acquire and treat petroleum and to manufacture and sell products from it; and the necessary clauses were, in fact, included in the Coal Industry Act 1977. Late in 1976 the NCB and BSC also agreed to set up a partnership, British United Chemicals, in which they shared equally, to conduct their joint interests in petrochemicals. In November the NCB considered proposals for the whole possible range of new activities. They included, among others, a new plant on a greenfield site for British United Chemicals, with a 63 MW power plant and facilities to produce chloralkali, vinyl chloride monomer, styrene, and PVC in large quantities. There were forecasts of large and growing markets, an estimate of £201 million capital cost shared by NCB and BSC, with the finance raised mostly outside, and the suggestion that it was a logical development of the NCB's chemical interests and need not be affected by the loss of the oil interest to BNOC. It was probably the wildest scheme ever put before the national board. It received appreciable support, but also met effective resistance. All but one of the part-time members expressed great scepticism, and they included Sir Jack Wellings, the chairman of NCB (Coal Products) Ltd., the body which would be immediately responsible; and the whole scheme received a comprehensive application of cold water from the NCB Finance Department. The plans were not formally rejected and further study of them was permitted, but after this they got nowhere.[1] Neither NCB (Hydrocarbons) Ltd. nor British United Chemicals ever traded. A report in November 1977 put things succinctly:

> The transfer of NCB (Exploration) Ltd. to BNOC has deprived Hydrocarbons of the anticipated secured source of crude oil supplies. At the present time the cost of crudes on the open market together with the tolling fees enables only narrow margins, which would not cover the funding costs on working capital. For the time being, therefore, toll refining of crude oil is being deferred until circumstances and timing are more propitious.

And, since the first step could not be taken, neither was the expensive sequence which might have followed.

It was a merciful deliverance. The market forecasts were hopelessly wrong. Within a few years the major oil companies and chemical companies were finding that petrochemicals had become their nightmare and were involving them in heavy losses which dragged down the results of their business as a whole. If the NCB had allowed themselves to be drawn into that miserable experience, they would have been severely distracted and would have produced indefensible losses. Fortunately, there was enough good sense about to save them from that fate. And there was ample time to prepare for the eventual use of coal as a major chemical feedstock, and there were more gradual routes to follow.

It was, indeed, the case that the NCB had in the sixties begun to take a greater interest in chemical manufacture, partly to get more value from the further processing of coal products, and partly to preserve an outlet for the by-products of carbonization in the face of greater competition from oil-based products. But the development of these activities was neither very rapid nor very rewarding. The failure of Phthalic Anhydride Chemicals was one disappointment. The joint venture with Stanton and Staveley, which became part of BSC when steel was renationalized, was more successful. When the two undertakings formed Staveley Chemicals Ltd. they were anxious to get more value out of crude benzole by refining it further, and they agreed to build new plant

[1] NCB *Ann. Rep.* 1975–6, 56; 1976–7, 52 and 54; 1977–8, 5–6.

for this purpose. The NCB insisted that the products of the benzole refinery should be processed in a new plant to make cyclohexane. Stanton and Staveley insisted that the new joint firm should manufacture PVC. Conoco were invited to take a half share in the PVC plant and thereby came to own 10 per cent of Staveley Chemicals Ltd., with the shares of the NCB and BSC reduced to 45 per cent each.[1] The plans were put into operation with fair speed and the company began to earn some profits in 1969–70. In shared profits and interest on loans Staveley Chemicals Ltd. gave the NCB a reasonable, though rather erratic, return in most years until 1980–1, when there was a substantial loss.[2] But the desire to go one step further with the products of this firm, by turning its cyclohexane into caprolactam (the raw material of nylon 6), lured the NCB into their other major chemical investment, which was disastrous.

Caprolactam was made by Nypro (UK) Ltd. In 1969 the principal shareholder, Dutch State Mines (which also manufactured caprolactam in continental Europe) approached the NCB with an invitation to share in the building of an extended plant at Flixborough in Lincolnshire, which was wanted partly in order to overcome a projected imposition of UK import duty on caprolactam. After months of consideration, and discussion with the Ministry of Power, the NCB, relying on the technical knowledge and high reputation of Dutch State Mines, in February 1970 bought a 45 per cent share in Nypro (UK) Ltd., the same proportion as Dutch State Mines, with Fisons holding the other 10 per cent.[3] The shareholders provided most of the finance by loan and waived all interest until after the completion of the extended plant, which was commissioned early in 1973. Nypro nevertheless produced a loss in 1973–4, though it operated at a profit in the early months of 1974. On 1 June 1974 the plant was destroyed by a huge explosion, followed by fire, which killed twenty-eight people and damaged property for miles around.

The NCB were told that if the company were liquidated, their losses (an investment of £3.15 million) would probably be covered by the insurance, which was adequate. But, although Fisons withdrew and sold their holding in equal parts to the other two shareholders, the NCB

[1] In October 1978 it was agreed that the 10 per cent holding by Conoco would be transferred to Norsk Hydro.

[2] See the schedule of investments in associated companies, published in NCB accounts each year.

[3] NCB *Ann. Rep.* 1969–70, II, 38.

chose to continue, and it was decided to rebuild the plant, though caprolactam was to be made from phenol (which could also be derived from benzole) instead of by oxidation of cyclohexane. Courtaulds had expressed anxiety for a UK source of caprolactam for their nylon 6 production, and a readiness to support Nypro. The NCB consulted the government and the Treasury gave explicit approval. So a course of action was followed which led the NCB into large losses that need never have been incurred if the explosion had been allowed to terminate the venture. By March 1976 prospects looked less rosy than they had done, but estimates of sales and profits were still judged adequate and the NCB committed themselves to the necessary investment. Even in July 1978, when the nylon market was clearly depressed, the two relevant customers (Courtaulds and British Enkalon) were still using more caprolactam than the rebuilt Nypro plant could produce and the project was still calculated to give a DCF return of 9½ per cent. The market, in fact, only got worse and worse. The rebuilt plant came into production in 1979, with few physical problems but with its product facing low prices and progressively diminishing sales in an overstocked market.

The NCB share of Nypro losses was £4.1 million in 1979–80, £6.3 million in 1980–1, and £2.5 million in 1981–2, and these losses came on top of others in the preceding years when it had been thought prudent to write off deferred revenue expenditure. In 1981 the UK production of nylon 6 virtually ceased. Nypro was closed down and the company was put into voluntary liquidation in December 1981. All these losses fell directly on NCB (Coal Products) Ltd. The NCB themselves also had to complete the financial story by setting aside £20 million to meet the guarantees they had given to loans raised by Nypro from third parties.[1] All the would-be enterprise in trying to get a bigger financial yield from coal's products had turned into a quick way to lose money, and it was not much comfort, or a very strong excuse, that the NCB had not acted alone. The hydrocarbon and chemical interests are among the clearest illustrations of the business strengths and weaknesses of the national board in the seventies. North Sea gas and Nypro show the effects of the range of quality in judgement.

It was left mainly to the carbonization activities to determine the success of business in coal products after 1975. Circumstances, however, became increasingly unpropitious and reaction to them was rather slow. Change within the undertakings was less drastic than change in the

[1] Ibid. 1974–5, 53; 1977–8, 50 and 62; 1979–80, 56 and 71; 1980–1, 49–50 and 79; 1981–2, 49,58, and 79.

business environment within which they operated. The main problem was the depressed condition, and eventual sharp contraction of the steel industry, which greatly reduced the demand for metallurgical coke from any ovens other than those of the BSC. Technological changes worked in the same direction. Improved varieties of steel, more cheaply made, reduced the demand for foundry products, and thus for foundry coke, which had been the least unstable element in the NCB's coke market. The domestic market held up better, but the trend nevertheless was downwards, both for coke and manufactured fuel. 'Phurnacite', which had been so successful for so long, produced an operating loss in every year but one from 1973–4 onwards, and its troubles were as much in production as in the market. Some of the old plant created serious pollution, so much so that there were threats of closure by the Alkali Inspectorate, though this was avoided. 'Homefire' did as well as it had been doing since 1971, but only for short periods was it possible to have the two-stream production necessary for steady profit, and this fuel remained a loss-maker. The problems of the carbonization market were not peculiar to Britain, so even though in some years the quantity of exports increased, this was accomplished only by selling at lower prices than at home. Secondary by-products fared rather better, though there was the difficulty that if less coal was carbonized, a smaller throughput of by-products was available. Thomas Ness Ltd. generally operated profitably, but the profits were small and came mainly from tar distillation. The recession in the building industry at the end of the seventies hit the market for the various synthetic building and roofing materials manufactured from tar products.

Coke suffered from all the market problems which have already been examined in relation to coking coal and its pricing.[1] They were not solved even though the rebates which the BSC secured on the price of coking coal had to be extended to all producers of coke and carbonized manufactured fuels, including the NCB's own subsidiary, National Smokeless Fuels Ltd. Although the cost of coal was about 65 per cent of the cost of coke production, the rebates on coal were insufficient to lower the cost of inputs as fast as the revenue from sales. This was partly because plant had to run below full capacity, partly because accumulating stocks and over-supply created and maintained a buyer's market. National Smokeless Fuels also had some disadvantage from the retention of old and inefficient capacity. In 1977 the price which the BSC charged in their internal accounts for blast furnace coke from their own

[1] Chap. 7 above, pp. 400–2.

ovens was £12 to £15 a ton less than the price charged by National Smokeless fuels. The NCB believed this to be a distortion, but they nevertheless admitted that there was a genuine difference averaging about £5 a ton.

Figures of sales and production show what was happening. In 1973–4 National Smokeless Fuels sold 3,406,000 tons of coke, 387,000 tons of breeze, and 1,105,000 tons of manufactured fuel. The total of 4,898,000 tons (4.98 million tonnes) exceeded production, which was reduced by the 1974 strike, by 172,000 tons (175,000 tonnes). The next year showed a great improvement in sales, with a large increase in production, but then sales returned to around the level of 1973–4, with the result that stocks began to pile up. Total sales in 1976–7 (on a comparable basis) were 4,734,000 tons (4.81 million tonnes), or 5,122,000 tons (5.20 million tonnes) on the disposals basis used thereafter.[1] But production was 5,877,000 tons (5.97 million tonnes). Sales of blast furnace coke fluctuated wildly, and those of foundry coke were gradually declining and were only 646,000 tons (656,000 tonnes) in 1976–7. After this the general decline became severe. Total sales in the following year fell by over a million tonnes, with the sharpest fall in sales of blast furnace coke, and though the drop in production was even larger, stocks were still rising. Sales were rather better in the next two years, but this was almost entirely due to temporary rises in the erratic sales of blast furnace coke, together with some additional exports. The trend of most items was downwards, and a diminution of stocks by over a million tonnes in two years was achieved only by operating ovens below capacity, which added to costs. Sales of domestic manufactured fuel fell below a million tonnes in 1978–9 and went on falling.

The year 1980–1 was disastrous. During the steel strike in early 1980 the BSC coke ovens had gone on working and had built up stocks. The sequel was that in 1980–1, for the first time, no NCB blast furnace coke was sold. Sales of domestic coke fell by over 200,000 tonnes and sales of manufactured domestic fuel dropped below 900,000 tonnes. The slight recovery in 1981–2 was attributable mainly to a three-year contract to supply 250,000 tonnes of blast furnace coke annually to BSC, and to increased exports (which became the largest single category of sales and were equal to more than a quarter of total production). Government payment of part of the price of foundry coke failed to revive demand, and sales, at 311,000 tonnes, were the lowest ever. National Smokeless

[1] From 1976–7 the published figures for both disposals and production included breeze and char used internally. In previous years these were excluded.

Fuels disposed of 4.1 million tonnes of coke, breeze, char, and manufactured fuel and produced 3.3 million tonnes. There were still 1.6 million tonnes in stock at the end of the financial year, and the finances were hampered not only by the increased reliance on exports but by the rise in the proportion of breeze and char, which were also low-priced products. Nor was the achievement of a cutback in production and consequent appreciable lowering of the excessive stocks a sign that a new balance had been attained at a lower level. The slide continued through 1982, and in 1982–3 as a whole there was a further drop of over 400,000 tonnes in disposals. Only manufactured domestic fuel showed any significant recovery of sales. For nearly everything else the trend continued downwards.[1]

One of the serious problems was the difficulty and slowness of the adjustment of productive capacity to demand. This appears to have arisen for four principal reasons. One was that, even as late as 1977, hardly anyone seems to have recognized the strength of the forces making for a *permanent* reduction in the use of coke. The second was perpetual worry about trade union attitudes. Every time any suggestion of the closure of some capacity was made, expectation of union resistance was expressed as an argument for proceeding cautiously, or not at all. The third was a desire to preserve outlets for the production of coking coal collieries. The fourth was the division between the NCB and the BSC. The latter consumed around 1977 about 10 million tons of coke a year, of which National Smokeless Fuels supplied only a small and fluctuating proportion. The BSC found advantages in having coke ovens sited close to steelworks, and even at this time of over-capacity in the coke industry as a whole, they were still rebuilding coke ovens, with some increase in capacity, with the object of self-sufficiency. There was a joint NCB/BSC working party, but it was difficult to control this tendency to add to capacity when the BSC were both the larger producer and much the largest user. In the end, the limiting factor was the parlous financial position of the BSC, which caused a shortage of capital for rebuilding coke ovens.

National Smokeless Fuels in 1977 had thirteen coke works, employing 5,500 men, and capable of producing about 4½ million tons of coke a year at full throughput. But in 1977–8 disposals of coke and breeze only just exceeded 3 million tons, and in all subsequent years but one were below (and usually well below) 4 million. Yet in 1977 the only proposed

[1] Figures of production and sales are given in NCB *Ann. Rep.* each year in the chairman's statement for NCB (Coal Products) Ltd. and subsidiaries.

change was the closure of the Glasshoughton works (mainly used in recent years for blast furnace coke), which reduced capacity by only about 200,000 tons. This was accomplished in 1978; but not until 1980, when the foundry plant at Norwood and the domestic and industrial plant at Manvers were closed, was there adjustment on the scale already needed in 1977. But by 1980 the market indicated the need for further adjustment.[1] It was these delays that prolonged the situation in which there had to be repeated judgements on whether losses would be bigger by running plants well below capacity or by stocking more coke.

National Smokeless Fuels was much the largest subsidiary of NCB (Coal Products) Ltd., and (though Nypro was for several years also a strong adverse financial influence) it was the principal determinant of its fortunes.[2] Once coke ran into difficulties, NCB (Coal Products) and subsidiaries became rather a tottery business. It paid its preference dividend in its first three years and also paid a dividend of 15 per cent on the ordinary shares in 1974–5 and 1975–6. After that there were no dividends of any kind. The government began to pay a coke stocking grant in 1977–8. Without that there would have been an operating loss. Even with the continued grant there were operating losses from 1980–1 onwards, and overall losses in each year from 1977–8 except 1980–1, when there was nominally break-even. The results would have looked still worse had not the benefits of group taxation relief (which were available because of the profitable operation of NCB (Ancillaries) Ltd.) been used from 1980–1 entirely to improve the accounts of NCB (Coal Products) Ltd. In the first ten years of existence the latter had losses of £16 million, after receiving £71 million in government grants and improving its results by £35 million through granting group tax relief. In the last three of those years, despite benefits of these kinds, losses totalled £39 million. These were large losses for a company whose total capital employed never exceeded £154 million and dropped to £115 million in 1982–3.[3] They led to a large increase in borrowing, and by the

[1] NCB *Ann. Rep.* 1977–8, 62; 1980–1, 78. One-third of the plant at Manvers was closed in 1978. Further closures were proposed in 1982 (*Ann. Rep.* 1982–3, 66).

[2] Apart from the undertakings discussed in this chapter, there were a few small subsidiaries of subsidiaries or of associated companies, and there was one venture into the field of specialist advisory services. This was Coal Processing Consultants, which was started in 1976–7 as a 50:50 partnership bewteen NCB (Coal Products) and Babcock Woodall Duckham. Its activities were taken over in 1979 by a limited company of the same name, in which British Petroleum Ltd. took a one-third interest. In 1982 the latter withdrew and the company was again a 50:50 venture of the original partners. It was profitable in 1982–3 but previously made only losses. It was too small to have much effect on the finances of NCB (Coal Products).

[3] NCB *Ann. Rep.* 1982–3, 76.

summer of 1980 the company's bankers were refusing to renew overdrafts without the guarantee of the parent organization. The NCB found it necessary to waive interest on their own loans to NCB (Coal Products) and they had to start thinking about putting in additional capital to prop up the subsidiary, which they did by way of both equity and loan in 1983.[1] It was a melancholy story.

### iii. Ancillaries

Although the non-mining activities taken over by the NCB from the colliery companies were numerous and varied, apart from coal products they were mostly small both absolutely and in relation to the industries in which they operated. The range and character of these other operations did not change very much during the first twenty or so years of nationalization. Of those that were genuinely commercial the most important were the brickworks, the coal selling depots, and the estates and farms.

The brickworks, taken together, formed a business much above the average size for the brickmaking industry, but most of the individual plants were small and many were old and not very efficient.[2] But the brickworks, as a whole, were modestly profitable despite the existence of price control until June 1948; their market prospects were fair; and in their earlier years they fitted in well with colliery needs. There was a good case for pruning the worst but keeping and improving the rest and seeking to get more out of the assets. This was what was attempted. The original eighty-five brickworks and pipeworks had been reduced by 1951 to seventy-five brickworks and five pipeworks. Output of bricks rose from 389 million in 1947 to 473 million in 1951, and about a quarter were consumed by the NCB.[3] Despite the absence of any strong enthusiasm for further expansion at this time, the improvements in management and efficiency that were achieved made some further growth practicable. In 1956 seventy-four brickworks produced 522 million bricks, and the NCB output temporarily reached about 12 per cent of the national total. In the sixties output tended to run a little lower, though 513 million bricks were produced in 1964–5. Security in the

[1] NCB *Ann. Rep.* 1982–3, 68 and 73.

[2] The NCB thought them 'a mixed bag of assets' and quickly introduced a standard system of costing so as to assess their relative efficiency and profitability (NCB *Ann. Rep.* 1948, 81).

[3] Ibid. 1951, 47–8. Some of the figures published earlier are higher than those cited here, because they include firebricks and other fireclay products.

market was sought by closing more works, reconstructing others and adding a couple of new ones, increasing the proportion of the more profitable facing bricks, and (in 1966) further increasing the range by buying the Whittlesea Central Brick Company Ltd. Nevertheless, the Brickworks Executive found itself operating in more difficult conditions than previous management and the difficulties increased. Output went down to 472 million in 1965–6 and 441 million in 1966–7. By 1969–70 production had fallen to 278 million, but this excluded Scottish production as the NCB brickworks in Scotland were combined in February 1969 with those of Alexandra Building Services (a subsidiary of Thomas Tilling Ltd.) to form a new associated company named Scottish Brick Company Ltd. At the time of the merger the NCB brickworks in Scotland were producing at the rate of about 160 million bricks per year.

The management of the NCB brickworks in England had also been reorganized in March 1968, so that, besides the Whittlesea Central Brick Company, there were two other units. Under the Brickworks Executive, the works in Northumberland and Durham were run by Northern Brick Company and the rest by Midland Brick Company. But despite these changes, and an internal shake-up in Midland Brick Company, which was earliest in difficulties, the brickmaking activities ran into loss at the end of the sixties.[1] Lack of profit came at a time when contraction of the coal industry had reduced both the colliery output of brickmaking material and the colliery demand for bricks. So the question began to pose itself whether operating brickworks was any longer a very appropriate activity for the NCB.

When the Heath government began to press for a reduction of ancillary activities and their subjection to closer governmental control, sale of the brickworks looked to be the obvious way of simultaneously satisfying the government and perhaps serving the interests of the NCB. But in fact, very divided views emerged in the national board about the relative advantages of alternative courses, and, though pressure from the government was renewed from time to time, it was not continuous and the board were left free to make their own decision. The reasons why some members of the board wanted to retain the brickworks were twofold. Thanks partly to the reorganization of management and still more to a building boom, the brickworks as a whole returned to profit in 1971–2, though many individual works ran at a loss; and there was concern about industrial relations difficulties, which might affect the collieries,

[1] Ibid. 1956, I, 29; 1964–5, I, 27; 1966–7, I, 28; 1968–9, I, 27; 1969–70, I, 30.

because the NUM submitted written objections to the sale of the brickworks and represented the workers at some brickworks. On the other hand, the opposite view was supported by the arguments that the brickworks were no longer integral to colliery business, that additional capital expenditure (which could be more usefully directed elsewhere) would be unavoidable in the next few years, that in such a cyclical industry as brickmaking the best time to sell a business was during a boom, and that the NUM objection was explicitly on political grounds and acknowledged the possibility of a commercial case for sale. The former view prevailed at first, the latter in the end.

Early in 1972 agreement was reached to sell Midland Brick Company, but the implementation was held up by the coal strike of that year, and then the NCB, mainly because of worry about industrial relations, went back on the agreement. This, naturally, damaged confidence in their reliability in sale negotiations. It was another year before renewed indications of willingness to sell began to show signs of producing results. In the meantime all the brickmaking activities had been brought under NCB (Ancillaries) Ltd., with the Northern and the Midland companies replaced by limited liability companies of the same name. Most of the rapid closure of small uneconomic works had been accomplished by 1970 but one or two more were closed later. In 1973 only nineteen were operating and a good part of the closure costs had come to an end. Brickmaking still made reasonable profits for the NCB in 1973–4, £1 million in eight months until, on 25 November 1973, the sale of all three companies was completed. In the end, the sale went well. The NCB's Finance Department had thought they would be lucky to obtain the equivalent of the gross book value of £4.5 million, a third of which was the amount included for goodwill in the purchase price paid for the Whittlesea Company, and only £2.2 million of which represented fixed assets. In fact, the sale realized £5.6 million, and the only union reaction was a request that the NCB would remind the purchasers of the Labour Party's policy of renationalizing without compensation any public sector undertakings sold to private enterprise.[1]

Even though the brickworks were currently profitable after meeting their share of interest charges, it was a well-timed sale of undertakings that had become something of a distraction from the main business. After this the only involvement of the NCB in brickmaking was through their 50 per cent share in Scottish Brick Company Ltd. This undertaking, after being initially profitable, ran into loss in 1974–5, and

[1] NCB *Ann. Rep.* 1973–4, 58 and 59.

the losses became persistent and heavier, with the result that in 1976–7 the NCB wrote down their investment to a nominal value of £1.[1] The sale year of 1973–4 was the finale of profit from bricks.

The coal distribution activities remained fairly small and were not much changed before the sixties. The collieries which incorporated a selling depot continued to operate it, and the selling depots away from the collieries were managed separately and regularly produced a few hundred thousand pounds profit each year. The sixties saw various attempts to preserve business through new distributional facilities and services. There were some additional retail services operated by the NCB sales regions on much the same lines as those of any other coal merchant. A few packaged fuel plants were set up. There were the new mechanized coal concentration depots in southern England, which were operated as a separate management unit. The biggest changes of scale came through association with firms in the private sector. In 1965 there was the purchase of a majority shareholding in J. H. Sankey and Son Ltd. (which became a partly owned subsidiary of the NCB) and thus the entry into the sale of domestic solid fuel heating appliances. Then in February 1966 there was agreement with Amalgamated Anthracite Holdings Ltd. for the NCB to share in its distributional business. To implement this agreement, a new partnership, British Fuel Company, was established and began trading early in 1967. The NCB originally put in one-sixth of the capital, a proportion subsequently built up in stages to 40 per cent and then to 49¾ per cent. These two ventures soon came to be among the chief contributions to ancillary profits.

When the reorganization of 1973 took place it was decided to bring together most of the wholly owned fuel distribution activities—retail sales service, coal handling depots, packaged fuel plants, and the Ayr shipping services—into one unit. For this purpose an existing wholly owned non-trading company, National Fuel Distributors Ltd., was activated. The mechanized coal concentration depots were placed in a new company, Southern Depot Company Ltd., which also took over a depot near Cheltenham, hitherto run by the NCB's Welsh sales region. J. H. Sankey and Son Ltd. continued as a separate partly owned subsidiary company. NCB (Ancillaries) Ltd. also took charge of the NCB investment in associated companies engaged in fuel distribution. The British Fuel Company was the chief, but there were a few much smaller

[1] Ibid. 1976–7, 40, 65, 71, and 72. Scottish Brick Corporation Ltd. (the successor to Scottish Brick Co. Ltd.) made a small profit in 1982–3 (NCB *Ann. Rep.* 1982–3, 47).

undertakings, and subsequent disposals and acquisitions.[1] Distribution became the largest and most profitable activity of NCB (Ancillaries). The domestic solid fuel market was no longer very big, but the NCB were involved in a substantial share of it. In 1977–8, of the £11.9 million operating profit of NCB ancillaries, £7.5 million came from distribution activities, of which J. H. Sankey and Son contributed £3.0 million, and the other £4.5 million came from fuel distribution. In 1980–1 the contribution was £16.9 million (of which Sankey produced £4.3 million) out of total profits of £22.2 million. In 1981–2 Sankey ran into loss, which led in 1982 to the disposal of the retailing division and the electrical wholesaling division, but solid fuel distribution was still able to produce £12.1 million of the total 1981–2 profit of £15.4 million.[2] It all suggested that life is often easier as a middleman than as a producer.

The land owned by the NCB amounted in the earlier years to some 360,000 acres and the total changed little in the first decade. About one-sixth was then needed for colliery and other operations, about one-sixth for housing, and the remaining two-thirds was not immediately needed for operational purposes and was mostly let in farms.[3] It needed to be managed commercially, but also in such a manner as to ensure that land was available for transfer to operational purposes when and where required. The management was such as regularly to produce a small profit. In the sixties, as the coal industry contracted, the future needs of land declined, though there was a larger turnover as land was bought for opencast mining and resold after restoration. Although the total area of land tended to fall, land values and rentals were rising and the operating profits of estates and farms were higher, getting up to over half a million pounds annually.[4] In 1973 a new company, Coal Industry Estates Ltd., was set up to manage the non-operational land, as a subsidiary of NCB (Ancillaries) Ltd. Initially it had about 180,000 acres of estates, farms, and woodlands and produced operating profits each year of £700,000 upwards, passing £1 million for the only time in 1979–80. It was also able to produce capital profits several times as large, as it could identify estates each year which were never likely to be needed and sell them at good prices. In 1980, however, the NCB took back from the company 111,000 acres which, it was suggested, would probably be

[1] NCB *Ann. Rep.* 1966–7, II, 29; 1973–4, 58.

[2] Ibid. 1977–8, 51 and 77; 1981–2, 59, 90, and 92. The partial disposal of Sankey was only the first step in disengagement from this business, which was completed in 1984.

[3] Ibid. 1951, 48. 1957, I, 29.

[4] Ibid. 1970–1, I, 28–9 summarizes the arrangements for estate management before 1973.

needed for operations in the future. This was a remarkably high proportion, indicating extreme caution about operational contingencies far ahead. Coal Industry Estates Ltd. was left with only 37,000 acres and its activities and profits were reduced.[1]

As new techniques emerged, with applications to the use of coal or the administration of the coal industry, so a few additional ancillary activities were introduced to make use of them. The two most prominent concerned the use of computers and the adoption of district and other large-scale heating schemes. During the sixties the NCB began to make increasing use of computers, and not all of their facilities were fully employed for anywhere near the whole time. The wish was to make some money out of the spare computer time. The first attempt, stimulated by the Economic Development Council ('little Neddy') for the Hotel and Catering Industry, involved association with International Reservations Corporation, a wholly owned subsidiary of Planning Research Corporation of California. The two parties established International Reservations Ltd. in 1969. The purpose was to use the NCB computer installations to operate a computer based hotel reservation system. It was never a success and made no money. The NCB sold out in 1972–3.[2] The second attempt did better. When NCB (Ancillaries) Ltd. was formed in 1973, the NCB's six computer centres were made the responsibility of a new subsidiary, Compower Ltd. Though the first duty was to serve the NCB's own activities, Compower Ltd. had an external bureau to market the available spare computer time. It was not tied to one associate and could make whatever commercial contracts were practicable and advantageous. In the later seventies and early eighties the external activities were growing faster than the internal. Profits increased from under £1 million in 1973–4 to £4 million in both 1981–2 and 1982–3, on a turnover of £25 million.[3]

Several district heating schemes were started in the sixties under various arrangements made by the NCB Marketing Department. But the most successful venture was Associated Heat Services Ltd., which was formed in 1966 by Solar Industries Ltd., Compagnie Générale de Chauffe, and the NCB, who took equal shares. Its purpose was to promote and operate district heating schemes and to provide consultancy,

[1] Ibid. 1979–80, 82 and 84. Other figures from the accounts of NCB (Ancillaries) Ltd. in *Ann. Reps.* for successive years.

[2] NCB *Ann. Rep.* 1969–70, II, 39; 1972–3, 30.

[3] Ibid. 1973–4, 58 and 61; 1982–3, 78, 80, and 88. There were, in fact, seven computer centres when Compower Ltd. was formed, but those at Doncaster and Mansfield were merged almost immediately.

advisory, and management service for large heating schemes. Some changes took place in ownership. William Cory and Son Ltd. came in as fourth equal shareholder, and then Solar Industries dropped out, so that once again there were three. It took several years to get the undertaking going, but it became profitable in 1971–2 and quickly went from strength to strength. The annual profits were soon many times the book value of the assets. The NCB's share of profits passed £750,000 in 1979–80 and was £971,000 in 1981–2, while the book value of their investment was only £64,000. With another Conservative government expressing doubts about the NCB retaining marketable ancillaries, Associated Heat Services Ltd. became in May 1982 the only NCB subsidiary or associate to seek and obtain a listing on the Stock Exchange. When that was done the NCB and Compagnie Générale de Chauffe each retained 30 per cent of the shares and the remaining 40 per cent were sold to the public. The undertaking became a public limited company. Despite the reduction in the NCB's share, they earned over £1 million from this business for the first time in 1982–3.[1]

Various other small ventures of different kinds were undertaken. Two or three flopped after a few years; others became profitably established. Among the successes were PD–NCB Consultants Ltd., a 50:50 undertaking with Powell Duffryn, which the NCB entered in 1974–5. Its business was to provide an international mining consultative service, and its name was changed in 1979–80 to British Mining Consultants Ltd. Another was Horizon Exploration (Holdings) Ltd., in which the NCB and English China Clays Ltd. took equal shares in 1979–80. This company undertook marine and land seismic surveys for minerals and hydrocarbons both at home and abroad, and used new techniques which the parent undertakings had developed. In 1981–2 it gave the NCB a £650,000 profit on an equity investment of rather less than £2 million, profit having more than doubled over the previous year.[2]

NCB (Ancillaries) Ltd. was never a very large undertaking—its labour force started at about 6,300 and after ten years was just over 4,000—and so could not have great weight in the NCB organization. But it operated with steady success and continued to do so when many of the larger NCB operations were financially unrewarding. It became progressively more capital-intensive, but was able to generate internally all the additional capital needed and produced a satisfactory return on capital. In its

[1] NCB *Ann. Rep.* 1966–7, II, 30; 1969–70, II, 38; 1971–2, I, 21; 1979–80, 57 and 83; 1981–2, 59 and 91; 1982–3, 47 and 79.

[2] Ibid. 1974–5, 62 and 63; 1979–80, 57 and 83; 1981–2, 59 and 91.

first ten years the ratio of operating profit to capital employed only once fell below 20 per cent (in 1981–2, when it was 19.0 per cent), and it was usually in the range 25 to 35 per cent. On this basis its worst year was better than the best year of NCB (Coal Products) Ltd., which it also surpassed in turnover from 1977–8 onwards.[1]

The formation of NCB (Ancillaries) Ltd. and its subsidiaries was intended to be, among other things, a means of managerial reform. The consultant who examined the activities in 1972, before the new structure was devised, reported that 'many times it was found that a dichotomy of responsibility existed, and, furthermore, many line managers were content to leave the final mechanics of budgeting and explaining variances to Finance Department, thus blunting the edge of management responsibility'. The remedy he proposed required that the 'axis of management must be clear with no division of control allowed to dilute its authority or responsibility'; and he added that 'the adoption of the terminology and methods used in commerce should assist the process of change to a marked degree by creating the right environment for management thinking and action'. The subsidiary structure created below NCB (Ancillaries) Ltd. differed in various details from the proposals of the report, but the essential principles were adopted. It was thus not inappropriate that the consultant who made the proposals, Gerald Jackson, should have become the first Managing Director of NCB (Ancillaries) Ltd., with the task of carrying them out. He remained for five and a half years, until his retirement.

In one or two respects (for instance, in relation to the management of land) some divisions of responsibility remained, but a strengthening of a commercial approach was undoubtedly achieved and was necessary as a basis for the financial results produced. The ancillary activities had an internal purpose, to provide goods and services needed in the NCB's main activities or to make fuller use of their products. But most of the activities were extended to the competitive world outside. In most years internal transactions amounted to only 20 to 25 per cent of total turnover, and the results depended on commercial effort.[2] They were good enough for the regular payment of high dividends after the retention of enough profit to provide for internal needs. In its first five years, NCB

[1] Ibid. 1982–3, 76 and 88.

[2] Relevant figures are in the company's published accounts in NCB *Ann. Reps.* In 1982–3 the proportion of turnover which was internal to the NCB rose to around 30 per cent, mainly because of the disposal of parts of J. H. Sankey and Son Ltd. which had sold almost entirely to the public.

(Ancillaries) Ltd. paid to the NCB £27.0 million in dividends plus £16.1 million by way of group relief.[1] The process went on steadily, with increased amounts of group relief being used specifically to lessen the difficulties of NCB (Coal Products) Ltd. from 1979–80 onward. Even in a year of relatively difficult trading conditions, 1981–2, it was still possible to pay a $22\frac{1}{2}$ per cent ordinary dividend after adding to reserves,[2] and in other years the dividend was appreciably higher. It was a useful, though small, benefit and a good example to the NCB as a whole.

### iv. The contribution to the coal industry

Such information about financial results is the obvious kind of evidence from which to start when trying to assess the usefulness of attaching ancillaries to the main business of the coal industry. For the first twenty-five years the NCB regularly published a combined total figure for the annual operating profits of what were then classed as 'ancillaries', which included all the non-mining businesses and some internal services.[3] A more realistic figure, and one more nearly comparable with later data, is obtained by excluding two easily identified components which are clearly inappropriate to the category and were removed from it in 1973. These are housing, which was not run as a commercial service, and the net revenue derived from the owners of licensed mines, as this properly belongs to the mining rather than the non-mining business. It happens that these two items almost cancel each other for the period as a whole, for the housing account recorded a deficit in seventeen of the twenty-five years, but their exclusion alters the figures significantly for shorter periods, as appears from Table 9.1. All the figures exclude income from associated companies, but participation in joint enterprises was little more than nominal before the later sixties.

In 1972–3 the accounting categories were changed to conform more closely to those to be adopted when the two holding companies for non-mining activities began operations in 1973. The categories, while not identical with those of column (5) in Table 9.1, were not so wildly different as to make comparison seriously misleading for general purposes. On the new basis the 'ancillary' operating profit was £9.2 million in 1972–3,[4] and for the ten years 1973–4 to 1982–3 the two

[1] NCB *Ann. Rep.* 1977–8, 75. [2] Ibid. 1981–2, 91 and 92. [3] Ibid. 1971–2, II, 4–5.

[4] The figure for 1972–3 is a revision which appeared in NCB *Ann. Rep.* 1973–4, 40. Compared with the original version (*Ann. Rep.* 1972–3, 32), the profits of chemical secondary by-products have been revised upwards and the apparently accidental omission of the profits of LS Leasing Ltd. from the figure for offshore activities has been corrected. The profits and losses of associated companies (which were also revised) are excluded from the total, as in all other years.

Table 9.1. *Operating profits (losses) of NCB 'Ancillaries', 1947 to 1971-2*
(£ million)

| (1) Period | | (2) 'Ancillaries' as published | (3) Housing | (4) Licensed mines | (5) 'Ancillaries' narrowly defined (2) minus (3) minus (4) | (6) NCB operating results inclusive of (2) |
|---|---|---|---|---|---|---|
| 1947–1956 | (10 years) | 26.2 | (7.8) | 0.7 | 33.3 | 173.7 |
| 1957–1966/7 | ($10\frac{1}{4}$ years) | 16.2 | (4.3) | 3.4 | 17.1 | 290.3 |
| 1967/8–1971/2 | (5 years) | 42.5 | 7.1 | 3.0 | 32.4 | (14.1) |
| TOTAL | ($25\frac{1}{4}$ years) | 84.9 | (5.0) | 7.1 | 82.8 | 449.9 |

*Source*: NCB Annual Accounts.

holding companies together made operating profits of £155 million, exclusive of £21 million derived from associated companies and partnerships and £40 million of capital profits. If the figures for these years are added to the total of column (5) of Table 9.1, they give a total 'ancillary' operating profit of £247 million for the thirty-six and a quarter years from 1947 to 1982–3. With the additional eleven years the total NCB operating profits for the whole period go up to £715 million, though the figure is somewhat confusing because of the inclusion of government grants in some of these later years. But the figures at least give a rough indication that over the first thirty-six years, the NCB derived rather more than one-third of their operating profits from non-mining activities, and the effect of government grants was slightly to understate the proportion. The dependence on non-mining profits was least in the late fifties and early sixties, mainly because, in five of the ten years starting with 1957, the 'ancillaries' operated at a loss. The apparent dominance of non-mining profits in the quinquennium beginning in 1967–8 is a little misleading because the NCB as a whole made operating profits except in the strike year of 1971–2: a total of £106.1 million in the first four years, during which the whole of the 'ancillary' profits in column (5) of Table 9.1 were earned. The largest non-mining contribution came after 1971–2. In the next eleven years non-mining activities provided more than half of all operating profits; in fact, about two-thirds if the share of profits of associated companies is included.[1]

It might be contended that, as the NCB were required only to break even, the higher operating profits of the non-mining activities were supererogatory, but operating profits were required in order to provide at least for the service of loans. The operating profits of the main activity, the collieries, were less and less adequate for this purpose, so increasing reliance had to be placed on other surpluses in order to cover interest charges completely and to achieve break-even on the business as a whole. Non-mining profits were both needed and available to help to cover interest charges in 1947, 1952, 1954, 1955, 1957, 1960, and regularly from 1964–5 onward.

The profit contribution of non-mining activities was made without them taking up a large proportion of NCB business resources. There are problems of changing definition which make comparisons less exact than is desirable, but the general picture is clear enough. In 1954 the sales by 'ancillaries' were £78.2 million out of a total NCB turnover of £771.7 million, which included £58.4 million of intra-group transactions

[1] All figures are from the accounts published in NCB *Ann. Reps.*

that were not separately specified by type of activity. In 1981–2 the non-mining turnover was £688 million (£541.9 million if sales within the NCB are excluded) out of the NCB total of £4,727 million. Figures of assets employed can be a somewhat misleading basis for comparison because the book values of some assets (e.g. land and some classes of buildings) become very unrealistic with the passage of time; but here also, though fine distinctions cannot be trusted, wide differences are significant. Because of the long delay in settling the value of assets transferred at nationalization, there are no comprehensive asset figures for 'ancillaries' until 1954. At 1 January in that year the net fixed assets of 'ancillaries' were £44.3 million out of the NCB total of £480.1 million (or £409.3 million if the value attributed to minerals is excluded, as it should be for long-term comparisons). In 1982 the net fixed assets of non-mining activities were £59 million (£41 million for coal products and £18 million for ancillaries) out of the NCB total of £2,788 million. Of course, the capital needs of deep mining were disproportionately concentrated on fixed assets; but the total assets employed by the non-mining activities were only £202 million (£123 million for coal products and £79 million for ancillaries) out of £3,553 million for all NCB activities.[1] The share of non-mining activities in fixed assets appears to have fallen from around 10 per cent to less than 3 per cent, and even their share of total assets dropped to 6 per cent or less; but they increased their share of turnover from around 10 per cent to a little under 15 per cent, and their contribution to profits quite disproportionately, though the proportion varied widely from year to year and according to the method of accounting for government grants.

The particular sources of the main non-mining contribution to NCB revenues and surpluses changed over time. For most of the period the fortunes of the coke ovens were decisive. In the first ten years they contributed £18.1 million to operating profits, and in the quinquennium starting with 1967–8, £15.2 million; i.e. when the coke ovens were faring reasonably well they could be expected to provide directly[2] about half the total non-mining profits. Likewise, if the coke ovens fared badly there was not a lot that could compensate. It was because the coke ovens made an operating loss of £16.4 million in the second ten years that the

[1] NCB *Ann. Rep.* 1954, 50–1 and 58–9; 1981–2, 46, 47, 56, 82, 83, 95, and 96. In 1954 the figure for fixed assets of 'ancillaries' is the total of all items other than collieries, opencast, houses, central services, and general offices. Minerals are left out of account because their book value was written off in 1973.

[2] The figures relate only to the immediate products of the coke ovens, i.e. coke, breeze, and primary by-products. They exclude all other coal products.

non-mining contribution was then so slight. But despite a brief revival in 1974–5, the good days for coke ovens had virtually ended by 1971. Under NCB (Coal Products) Ltd. the coke ovens and manufactured fuel plants were grouped together as a single activity. In the first ten years of that undertaking they recorded an operating loss of £56 million after receiving £71 million in grants and benefiting for four years from group tax relief. By this time, however, there were other activities that could more than compensate. North Sea gas was important until it was taken away. Thereafter, distribution of coal and heating appliances, computer services, and heating services were what chiefly mattered.

The size of the role of particular activities is relevant to a consideration of the indirect effects on the coal industry. The answers to some of the questions, particularly those about sales and output of coal, must be speculative, because the questions cannot be removed from the category 'What might have happened if . . .?'. It is probable that the coke ovens and manufactured fuel plants increased the use of coal a little for a few years, simply by having available more of an acceptable product and turning out more of the product in modernized plants. But the greater effects were probably defensive: the slowing down of the decline in the use of solid fuel in the sixties, as is suggested, for instance, by the retention of a high level of sales for domestic coke and 'Phurnacite'. Even this success was restricted, however, by the slowness in bringing improved manufactured fuels into commercial production. There was probably a greater influence on the output and sales of coal of particular qualities from particular collieries. As long as the NCB had their own large facilities for making coal products they could decide for themselves what material to use, from what source, subject only to the technical capabilities of their plant and the saleability of the products. In the absence of the NCB coal products plants, some collieries would have found it harder to dispose of their output, while others would have been faced by even more pressing demand. This alternative would, in some cases, have been more profitable for the NCB; but if coal products plants helped the NCB to maintain coal output rather than to improve profits, this is just one of many illustrations of the policy dilemma implicit in the statutory duties of the NCB.

In some cases there was little conflict between output and profit. For many years coke ovens made profits which more or less made up for losses on the mining of the coal which they used; and there were not many other ways of making money out of coal in South Wales, as 'Phurnacite' generally did until the end of the sixties. But by the seventies the coke ovens and manufactured fuel plants were not achieving much for coal except to preserve some home output, with the penalty of increasing financial losses, from the competition of imports of various kinds. And there were always some projects for which the commercial argument turned out weak when put to practical test. Types of manufactured fuel might be devised which were made from coal for which the market was difficult. But when the coal was mined at a loss and the fuel was manufactured at a further loss, there is a strong possibility that the general benefit might have been greater if nothing of this kind had been done. The happier state of affairs, which applied quite widely for many years, was where profitable Yorkshire and East Midlands coal was turned into coke, with an additional profit. Thus was maintained an assured demand which justified the development of colliery capacity of above average efficiency.

Other coal products activities had less to offer the collieries. In particular, it has to be remembered that it does not require large quantities of coal to serve as material for large values of coal-based chemicals. Any chemical activities had to justify themselves by the profits to be derived from their own manufacturing and selling processes, or, on a small scale, by the argument that, since primary by-products existed anyway, it was better to make a little more money from them, if possible, by turning them into secondary by-products.

But the distributional activities, which became increasingly important and successful, were of more direct service to the collieries, even if they did less to expand the market than to limit its erosion. The increased attention to distributional service was a feature of the sixties that was a direct response to the need to go out and sell coal after the captive market had disappeared. It seems clear that if the NCB had not done a great deal to improve distributional facilities, particularly at the wholesale level, and to raise the quality of service, nobody else would have done as much and the attractiveness of coal would have declined faster and further. And the discovery, by experience, that a well-run merchanting business could be a profitable investment helped to make the interests of production and distribution reinforce each other.

There remains the problem of possible diversion of resources and effort from more central activities. It has been shown that, in proportion to output and revenue, the demands of the non-mining operations for resources were small in comparison with those of collieries. But the possible distraction of managerial time, effort, and skill, about which so much concern was being expressed in 1951,[1] was a real danger. The variety of functional responsibilities, many of them rather small in any particular unit, was very wide for divisional and lower tiers of administration. The weaknesses of excessive proliferation were only gradually reduced by greater specialization of function, but by the sixties they were probably not serious, especially after the formation of Coal Products Division. By the late sixties, indeed, there may have been some new positive strengths. This was mainly because, from then on, it was some of the non-mining activities that began to bring the NCB into closer relations with firms of different kinds in the private sector. Larger and more numerous investments in associated companies were the commonest sources of these new links, but there was a comparable influence from the establishment of additional subsidiaries through the purchase of going concerns in the private sector. Still more cross-fertilization was possible when wholly owned subsidiaries entered new fields by acting jointly with firms already established there and relying on their specialized experience, as happened with the offshore activities in gas and oil.[2]

All these various matters influenced judgement on the rationality of the retention of non-mining activities within the NCB, whose *raison d'être* was coalmining. Initially, any other course would have reduced the smoothness of the transition to public enterprise, and little change could have been sensibly made for several years. Subsequently, most of the activities still fitted in fairly closely with the main business of coalmining and probably brought more strengths than weaknesses, though they were not all necessarily kept within the best possible boundaries, and adaptation to changed circumstances was not easy for undertakings that had put down deep roots. Brick, tile, and pipeworks became decreasingly relevant to the coal industry as it contracted, and the NCB might well have withdrawn from them earlier, though the eventual sale turned out to have been propitiously timed. Yet even after that, it might be claimed that in coalmines there was plenty of good fireclay and other clays that went unused for lack of a closely integrated outlet.

The NCB did moderately well out of coke ovens for twenty-five years and there were relationships, based on plant location, coal supply, production experience, and established markets, that it would have been unwise to dissipate. But the NCB may have clung too long to too much.

[1] pp. 172 and 480–1 above.

[2] The mining activities also got some cross-fertilization as they made and acted on their various agreements to exchange information with mining undertakings abroad, e.g. RAG; but this was not so close a relationship as a shared responsibility for output and profits.

The warnings of 1951 that NCB output of blast furnace coke and coke oven gas could never hope to serve anything other than an unreliable residual demand were soundly based. It might have been in both the general and the NCB's interest if production of blast furnace coke had been left to the steel industry and any independent ovens that were willing to stay in that branch of the business. It has been noted that in 1951 a minority of the national board would have been prepared to put all publicly owned coke ovens under a joint board of the coal, steel, and gas industries. It is not possible to see that subsequent experience completely destroyed their case, though every conceivable course had some disadvantages. The events of the later seventies certainly indicated that even more co-ordination than existed was needed. By that time it seemed that the benefits to the NCB from still continuing in the carbonization business were limited to the retention of a slightly greater influence on the use of British, rather than imported, coal than would otherwise have been the case.

But such considerations raise wider questions that go outside the present context. They are related to the general problem of defining and conducting the relations between nationalized industries. Can these relations be achieved, with the greatest public advantage, by mutual co-ordination between their boards, or does there have to be a continuing series of governmental decisions on the delimiting of function? A problem of much the same kind emerged from the entry of the NCB into North Sea gas and oil operations. On the whole this proved to be an excellent investment, and there was a feeling of unfairness at least about the financial terms on which it was compulsorily transferred to BNOC. It was a coherent redirection of resources for a fuel undertaking that had long been closely involved in the gas market to renew that involvement in novel conditions of technique and supply; and it was a practical quick expedient to establish a public interest in the offshore energy resources by using the existing fuel undertakings. But it would not have been practicable to continue on the lines that had been begun. When the NCB had started to produce from all the proven oilfields in which they had an interest, and still more if they had continued with their associates to bid for more licences in each successive round and had consequently shared in further oilfield discoveries, the nature of their business would have been transformed. They might well have become more dependent on gas and oil for revenue than on coal, with traumatic consequences for industrial relations, management, and the allocation of resources. The gas and oil interests had to be put in another institution.

The other non-mining activities gave rise to no such doubts. They neither threatened to match coalmining in size nor were they fashioned into a Trojan horse filled with competing fuels. There were occasional mistakes. It was enterprising to seek profit from spare computer time, but eccentric to tie it all to the booking of hotel rooms. This, however, was only a small and short-lived aberration. The activities carried on under NCB (Ancillaries) Ltd. were much more useful. They were related to the coal industry and created business for it. They earned profits when every little helped. They brought an increased variety of managerial and other business contacts. In all these ways they were beneficial to the NCB, and as they appeared to be as successful under NCB operation as they were likely to be anywhere else, there was no obvious reason why the NCB should not continue to receive that benefit. Non-mining activities had been contributing to the coal industry since long before nationalization. All the signs were in the early eighties that, even though the nature of the ancillaries that could be most helpful had changed, they still made a contribution which brought the coal industry a modest supplement of strength.

CHAPTER 10

# Welfare

## i. The organization of welfare provisions

There are many things in a mineworker's life, sometimes even the security and length of life itself, that can be deeply affected by the terms of his employment, his methods of work and the legal provisions applicable to them, the involvement of his employer with the community in which he lives. The strongest influence has almost certainly been the normal relation of employer and employee through the wage system, all the more so because for most underground workers there were for so long so many varying constituents of wages that were related to the precise location and physical condition of workplaces and the detailed way in which particular jobs were carried out. The settlement of all such matters was continuously central to the operation of the coal industry, and there is no need to go over them again; but it is important to remember that a worker's well-being depended on how much he was paid, how regularly, and in return for what effort involving what long-term physical cost. Welfare is not created by a few supplements to daily working life.

But there were other items besides the wage system that had to be taken into account. Some of them were wholly or partly monetary, such as provision for sickness, injury, disablement, retirement, or death. Some of them involved buildings and services, such as housing and recreation. Some of them related to guidance and limitation on methods of work, mainly in the interest of safety and health.

One of the noticeable features of pre-nationalization arrangements, which contrasts with what happened afterwards, is the rather small extent to which most individual employers took any direct initiative on these matters. The principal exception to this generalization is housing, which many mineowners undertook for essential business reasons. Collieries often had to be started in sparsely inhabited rural districts where employers could locate and retain enough workers only by building houses for them. This was frequently done quickly and cheaply, so that the poor quality of much colliery housing became notorious. Even so, it was no more than a substantial minority of mineworkers who were housed by their employers, but the NCB took over, and kept in use for many years, some 140,000 pre-nationalization houses. The rest of the provisions came from a miscellany of sources. Payments for sickness, injury, and retirement depended mainly on whatever schemes the state introduced from time to time, together with friendly society and local trade union benefits. Some colliery firms, but by no means all, had superannuation schemes for their salaried staff, but they did not provide in this way for their mineworkers. The setting of standards for methods of work, in the interests of safety and health, also depended primarily on the state, through mining legislation and the work of the Mines Inspectorate. Employers applied the standards laid down and the trade unions, through their statutory right to carry out their own mine inspections, provided a continual check. But it was only a very few enlightened colliery owners who worked out better and safer methods which went well beyond the current statutory minimum and contributed to the setting of improved standards for general use. Most improvements came through regulations and new legislation based on conclusions reached by the Mines Inspectorate when investigating mining accidents.[1]

There were a few outher sources of progress. The Mining Association of Great Britain had given a little support to safety research since 1908, when it made an initial grant of £10,000. The Safety in Mines Research Board, which depended on outside funding, had operated a field research station at Buxton since 1926 and had had headquarters and laboratories at Sheffield since 1927. It also conducted some research in London, at Imperial College. But it got into great financial difficulties in the Second World War and consequently, in 1946, on the day that CINA received the Royal Assent, it was taken over by the Ministry of Fuel and Power.[2]

Similar reliance on small external provision characterized medical matters. Very few doctors and nurses were employed by the pre-nationalization coal industry, though colliery companies sometimes retained a doctor to attend in case of accident and to advise on medical claims for compensation, and all collieries had by law to provide a first aid room. In 1943 the Ministry of Fuel and Power inaugurated a Mines

[1] One former colliery manager told the author of the lasting impression made on him by the instruction of his father, a mines inspector: 'Remember, every one of these regulations is there because somebody got killed.'

[2] PRO, POWE 10/395; SMRE, *An Introduction to the Safety in Mines Research Establishment* (1952 edn.). Though the Ministry formally took over in July 1946 there were delays in working out new arrangements and the Board and its governing committee continued to function until 30 June 1947 when they were replaced by a branch of the Ministry.

Medical Service, but it was still only very small in size in the late forties. There was also a little government sponsored research. The Medical Research Council had conducted an investigation of pneumoconiosis in South Wales in 1942 and 1943, which led to the introduction in 1943 of the Coal Mining Industry (Pneumoconiosis) Compensation Scheme, under the terms of which 20,000 miners had been withdrawn from the industry by the end of 1947. Yet, in view of its seriousness, it is remarkable how little had been done to establish even the incidence of the disease. For such a dangerous industry as coalmining, medical services and medical research remained poor at the time of nationalization.[1]

There was, however, one substantial organized contribution by the coal industry to the welfare of its workers, though this also originated from governmental pressure rather than spontaneously. The Mining Industry Act 1920 had established a Miners' Welfare Fund, financed by an output levy on coal which (except for seven years in the thirties, when it was halved) remained constant at 1*d.* per ton. The Mining Industry Act 1926 added a 5 per cent levy on coal royalties. The fund was originally administered by a committee, which in 1939 was superseded by the Miners' Welfare Commission. This body consisted of representatives of employers, royalty owners, and trade unions, together with independent members, and all members were appointed by the government. It operated through twenty-five district committees, each with a roughly similar balance of membership. The Commission was not allowed to provide housing, but this was the only specific limitation on the interpretation of the statutory duty to administer the fund in the interest of 'the social well-being, recreation and conditions of living of workers in or about coal mines'. Mining education and research were explicitly allowed. In practice, most of the expenditure (which totalled over £30 million in the thirty-one and a half years the fund existed) went on pithead baths, though part of the running costs was met from payments by miners for their use. Baths, however, were not added to during the war, but from 1941 the Commission was allowed to build colliery canteens, and in 1942 it launched a scheme for rehabilitating injured miners. By the beginning of 1947, pithead baths had been built at 366 collieries, with provision for 450,000 men; there were over 900 canteens, over 1,500 clubs, institutes, and recreation grounds, and 18 convalescent homes; there were several schemes to assist the education of

[1] Sir H. Roberts, 'Health and Safety', in *Colliery Guardian*, *National Coal Board*, 1957, 52; J. M. Rogan, 'Medical Development and Medical Problems', in ibid., 95, 96, 98, and 100; NCB *Ann. Rep.* 1947, 55–6; PRO, POWE 10/273.

mineworkers and their families; and there were contributions to research, mainly on safety, the Commission having become a statutory contributor of £20,000 a year to the Safety in Mines Research Board.[1] The Commission had obviously done a lot of very valuable work, though it might be thought that much was of the kind which, in some other industries, would have been undertaken directly by the good employer.

Nationalization greatly changed both the scope of welfare arrangements and the allocation of responsibility for them. It was laid down by statute that the policy of the NCB, consistently with the proper discharge of the prescribed duties, must be directed to securing, amongst other things, 'the safety, health and welfare of persons in their employment',[2] which went far beyond the practice of the previous employers. The power of the Miners' Welfare Commission to finance research on health and safety was abolished, and a duty was laid on the Ministry of Fuel and Power to prosecute such research directly, or by engaging others, or jointly with others, and to meet the cost from moneys voted by Parliament. The composition of the Miners' Welfare Commission was slightly changed and the Minister's authority in relation to the Commission was strengthened. But despite the additional responsibilities given to the Minister and the NCB, the Commission and its source of revenue, the Miners' Welfare Fund, were kept in operation. Indeed, the Commission was empowered to act as agent of the NCB in any matter concerning the health or welfare of NCB employees.[3]

There was inevitably some redistribution of welfare functions among the Ministry, the NCB, and the Miners' Welfare Commission, and initially there were some rather awkward overlaps. The NCB in 1947 immediately began to create their own safety organization on a much bigger scale than the industry had previously known. They set up a Safety Branch, directed by a Chief Safety Engineer, in the Production Department and gradually extended it as a service to line management at each level of the administration. In a sense the employer was still applying standards laid down in legislation and by the inspectorate, which had some guidance from research conducted by the Ministry; but now the employer was making a more positive contribution.[4] The NCB's

[1] NCB *Ann. Rep.* 1947, 34–5; 1951, 64–5; CISWO *Ann. Rep.* 1952, 11; PRO, POWE 1/16, 1/44, 10/395.

[2] CINA, s. 1 (4)(a).

[3] CINA, ss. 40 and 42.

[4] Roberts, in *Colliery Guardian*, *National Coal Board*, 1957, 45–6; J. L. Collinson and J. S. McLintock, 'Safety and Health', *Colliery Guardian*, Nov. 1976.

staff were seeking to do much more than before in going beyond minimum statutory standards and devising working practices, sometimes needing modification to the specifications of equipment, which contributed directly to the increase of safety. Such an extension of activity could sometimes create a conflict with existing regulations. Nevertheless, there still was a basic division of function among the various parties, and satisfactory informal contacts could reduce the difficulties where the division became blurred or the progress sought by one party seemed to be resisted by another.

The organization of medical services involved, for a time, rather more confusion of responsibilities. The NCB began the creation of their own medical service in June 1947 by appointing a Chief Medical Officer at headquarters and then going on quickly to appoint Divisional Medical Officers and, more gradually, some Area Medical Officers. In 1947 they also appointed a Director of Research in Medical and Human Problems, whose office was united in 1951 with that of the Chief Medical Officer. The medical organization set out from the beginning to increase the medical treatment facilities available at the larger collieries. But the Mines Medical Service of the Ministry of Fuel and Power remained in existence, although it lost four of its senior staff who became NCB Divisional Medical Officers. One of its innovations had been, since 1945, to instruct the Miners' Welfare Commission that every new pithead bath should have a medical centre attached, and the Ministry bore the extra cost. These medical centres were taken over by the NCB, but they used the Ministry's Mines Medical Service to run them as agents until the end of 1947. The Ministry's medical officials maintained that, long after that, it was necessary for them to exert direct influence on colliery management in relation to the latter's responsibilities for health. Not until late 1951 did they agree that the NCB medical service had developed sufficiently to do this. The NCB would almost certainly have disagreed and would have claimed that they did exert influence directly on colliery management. But they would have agreed that their service developed more slowly than they wished, for reasons outside their control.

In 1949 the government set up a Committee of Enquiry on Industrial Health Services (the Dale Committee), with the object of establishing their best relationship to the new National Health Service, and asked industry to defer any major development of medical services. So the NCB practically ceased recruitment of medical and nursing staff. The

Dale Committee reported in 1951 in favour of the provision of medical services by employers. The NCB then resumed recruitment and expansion of medical services. They did not transfer any of their facilities to the National Health Service, and it was recognized that the NCB and not the Ministry of Fuel and Power should be the body to provide medical service for the mines. The medical officers of the Ministry had then an advisory and consultative role to meet the needs of the Mines Inspectorate and other branches of the Ministry. From the end of 1951 the division of functions was unquestioned.[1] There were still various different bodies engaged in research, but everybody concerned was clear about who was doing what, though it was probably not ideal that the Medical Research Council, the Safety in Mines Research Establishment, and the NCB were all simultaneously looking at aspects of pneumoconiosis.

One other feature of the developing organization was the increased involvement of workers' representatives. The Coal Industry National Consultative Council established a Safety and Health Committee. Other bodies were the National Pneumoconiosis Joint Committee and the Safety in Mines Research (Advisory) Board. All these brought together representatives of the NCB, the trade unions, managers and officials, and HM Inspectors of Mines.[2] There was thus more opportunity for positive suggestions from all those directly involved, at every level, instead of most of them passively receiving the instructions of one small group.

The wider concerns pursued through the Miners' Welfare Commission had always brought employers and unions together for joint action, and continued to do so. It was here, however, that the changes of responsibility which accompanied nationalization created the most difficult problems of reorganization, even though the ending of the Commission's obligations to safety research was a step towards clearer demarcation and left a little more money available for other welfare objectives. One problem was that the NCB had obligations for the welfare of all their staff, including non-miners, while the Commission had wider powers than the NCB to provide for miners' dependants; yet their functions mostly overlapped and it would have been wasteful to duplicate each other's work. The first attempted solution came through the establishment of a joint authority called the National Miners' Welfare Joint Council, which consisted of all members of the Miners'

[1] Rogan, in *Colliery Guardian*, *National Coal Board*, 1957, 95; PRO, POWE 10/273; NCB *Ann. Rep.* 1947, 55–6; 1948, 29; 1951, 59–60; NCB 62nd meeting, min. 749, 7.2.47.

[2] Roberts, in *Colliery Guardian*, *National Coal Board*, 1957, 45.

Welfare Commission plus two more representatives of the NCB, with an administration serving both bodies, accommodated and staffed by the NCB. This body was supported by Joint Welfare Committees in divisions and areas, which replaced the District Welfare Committees of the Miners' Welfare Commission. Some NCB members were unhappy that such a scheme involved the NCB in paying twice over, and the national board agreed that their direct payments would mostly go on buildings. But the details of the new system were worked out by mutual agreement during 1947 and the scheme came into operation at the beginning of 1948.[1]

On the whole, the new arrangement enabled agreed programmes to be carried out smoothly, but it could not remove all conflicts of objective. For instance, the Miners' Welfare Commission allocated new pithead baths to localities in accordance with the proportion of miners who lacked bathing facilities. The NCB wanted to give highest priority to those pits without baths to which they wished to transfer men because there were immediate opportunities of expanding output. In the end, where the NCB wanted pithead baths provided 'out of turn' they bore the whole cost themselves. They were the more ready to do this because they had already taken over the running costs of all pithead baths (except that miners still paid a charge for soap and towels) and, unlike the Commission which had no capital and had to finance everything out of revenue, they could charge additional baths as capital investment. As Lord Hyndley said, in dealing with baths they were performing a function of production under the guise of welfare.[2]

Such needs to act outside agreed programmes emphasized the continuation of unresolved problems. Nobody was altogether satisfied. Moreover, the output levy on which the Miners' Welfare Commission depended for revenue was always granted for a few years at a time and, unless renewed, was due to end on 31 December 1951. So, late in 1950, Arthur Horner proposed consideration of future organization after 1951. A committee was set up, its main conclusions were accepted and formed the basis of an agreement between the NCB and the NUM in July 1951, the necessary statutory authority was given by the Miners' Welfare Act 1952, and new arrangements operated from 1 July 1952.

[1] NCB *Ann. Rep.* 1947, 34–5; NCB 55th meeting, min. 656, 14.1.47; 57th meeting, min. 682, 21.1.47; 59th meeting, min. 709, 28.1.47; 136th meeting, min. 1725, 14.10.47; 139th meeting, min. 1764, 24.10.47; 144th meeting, min. 1833, 7.11.47.

[2] NCB 196th meeting, min. 533, 13.4.48; 256th meeting, min. 120, 25.2.49; 273rd meeting, min. 361, 15.7.49; NCB *Ann. Rep.* 1951, 65.

The essential features were that the Miners' Welfare Commission and the National Miners' Welfare Joint Council were abolished, and separate provision was made for 'colliery' welfare, which would be the responsibility of the NCB, and 'social' welfare, to deal with which the Coal Industry Social Welfare Organisation (CISWO) was established, with representation from the NCB, NUM, NACODS, and managerial grades. Colliery welfare was defined to include pithead baths, canteens, cycle stores, bus shelters, industrial medical services and colliery treatment centres, the encouragement of vocational education and training, the provision of suitable accommodation for workers, the welfare of those off sick or injured and their resettlement at work. The corresponding definition for social welfare explicitly referred to 'persons in the coal industry and other NCB employees, and their families'. It included provision of recreation grounds for sports and games, recreation or community centres, youth clubs and camps, the organizing of holidays and holiday camps, the assistance of the sick or disabled and the provision or support of convalescent homes and funds, and the encouragement of education in all fields other than vocational training for the industry. The functions, assets, and liabilities of the Miners' Welfare Commission passed to the NCB or CISWO as appropriate, and all NCB assets used for 'social' welfare were made available to CISWO. The cash balances and investments transferred to CISWO on 1 July 1952 amounted to £1,461,490, and the NCB agreed to make £1 million available to CISWO as required. The future financial needs of CISWO were to be met by grant by the NCB from time to time to cover the estimated cost of authorized social welfare activities. CISWO was incorporated on 19 February 1952 as a private limited company with up to six directors appointed by the NCB (one of whom must be a person in a managerial grade), and up to six appointed by the NUM (of whom one was to be nominated by NACODS). The chairman had a five-year term and was drawn alternately from the NCB and the NUM. CISWO had a Divisional Welfare Committee in each NCB division, with a similar alternation of chairmanship. The allocation of funds was determined centrally, but the approval and execution of projects was delegated to the Divisional Committees. The NCB provided administration without charge.[1]

The new system was based on a more rational division of functions and responsibilities. It stood the test of time by continuing without

[1] Minutes of meetings of National Miners' Welfare Joint Council: 18.10.50, min. 20; 15.11.50, min. 2(c); 20.6.51, min. 3; 18.7.51, min. 3. NCB 347th meeting, min. 86, 16.5.52. NCB *Ann. Rep.* 1951, 66; 1952, 67–8. CISWO *Ann. Rep.* 1952, 4, 13–14. PRO, POWE 10/469.

fundamental alteration, apart from some changes in local structure after the abolition of NCB divisions. At its beginning CISWO was already responsible for 700 halls and institutes, 600 recreation grounds, holiday centres at Rhyl and Skegness, and twenty convalescent homes. (The Miners' Welfare Commission had transferred all its rehabilitation centres, except that at Uddingston, to the National Health Service in 1951.) It operated a national Scholarship Scheme and a Students' Exhibition Scheme which then awarded an average of 12 new scholarships and 10 new exhibitions each year.[1] These educational schemes were altered in 1955, when the field of beneficiaries was broadened, and again in 1956 when full scholarships were restricted to those unable to get public grants. The rest (a growing number) got non-recurring awards of £100, the maximum allowed without reduction of the grant from public funds. In 1969 the two schemes were combined to form a new trust, the Miners' Welfare National Educational Fund, which provided grants for study of a wider range of courses than before.[2] By 1980 the number of new awards made in the year had risen to 191, the recent increase having been rapid. At the start of the 1979–80 academic year the total number of awards held was 149, of which 117 were non-recurring; and the total number of students who had held awards and who completed their courses in that year was 77.[3]

Another addition to the schemes associated with CISWO resulted from a recognition that, though newsworthy disasters aroused generous public aid to the victims, there were many obscure misfortunes for which no extra provision was made. The new scheme was the Coal Industry Benevolent Trust, set up in March 1976 to assist with money or services the families of mineworkers suffering hardship through bereavement, sickness, or disablement. It was administered by a committee consisting of the directors of CISWO and eight other persons, three nominated by the NCB, four by the NUM and one by NACODS. Its funds came from payments of 3p per week from employees in the coal industry and equal contributions from the NCB and some other employers. Grants were made to relieve hardship from deaths caused by industrial accidents or industrial disease, to dependants of persons admitted to hospital as a result of accidents at work, and to mining families with problems in caring for physically or mentally handicapped children. In five years the annual income of the fund

[1] CISWO *Ann. Rep.* 1952, 22–5.

[2] Ibid. 1969, 37.

[3] Ibid. 1980, 33.

rose from under £200,000 to nearly £1,400,000 and its accumulated funds to nearly £4 million.[1]

The cost and extent of the activities supported by CISWO (i.e. excluding the Coal Industry Benevolent Fund) did not grow significantly in real terms, except that the cost of administration was taken over by the Organisation instead of being borne by the NCB, though as the NCB provided nearly all the income, this was not much more than a book-keeping change. The lack of growth (and for a good many years a real decline) came mainly because the number of workers in the coal industry fell so much, because the need to provide additional buildings out of revenue had become small, and because there had been so large an increase in the services and benefits provided by the state. Total expenditure by CISWO amounted to £1.44 million in 1979 and £1.86 million in 1980, and around 30 per cent of this was going on administration, mainly salaries. But if the scale of activities did not grow there were some changes in the range. By the later seventies there were such things as an annual miners' week-end and festival, a Coal Queen competition, art and photographic competitions, grants to provide new instruments for colliery brass bands, as well as attempts to tackle some neglected personal difficulties through experiments with a social worker counselling scheme and with a block insurance scheme for seriously disabled drivers.[2] But the impact of CISWO on miners' lives was probably less than had been expected when it was set up.

## ii. Pensions and other non-wage payments

Although payments of various kinds outside normal wages were very often treated as supplementary items for negotiation along with wages, there were some which, in some respects, had to be treated apart, if only because of the specialized (and often large) financial provisions which they involved. The most important were the superannuation schemes, for security in retirement is of concern to all workers, and the operation of a funded pension scheme for a very large labour force is a complex business handling huge capital sums and requiring continual adjustment to actuarial needs.

In 1947, in accordance with regulations made by the Minister of Fuel and Power under CINA, the NCB established a superannuation scheme for all their salaried staff. Some of these had not previously been in any

[1] CISWO *Ann. Rep.* 1980, 55–61.

[2] Ibid., 9–19 and 26–9.

scheme, and some had been with employers whose schemes were inferior to that now started by the NCB. For such people prospects on retirement were better than before, though if they were near retiring age when they came into NCB service, their retirement income would still be low. It was only with the passage of time and the opportunity to get in long service that the scheme came to provide real security in retirement. It was a funded, contributory scheme, to which the employee originally contributed 4 per cent and the employer's ordinary contributions were initially equivalent to 8 per cent of annual salary. But the NCB had the statutory obligation to make good any deficiency disclosed on periodic revaluation. Such a deficiency was expected from the beginning, mainly because the NCB had to provide for rights, in respect of pre-nationalization service in the coal industry, which were protected by CINA (s. 37) but were not covered by funds transferred to the NCB. So although the first valuation was not due until 1953, the NCB were already setting aside £900,000 a year before that, in addition to their ordinary contributions.[1] In fact, deficiency payments became a regular feature and remained so even when pre-nationalization pension rights were ceasing to be a serious burden. This was because of improvements in the terms of the scheme and because of the eroding effects of inflation on any scheme where pensions were related to earnings. The NCB were regularly advised by the government actuary of the level of payments over a stated period that would make good the deficiency in the fund, and had to make provision accordingly. Membership of the scheme was compulsory for all eligible staff, and when the state introduced a new graduated pension scheme in 1959 they were contracted out of it. By 1960 the Staff Superannuation Scheme Fund had risen to £120.8 million, and £2½ million a year was being paid to 15,400 pensioners. Twelve years later the fund was £286.5 million, but its growth had failed to keep pace with pension entitlements, for £20.7 million was paid in the year to 42,401 beneficiaries.[2]

In 1971–2 the scheme was amended so that pensions were related to final salary instead of the average of the last three years, and from 1975 pension increases were paid annually in proportion to the rise in retail prices. By then the burden imposed by inflation on the finances of pension schemes was causing great concern. The NCB had already

[1] NCB *Ann. Rep.* 1947, 27; 1952, 149. The scheme covered all administrative and office staff and colliery officials down to deputy and provided for death in service as well as retirement (NCB *Ann. Rep.* 1946, 7).

[2] Ibid. 1960, I, 32; 1971–2, I, 36–7.

undertaken to make special additional contributions to the staff scheme as a consequence of recent large salary increases of contributors, but the government actuary reported that a further deficiency payment was necessary on top of those. By 1982 the additional payments had become so regular that they were replaced by an increase of approximately 1 per cent in the standard contribution paid by the NCB.[1]

There were few occupational pension schemes for industrial workers in 1947 and the coal industry was no exception, but it was exceptional in getting one relatively early, mainly because this was one of the claims by which the NUM had set great store from the beginning. In 1951 a scheme was negotiated and approved (as statutorily required) by the Minister of Fuel and Power. It started operation at the beginning of 1952. It was a contributory scheme, paying flat-rate pensions at levels determined by length of service. The original range was from 10*s.* (50p) to 30*s.* (£1.50) per week, though nobody who retired in the first year qualified for more than 15*s.* (75p). Initially the employee's contribution was 1*s.* 6*d.* (7½p) and the employer's 2*s.* (10p) per week, with slightly lower rates for surface workers. Workers who were too near retirement to contribute for the minimum pensionable period were given back service credits, which enabled them to qualify for the minimum pension of 10*s.* per week at age 65. In order to cover the cost of this concession, the NCB made a special initial contribution of £2 million, and undertook to make annual deficiency payments for 25 years. Membership of the scheme was obligatory for new entrants but voluntary for those already in the industry, though both the NUM and the NCB made great efforts to persuade everyone to join. By the end of 1952 about 80 per cent had done so.[2] In 1959 membership had become nearly complete, and the NUM agreed that membership should be compulsory, in order that mineworkers could be contracted out of the new state graduated pension scheme. The number of beneficiaries was rising very fast. In the year from 5 April 1959 it went up from 79,400 to 113,200, and the annual cost of payments to the latter number was £5,200,000. By September 1971 the number of pensioners was 228,000 at an annual cost of £14.1 million. The pension fund had risen to £124.9 million.[3]

It will be obvious that the rate of pensions was low. One of the recommendations of the Coal Industry Examination in 1974 was that the NCB and NUM should negotiate a better scheme. As a result, an

[1] NCB *Ann. Rep.* 1974–5, 9 and 35; 1982–3, 38.
[2] Ibid. 1951, 77; 1952, 56.
[3] Ibid. 1960, I, 32; 1971–2, I, 36–7.

earnings related scheme, which also provided increases in widows' benefits and retirement lump sums, came into operation on 6 April 1975. Though the problem was not usually as severe as with the staff superannuation scheme, the NCB were accustomed to making actuarially required deficiency payments, and the adoption of an earnings-related scheme imposed additional burdens of this kind. The government, however, agreed to cover about half the initial capital liability imposed by the new scheme. The government also made repayments in subsequent years to meet charges arising from cost of living increases in pensions. In effect, in the early eighties, the government were meeting the costs arising from the application of improved pension terms to those who retired before 6 April 1975, and the NCB were making payments to cover other deficiencies disclosed by subsequent valuations.[1]

The operation of these large pension schemes presented substantial managerial problems. Apart from such initial worries as the absorption of over 200 pre-nationalization pension schemes into the NCB Staff Superannuation Scheme in the late forties, there was the permanent task of managing an ever-growing volume of investible funds. Each pension scheme had a managerial committee with equal numbers of NCB and union representatives, each management committee had an investments sub-committee, and there was a joint sub-committee combining the investments sub-committees of the two funds. This latter body met regularly with an advisory panel of outside investment experts which was created in 1950. An external consultant on property was appointed in 1953 as the funds prepared to invest in property and property development in addition to portfolios of securities. The NCB Estates Branch was also much involved, but the scale of investment increased and a separate Superannuation Estates Branch was created in 1960. In 1972 a new post of Director of Superannuation Investments was created. In 1977 his status was raised to Director-General, with a unified Superannuation Investments Department divided into three branches: marketable securities, property, and industrial finance. At the end of 1975 the two funds together totalled £685 million, of which £206 million was in property. But the conversion of the mineworkers' scheme to an earnings related basis, together with the high rate of inflation, very rapidly increased the cash flow into the pension funds. In 1977 their combined book value first surpassed £1,000 million, but the book value greatly understated their market value.[2] By this time the pension funds

[1] Ibid. 1974–5, 9; 1977–8, 41; 1982–3, 38.

[2] H. R. Jenkins, 'Superannuation Investments Department', *Colliery Guardian*, Jan.1978.

had become important investment institutions in their own right. Their combined weight made them a major influence in the securities and property markets, and their industrial finance branch was wealthy enough to permit a worthwhile policy of investing to promote new enterprise in mining districts.

The other important scheme for payments outside the wage system affected far fewer people, as it dealt with those affected by industrial injury. It arose out of the fundamental change in the law on the payment of workmen's compensation, which was effected by the National Insurance (Industrial Injuries) Act 1946. Workmen's compensation payments ceased to be related to earnings. Instead there was a flat rate injury benefit payable for 26 weeks, followed by disablement benefit at rates dependent on the degree of disablement. This could be supplemented, where necessary, by dependants' allowance and a constant attendance allowance or a special hardship allowance. The new arrangements were brought into operation on 5 July 1948. A clause in the Act allowed the Minister of National Insurance to approve supplementary schemes (not financed out of public funds) for any industry whose employees and employers made out a case that the conditions in their industry justified greater benefits. The NUM argued that coalmining was the supreme example of such an industry and asked the NCB to join them in designing a supplementary scheme and submitting it for approval. This was done and the scheme came into operation on 2 August 1948, though comparable payments were made outside the scheme from 5 July to 1 August.

The supplementary scheme gave a flat rate weekly supplement to injury benefit and normally a one-third supplement to disablement benefit (but not to dependants', constant attendance, or special hardship allowances). It also paid a small supplement to a widow who benefited under the Act, and a larger one if she received a children's allowance under the Act or were over 40 or were incapable of supporting herself. Mineworkers paid weekly contributions to the supplementary scheme, but nearly all the cost (six-sevenths at the outset) was borne by the NCB, and they regularly adjusted the financial provision they made for expected liability. The supplementary scheme worked smoothly from the beginning and committees established to consider disputed cases never needed to meet. In the early years it was costing about £1.3 million annually. Twenty years later the annual cost was down to about half that in cash terms, still less in real terms. Because of a further change in the national insurance provision for industrial injury, the supplementary scheme could not admit any new cases after 30 March 1970. The claims on it therefore steadily diminished and it was wound up in December 1977. Adequate funds remained to cover all terminal liabilities, and the small number of continuing claims were transferred to the mineworkers' pension scheme.[1]

In fairness, it seemed that those injured before 5 July 1948 should not be worse off than the rest, but no supplementary scheme was permissible for them under the National Insurance (Industrial Injuries) Act. It was possible, however, for the Minister of Fuel and Power to approve such a scheme under powers conferred by CINA. He gave his consent to a scheme which came into force on 5 July 1948 and which also applied to beneficiaries of a pneumoconiosis scheme started in 1943. This scheme, inevitably, had to be non-contributory. In general, subject to an overriding maximum income from all sources, it made a weekly payment at full rate for men totally incapacitated, at half rate for men partly incapacitated and out of work, and at quarter rate for men partly incapacitated and in work. In the latter half of 1948 some 34,000 men were receiving payments under this 'old cases' scheme, which was costing over £1 million a year until 1952. Some changes of detail were made in the scheme from time to time to ensure that the total position of beneficiaries was not worsened by changes in the methods of reckoning eligibility for national insurance benefits. As the 'old cases' scheme was outside the national insurance legislation, it could not be superseded as the 'new cases' scheme was. It had to go on as long as any of the beneficiaries survived. Even in 1979–80 it was still costing about £200,000 a year. The NCB, of course, also had liability for workmen's compensation in cases arising before 5 July 1948, most of the liability having been taken over from the pre-nationalization colliery companies. These liabilities, too, continued as long as the beneficiaries survived. Around 1980, in workmen's compensation and supplementary injury payments together, these cases from before July 1948 were still costing the NCB about £500,000 a year.[2]

### iii. Housing

For a very long period, both before and after nationalization, the housing of mineworkers presented employers with a number of interrelated dilemmas. It had long been the practice for many mineowners to build

[1] NCB *Ann. Rep.* 1948, 40–2; 1951, 76; 1952, 56; 1971–2, II, 30; 1977–8, 44.

[2] Ibid. 1948, 42–3; 1952, 56; 1971–2, I, 37; 1977–8, 44; 1979–80, 51.

and retain ownership of the houses occupied by a large proportion of their workers. This had the advantage of helping them to attract and retain enough workers in districts which had previously been sparsely inhabited and to have them conveniently close to the colliery to suit the awkward hours that shift working requires. But it also had great disadvantages. What was sought and provided was usually housing that was cheap to both employer and employee. It was cheap both because it was built to poor standards and because, to attract workers from elsewhere, employers charged very low rents, often well below a remunerative level for even the poorest of the houses that were built. Once they had got the cheapness, the employees were determined to keep it, but the employers were unable to prevent their housing becoming a serious expense. Houses cheaply built soon deteriorated and were costly to maintain and repair, and the employers could not recover much or any of the cost by raising rents without provoking strong collective resistance which could create a crisis in industrial relations throughout their business. And changes of occupation began sometimes to prevent employers getting possession of their own houses for their own workers.

Moreover, the provision of colliery housing in this way often created communities in which everybody had one occupation and one landlord who was also the one employer. Such conditions were socially restrictive and encouraged a very insular outlook. But they also made for great cohesiveness, in which pit and village were united, and both the pleasures and quarrels of the one regularly involved the other. Goodwill could be quickly spread in such communities, but industrial questions could also be hopelessly tangled in disagreements originating in other community experiences, and rational business management was all the more difficult.

By the nineteen-forties it seemed there was more and more emphasis on the disadvantages of colliery provision of workers' housing. When the regional surveys of the coal industry were made at the end of the war, one of the subjects of enquiry was the housing need in the immediate future. From many districts there came indications of the need for more men if output potential was to be achieved, and estimates of the number and location of additional houses that would be needed if those extra men were to be attracted. But almost everywhere it was said that colliery companies had no plans to build any more houses, and local authorities were seen as the only possible substitute. And over and over again it was argued that future building should be in mixed development, not in colliery housing estates. Thus, from Northumberland came the unqualified statement that the policy of building villages solely for miners should be abandoned and bus services had made it unnecessary. In Yorkshire it was stated that the housing needs of miners should be considered as an integral part of the needs of the people as a whole, and this would tend to relieve the insularity of pit communities. The North Midlands committee unanimously recommended the aim of establishing substantial communities with a variety of subsidiary industries, and suggested that, in general, new mining communities should be grafted on to existing centres of population and not established in isolation. In the region of the Midlands Committee, which covered North and South Staffordshire, Cannock Chase, Shropshire, and Warwickshire, it was noted that there were very few separate pit communities and that this state of affairs ought to be repeated in any new developments.[1]

The NCB were in no position to introduce a new policy quickly. They took over the 140,000 or so colliery houses because it would have been impossible to maintain continuity in the operation of the coal industry without doing so. But they soon had to face all the old dilemmas. They considered the suggestion of the Minister, Gaitskell, who thought it would be good social policy for them to sell houses to their own employees, but they agreed in March 1949 not to do this. They would not have been likely to find many willing and able buyers. But the houses were physically an unattractive inheritance which constituted an unwelcome financial drain. At the end of 1948 the total stock of NCB houses was 141,912, of which 37 per cent were classed as 'poor' and only 34 per cent were classed as 'reasonably modern and in fair condition'. Only in Yorkshire, the East Midlands, and Kent did a majority of the houses qualify for the latter category. So the NCB told the Ministry of Fuel and Power in August 1949 that they favoured the transfer of their houses, on suitable terms, to the local authorities. But the Minister of Health, Aneurin Bevan, pointed out to Lord Hyndley that the houses were still unvalued for compensation for nationalization, and as the NCB did not know how much they would have to pay for them, there was no possible financial basis for an approach to the local authorities.[2]

So the NCB were stuck with their houses, and soon had to consider

[1] Ministry of Fuel and Power, *Regional Survey Reports*; e.g. *Northumberland and Cumberland*, 4; *Durham*, 31–2; *North Eastern*, 57; *North Midland*, 21; *Midland*, 26–7.

[2] NCB 235th meeting, min. 1149, 15.10.48.

whether they should build more. There were some areas to which they needed to attract more men, and they could not do it if the men had no assurance of somewhere to live. From 1950 they were able to persuade the Ministry of Health to approve extra house building by local authorities in mining districts. By the end of 1952, 2,650 of these additions to the programme had been completed, 3,800 were under construction, and sites had been found for another 3,200. This was helpful but not enough. So in 1952 the NCB set up the Coal Industry Housing Association, with an original programme to build 19,842 houses, of which seven-eighths would be in the 'manpower deficiency' regions, Yorkshire and the Midlands. Practically all these houses were completed by the end of 1955, and it was estimated that this original programme involved a loss of £1 million a year. But it was considered essential in order to prevent labour shortage in the most productive coalfields, and smaller supplementary programmes were subsequently undertaken, with 469 out of 1,500 completed by the end of 1960.

In the main programme the houses were of non-traditional types, chosen to save on bricks and labour, which was perhaps unwise if long-term maintenance costs were considered. Efforts were made to avoid some of the past mistakes and the aim was to create mixed communities. On fifty sites (about half the total) there was combined development by the Coal Industry Housing Association and the local authority. But inevitably such rapid and specialized development tended to add to communities which were dominated by mining interests, and the weaknesses of the exclusively residential estate, so glaring in the inter-war years, were not absent. In the original programme the Coal Industry Housing Association built only 48 shops and 620 garages.[1] There were also some initial difficulties. There were places where the local miners objected to the allocation of new houses to transferees from other coalfields, although the encouragement of transfer was one of the objectives of house building; and on some estates there was difficulty in filling all the houses as soon as they were completed. But these were not lasting problems.

It had never been expected that there would be need for much more house building after the completion of the initial programme,[2] and any need was made less acute by the reduction in manpower from the late fifties onwards. Even then, the effort to encourage long distance

[1] NCB *Ann. Rep.* 1952, 13; 1954, 13; 1956, I, 75; 1960, II, 53–4.

[2] NCB 366th meeting, min. 57, 15.3.53; 372nd meeting, min. 113, 25.9.53; 395th meeting, mins. 152 and 153, 19.11.54.

transfers to the more productive coalfields was hampered in a few localities by housing shortages, so there were spells of house building from time to time, but on a much smaller scale than between 1952 and 1955. The main instrument continued to be the Coal Industry Housing Association, though there were six other very small housing associations, each concerned with only one locality, whose borrowings from public authorities were guaranteed by the NCB, after the latter had acquired their share capital, which was little more than nominal. Loans to the Coal Industry Housing Association were provided, interest free, directly by the NCB, but the financial burden was lessened a little because most of the home building, like that of local authorities and other housing associations, qualified for Exchequer subsidy. By March 1966 there had been supplementary programmes which increased the original programme by about a quarter, and altogether 24,070 houses had been completed by that date. The NCB had then provided £25 million in share capital and £19.3 million in loans to the Coal Industry Housing Association. The NCB charged no interest, but took as dividend whatever small surplus the Association achieved. The NCB also took on a full repairing lease, without paying rent to the Association, all the houses built by the Association, and provided in their own accounts for depreciation. The other six housing associations received only £500,000 in share capital and loans, and made no profits. A few hundred planned houses still remained to be completed, but this was virtually the end of NCB house building. At the end of March 1971 the NCB investment in the share capital of housing associations was unchanged and the total of loans to all of them was £20.7 million.[1]

By the late sixties the NCB were reducing their stock of houses. Many of those built long before by colliery companies had always been poor and had become worn out. Every year some were demolished under slum clearance schemes, and if there were tenants to be rehoused, this was usually done by the local authority. But there were various districts where the large reduction in the number of colliery employees had made some of the NCB houses superfluous. Area managements were from time to time asked to identify houses that were no longer required, and if they were not ripe for demolition attempts were made to sell these. They were usually offered to the local authority and then, if the

[1] NCB *Ann. Rep.* 1960, II, 53–4; 1965–6, II, 24; 1971–2, II, 34–5. In 1965–6 some of the NCB's loans to the Coal Industry Housing Association were replaced by additional shares. There are some slight discrepancies, which it is impossible to reconcile, in the NCB statistics for houses completed by the Coal Industry Housing Association at particular dates.

council did not want to buy them, to the sitting tenants. By March 1971 the stock of houses vested at nationalization had fallen to 83,744 and the number of Coal Industry Housing Association houses still retained was 24,189. A year later there had been a further reduction of over 4,000 to a total of 103,498.[1]

There were many good reasons for the NCB to reduce their housing activities as soon as there was no overwhelming need to stimulate labour recruitment and transfer by means of direct housing provision. The strongest reasons were the disrepute (not least in the eyes of the NUM) of being a slum landlord, and the high cost. The NCB prepared profit and loss accounts for housing, and down to 1966–7 these showed an operating loss in all but three years, mainly because of heavy costs of maintenance and improvement. But the true financial position was much worse. Houses involve large payments of interests, which easily turn an operating profit into a loss. And the housing account was credited with notional rents, although in many cases the tenants paid much less. The general policy was to fix the rents of the houses built by the Coal Industry Housing Association at the level of the rents for comparable council houses (except where this level was higher than the maximum permitted for a particular estate by the housing subsidy agreement), and the rents of the vested houses at the level permitted by the Rent Acts; but most tenants who were employees, retired, or widows of employees were allowed concessionary rents. In the extreme case, in Northumberland and Durham, under union agreements going back to the years just after the First World War, serving miners were entitled to free housing or to a fixed payment in lieu if their employer did not house them. The housing account was credited with the full rent, and the difference between that and the rent actually paid was charged against the activity in which the tenant was employed, usually the colliery.[2] So, although the housing account showed an operating profit in every year from 1967–8 to 1971–2 (rising to £1.87 million in 1970–1), when interest and the charges to collieries and other activities for free and concessionary houses are taken into account, the true result was a loss of between £4 million and £5 million in each of these years.

The concessionary terms added to the problems of house ownership for the NCB, as they were regarded as one of the benefits of employment, which, if altered, could be a subject of *industrial* dispute. Thus in late 1968 the NCB decided to bring the actual rents paid for vested

[1] NCB *Ann. Rep.* 1970–1, I, 28; 1971–2, I, 21.

[2] Ibid. 1960, II, 52; 1971–2, I, 21.

houses up to the maximum permitted by the Rent Acts, and to do this by means of annual increases of 37½p per week. Similar annual increases were to apply from late 1969 to Coal Industry Housing Association houses, wherever rents were below the council house level. But the outcry was such that, after one increase of 37½p per week, the rents were frozen again.[1]

There were further embarrassments that discouraged the continuation of house provision. For instance, the employer-landlord was liable to get the odium for anomalies which were caused, in fact, by legislation. Thus the rent for identical houses often differed because one was on a controlled tenancy (established before 1957) and one on a regulated tenancy, created since June 1957 and subject to the Rent Act 1965 and the Housing Act 1969. There were also anomalies caused by the fact that only a minority of workers could be provided with a house, and there were difficulties because the houses could not always be made available for employees even when they were wanted. In 1971 only about 20 per cent of NCB industrial employees lived in an NCB house. Of the occupied NCB houses, 32 per cent were occupied by retired workmen and widows and nearly 10 per cent by persons unconnected with the coal industry or the NCB's non-mining activities.

The Housing Finance Act 1972 was a suitable occasion for a reconsideration of housing policy by the NCB. The Act provided for the staged conversion of controlled to regulated tenancies, extended the principle of fixing a fair rent to local authority as well as private dwellings, and shifted eligibility for rent allowances or rebates from the dwelling to the tenant, in both the public and private sectors. So, by copying the newly generalized principles, the NCB could move towards the registration of a fair rent for nearly all their houses and actually charging that rent. Indeed, the transfer of rent rebates by local authorities to the tenant instead of the house meant that subsidies received by the NCB in respect of Coal Industry Housing Association houses were being phased out, so there would be a further financial loss unless there was an offsetting increase in rents. Nevertheless, it was thought prudent, though rent increases were made, to limit them to only a proportion of what

[1] In 1970–1 the cost borne by collieries and other activities for free and concessionary houses was £2,725,000, of which £1,245,000 was attributable to the provision of houses in Northumberland and Durham. The cost of cash allowances in lieu of houses in Northumberland and Durham was £400,000, which was *additional* to all the figures cited. 'Free' housing in Northumberland and Durham meant rate-free as well as rent-free, but the tenants paid the water rate. If their houses had been improved they also paid an improvement charge, either in cash or by reduction of concessionary coal.

could legally have been charged. And the NCB continued to be embarrassed by their association with a large stock of houses, many of which were obviously slummy.

By 1974–5 there had been little essential change in the NCB's policies and position as a housing landlord. The proportion of NCB workers housed by their employer was about the same, but the proportion of NCB tenants who were outside the industry was rising as the proportion of widows and retired miners fell. In March 1974 the NCB still had 90,000 houses, of which 56 per cent were more than 50 years old. There were 3,000 houses which were regarded as unfit and 27,000 others which were incapable of improvement to a good standard with adequate amenities. The NCB would have classed many of the latter as unfit, had it not been that the local authorities did not expect to be able to include them in slum clearance programmes in the foreseeable future. There were 15,000 houses which were agreed to be surplus to operational needs. Disposals tended to be slow because local authorities often delayed a decision about their willingness to buy, and because some NUM local branches were opposed to the sale of NCB houses. The one encouraging sign, however, was that, where houses had been sold to local authorities, colliery employees showed little hostility to the prospective loss of concessionary rents. Apparently, the council was widely accepted as the 'proper' housing authority, and the NUM local councillors, who were numerous in many mining districts, probably thought it better to municipalize houses rather than risk their sale to private owners.

The basic question for the NCB was whether to try to get out of housing altogether. There were sharply divided views about this and discussions went on for more than a year in 1975 and 1976. Most people thought the existing method of identifying surplus houses and seeking to sell them worked too slowly to answer any of the perceived problems. But whereas some would have liked to seek government support for the sale of all NCB houses to local authorities, others thought that such a change of policy might provoke industrial disputes and that the retention of a substantial supply of NCB houses was a necessary aid to the maintenance of adequate mobility of labour.

In July 1976 the question was at last resolved by a decision to dispose of all workmen's houses as rapidly as possible, except for a tiny number essential for operations, such as those occupied by rescue brigadesmen. In fact, the practicable speed was such as not to defeat entirely the preferences of those who wanted to retain some core of housing for the time being; and the supply of labour was becoming less difficult, so that tied houses were a less necessary lure. Under the new policy, houses were first offered to sitting tenants at a price taking account of their security of tenure, i.e. usually about half the estimated vacant possession value of the house. If the sitting tenant did not want to buy, then the house was offered to the local authority at a price negotiated between the NCB estate managers and the district valuers of the Inland Revenue Department. If the local authority chose not to buy, then the house could be offered to another purchaser, with a preference for a registered housing association.[1] In order to facilitate the new policy the ownership of all Coal Industry Housing Association houses was transferred to the NCB in 1977. The Association reported for the last time in 1978, having ceased operations.

When the new policy was adopted the NCB still owned a little over 85,000 houses. By March 1978 the number was down to 76,800. Four years later the figure was 41,000, and in March 1983 it was 36,000. Of the 47,000 sales from the end of 1976 to early 1983, 32,000 were to sitting tenants. The NCB still had a significant financial involvement in housing—rents collected by collieries rose in the early eighties to over £20 million a year—but it was a proportionately less serious drain than it had been and the housing commitment gave rise to less friction.[2] It did not appear that operational needs any longer required the NCB to keep large numbers of houses available, nor was the diminution of their housing role any matter for general regret in the communities around the collieries. Communal life changed greatly, but the change had more to do with levels and security of income and employment, the changing mix of occupations, and the mobility created by car ownership, than with anything that the identity of the housing landlord could determine.

### iv. Safety

Because coalmining was such a hazardous occupation, attention to safety needed to be as pervasive as possible in the operation of the industry. Nationalization made it easier to meet this requirement. It placed a greater obligation for safety on the employer, it created an employer with sufficient resources to be able to contribute to the initiation of improved practices, and it established a breadth of ownership which could accelerate the general adoption of improvements as soon as

[1] NCB *Ann. Rep.* 1976–7, 16.

[2] Ibid.1976–7, 16; 1977–8, 17; 1982–3, 15, 50, and 54. Some rents came from sources other than workmen's houses.

they had been proved. The main responsibility had fallen on the mines inspectors, enforcing the law and suggesting amendments to it, and the colliery managers, on whom statutory duties were placed personally. The employers, of course, had had to appoint properly qualified persons—there were statutory qualifications and duties for other colliery officials in addition to managers—and some colliery companies had given a lot of attention to safety, beyond any formal requirement, in their arrangements for training, supervision, and working practices. But the creation of a statutory obligation for the NCB, in the direction of their policy, to secure the safety of their employees, was new. Other features, too, were new. There was common responsibility for all the central rescue stations. There were thirty-seven of these, which previously had been controlled by thirty-one separate bodies. All the stations and their staffs were transferred to the NCB on vesting date and were placed under the control of the divisional boards. It was soon demonstrated that, in dire emergency, rescue teams could be brought from several parts of the country to act as one body.[1] It was new also that the NCB had a scientific department with functions applying to the whole industry. Once it had had time to get its routine activities well established, it was able to devote some of its resources to safety problems and was helpful in securing the more rapid application of the results of research, wherever it was carried out.

The kind of safety organization which the NCB initiated in 1947 continued without a lot of change in its fundamentals, though with some expansion and many adaptations of detail. This was the first time in the history of the coal industry that specialist safety practitioners had been generally appointed in the collieries. Both the generality of such appointments and the gradual recognition of the need to improve the qualifications of the personnel were important. From the beginning, safety engineers down to area level held the statutory first class certificate of competency which qualified them as mine managers, but the safety officers at collieries were mostly drawn from the underofficials and few had a statutory qualification higher than a deputy's certificate. This began to change in the late fifties, when safety engineers were first appointed at some large collieries and collieries with bad accident records. From 1965 it became standard policy for every colliery with an output of over 500,000 tons a year to have an established post of colliery safety engineer, whose holder should have a first class certificate of competency, with at least one colliery safety officer in addition.[2]

[1] NCB *Ann. Rep.* 1947, 53 and 56.

[2] Collinson and McLintock, *Colliery Guardian*, Nov. 1976.

The location of the safety branch in the Production Department (renamed Mining Department in 1971) of the NCB was sound practice, for the methods of production had to be the basis of safe working. In most respects safety and efficiency went hand in hand. Safer conditions permitted greater continuity of operation. Measures to facilitate the efficient execution of jobs lessened the resort to short cuts and improvisation that might be risky. The introduction of mains lighting underground, for instance, once ways had been found to make it safe and reliable, removed the temptation to disregard safety rules in the effort to see properly what had to be done. Dust control in the interest of health was also a contribution to the lessening of one potential factor in explosion and fire in some circumstances. In addition to day-to-day responsibilities for mining practice, the Production Department of the NCB had a significant role in research. In 1947 it took over the work of the Safety in Mines Research Board on strata control, which has a direct relevance not only to the maintenance of conditions physically stable enough for the driving of roadways and the extraction of coal to continue, but also to the risk of accidents from falls of ground, to the amount of dust produced in mining, and to the emission and accumulation of gases. This work thus bore on safety at critical points, for falls of ground and explosions were major hazards; and the great safeguard against explosions and related fires was to adopt methods of mining, and the monitoring of conditions, that would prevent the coexistence of methane (firedamp) in explosive proportions, accumulated coal dust, and any source of sparking.[1]

The role of the NCB, not only in research but still more in its development and application, remained central to the improvement of standards of safety and probably became more so from the late fifties onward. This was not just a matter of organization, with a concentration on specific problems in the Central Engineering Establishment and, later, in the MRDE. It was a consequence of the drastic changes in mining technology that took place then, and took place without the existing body of legislation offering much guidance about the best way to apply them. It was the engineers responsible for equipment design and for production who had to work out the methods of new technology that combined efficiency with safety, and hope that what seemed

[1] NCB *Ann. Rep.* 1947, 51; Roberts, in *Colliery Guardian*, *National Coal Board*, 1957, 48. Methane is explosive when it forms about $5\frac{1}{2}$ to 13 per cent of the atmosphere and the risk from it can therefore be reduced both by strong ventilation and by draining it away to some point where it can be safely stored for sale or use, or where it can be dispersed in an atmosphere that effectively dilutes it.

the ideal methods would not be frustrated by existing regulations. The significant thing was that from the late fifties there was more and faster innovation than for a very long time past, and the safety aspect was an integral part of the innovations, not something that had to be worked out afterwards and applied by legislative pressure, as had tended to happen in earlier periods. With the thoroughgoing adoption of more complex technology, organization for safety purposes was made both wider ranging and more cohesive. From 1974 responsibility in the NCB for safety, the mines rescue service, ventilation, and dust control was united in a Directorate of Mining Environment within the Mining Department.[1]

There continued to be a major governmental contribution to safety research and the testing of equipment, and the overriding responsibility for safety policy and its enforcement remained with the Mines Inspectorate. The initial arrangements for the Safety in Mines Research and Testing Branch (the former Safety in Mines Research Board) in the Ministry of Fuel and Power do not appear to have been very happy. In 1950, at its own request, it was transferred from the Safety and Health Division to the Chief Scientist of the Ministry and renamed the Safety in Mines Research Establishment. It was a substantial organization with a staff of 260 in 1952, 160 of them in scientific or technical posts. Its research was in three broad categories: engineering (which did not include coalcutting and loading machinery), explosions and allied hazards (which involved a lot of testing of equipment and substances), and health research (which was mainly on dust and its control). The increasing quantity and variety of equipment used underground, and especially the general electrification of collieries, gave added importance to the work of the Safety in Mines Research Establishment, both in research and testing, perhaps above all in relation to flameproofing and the containment of explosions and fires. A close relationship with the Mines Inspectorate gave the Establishment an important function in helping to discover the causes of some accidents and to devise remedies to lessen the risk of their repetition. This gave it an influence on the form and content of new safety regulations and, through them, on some methods of working. There was less in the way of direct and general effect on working practices, as the Establishment did not have a continuing involvement in the daily activities of production, such as

[1] R. B. Dunn, 'Mining Department Organisation and Objectives', *Colliery Guardian*, Oct. 1976. This arrangement, of course, did not and could not detract from the statutory responsibilities placed on colliery management.

determined the emphasis in the NCB's searches for innovations of method; and the Establishment's programme at any one time did not extend across the whole field of mining in an integrated way, but concentrated in detail on a variety of specific projects.

When a revised national system for industrial safety came into operation at the beginning of 1975 the Establishment became part of the Research and Laboratory Services Division of the Health and Safety Executive. It was not so self contained as before but formed the main part of the Safety Engineering Laboratories and the Explosion and Flame Laboratories, as well as contributing to the Occupational Medicine and Hygiene Laboratories. Its functions in relation to coalmining were not greatly changed. It remained responsible for testing and certifying electrical equipment and breathing apparatus and certain other types of plant, equipment, and materials. It also continued to carry out tests required as a result of inspections by the Mines and Quarries Inspectorate.[1]

Though so many different people contributed to the achievement of mining safety, and an increasing initiative came from those involved in production, the last word could come from the mines inspectors.[2] In some ways they soon became able to do a better informed job. At the time of nationalization they were operating through a newly revised organization, which included specialist inspectors for roof control, dust suppression, and mechanical engineering (additions to the earlier specialist electrical and medical inspectors), and which also sub-divided the inspection divisions into smaller districts in each of which a district inspector was given a lot of autonomy. It was thus possible for the inspectors most immediately responsible to act on the basis of closer knowledge of local conditions.[3] This revised organization worked well and was little changed (apart from the addition of responsibility for quarries) until 1967. Then, over a period of seven years, there was a long series of boundary changes, first mainly to match the new area organization of the NCB, and then to merge districts and combine divisions as the workload was diminished by the great reduction in the number of mines. In the end divisions were abolished and a reduced number of

[1] PRO, POWE 10/395; SMRE, *An Introduction to the Safety in Mines Research Establishment* (1952 edn. and 1957 edn.); Roberts, in *Colliery Guardian*, *National Coal Board*, 1957, 46; Health and Safety Executive, *Health and Safety: Mines*, 1980, 7; *Guide to the Coalfields*, 1981, 55.

[2] For a summary of the related roles and duties of the Minister (making regulations), the inspectors (enforcing them) and colliery managers (complying with them) see NCB *Ann. Rep.* 1947, 50.

[3] Roberts, in *Colliery Guardian*, *National Coal Board*, 1957, 45.

districts all reported direct to national headquarters. In 1968 there was also the innovation of adding specialist civil engineering inspectors to the establishment. This was a response to the Aberfan disaster and the legislation which followed it. For the same reason district inspectors were sent on courses of instruction about tips and lagoons. After 1974 the organization entered another period of stability, but now within the new Health and Safety Executive.[1]

The inspectors were able to exercise a strong influence on safety in a variety of ways besides the formal application of regulations and their suggestions for amendments of them. Much depended on the informal discussions with colliery management which accompanied their daily work. This was a two-way process, for the inspectors were not just imparting their experience and knowledge of the regulations, they were having to react to the use of new equipment and to the ideas of the men on the spot about the best way to conduct certain operations in particular local circumstances. The inspectors also joined with management and union representatives in the safety committee of the Coal Industry National Consultative Council and in the conduct of safety competitions.[2] The requirements of the inspectorate, the mining acts and regulations also had a continuing influence on the education of mining engineers.

Yet in some respects, arising particularly from the timing of legislation, the influence of the inspectorate may have become less valuable than it could have been. At the time of nationalization the main controlling statute on management and safety in coal mines was the Coal Mines Act 1911. This was not replaced by a new consolidating measure until the Mines and Quarries Act 1954, which was brought into operation on 1 January 1957 and was obviously a stimulating influence, for a good deal of action had to be taken in the mines to conform with the more stringent safety requirements it imposed.[3] But in important ways the new measure was unfortunately both late and early. Much of its content, and the thinking behind it, came from the report of the Royal Commission which had sat from 1935 to 1938 and whose recommendations had been pushed aside by the preoccupations of war and nationalization. And the Act came just before the great spurt of mechanization which

[1] The changing structure and staffing of the inspectorate can be followed in the published annual reports of HM Chief Inspector of Mines and Quarries or in the successive annual editions of *Guide to the Coalfields*.

[2] Collinson and McLintock, *Colliery Guardian*, Nov. 1976.

[3] NCB *Ann. Rep.* 1957, I, 51.

transformed mining methods and many of the physical conditions of work in mines. So the Act was, to a considerable degree, an instrument for regulating the conditions of a mining age that was passing. Yet, although there was, from time to time, supplementary legislation on limited aspects, the Mines and Quarries Act continued in force, and thirty years later there were still no steps to prepare for its replacement. This was an unusually long period in the history of mining legislation and was specially inappropriate because it was a time of exceptionally comprehensive technological change. It meant that the mining industry in the seventies and eighties had to operate within a legislative framework that was, in important ways, obsolescent.

A good deal of adaptation was achieved. Much of the Act was concerned to lay down principles that must be observed and to provide that their detailed application should depend on subsequent regulations. The specific details could change as conditions changed without necessarily involving a departure from principles, but no principle can be stretched indefinitely. Where it was impossible to match regulations to the nature of new equipment or techniques there was discretion for inspectors to prescribe what might be done as an authorized departure from statutory regulations for a stated period. There had to be a lot of resort to this device, which was unsatisfactory as a means of dealing with practices and equipment that had been established as sound by careful testing and prolonged experience. Standards of safety were not weakened by the obsolescence of the legislation. Compromises erred on the side of caution and did not always take much account of what was operationally preferred. Innovation in mining practices should be applied in ways that seek to combine optimal safety standards with optimal operating efficiency, but standard regulations within the existing statute sometimes made that impracticable, to the detriment of efficiency, and they caused the diversion of inspectors' and mining engineers' time and effort into too frequent consideration of cases for special exceptions and their renewal.

There was, however, one change in law and administration which probably most people in the coal industry would rather not have had. This was the transfer of the mines inspectorate and of responsibility for safety and health legislation from the Department of Energy (which had to deal with all aspects of the coal industry as one of its chief concerns) to the new Health and Safety Executive (which had mining matters as only a small part of a field of responsibility which covered occupational health and safety everywhere). There was a feeling that a high proportion of the safety requirements of mining are very different from those of manufacture or commerce, and that people in the upper levels of the Health and Safety Executive were neither sufficiently aware of the extent of the differences nor sufficiently interested in them.

There were several episodes which infuriated the NCB and some of the unions. Among the most serious was the decision of the Executive, early in 1976, to bring prosecutions against members of BACM and NACODS, but not against a member of the NUM, who were criticized by a Court of Enquiry into an accident at Houghton Main Colliery. The result of this decision and related actions by the Health and Safety Executive was a policy of non-co-operation in certain important respects by members of NACODS, at first in Yorkshire and then, for a time, nationally, with appreciable losses of output. Members of NACODS were incensed both by what they regarded as the unfairness of the prosecutions and by the Executive's interpretation of regulations in such a way as to make a deputy statutorily responsible for things that happened when he was legitimately absent. Everybody was upset that, contrary to precedent, the Executive used as prosecution evidence the proceedings of the previous Court of Enquiry at which, in accordance with long-standing agreed practice, no one had been legally represented. It did not make the Executive any better respected that this affair ended with the quashing of most of the convictions on appeal. There were also complaints by the NCB of attempts to introduce 'across the board' regulations without regard to their implications for coalmining. Protests were aroused, too, by erroneous and detrimental press briefings which the Executive gave when statistics on health and safety in the coal industry were published. Late in 1977 it was privately indicated to the Department of Employment that, if there were any more errors of this kind, the NCB would make their concern public.

After this, the worst of the irritations seem to have been somewhat reduced. But two years later there were still many worries. The NCB found it necessary to contradict a press statement by the Health and Safety Executive about an accident at Bentley Colliery, and representatives of the Coal Industry National Consultative Council met the chairman of the Health and Safety Executive about the same subject. The NCB were complaining that the Executive's preoccupation with regulations which applied generally to industry delayed the making of urgently needed new regulations on, among other things, ventilation and explosives in mines. They asked for the Mines Inspectorate to be allowed more autonomy within the Health and Safety Executive. It was a state of affairs that made remedies for the obsolescence of mining legislation harder to achieve.

But whatever flaws there may have been in the legislation and administration, great improvements in safety were achieved and maintained. There were improvements in the training of mineworkers. There were frequent safety campaigns, with the form varied from year to year. There were safety competitions: in the early eighties collieries were grouped in ten divisions of a 'safety league' with promotion and relegation between divisions. There were constant reminders from new devices. Perhaps this was the most important influence of the development of the self-rescuer between 1965 and 1970 and its eventual ubiquity. This was a small piece of apparatus to be attached to an underground worker's belt. It enabled him to breathe properly in an atmosphere contaminated by noxious gases, especially carbon monoxide, and to sound an alarm if he was trapped. It became universally available by early 1972, and from October 1973 it became a condition of service, agreed with the NUM, that every mineworker should continue to carry the self-rescuer on his person all the time he was underground.[1] It was one perpetual reminder of safety precautions nobody could forget. But the greatest influence was always the working out of improved mining practices in which safety was an integral consideration.

The results showed in the statistics of accidents. The figures, especially of fatalities, for any year could be distorted by the irregular timing of major disasters. This kind of distortion became much rarer. In the first thirty-five years of nationalization there were fifteen disasters which each involved at least 10 deaths. Seven of these disasters, accounting for 326 deaths (almost exactly half the death roll for the fifteen accidents) occurred in the first five years.[2] After 1960, apart from the appalling tipslide on to the village of Aberfan,[3] there

[1] Progress with the self-rescuer is described at some point in each annual report of HM Chief Inspector of Mines from 1965 to 1971 inclusive. See also NCB *Ann. Rep.* 1970–1, I, 11; 1971–2, I, 8 and 51. At collieries where self-rescuers were provided it was from 1968 a condition of service to keep them readily available at the place of work. The change in 1973 was to compel them to be carried on the person at all times underground.

[2] Four of the disasters, all from explosions, occurred in 1947, in the first nine months of the nationalized industry. They included the explosion at William Pit, Whitehaven, which killed 104, the largest death toll in any mining disaster (apart from Aberfan) under the NCB.

[3] Chap. 6, section vi above for an account of this disaster.

were only four such large disasters, with a total of 75 fatalities. Most of the large disasters were caused by explosions, and the great improvement was the prevention of the build-up of the combination of conditions liable to cause explosions, and the development of measures to contain the spread of fire if those precautions failed. So great was the improvement that the statistics were exposed to a different source of distortion. The average incidence of fatal accidents was sufficiently reduced for quite small random variations to produce peculiar figures for individual years. But over any period of several years, the trends were clear.

In every year down to 1952 there were over 400 fatalities, including 612 in 1947, but thereafter 400 was never reached again except in 1955, when there were 408. After 1960 the total never again reached 300; and 1965–6, with 217, was the last year when the figure was over 150. The total fell below 100 in 1969–70 and below 50 in 1976–7. It remained below 50 except in 1978–9 when there were 72 deaths. A clearer indication is given by the rate of fatalities per 100,000 manshifts worked. This figure was 0.34 in 1947, but otherwise exceeded 0.25 only in 1950 and 1951. With these exceptions the rate was usually between 0.20 and 0.25 down to 1965–6, though in four years it was slightly lower. In 1966–7 it dropped sharply to 0.15 and never again exceeded that level. It fell below 0.10 for the first time in 1976–7 and remained there except in 1978–9.

The trend in the more serious non-fatal accidents is much more difficult to establish because of changes in 1959 and 1981 in the definition of what had to be reported in this category. The 1959 change ought to have caused a reduction in the figures, because minor reportable accidents were thenceforward excluded from the category. But as no such trend appears in the figures, the true incidence of serious accidents must have been rising, and this is suggested by other evidence. The rate of casualties had been highest from 1947 to 1949, at around 1.3 per 100,000 manshifts, but, after running significantly lower, was rising again in the late fifties, with a maximum in 1959. There was a gradual downward tendency in the late sixties, and in the seventies the rate fluctuated around 1 per 100,000 manshifts, with a minimum of 0.87 in 1979–80. The new definition in 1981 increased the figures so much, by including for the first time everyone who spent 24 hours in hospital as a result of an accident, that comparisons with previous years are impracticable.

Most accidental injuries were not serious, but all that caused absences of three days or more were recorded. When combined with the statistics of fatal and serious, non-fatal accidents, they provide an indication of of the general trend of safety. It is not a completely objective record because the figures are influenced by the readiness with which individuals seek to get back to work, and that in turn is influenced by the provision of income during time off work. But the statistics should give an acceptable approximation, in which large changes are clearly indicative of real trends. The low figures for 1947 and 1948 may be attributable, at least in part, to differences in the conditions of national insurance payments. Except in those years there were usually around 130 casualties per 100,000 manshifts until a sharp rise set in from 1958. The rate rose rapidly to 210.3 per 100,000 manshifts in 1965–6. After that it fell in each succeeding year. It dropped below 150 in 1970–1, was down to 100 in 1976–7, and continued a sharp decline to reach 62.5 in 1981–2 and 52.8 in 1982–3.[1]

These figures do not suggest that everything was becoming easy. It should be remembered that an accident rate of 100 per 100,000 manshifts means that, at a colliery with 1,000 men (not an enormous size), on average on every working day someone is sufficiently injured to be off work for at least three days. Nevertheless, the statistics show that coalmining, which had been among the most hazardous of all occupations, had become very much safer, and that, in particular, the risk of being killed at work had been greatly reduced. Most of the improvement dated from the later sixties, i.e. from the time when modern mechanized methods had become general and familiar, and many of the most antiquated pits had been closed; and there was a further improvement in the later seventies.

The improvement was by no means evenly shared. In the first quarter of 1982 the accident rate per 100,000 manshifts at individual collieries varied from 12.3 to 134.7. In the early eighties the rate was generally better in Western Area than elsewhere, and worse in South Wales, South Yorkshire, and North East Area.[2] If the number of workdays lost (as a result of accidents), rather than the number of casualties, is related to 100,000 manshifts, the contrast becomes even sharper, though differently located. On this basis the rate for South Midlands Area (the best) was less than half that for South Wales (much the worst). But a rapid rate of improvement over the levels of the very recent past was

[1] NCB *Ann. Reps.* give figures. *Ann. Rep.* 1982–3, 20–1 summarizes the fatal and serious non-fatal accident statistics from 1970–1 onwards and at five-year intervals before that. For complete annual figures see Appendix 1, Table A2 below. All accident statistics for 1947 and 1948 include licensed mines, but thereafter the returns from them were separated, and all the later figures cited relate to NCB mines only.

[2] NCB *Ann. Rep.* 1982–3, 24–5.

general, and was by no means least in those areas, including South Wales, which obviously had most need of it. There were, too, some worries about the incidence of accidents in particular kinds of work. While the risk of large explosions, which in the past had often had such harrowing and devastating effects, had greatly decreased, and while there was a continuing decline and low level of serious accidents at the coal face, there was more difficulty in keeping down accidents in underground transport and haulage. In the late seventies there were several serious accidents of this kind which it was alarming should have happened at all. It was these activities which now persistently produced the largest number of fatal and serious casualties.[1] Yet even these accidents showed up so sharply because of the great general improvement in standards. They showed how much room for improvement there still was, and where extra effort needed to be directed. But they only slightly modified the achievement of the preceding fifteen years in making coalmining a much safer occupation.

## v. Health

There were two main types of service to be offered within the industry to improve the health of mineworkers. One was the provision of staff and facilities on the spot to deal with the immediate effects of illness and injury at work, and also to check the health of new recruits. The other concerned the main occupational diseases of coalmining. It was necessary to carry out research to establish their incidence and to seek their causes and characteristics, as a guide to the alteration of working methods and the working environment in such ways as would help to reduce the incidence of the diseases and, as far as possible, eliminate them. It was also necessary to provide for the observation of the progress of the diseases in those who had contracted them, and to adjust the type and location of their work accordingly.

The early difficulties of building up a day-to-day medical service—the smallness of the foundation laid by the colliery companies, the uncertain division of function between the Ministry and the NCB, the cessation of new plans while the Dale Committee was sitting from 1949 to 1951—have already been touched on. But there was a clear appreciation of

[1] NCB *Ann. Rep.* 1976–7, 14–15; 1977–8, 17–18; 1978–9, 20; 1982–3, 14 and 24–5. The achievement of greater safety below ground encouraged additional efforts to reduce danger on the surface, where increases in the amount of mobile plant and heavy machinery had created some new hazards.

what was needed and a willingness to provide it; and from 1951, when the earlier uncertainties were removed, there was rapid progress. From the early thirties there had been a statutory duty to provide first aid rooms on the surface at collieries, and a few of these were provided with a little more than the basic minimum of facilities and attendance, so that they were able to serve as a model for what the NCB developed into simple 'medical units', which became standard at smaller collieries. A very few colliery companies had established medical treatment centres, each of which was in charge of a state registered nurse. The pattern which they introduced consisted of a reception room, treatment room, consulting room, and recovery room for serious accident cases, and this remained standard, though in later years baths or showers, for patients to use before going to hospital, and physiotherapy facilities were added. But immediately before nationalization the coal industry had only eight medical centres and twenty-two state registered nurses. The Ministry of Fuel and Power, however, had plans to build 200 medical centres. During 1947 the NCB agreed to take over these plans, but until 1951 progress could be only slow. In 1950 there were still only fifty-two medical centres and ninety-eight state registered nurses. But the original target of 200 medical centres was reached in 1953, and more and more were built until a maximum of 373 was reached in 1960. The number of medical units increased even more and was 473 in 1960, some collieries having a medical unit as well as a medical centre. After this the closure of collieries caused a very great reduction in the number of medical units, but the number of medical centres declined much more slowly,[1] as it was standard policy for all but the smallest collieries to have a medical centre. The employment of state registered nurses, which reached a maximum of 379 in 1958, also fell rather quickly after 1963. After dropping to 156 in 1975 the number was kept around 170 in subsequent years.

Responsibility for the organization of treatment services rested with the medical officers, who were also responsible for advising colliery managers about first aid and morphia supplies underground, for administering the Pneumoconiosis Supervision Scheme, and for holding consultations with men found to be suffering from any industrial disease. Colliery treatment centres never normally gave more than initial treatment, and it was the responsibility of the medical staff to ensure effective co-ordination with hospitals, family doctors, and the social services in relation to both the treatment and the rehabilitation of

[1] For example, between 1960 and 1962 the number of medical units was reduced by 205, but the number of medical centres by only 25.

sick or injured workers. Most of the NCB doctors were deployed at divisions and areas (from 1967 at areas), whereas the nurses were at collieries. The number of doctors was built up from seven in 1947 to eighty in 1957, after which the establishment contracted along with the rest of the industry. After 1973 the total number was kept between thirty and forty.[1]

Apart from giving initial treatment, the main routine function of the medical service was the examination of new recruits. At the time of nationalization the statutory position was that all new entrants under 18 must be medically examined, and in South Wales, because of the prevalence of pneumoconiosis, all entrants of any age must be both medically examined and given a chest x-ray. The examinations were usually carried out by doctors employed by the Ministry of Labour and National Service. Very few employers had ever arranged for medical examinations not statutorily required. The statutory medical examinations were taken over by the NCB medical service in 1952, when the statutory age limit was raised from 18 to 21; and the radiological examination of new entrants in South Wales also passed to the NCB in 1953. Although there was no statutory requirement, the NCB quickly developed the practice of examining adult recruits. By 1956, 85 per cent of adult entrants were examined by NCB doctors. The rest were those recruited from non-mining towns and, by arrangement, they were examined by local practitioners in accordance with NCB medical standards. Juvenile recruits to NCB non-mining industries, who were required by the Factory Acts to undergo medical examination, were also examined by NCB doctors. Some risks of individuals being placed in particular jobs for which they were not physically fit were thus avoided. Some special classes of worker were also regularly re-examined. For instance, there were annual medical examinations for all members of mines rescue brigades and for some workers in ancillary industries who were exposed to risks of skin cancer.[2]

The effects and value of all these routine medical activities[3] are impossible to measure, but, having got them, nobody wanted to be without them, and the extent of their voluntary use is some indication of the value set on them. The other main medical function, the control and

[1] Rogan, in *Colliery Guardian*, *National Coal Board*, 1957, 95–8; Collinson and McLintock, *Colliery Guardian*, Nov. 1976. Statistics were given from time to time, but irregularly, in NCB *Ann. Reps.* and in NCB Medical Service, *Ann. Reps.*, which began in 1957.

[2] Rogan, in *Colliery Guardian*, *National Coal Board*, 1957, 95–6.

[3] The specific functions of the NCB medical service are defined and listed in NCB Medical Service, *Ann. Rep.* 1957, 16–17.

reduction of occupational diseases, is more susceptible of measurement, though some of the statistics need a lot of care in interpretation.

The principal occupational diseases of coalmining have been dermatitis, the beat conditions of elbow, hand, and knee (with which tenosynovitis may also be grouped), nystagmus, and coalworker's pneumoconiosis. There were a few others with a much lower incidence, of which Weil's disease was the most serious.

Two of these diseases were fairly readily overcome by changes in the mining environment associated with technical conditions rather than by medical measures. Weil's disease, which is usually characterized by jaundice and attacks the muscles, sometimes with fatal consequences, is caused by an infective organism carried in the urine of rats. It occurred from time to time among miners but only irregularly. There was an outbreak in Scotland in 1951 and 1952 when twenty-three miners caught the disease and nine of them died. There were occasional cases later but nothing as serious again. The obvious remedy was to get rid of rats. Besides various rat extermination campaigns in the fifties and sixties, two other factors contributed. One was that rats were commonest in drift mines and fairly shallow pits, and these provided a decreasing proportion of the coal. The other was that rats found it congenial to share the well-strawed environment provided for ponies, and the removal of ponies from nearly all collieries destroyed this favoured habitat. So the rats had a hard time of it and by the seventies Weil's disease was virtually unknown among miners.

Nystagmus is an affliction of the eyes characterized by involuntary oscillation of the eyeball. At one time it was fashionable to seek explanations in the psychological condition of the sufferer, but poor lighting was almost certainly the cause, as indicated by both circumstantial and experimental evidence. When cap lamps and underground mains lighting came into regular use, nystagmus began to disappear. It is unlikely that all cases were ever medically certified, but the drop in the medical statistics indicates a real change. The number of new spells of benefit each year on account of nystagmus averaged 178 in the early fifties, fell below 50 from 1962, below 10 from 1970, and was so low as not to be separately recorded after 1974. Even by 1954 the rate of new cases was only about one-eighth of what it had been in the late thirties.[1] The rate per 100,000 manshifts worked fell from an average of 0.10 between 1951 and 1954 to 0.01 by 1968.

[1] Rogan, in *Colliery Guardian*, *National Coal Board*, 1957, 99–100; Collinson and McLintock, *Colliery Guardian*, Nov. 1976; NCB Medical Service, *Ann. Rep.* 1971–2, 22–3.

The other occupational diseases presented rather more difficulty, very much more in the case of pneumoconiosis. They called for more research, both into the characteristics and treatment of diseases and (by observation and statistical analysis) into the incidence of particular medical conditions in different mining environments and in association with different working arrangements. They needed also the adjustment of mining technology to take account of the results obtained. More of the research was done within the NCB organization than anywhere else, and the results were presented to professional notice by publication.[1] The research facilities were augmented in 1969 by the opening of the Institute of Occupational Medicine. This Institute, though an independent charitable foundation, was established on the initiative of the NCB and gave priority to the health problems of coalmining. It was financed mainly by the NCB, but it was given associate status by the University of Edinburgh and undertook work on contract for other industries. In the mid-seventies it was receiving support from the BSC and the Asbestosis Research Council.[2] Pneumoconiosis, as the widest problem, was investigated in many ways by different bodies. Although from the late fifties onward the NCB were the most heavily involved organization in the UK, the Medical Research Council conducted and supported field research over a long period, and one branch of the Safety in Mines Research Establishment was continuously involved through investigation of the effects of dust and the improvement of dust control. Pneumoconiosis was also a matter of international concern and enquiry. There was comparison and discussion of research and, with the encouragement of the International Labour Office, there was the establishment and improvement of an international radiological classification, to which the results of observations could be related. But efforts to understand, and measures to deal with, the characteristics of pneumoconiosis in the forms specifically associated with coalworking were heavily dependent on British investigations.

Though it was somewhat delayed, there was most success in dealing with the beat diseases. These were characterized by inflammation affecting the tissues near to certain joints, especially the knee, elbow, and hand. They were attributed to infections entering tissues which had

[1] The NCB published *Reports on Research on Medical and Allied Problems* from 1952, five years before their Medical Service started to publish Annual Reports. The two reports were combined from 1962 to 1970–1, after which the Institute of Occupational Medicine began to report on research.

[2] Collinson and McLintock, *Colliery Guardian*, Nov. 1976; NCB Medical Service, *Ann. Rep.* 1968–9, i.

been weakened or damaged by continual pressure or by continual use in striking heavy blows. Beat knee was particularly prevalent, and down to the mid-fifties showed no sign of reduction, a state of affairs attributed to the large proportion of coal seams of a thickness which required face-workers to do everything in a kneeling position.[1] The original ideas worked well for some beat conditions but not for others. Beat hand had almost disappeared by 1970 because abrasions were much reduced by the use of protective gloves and by the great decline in pick and shovel work at the coal face, and because infections were promptly treated with antibiotics.[2] But in 1961 it was concluded that infection was of slight importance in beat knee. The effects of pressure were what mattered and research was redirected accordingly.[3] It was the long and eventually successful search for a design of knee pad that provided an effective cushion against heavy pressure which brought down the incidence of beat knee. Mechanization of coal getting also helped by superseding many of the manual operations that had to be done kneeling. After 1963 there was a continuous and rapid decline in cases of beat knee per 100,000 manshifts: the rate fell from 4.55 in 1963 to 1.88 in 1971. The rate thereafter continued a general decline, but more slowly and erratically.

Beat elbow probably depended most for improvement on a reduction in the number of forceful percussive movements made with a bending of the arm. After a sharply reduced incidence in the mid-fifties, there was no more sustained improvement before 1963, after which the rate per 100,000 manshifts was approximately halved in eight years, to just over 0.25. Tenosynovitis, a painful condition of the tendons and tendon sheaths just above the wrist, attributed to the arduous use of the forearm muscles (especially in non-habitual movements), had been less common, but declined later and more slowly. This may have been because it was more likely to arise in heavy work in roadways or in moving equipment than in coalcutting, and there was not so much early lightening of this sort of work by mechanization. But a reduction set in during the late sixties. The rate of 0.25 per 100,000 manshifts in 1971 was still just over half that of twenty years earlier. The gradually extending influence of mechanization continued the downward trend, especially in the late seventies. The combined incidence of beat elbow and tenosynovitis was halved between 1976 and 1980, when it was about one-third that

[1] Rogan, in *Colliery Guardian*, *National Coal Board*, 1957, 99.

[2] NCB Medical Service, *Ann. Rep.* 1971–2, 28.

[3] Ibid. 1961, 14.

of beat knee, even though that was now low enough not to be a serious problem.[1]

Dermatitis remained about the most obstinate of the occupational diseases. This was probably because there were so many different substances that could irritate, so many sources of friction and warmth that could keep them adhering to the skin. It may also have been partly because dermatitis did not always seem a serious enough complaint for a sufferer to take adequate precautions. For a long time the recorded incidence showed no signs of falling and, indeed, reached a peak as late as 1969, and in subsequent years it was well above that for beat knee. Thereafter, better ventilation, reductions in dust, better designed workwear must all have helped. On the other hand, there were more numerous chemicals and oils used underground that were suspected of sometimes having bad effects on the skin, and cement-based materials used in strata control were known to be a source of skin problems. Particularly after 1972 the statistics show quite a sharp reduction. The number of new spells of benefit per thousand employees fell from 3.41 in 1976 to 2.22 in 1980. But there were strong suspicions that the problem was more widespread. The statistics record only men who received national insurance benefit and were off work for more than three days, whereas many sufferers from dermatitis simply had first aid medication and carried on.[2] Dermatitis is not a disabling disease but could cause intense irritation that was bad both for equanimity and for quality of work. There remained a need for careful and perhaps increased attention to hygiene and for prescriptive rules about methods of handling potentially irritant substances.

By far the most serious of the occupational diseases of coalmining was pneumoconiosis, a condition of the lungs caused by the retention of dust as a result of working in a dusty atmosphere for a very long period. It has both simple and complicated forms, the latter (which develops only after the simple form is present) known also as progressive massive fibrosis (PMF), each form subdivided into categories according to the interpretation of radiological evidence. The method of classification,

[1] NCB Medical Service, *Ann. Rep.* 1971–2, 24–9; 1980–1, 24. The incidence was calculated by counting the number of new spells for which national insurance benefit was paid. Until 1974 the rate was expressed in proportion to 100,000 manshifts worked, but thereafter it was in proportion to 1,000 men employed. Thus the figures of the rate were roughly double those on the earlier basis. On this new basis the rates fell between 1976 and 1980 from 1.81 to 1.16 for beat knee and from 0.79 to 0.39 for beat elbow and tenosynovitis together.

[2] Ibid. 1971–2, 30–1; 1980–1, 23–4; Rogan, in *Colliery Guardian*, *National Coal Board*, 1957, 100; Collinson and McLintock, *Colliery Guardian*, Nov. 1976.

first adopted in 1953, was refined and extended from time to time, in 1958, 1968, and 1971, under the auspices of the International Labour Office. The NCB, for their own work, further subdivided the classification of 1968 in order to indicate more accurately the degree of progression within the three categories of simple pneumoconiosis. Category 1 of simple pneumoconiosis, which showed the first signs of dust retention, with a few very tiny round shadows on the x-ray film, was not considered a disabling condition in itself, but the subject needed to be watched to see whether the condition progressed at all. Official certification of pneumoconiosis and assessment of disability was done independently by panels of doctors employed by the Ministry of Health and Social Security. It was NCB practice to refer to the panels anyone who had progressed into or near to category 2 of simple pneumoconiosis, with more or slightly larger round shadows on the x-ray film, for it was at this stage that the panels were most likely to certify a man as having pneumoconiosis. Anyone who was so certified, or who, though not certified, was judged by the NCB doctors to be comparably affected, was normally moved, unless he objected despite advice, to work where he would be less exposed to dust. Such a change of environment could help to stop the progress of the condition and preserve a standard of health in which the deterioration of lung function was not greatly accelerated nor life expectation much reduced, though these results were not always achieved.

Complicated pneumoconiosis, in which confluent shadows more than 10 millimetres in diameter appeared and there was consolidation of fibrous tissue, was very serious and could become badly disabling. In some sufferers PMF did not progress, in others it did so persistently. Even in the seventies nobody knew why there was the difference, and if PMF was going to progress in a particular sufferer, nobody knew any way to stop it, though there was some evidence that an abnormally high proportion of quartz in coalmine dust was a factor in the causation of PMF.[1] But this could not be a general explanation as so few mines had much quartz in their dust.

In the forties, though there was much concern about pneumoconiosis, the extent of its incidence was unknown and there were only very imprecise ideas about the influence of different types and quantities of dust and different periods of exposure to dust hazards as factors in its

[1] NCB Medical Service, *Ann. Rep.* 1974–5, 5, and 1975–6, 5–6 give clear accounts of the criteria and methods of classification of pneumoconiosis. *Ann. Rep.* 1980–1, 9 for the possible influence of quartz in dust.

causation and progression. It was known to be a serious problem in South Wales, but its extent elsewhere was almost certainly underestimated. There were some indications that the pre-war introduction of mechanical coal cutters had been associated with some increase in cases of pneumoconiosis, and therefore there was particular concern to suppress more of the dust created by new equipment. Perhaps because attention was usually drawn to the disease only after it had progressed some way in a sufferer, a gloomy view was taken of its physical effects, and until 1948 men who were certified to have pneumoconiosis were required to leave the coal industry. This not infrequently caused them to become unemployed, so there was some disincentive to seek certification. In 1948 it became permissible for such men to remain in the industry, provided they were employed in dust approved or dust free conditions.[1]

There was clearly a need to acquire and apply much more knowledge and give workers better protection. The Ministry of Fuel and Power in 1947 set up a National Pneumoconiosis Joint Committee, with Sir Charles Ellis, the board member for science, as NCB representative. The Joint Committee came up with the suggestion that the medical examination of all mineworkers could be a means of prevention and control of pneumoconiosis. The NCB agreed that this would be helpful, at least to the extent of having initial examinations, though periodic examinations could be introduced only gradually. But they argued that the work and the cost should be the responsibility mainly of the National Health Service, not the NCB.[2] So there was a long delay before the suggestion was acted on. It was not until 1956 that the NCB decided that, as a matter of policy, a chest x-ray to establish whether or not there was pneumoconiosis should be offered to every mineworker, and it was 1958 and 1959 before the steps to put that policy into practice could be taken.[3]

In the meantime, research on pneumoconiosis and attempts to reduce levels of dust in mines continued. The most notable was the Pneumoconiosis Field Research begun by the NCB in 1952 and continued for many years thereafter. Its aim was defined thus: 'to determine how much and what kinds of dust cause pneumoconiosis and to establish what environmental conditions should be maintained if mineworkers

[1] Rogan, in *Colliery Guardian*, *National Coal Board*, 1957, 98–9.

[2] NCB 106th meeting, min. 1327, 4.7.47; 169th meeting, min. 117, 20.1.48; 234th meeting, min. 1124, 8.10.48; 235th meeting, min. 1138, 15.10.48.

[3] NCB Medical Service, *Ann. Rep.* 1957, 2; 1958, 3.

are not to be disabled by the dust they breathe'. It was carried out at twenty-fve collieries with a combined labour force of about 30,000. They were selected from all the main coalfields, and the effort was made to constitute a representative sample of underground conditions in British collieries as a whole. The plan was to build up a cumulative record of dust exposure for each man employed at these collieries, x-ray every man at regular intervals, and try to demonstrate the relationship between dust exposure and the development of pneumoconiosis.[1] The task was extremely difficult. Much of the evidence must be statistical and the problems of interpreting pneumoconiosis statistics are formidable. The individual membership of the population examined undergoes continual change to an appreciable degree, and individuals move between environments of differing dust characteristics for periods of varying length; the age structure of the whole population changes over time and the age structures of the populations exposed to different types of dust conditions cannot be accurately standardized; and pneumoconiosis is related to cumulative influences of dust exposure over a long period, so a change in the incidence or progression of pneumoconiosis cannot be clearly related to a particular change in dust exposure, except after many years. It is no help to look only at the evidence from the condition of those who entered after a particular change in dust conditions began, for, whatever the dust conditions, it is rare for anyone under the age of 35 to show any indications of pneumoconiosis. Results cannot be strongly suggestive until mineworkers have been employed for a good many years and grown older.

So the research was necessarily slow in becoming fruitful. Nevertheless there were some results which suggested the need for new approaches to the problems. Medical Research Council investigations in South Wales in the thirties had indicated that the prevalence of pneumoconiosis varied with the number of particles of respirable dust in a given volume of air. The only practicable way of measuring the amount of respirable dust was by using a standard thermal precipitator, which enabled a count, under the microscope, of particles in the range 1 to 5 microns. The standards for acceptable dust levels in the coal industry were also based on a count of particles in the 1 to 5 micron range. (A micron is a thousandth of a millimetre.) Yet an increasing amount of the evidence seemed not to fit the previous assumptions relating pneumoconiosis to the quantity of dust. Clearer results did not emerge until in

[1] The best summary of this prolonged research is Jacobsen, Rae, Walton, and Rogan 1970, which was reprinted as a supplement to NCB Medical Service, *Ann. Rep.* 1969–70, 30–6.

the mid-sixties it became possible to design effective elutriators which would screen out the coarse non-respirable particles when a sample of the atmosphere was taken. All that then needed to be done was to weigh and analyse all the dust in the sample. This new method of gravimetric sampling was thus much simpler to operate. It revealed that, if two samples contained identical numbers of respirable particles, the dust in one could have a mass up to five times that of the dust in the other. Re-examination of the research results showed that if the environmental hazard were measured by mass of dust instead of by particle count there was a much better correlation with the radiological response. So from 1969 all regular monitoring of dust in mines was done by gravimetric methods, and dust standards were set in terms of milligrams per cubic metre. From that time onwards it began to appear that there was a firm grip on the problem and that there was unmistakable progress in reducing pneumoconiosis. It was a triumph of improved instrumentation. The gravimetric sampler was invented in the course of the NCB researches and was sold and adopted throughout the world.

In 1958 the NCB offered all their mineworkers a radiological examination for pneumoconiosis, to be repeated every five years. The examinations began early in 1959, and it took between four and five years to complete examinations at every colliery. In 1974, when the number of men had been greatly reduced, it was decided to increase the frequency of examination to every four years. There was no compulsion, but in the first round just over 85 per cent of those eligible elected to be examined,[1] and the response remained around that sort of level. In 1980 the response was nearly 90 per cent.

This comprehensive programme of examinations had several useful results. It provided a lot more data and gradually accumulated enough evidence to make it less uncertain whether changes in conditions and methods were reducing the incidence and progression of pneumoconiosis. It also identified the men who were most affected and most at risk, and made it much easier to look after them. For many years the NCB medical service supervised the health and dust exposure of known certified pneumoconiosis cases working at collieries. These arrangements were made formal in 1975 by the Coal Mines (Respirable Dust) Regulations. Miners already certified and those uncertified who were considered by the NCB medical service to have category 2 pneumoconiosis or worse, or who were under 35 and showed evidence of

[1] The response and results of the first round of examinations are summarized in NCB *Ann. Rep.* 1963–4, 38–9.

category 1 pneumoconiosis, or who showed rapid progression of the disease between two successive examinations, were offered clinical examinations, lung function tests, and chest x-rays every two years, and the results were sent to their general practitioners. Miners under the scheme were recommended to take jobs in low-dust conditions (less than half the recommended maximum for coal faces). Allowances from the Department of Health and Social Security were available to help compensate for loss of earnings as a result of a change of job in these circumstances. The whole scheme was voluntary but most of the men concerned made use of it and acted on the recommendations.[1]

Long-term comparisons from both the field research and the periodic medical examinations are limited by colliery closures and mergers, changes in the population of the sample, and changes in the age of retirement. But comparable results can be given for fifty-three collieries, drawn from all areas, and surveyed in 1962, 1967, 1972, 1976, and 1980. Some of the principal results are shown in Table 10.1.

Table 10.1. *Prevalence of pneumoconiosis at 53 collieries, 1962-1980* (percentages of men x-rayed)

| Year | All categories | | Category 2 or more | | PMF | |
|---|---|---|---|---|---|---|
| | All ages | 55 to 59 | All ages | 55 to 59 | All ages | 55 to 59 |
| 1962 | 8.8 | 22.2 | 3.8 | 10.2 | 0.9 | 2.5 |
| 1967 | 7.8 | 19.6 | 2.8 | 7.6 | 0.7 | 2.3 |
| 1972 | 8.0 | 19.1 | 2.5 | 7.1 | 0.7 | 2.1 |
| 1976 | 5.4 | 14.9 | 1.5 | 4.7 | 0.5 | 1.5 |
| 1980 | 3.4 | 11.9 | 0.8 | 3.2 | 0.3 | 1.3 |

*Source*: NCB Medical Service, *Ann. Rep.* 1980–1.

The figures suggest a general gradual improvement, which was much greater in the seventies than the sixties. (The apparent increase in the total for all categories at all ages in 1972 is caused by a rise in the average age of the sample; there is a slight fall in the rate specific to each age group.) But the statistical change between the first and third dates is

[1] NCB Medical Service, *Ann. Rep.* 1980–1, 17.

slight enough for it to be possible (though unlikely) that the improvement was illusory. It could, for instance, have resulted from differences in the rates at which pneumoconiosis sufferers and others left the coal industry, in a period when manpower fell rapidly. It is at least likely that things were not going as well for much of the sixties as was believed at the time. This was a difficult period. Mechanization was causing more dust to be created; more water spraying during coalcutting and more equipment for dust extraction were removing much of that extra dust; manpower savings caused by mechanization were reducing the number of men exposed to any dustier atmospheres that may have been created. It was hoped that the favourable would more than offset the unfavourable changes. Stricter dust standards were set to try to ensure that this hope was realized, and on the whole the stricter standards were met. But in the absence of any gravimetric methods of monitoring respirable dust, these standards did not everywhere provide conditions as good as was believed.

Even at face value, the earlier statistics contained some worrying components. All the additional evidence confirmed the old belief that the prevalence of pneumoconiosis was much worse in South Wales than elsewhere, though it was serious in some other places, and the rates of progression in South Wales were exceeded in several other coalfields. In the collieries whose results are tabulated in Table 10.1, those in South Wales showed 40.7 per cent in the 55 to 59 age group as having pneumoconiosis in 1962, and, at 40.1 per cent, the figure for 1972 was higher than that for 1967. The figures in that age group for cases of category 2 pneumoconiosis or worse were higher (at 17.2 per cent) in 1972 than 1967, and for PMF the percentage rose at each examination down to 1972, when it was 7.3. The same trend for PMF among older men appeared in the Western Area collieries in the sample, and in the Barnsley Area collieries in 1972 there were still 10 per cent of the men aged 55 to 59 who had category 2 pneumoconiosis or worse. It was possible that all these unsatisfactory figures might be explained by the dust conditions to which the sufferers were exposed before periodic examinations were instituted, though it would be hard to explain in this way the rise between 1962 and 1972 in the prevalence of category 1 pneumoconiosis in the age group 35 to 44 in the South Nottinghamshire constituents of the sample.[1] Collieries in other areas showed more

[1] NCB Medical Service, *Ann. Rep.* 1980–1, 8–16 presents and comments on the data for the fifty-three collieries in much more detail. Data showing at various dates higher progression indices in several other areas than in South Wales are given in *Ann. Rep.* 1967–8, 13 and 1968–9, 12. Durham, the North West, and Kent showed the highest progression indices at this time.

favourable trends. It is probable that, on the whole, there was some general improvement, but it was slow and far from universal.

Any genuine improvement in the sixties would continue to help a further downward trend in the next decade, but the strongest subsequent influence was the application of more certain and relevant knowledge to the establishment of new dust standards, together with the adaptation of techniques to ensure conformity with the standards. In 1970 the NCB recommended a respirable dust standard of 8 milligrams per cubic metre of air in the return airway leaving the coal face, and this was accepted by the Ministry of Technology and became statutory.[1] In September 1975 the Coal Mines (Respirable Dust) Regulations were introduced with provisions for progressively severe sanctions for continued operations in workplaces which failed to meet the dust standards. There were a few cases in the next two years of work having temporarily to stop on faces and drivages because of such failures. But dust suppression was going so well that in 1977 the NCB voluntarily adopted more stringent standards of 7 milligrams per cubic metre at coal faces and 5 milligrams elsewhere, and these new standards were made statutory in 1978.[2] In March of 1979, 1980, and 1981 there was no class of workplace where the percentage meeting dust standards ever fell below 97.5 per cent (the figure for drivages in 1980), and for most it was 100 per cent. This was a huge improvement in meeting standards if it is compared with the mid-sixties when standards, expressed in numbers of particles, were much less satisfactory.[3]

It was in such conditions that the risks of developing pneumoconiosis among all but the oldest miners were greatly reduced. The total number of new certifications by the Pneumoconiosis Medical Panels of the Department of Health and Social Security fluctuated about an only slightly declining level, but a constantly growing proportion consisted of ex-miners, much of whose dust exposure was well in the past. Certifications among working miners were more than halved in the decade after 1970. By 1980 there were nineteen collieries (9 per cent of the total) which did not have a single worker registered under the Periodical Medical Supervision Scheme for pneumoconiosis.[4] The statistics already

[1] NCB Medical Service, *Ann. Rep.* 1969–70, 36.

[2] NCB *Ann. Rep.* 1975–6, 16; 1976–7, 15; 1977–8, 18.

[3] NCB Medical Service, *Ann. Rep.* 1980–1, 21. Of 24 separate entries for different classes of workplace (8 for each year) 17 show 100 per cent meeting dust standards. Cf., e.g., *Ann. Rep.* 1967–8, 7, which gives figures for working places meeting approved dust standards at the end of September 1965, 1966, and 1967. For drivages the figure was always below 75 per cent, and at power loaded coal faces only on cutting shifts in 1966 did it reach 90 per cent.

[4] NCB Medical Service, *Ann. Rep.* 1980–1, 8 and 19–20.

cited all indicate a rapid improvement. There was still plenty of work to be done in investigating and trying to check PMF among older men, even though it was already much less prevalent and its prevalence was bound to go on diminishing as fewer men contracted pneumoconiosis. But pneumoconiosis was sufficiently under control for it no longer to need to dominate research in the occupational health of coalminers to the extent that it had done for several decades. So it was that from about 1980 the medical research of the NCB was able to shift its emphasis to pay more attention to complaints that were significant but less prevalent, or common though apparently less serious. Thus there were investigations into the possible influence of dust on bronchitis and emphysema, there was an increasing amount of research on back pain, and a development of ergonomic research on mining systems and the mining environment as well as on machine and vehicle design.[1]

Such shifts of interest were signs of an important and welcome change in the coal industry. It had for a very long time had a fearsome reputation as just about the most dangerous of all industries. It still had more hazards than most, but particularly in the sixties and seventies it became very much safer, and of the immediate contributions to that result, the declines in fatal accidents and in the prevalence of pneumoconiosis were the most impressive.[2] But behind them, and most of the other immediate influences, was the most fundamental factor: the revolution in mining methods and equipment.

[1] NCB Medical Service, *Ann. Rep.* 1980–1, 38–42; NCB *Ann. Rep.* 1981–2, 28; 1982–3, 17.

[2] As early as 1969 it was noted that more fatalities underground came from sudden illness, notably coronary thrombosis, than from accidents, and that first aid services underground had to be adapted accordingly (NCB Medical Service, *Ann. Rep.* 1969–70, 23).

PART D

# THE RECORD

CHAPTER II

# Functional Problems and Performance

When the coal industry was nationalized both specific and general objectives were set for it. Throughout the years of nationalized operation there were some important features that were continuous or constantly recurring. Many of them had a strong influence on the way the objectives could be approached, and on the possibility of ever fully attaining them. Some of them revealed the persistence of difficulties that it had been hoped nationalization would remove. Some of them indicated problems which had not been foreseen and which provoked argument about their origin and about their possible association with nationalization itself. Circumstances change, and the detailed manifestation of these various characteristics changed with them. Much of the detail has already been displayed in chronological exposition, but it is worth while to seek from it the underlying common elements and look at them again in a more thematic fashion. In this way it may be possible to get a better idea of the choices that were available, the forces that sometimes frustrated preferred choices, and the range of possibilities against which alone it is realistic to attempt some assessment of success and failure.

## i. Commercial and financial objectives

The commercial and financial objectives of the NCB were laid down in general, but fairly precise, terms in the original statute, and the intentions of the wording were plainly presented in the debates which preceded its enactment. The NCB were to work coal and make it available in such a way (including quantity, quality, and price) as to further the public interest in all respects; and they were so to conduct their business that financially they would at least break even on revenue account on an average of good and bad years. The capital they needed in order to meet these objectives would be obtained by borrowing, mainly from the Minister, and with repayments of principal and interest at rates directed by the Minister.[1] Thus they were expected to pay their way but they

[1] CINA, ss. 1 (1), 1 (4)(*c*), 26, 27, and 28 (1)(*b*).

were not required to make a profit, though they were not statutorily precluded from doing so. It was possible that the powers of the Minister to give directions to the board, though these were limited, could, on occasion, require them to perform their functions in a way that might be expected to produce a surplus. Indeed, they were statutorily compelled to establish a reserve fund (which cannot sensibly be done except out of a surplus), and they were explicitly required to restrict the use of any revenue surplus to their own purposes, subject to any overriding direction given by the Minister with the approval of the Treasury.[1] So the possibility of earning some surplus was clearly envisaged, though it does not appear to have been expected that it would ever be a major financial item. The wording of the provisions about a reserve fund and the application of any surplus was, in fact, the subject of a good deal of debate and amendment (generally in a limiting direction) before the bill was enacted. The tenor of the discussion and of proposed amendments, which eventually led the government to bring forward its own revised form of words, was that the NCB should create their own reserve fund, as opportunity permitted, in the normal way of any prudent business organization, and use it for particular purposes of their own; and that the coal industry should not be used to build up surpluses which the government could raid for quite different uses.[2]

The existence of a reserve fund could conceivably provide an internal source of finance for investment by the NCB in the expansion or improvement of their business, though that could not be significant unless there was the opportunity to build up the fund to a large size. The other financial requirements made that seem unlikely, as, in fact, it turned out to be. The assumption was always that investment in expansion, like the original capital, would be financed by loans from the Minister. So it was laid down that programmes involving substantial capital outlay by the NCB must be on lines settled from time to time with the approval of the Minister.[3]

The basic principles underlying the original prescription of commercial and financial objectives remained in force. Later Coal Industry Acts altered the limits on borrowing by the NCB, and two of them, in 1965 and 1973, provided for capital write-offs and restructuring. From time to

[1] CINA, ss. 29 and 30.

[2] Chester 1975, 592–6 sets out in detail the sequence of discussion (both in the government and in House of Lords debates) and amendment on the subject of reserve fund and surpluses. The original instructions for drafting the bill suggested that accumulated reserves above a stated minimum might be used to purchase stock for cancellation.

[3] CINA, s. 3 (2).

time there were variations in the detail of the way the government required the financial objective to be interpreted. A White Paper of 1961 defined the break-even requirement for nationalized industries as requiring revenue surpluses at least to offset deficits over a five-year period, after making some additional provisions out of revenue; and the latter proviso involved the NCB from 1963 in setting aside £10 million a year towards the difference between depreciation at historic and at replacement cost.[1] But this requirement was insisted on for only fifteen months, and then suspended. From 1973 the break-even requirement was fundamentally modified to relate to the revenue position after the receipt of government grants, though the terms and size of the grants were continually changed. In 1980, in association with the Coal Industry Act of that year, the government introduced a new financial framework that required a bigger contribution from revenue to the finance of new investment, and a move to revenue break-even, after social grants only, to be completed by 1983–4. But this framework was abandoned after only one year and the financial principles reverted to what they had been before 1980, though with a large increase in the sums required from the government. Essentially, the requirements from the NCB remained what they always had been, with one major change. The NCB continued to be an undertaking with an all-loan capital structure, required to make available the supplies of coal needed in the public interest, and to break even on revenue account. But the revenue account was supplemented by government grants, which at the time of nationalization would have been anathema.

But there had been, for quite long periods, other factors which modified the character of the statutory objectives. The most important of these concerned prices. The fixing of coal prices, in accordance with the NCB's judgement of the public interest, was by statute left entirely to the NCB;[2] but in practice that provision was of little effect. For years the NCB could propose, but not determine, the prices they would charge for coal. This was because the Minister had other opportunities to overrule them. For many years these were provided mainly by the long continuance of arrangements made for the special conditions of wartime. Originally there was the authority under regulations for the Minister to decide the prices of domestic fuel. More general and more important was the non-statutory 'gentlemen's agreement' by the coal-owners and the government that coal prices would not be raised without ministerial approval. This agreement was still operating when the

[1] NCB *Ann. Rep.* 1962, I, 4. [2] CINA, s. 1 (1)(*c*).

coal industry was nationalized. The NCB were persuaded to continue it, and it was this informal arrangement that enabled ministers to keep limiting and delaying coal price increases at a time when a competitive level would have been higher than the prices currently charged. It also ensured that imported coal was sold at British prices despite its much higher cost in the forties and fifties.

Governmental action was thus able to keep down the NCB's revenue both directly, through the level of prices, and indirectly, by demonstrating that it was pointless for the NCB to seek still higher prices and, in particular, pointless for them to look to higher prices as a means of providing more internal finance for their investment programme. In the sixties and early seventies the previous restraints were largely superseded by those in successive prices and incomes policies, which applied to the economy as a whole. By this time, however, market conditions for coal were no longer such that these controls could enforce much loss of revenue, though there were, on a few occasions, delays in permitting increased prices which caused some loss. Only when the UK entered the ECSC in 1973 did the NCB gain in practice the complete responsibility for pricing that was nominally conferred by CINA, but it was only in the first ten years that this restriction was a significant obstacle to the raising of revenue.

Formal restraints on the quantities sold could also have affected revenue. As coal remained until 1958 subject to government allocation to consumers, under wartime powers, there was nominally a higher control on the NCB's performance of its statutory duty of 'making supplies of coal available', but this made no difference in practice. All the pressure, while this situation lasted, was for the NCB to produce and sell as much coal as possible. Subsequent reductions in sales came mainly because of market forces. These were reinforced, especially from 1967, by government energy policy, which reduced the purchasing needs of some of the largest users of coal, and indicated to others that coal was on the way out and they might as well start using something else. But none of this altered the statutory position of the NCB.

Before considering how successfully the NCB met the commercial and financial objectives prescribed for it, it is useful to consider whether the objectives, as modified in the ways just described, were mutually compatible, and whether they encouraged, or even permitted, the pursuit of optimal commercial policies. The statutory definition of the duties of the NCB included references both to 'securing the efficient development of the coalmining industry' and to making coal available in ways 'best calculated to further the public interest in all respects',[1] and it is not at all clear that both requirements are necessarily of the same kind. Further provisions may have encouraged a leaning in one direction rather than another. Consistently with the proper discharge of their duties, the NCB were statutorily required to direct their policies both to securing the 'welfare of persons in their employment' and at least to breaking even on revenue account over a period.[2]

It is fair to conclude that the statutory wording implied more emphasis on public service and welfare, with adequate supplies at popular prices, than on the normal commercial pursuit of a financial surplus. (It would clearly have been impolitic to include the word 'profits' in CINA in any reference to operations after nationalization.)[3] There is no doubt that the members of the national board were deeply conscious of their role as providers of a public service, and this view was made explicit from time to time in passages in the early published annual reports. Even though they wanted higher and quicker price increases than they got, they were reluctant price raisers and would never have contemplated charging at world price levels in their first ten years, even if the government had not excluded the possibility. These attitudes may have subsequently weakened in some respects, but they did not disappear. Even in the seventies, in difficult financial circumstances, the national board could still be found discussing whether there was an obligation to provide particular classes of coal which the public wanted, despite the inevitability of heavy loss on their production.

The Fleck Committee in 1953–5, anxious to improve the performance of the NCB through better management, showed full appreciation of the peculiar conditions imposed by nationalization, but argued that they neither need nor should alter the way things were done. The committee's report made these comments:

Instead of the profit motive there is a statutory obligation to the customers. The Board cannot make a profit in the sense of building up a surplus from which to distribute dividends to private shareholders, but is required so to supply coal as to 'further the public interest in all respects'. In short, the British coal industry is now a public service. But so far as we can see, there is nothing in the legislation which created the National Coal Board, and under which the

[1] CINA, s. 1 (1)(*b*) and (*c*).

[2] CINA, s. 1 (4)(*a*) and (*c*).

[3] The Act appears to have been drafted in a way that would avoid mentioning profits if possible, but references to 'profits and gains' of colliery concerns crept into s. 22 (3), which related to the calculation of interim income to be paid pending satisfaction of compensation.

National Coal Board operates, that prevents the industry from being managed in accordance with the best commercial practice. . . . Financial policy provides one instance of the point we are making. Given that the Board cannot make profits in the normally accepted sense, but must rather seek to balance their revenues and their outgoings 'on an average of good and bad years', we can think of no aspect of financial policy in which the Board should not pursue much the same course as would a large commercial undertaking.[1]

This passage contains important basic truths. Administrative structure and managerial methods and relationships need to be directed to securing the most efficient use of resources; and their design depends on the nature and size of the undertaking rather than its private or public ownership. But in some respects the comments of the Fleck Committee seriously over-simplify. The statutory position of the NCB discouraged, and could prevent, some types of strategic policy decision which a privately owned business would have thought desirable for its own long-term health. It is common for a large private undertaking to obtain very different rates of return from one division of its business and from another, and for one activity or unit to subsidize another for a time. But if it found itself in a period of continuing low returns in the business as a whole, it is most unlikely that it would not seek to deal with its seemingly perpetual loss-makers by contraction or closure or drastic measures to raise their rate of return. The NCB neither did nor could act so rationally or so ruthlessly, though they made continual efforts (some of them successful) to improve the financial performance of loss-making units.

This may be illustrated in various ways. It was known from experience of the recent past that the quantity of coal the country was believed to need could not be supplied without drawing on the immediately available output of every coalfield, however costly and commercially unprofitable. In the years immediately before nationalization it had been impossible to get this output without the industry's cross-subsidies being supplemented by some direct subsidy from the Exchequer. But the NCB were expected to break even without subsidy, yet to serve the public interest by producing a bigger quantity of coal than before; and down to the mid-fifties they kept being told that their targets for future output were not high enough. They could meet the public interest criterion in terms of quantity only by continuing to operate large amounts of productive capacity which no conceivable improvements could keep from loss unless the price of its output were raised so much as to provoke an immediate outcry that the financial aspect of the public

[1] Fleck Report 1955, 7.

interest was being ignored. In fact, in the twenty years in which the NCB operated with a divisional structure (nine divisions most of the time, though only eight at the beginning and end), only two, East Midlands and Yorkshire, consistently made profits after paying interest on capital, and a third, West Midlands, was profitable often enough to break even over the whole period. The quantity of coal that the others could supply was all urgently needed down to 1957, and a good deal was needed after that for special quality. But they all had to be carried financially. Scottish Division never made a profit after 1949, and none of the others achieved a profit in more than one year out of the twenty.[1] Yet little pruning could be attempted as long as government and public called for maximum output.

A more restricted illustration is provided by the maintenance of deep mining of naturally smokeless coals, at enormously high cost. The judgement was that the public needed this type of coal and, though it must command a premium price, it must not be put beyond the reach of the householders who wanted to use it. A commercially oriented firm would have found such a policy indefensible for more than a temporary phase and would have acted on the policy that the consumer must meet the cost or do without.

Another aspect of the difficulty of the NCB in acting like a large private undertaking concerned the financing of major investment projects. There was the same need to evaluate the likely return on the investment, using the fullest available data and the best available techniques (which changed appreciably in the sixties). But there could be in practice a significant difference between getting an adequate return on capital already accumulated in reserves, or a return no worse than that already achieved on the existing equity, and getting a return which would still be worth while after meeting likely loan charges over a period when interest rates would be volatile and unpredictable but certainly high and possibly rising. This was not a major problem in the fifties and sixties, but it became very serious in the seventies and eighties. The financial structure of the NCB was designed and created at a time which proved to be near the unforeseen end of an era of cheap money. Initially, the insistence on financing all capital needs out of loans

[1] Summarized figures in NCB *Ann. Rep.* 1971–2, II, 6–7. They should be compared with earlier summary figures (e.g. *Ann. Rep.* 1961, II, 4–5). The differences show up the separate contributions of Northumberland and Durham to divisional losses, and the adverse effects on divisional results caused by the inclusion of the persistently loss-making Cumberland field, at one time in Northern and at one time in North Western Division.

probably looked more like a benefit than a burden. But in the long run it was a source of serious financial distortion. Very few private firms, and probably no large ones, would have operated happily with such a structure.

So, despite the judgements of the Fleck Committee, the commercial and financial operation of the NCB was constrained by the statutory position to be somewhat different from that of large private firms. It remains to be asked what the performance was, in relation to the prescribed objectives.

In terms of quantity, the performance was moderately satisfactory but not as good as had been hoped. Output at its maximum, in 1952, was 19 per cent higher than in the last pre-nationalization year, and until 1960 it was consistently greater than home consumption. But the control of coal supplies to consumers until 1958 may have been a restriction, though it is doubtful whether it much reduced the total quantity of coal used, except by domestic consumers. It was, however, a matter of concern that output reached its peak before consumption did, and that consumers did not always get the type and quality of coal they wanted. So the mid-fifties were a worrying time, with a shortage particularly of large coal, stagnant output, falling stocks, and, consequently, increased imports. It was not quite the ideal service of the public interest that had been envisaged, and indeed, the persistence of coal rationing after that of all other commodities had ended has to be seen as disappointing, though the coal control scheme could have been safely ended at least 18 months earlier than it was.

Particular shortages at home were alleviated by imports which, from nationalization until they ceased in 1959, totalled 26.9 million tonnes. Three-quarters of them came in during the three years 1954–6, mainly as a result of contracts arising from the anxieties of 1953 and 1954. But in the period 1947 to 1959 exports, at 159.6 million tonnes, were almost six times as large as imports, and exports could always be squeezed to keep home supplies adequate. But that meant forgoing market opportunities, especially in the late forties, and to some extent later, at a time when exports could be somewhat more remunerative than home sales. To replace exports by imports, however, much more than swallowed up the financial gain.

From 1957 there was no longer any lasting problem of supply. Although output fell slightly short of home consumption in 1960 and 1961, that was only an adjustment to immediately preceding excess. This experience was repeated at the end of the sixties as a result of the large scale of colliery closures in response to government fuel policy, and in 1970 there was some difficulty about meeting all demands. The resumption of imports in 1971, though resulting mainly from governmental political ideas, also had a precautionary element. But, in general, there was adequate productive capacity from 1957 on, and it was only in years of major strikes that supplies fell short enough to damage consumers.[1] There was also in later years, because of the narrowing of markets, a sufficiency not only of coal in general but of coal of most particular types, though there were arguments with the steel industry about coking quality, and the very best coking coal, suitable for making foundry coke, had certainly become scarce.

The public may have got all the coal they wanted for most of the time, but their interest was also very much in its cost, and abundance at high prices was not very welcome. Coal got some popular reputation for becoming expensive and there were many figures to support this view, though they did not do so altogether consistently. Official statisticians in 1983 argued that over the previous twenty years fuel prices had risen faster than those of any other group of commodities. They instanced, among other items, domestic coal and smokeless fuel, which had risen in price by 48 per cent in real terms, though this was much less than the rise for heating oil, and differed little from the rises for rates and water charges and for public passenger transport.[2] In fact, though they were regularly made, comparisons of coal prices with indices of retail or wholesale prices were not closely precise indications of what was happening. This was particularly so because of the great number of different coal prices, according to type and place, and because of the long-term changes in the proportions of coal going to different markets. One set of statistics suggested that from 1970 to 1981 there was a real price increase of 36 per cent in domestic coal and coke, but 68 per cent in coal supplied to industry.[3] Moreover, by the end of this period even the real price figures, so far as they can be indicated, are a little less indicative of the achievements of the producer, because they could be more influenced than before by the

[1] Summaries of production and consumption each year in NCB *Ann. Rep.* 1971–2, II, 84–5 and 1982–3, 20–1.

[2] *The Times*, 13 Oct. 1983. The average money rise in prices for all commodities was put at 525 per cent for the 20-year period. From the beginning of 1956 to the beginning of 1983 the official retail price index rose from 100.0 to 734.5 for all items, but to 1150.3 for fuel and light. The index based on Jan. 1974 rose by Jan. 1983 to 467.0 for all fuel and light and to 456.8 for coal and smokeless fuels, which may be compared with 626.7 for oil (Dept. of Employment, *Retail Prices Indices 1914-1983* (1984), pp. 2, 4, and 31).

[3] MMC, *National Coal Board*, I, 57.

extent to which government grants could make possible prices that did not cover costs.

Nevertheless, the problems of data and methods do not prevent an appraisal of the approximate general trend of coal prices. It seems clear that, as in the seventy years before nationalization, coal was a commodity supplied at rising real costs and prices, but that, within the long period, there were shorter spells when price movements were more favourable. There were rapid coal price rises in 1947, but from 1948 until some time in 1951 there was a fall in real terms, despite the high rate of inflation in those years. The NCB claimed that average pithead prices in 1951 had risen above the average of 1947 by less than half as much as wholesale prices generally: 27 per cent against 70 per cent.[1] But in 1951 a persistent rapid rise began, above the rate of inflation, and this real rise (with a check between 1957 and early 1960 in response to the contraction of sales) went on for the rest of the decade. From 1961 to 1969 the trend of prices in real terms was downwards. From 1969 the trend was reversed, and remained so, though the rates of real increase varied a good deal. In 1971 and 1972 the large increases in the general price level offset much of the rise in coal prices, but in 1974 and 1975 the latter rose hugely in real terms, before their rise slowed down again. The hopes, often made explicit in 1946, that nationalization would bring cheaper coal were not fulfilled. In its first 35 years the nationalized industry was probably matching that expectation in spells which amounted to not more than one-third of the period. The other two-thirds more than wiped out all the gains.

In one sense, however, the NCB acted as a buffer protecting the public against higher prices for a time. This happened because in the forties and fifties imported coal was sold at British prices, i.e. far below cost. In 1955, when much more was imported than in any other year, the cost per ton of imported coal to the NCB was 70 per cent higher than average pithead prices at home, and in 1947 the difference had been larger still. On the whole import programme to 1959 the NCB sustained a loss of £74.1 million through selling imports at UK prices.[2] These figures put a different gloss on the earlier figures of price trends. They show that, even in a period of appreciably rising UK coal prices in real terms, such as most of the fifties, the public were paying a good deal less

[1] NCB *Ann. Rep.* 1951, 8.

[2] Ibid. 1955, I, 17 gives figures which contain slight errors for 1955 losses on imports. The loss that year was £200,000 less than was stated in the report. Aggregate losses for the whole period, taking account of the corrected figures, are in NCB *Ann. Rep.* 1971–2, II, 4–5.

than internationally competitive prices for coal. Any supplier would have had to estimate the effect of higher prices on the quantity of sales, but it seems likely that at least until 1956 revenue could have been substantially increased every year by charging prices nearer to world levels. Thus it appears that, partly by their own interpretation of their duty as public servants, and partly by government compulsion, the NCB acted less than commercially in their first ten years. Some people have argued on these grounds that, as part of the public interest approach, the coal industry was used in these years of post-war readjustment to subsidize the rest of the British economy. But there are problems of definition and interrelationship to be sorted out if that argument is ever to be sustained. From the end of the fifties onward there were always alternative sources of fuel which set lower international prices, and even though from 1973–4 a price advantage over imported oil was re-established, there was never again a general fuel price discrepancy in favour of British coal.

The level of prices is one of the main links with revenue, and if prices were set uncommercially low for a time, this was liable to make it more difficult to get enough revenue to meet the statutory break-even requirement. Up to a point, this was the effect of the policy, though commercial pricing would have created not break-even but large surpluses until the late fifties. The best way to look at the revenue results is probably in two phases, before and after the Wilberforce settlement of 1972, which led to an appreciable change in financial provisions and requirements. In the period of twenty-five and a quarter years to the end of 1971–2 the NCB made an operating profit in every year except 1947 and 1971–2. The accumulated operating profits totalled £449.9 million. But CINA required the break-even to be after payment of interest and contributions to reserve fund, and this was achieved in only eleven of the years. In this whole period the total of deficiencies less surpluses was £306.4 million. Of this sum only £41.5 million was accounted for by taxation, the 15-month requirement of a contribution to the excess of replacement over historic cost of fixed assets, and 'other items'. All the rest (£264.9 million) was interest that was not covered by the operating surplus.

This summary contains at least one serious distortion. It would have looked very different if terminated a year earlier, for more than half the total net deficiency of twenty-five and a quarter years occurred in 1971–2 (£157.0 million) and was almost entirely attributable to the overtime ban and national strike of that year.[1] Some would contend that the

[1] Annual figures in NCB *Ann. Rep.* 1971–2, II, 4–5. Apart from interest, 'other items' were the

figures show an artificially favourable result because of the reduction of interest charges as a result of the financial reconstruction of 1965. In the first year after this reconstruction interest charges fell by £17.7 million (and were £21.5 million less than they would have been without the reconstruction), and in this period they never regained the level of 1964–5. But this is where other elements of policy come in. The amount of capital written off was £415 million, and interest rates on most of the residue were lowered. But those members of the Ridley Committee in 1952 who favoured more commercial prices that would be closer to marginal costs were thinking of a level which would have produced a surplus of about £200 million a year at that time. Sums of that order could have replaced all the borrowings for expansion and reconstruction in the fifties, thereby reducing interest charges enough to turn annual deficiencies into surpluses, and reducing capital liabilities by much more than the amount written off in 1965. They could also have left enough in reserve to set against further deficiencies arising well into the seventies. Such an exercise in the fifties would not have been politically acceptable in a nationalized industry, but as an accounting exposition it presents a case for the economic condition of the coal industry that is arguably at least as realistic as that given by the figures of the 1965 reconstruction.[1]

Other arguments which seek to put a worse interpretation on the NCB's financial results relate to the government grants which were first paid in 1966–7. In the six years from then to 1971–2 inclusive the NCB received £55.9 million in grants, of which £2.4 million was reimbursement of the costs of delaying colliery closures in 1967–8.[2] If these sums represent items properly and exclusively chargeable to the business, then they should be added to the total of deficiencies to make it more realistic. But do they? The £2.4 million was the cost of meeting a government-imposed requirement, on grounds of employment policy, to delay implementing commercial decisions reached by the NCB. It would be misleading to charge the cost to them. The remaining £53.5 million is less clear cut. The costs involved related to provision for

largest deductions from operating profits. These included, in the first few years, compensation for loss of office as a result of nationalization. The largest single element was the transfer of £7 million in 1963–4 to the Central Obsolescence Fund, mainly as a result of concern arising from the closure of the nearly new Rothes colliery, which involved the write-off of £6.5 million of assets.

[1] Chap. 6, section v above for additional financial detail and discussion.

[2] NCB *Ann. Rep.* 1966–7, II, 35; 1967–8, II, 32; 1968–9, II, 32; 1969–70, II, 32; 1970–1, II, 33–4; 1971–2, II, 32.

the redeployment of mineworkers and the early retirement of older mineworkers so as to assist the accelerated closure of collieries. The costs were not wholly covered by the grants and were partly borne by the NCB, though the government paid the larger share. It can be argued that such items are among the costs which any business undertaking must bear as part of the demonstration of its viability by adapting to compete in new market conditions. But when, as in this case, the new market conditions had been induced partly by a deliberate change of government policy, and the nature of the adaptation depended partly on the needs of government social policy, it was rational to attribute some of the costs to the government. The sharing of social costs between the NCB and the government was economically realistic, though the division between them was arbitrary, not scientific, and may have been just a little favourable to the NCB. So, too, may the government payments to the CEGB to burn extra coal in power stations, but as they applied only to 14 million tons in three years, they were too slight to have much significance.[1]

Down to 1972 the conclusion must be that the NCB met their obligation to supply coal in the public interest fairly well, but they failed by a not very large margin to meet the requirement to break even on revenue account. The immediate reasons for this failure can be found mainly in restraints on freedom to operate commercially, especially in the first ten years, and on higher than expected charges for the servicing of capital. In the last year, however, it was demonstrated that national strike action could damage financial performance far more than anything else could.

The financial outturn from 1972–3 on has already been examined in close detail and no repetition is needed.[2] The NCB, over the ten years starting with 1972–3, met the government's requirements, which were to break even after payment of interest and receipt of government grants, and also, in the closing years, to keep each year within the government's imposed external financing limit, which was an annual total of loans plus grants.[3] In one sense, there was greater success than before in meeting stated requirements, but the requirements were in a less onerous form, because there was little prospect of the NCB meeting them if they continued strictly in accordance with the original definition. This pragmatic relaxation of policy was a response to the weakening of the NCB's financial position as a result of two national strikes in

[1] Chap. 6, section v above. [2] Chap. 7, section vi above.

[3] There was dispute as to whether the NCB breached the external financing limit set for 1980–1 (Chap. 7 (p. 416) above).

three years, which left them with large deficits and an increased level of costs. But the changes in market conditions quickly enabled prices to cover increases in operating costs. From 1974, despite all the changes in the framework of law and policy, the constraints on the ability of the NCB to better their financial performance resembled those existing down to 1971. In every year from 1974–5 to 1980–1 (with a tiny exception in 1979–80) there was an operating profit which exceeded grants to assist operations, but only twice was there an overall surplus in excess of such grants. As before, but to a greater degree, the main drag on financial performance was the huge and rapidly rising cost of servicing loan capital. Lack of commercial freedom in operations and sales had ceased to be a serious influence, though external commercial conditions were supremely important and after 1981 were leading to a great deterioration of results. So it looks as though the lasting questions that needed to be raised were those concerning the financial structure of the NCB and those concerning the possibility of earning a bigger revenue surplus even in difficult conditions of market competition.

The NCB sought to show that they had not become a financial dependency of the government by publishing from 1977–8 to 1982–3 a statement of value added and its application. This provided a demonstration that the NCB paid more to central and local government, in rates, taxes, employer's national insurance contributions, and interest, than was received in government grants. This was probably a useful exercise in the correction of popular error, but it did not demonstrate anything remarkable. It is normal for large undertakings, because they are ratepayers and taxpayers, to be substantial net contributors to government revenue. The NCB contribution was positive but much less substantial than would be customary for so large an undertaking. The total net payment to central and local government was £172 million in six years, and for one year (1981–2) it turned negative, though it recovered in 1982–3.[1] Perhaps the most interesting feature is that an undertaking which does not seek profit, and therefore pays very little corporation tax, should contribute even as much as this. It is a testimony to the size of interest payments, all the more impressive when it is noted that in 1981–2 5.1 per cent, and in 1982–3 4.4 per cent, of value added was applied to the payment of interest on loans from sources other than the UK government. Yet again there is a reminder of the dominant influence of the cost of borrowing on the financial outturn.

So there is a question whether the NCB needed, or invested reward-

[1] NCB *Ann. Rep.* 1977–8, 4; 1978–9, 4; 1979–80, 4; 1980–1, 8; 1981–2, 8; 1982–3, 5.

ingly, all the capital they borrowed, often at such high cost; and it is difficult to give a firm answer, partly because the data are sometimes incomplete and susceptible of various interpretations, and partly because there are disagreements about the appropriate tests. At one time in the fifties the Treasury became worried that the NCB were increasing their borrowings with the prospect of having to service them from the proceeds of a little changed output. The NCB countered with the argument, expressed in various forms at various times, that as the coal industry consumed 2 per cent of its capacity every year, the relevant comparison was between what happens with and what happens without new investment; and as long as there was a need to maintain maximum output, such an argument could not be ignored. Likewise, the Ministry of Power realized that the NCB could, without the Ministry's knowledge, carry out investment schemes at pits that would remain loss-making. The NCB would justify such investment, as long as there were general grounds for retaining the pit, if the project yielded a good positive return, even if that showed only in a substantial reduction of the overall loss on the colliery. When there was less pressure for maximum output, such arguments could be questioned more strongly but not always totally rejected. The expense of raising loans to finance investment to keep loss-making pits in operation remained a serious problem, and it was hard to justify it unless it could be shown that greater expense was likely to be incurred by not doing it.

To get value for money out of investment it was necessary for the NCB's procedures to be reliable, for governmental supervision was exercised in too general terms to have much influence on the choice of sound or unsound individual projects. It could, indeed, be maintained that the wording of CINA imposed a requirement only to obtain the Minister's approval for a general programme of investment, not to consult him about particular projects.[1] Certainly, the Ministry of Power thought that in the first ten years a tradition was built up whereby the NCB gave little information about specific items, though, in fact, the civil servants could have dug out more information about their progress by comparing the successive annual investment reviews which they received, or even the appendices to successive published NCB *Annual Reports*. And from 1958 they got agreement about the submission of details of the largest projects. But even governmental regulation of the investment programmes of the nationalized fuel industries in general

[1] CINA, s. 3 (2).

terms had been recognized as rather weak and irrational.[1] The available information was too limited and approximate and too subject to frequent revision in the light of changes in prices and the availability of resources; and the government had few instruments of control other than the arbitrary imposition of cash limits and steel quotas.

Within the NCB organization national headquarters also had for a good many years only a rather general and limited control over individual projects, even though the larger schemes had to be justified to, and approved by, the national board, which also regularly considered reports on their progress. Investment schemes were devised mainly in the areas, and to some extent in the divisions. If they were small they came within the discretionary authority of the divisional boards. If they were large, it was the divisions that were responsible for carrying them out, once they had national approval. So, in practice, there was a tendency for both the character and quality of investment schemes to vary between coalfields. Some divisions, notably Scottish, were particularly prone in the fifties to instigate large schemes, which had less success than most of the smaller ones. In the sixties investment was reduced so much that far less could go wrong. In any case, the amount of control from national headquarters increased, especially after the abolition of divisions, and the procedures for staged grant of approval and for monitoring progress, variations, and changes in prospective returns became tighter. There were new techniques for calculating returns. Some use of DCF techniques was adopted in the later sixties, and they were applied generally in the seventies.[2] When heavy investment was

[1] PRO, POWE 37/100. Note on capital investment, prepared for information of the Minister, 23.5.52.

[2] The White Paper, *Nationalised Industries: A Review of Economic and Financial Objectives*, Cmnd. 3437 (1967) stated (5) that DCF techniques should be used for all important projects and that 8 per cent was a suitable test discount rate. *First Report of SC on Nationalised Industries*, 1967–8, II (HC 371–II of 1967–8) published a great deal of information about NCB investment appraisal methods. A Treasury exposition of appraisal methods, and particularly of DCF techniques, is included (14–22) and the Treasury gives its conclusions about the purpose and method of the review of the investment programmes of nationalized industries by government departments. In the NCB oral evidence, Lord Robens and D. M. Clement (then Director-General of Finance and later a board member), explain (143–5) the NCB investment appraisal methods, including reasons why at that time they used DCF methods only for non-mining projects and a few small projects in collieries. They also indicate the small amount of checking done by government departments, and their personal belief that government departments were not adequately staffed to be able to appraise individual projects. Lord Robens remarked (144), 'So far every project that has been submitted has come back without change. This is because they have not the technical people to indicate any changes.' Vol. III of the same report (HC 371–III of 1967–8) included (27–8) a Treasury memorandum of May 1968 on 'The Test Discount Rate' and (47–55) a Ministry of Power

resumed in the mid-seventies it was subject to much more standardized appraisal and supervision than it had been in the fifties. Even so, there remained large possibilities of miscalculation, not least because of the speed with which prices changed and the length of time taken to complete capital schemes, which gave many opportunities for drastic changes in the market conditions on which yield estimates had been based.[1]

Despite all the difficulties, it seems unlikely that there was a serious and persistent waste of capital, though the evidence is too sketchy for results to be put in any continuous and precise form. In 1951 the Ministry of Fuel and Power attempted a statistical analysis of improvements in performance between 1946 and 1950. The conclusions were that every 1 per cent increase in capital produced an average increase of 0.25 per cent in output, that a similar percentage increase in labour increased output by an average of 0.75 per cent, and that in current conditions it was cheaper to seek greater output by increased investment rather than by increased recruitment. The best returns to increased inputs had been in Kent, South Wales, and the North West, i.e. generally where performance had been worst. Most of the investment then had been in small schemes, and it was later concluded that this investment had been responsible for most of the higher output in the early years of nationalization.[2]

Larger schemes were mostly started later and took longer to complete, and came to fruition when less coal was wanted. The Ministry of Power commented in 1963 that the recent poor financial results gave the impression that investment in the aggregate had not produced good returns, thought it agreed that this was largely because pits were having to be operated below planned output and because so much of the improvement had gone into benefits for workers, such as extra holidays, shorter working hours, sick pay, and pensions.[3] In fact, at this time

memorandum of Aug. 1963 on the 'National Coal Board's Procedures for New Projects'. The latter concluded (50–1) that the NCB methods of preparing projects for approval were very thorough and left little to be desired, but queried the adequacy of some of the criteria used as a basis for judgement. In particular, it doubted whether there was sufficient insurance in the practice of trying to cover imponderable risks by going for a high rate of return after depreciation and calculating that return on the gross new additional investment instead of on the much lower figure of its average written-down value.

[1] MMC, *National Coal Board*, I, 180–211 examines in detail the NCB methods of investment appraisal between 1974 and 1981 and their results. It suggested (210) that the NCB should carry out a separate sensitivity analysis of costs and prices in appraising all major projects.

[2] PRO, POWE 37/65, 37/68, and 37/100.

[3] *First Report of SC on Nationalised Industries*, 1967–8, III, 51.

operating profits were steadily increasing, and one independent enquiry suggested less unfavourable results. The estimates of W. G. Shepherd, cited earlier,[1] suggest that half the reconstruction schemes in Scottish and South Western Divisions in the fifties were financial failures, but that those elsewhere were giving a marginal return of 12 per cent by 1961. He calculated, however, that investment explained only 17 per cent of the changes in profits. Throughout the sixties it also seems plausible that investment, especially in new equipment, was partly responsible for the reduction of coal production costs in real terms, though also that its contribution was less than that of the new technology and of the elimination of high-cost capacity. After the resumption of large investment projects the earlier uncertainties about the extent of the positive results remain. It would be rash to go any further than the MMC, which concluded: 'While we cannot say whether investment has been misdirected on the basis of marginal returns, we note that much investment has gone into collieries which are either unprofitable or of doubtful potential profitability.'[2]

No large business gets all its investment schemes right, but in the private sector fewer of the disasters receive close outside scrutiny than in the public. Whether the NCB were more prone than others to miscalculate investment is impossible to say. In the earlier sixties an independent assessment after close study was that serious misallocation had not been evident and that 'most Coal Board investment has probably been as productive in a commercial and economic sense as investment in private industries', though there was a caveat about the investment in Scotland and Wales and, to a lesser extent, in the two northernmost English divisions.[3] Nearly twenty years later the MMC noted 'that, as a result of a tendency to be optimistic about physical and market-related factors, the NCB has tended to over-estimate the prospective revenue and under-estimate the prospective costs which a colliery will experience following investment'.[4] That characteristic was far from new in the NCB, but there is plenty to suggest that it is familiar elsewhere in both public and private enterprise.

So, while it is clear that the high cost of servicing capital was a major influence on the failure of the NCB's financial results to come up to

[1] Shepherd 1965 gives numerous other calculations about NCB investment and profits and the relation between them; see 70–2, 80–3, 112–17, and 139–41.

[2] MMC, *National Coal Board*, I, 211.

[3] Shepherd 1965, 140–1.

[4] MMC, *National Coal Board*, I, 211.

expectations, it is not clear whether these high costs were the NCB's own fault to any significant degree. They contributed to their difficulties in two ways. Capital was borrowed for some schemes which were quite unjustified by results. And many projects in both the fifties and the seventies appear to have taken an excessively long time to complete. The effects of this were that financial charges had to be paid for a longer period before any revenue was produced, and that the longer wait increased the risk that the estimates on which the project had been evaluated would be falsified. But it may be that the number and size of such errors were no greater than should be regarded as probable on the basis of experience throughout the economy, regrettable as that may be. In other respects the source of the difficulties was external. The NCB were unlucky that twice they had to undertake large capital programmes, partly because of preceding periods of minimal investment in coal, and on both occasions the programmes came to fruition in adverse busines conditions. And they were not responsible either for having to raise all their outside capital by loan, or for the heights to which interest rates rose. It seems almost certain that, whatever investment miscalculations occurred from time to time, they had much less effect on the cost of servicing capital than did the financial structure, which in turn depended on political status.

The other side of the arithmetic of success, how to get more revenue out of all the assets, material and human, was more complex. The constraints imposed by interpretations of the public interest have already been emphasized. They, too, have much to do with political status. But there were possibilities of at least partly relaxing these constraints, or being more closely bound by them, through managing the business in one way rather than another and through trying to elicit and sustain a more helpful human response. Performance in relation to commercial and financial objectives is not fully explicable without attention to all these other factors.

## ii. Industrial relations

One of the most widely shared hopes raised by the nationalization of the coal industry was that it would permanently improve the relations between management and workers which had been so bad for so long. Nationalization was the fulfilment of the most cherished aspiration of the mining trade unions, so it was expected that their members would show a new attitude to their work and greater co-operation with their employers. There were many who would, in general, have been politically opposed to nationalization but who were willing to accept it for the coal industry with little more than token complaint, because they did not see how a basic industry could run properly with so much internal feuding and they suspected that a move towards general peace was unlikely to happen without a change of ownership.

The government shared both the concern and the hopes about industrial relations. There was no provision for direct involvement of workers' organizations in the running of the industry. That had once been a much-favoured approach to nationalization and still had some support among active trade union leaders; but it was no longer preferred by the most influential Labour politicians or by union leaders in the coal industry.[1] Indeed, at the 1948 Labour Party Conference, a resolution 'to convert nationalisation into socialisation' and have trade union participation in the direction and management of nationalized industries at all levels was opposed by the president of the NUM.[2] But there was an attempt to establish a closer identity of workers and management by appointing trade unionists on a personal basis to membership of the national board, as was done at the outset with Citrine and Edwards; and by relying on the board to make a new approach to industrial relations and set up better arrangements for conducting them. Surprisingly, the government originally proposed to include hardly anything about this in the legislation, on the ground that the need was so obvious that it was inconceivable that the NCB would not respond to it appropriately. Almost everything in the statutory provisions resulted from amendments brought forward by the government in response to pressure from the Conservative opposition.[3] Besides being required to direct policy so as to secure the welfare of their employees, the NCB had to direct it so as to get the benefit of the knowledge and experience of the employees in the organization and conduct of the operations; and they were also required to establish joint machinery with employees' organizations for the settlement by negotiation of the terms and conditions of employment (with provision for arbitration) and for consultation on various matters, including the organization and conduct of operations.[4]

The setting up of this machinery, which for consultation was wholly new and for negotiation and conciliation was partly new, was done as

[1] Chester 1975, 383–4.
[2] Ibid., 845.
[3] Ibid., 783 and 790–1.
[4] CINA, ss. 1 (4)(*a*) and (*b*) and 46.

quickly as possible.[1] Over the years there were adjustments of detail to take account of changed circumstances (though perhaps not enough), and a few additional agreements with unions other than the NUM, which, for various reasons, were not practicable at the outset. But the basic principles and structure of the arrangements were not significantly altered.[2] The NCB's view of the continuity of the arrangements for consultation was expressed like this in 1982:

> The consultative machinery set up in 1947 has stood the test of time: it has been modified to match changes in the Board's organisation; and it has been extended to bring in additional parts of the enterprise. It has also been extended at national level to provide a high level forum for discussion of policy issues. The basic form of the 'consultative machine' as it is known, however, has remained constant and, as a result, it is firmly established and widely respected throughout the industry.

That view would not have been limited to the employer. If there were weaknesses in the arrangements for either conciliation or consultation they do not appear to have been in the machinery so much as in the ability to get those concerned to use it properly whenever it was needed, especially at the lower levels.

With correct and comprehensive use the machinery ought to have reduced quickly the misunderstandings that inevitably arise and to have made possible the peaceful settlement of disagreements about the terms and conditions of employment. So complete a contrast with what had been happening for so long was too much to hope for, but an appreciable improvement was not. It is evident, however, that improvements were patchy, intermittent, and subject to periods of reversal.

The obvious good sign was the absence of an official national dispute until the overtime ban and subsequent strike in the winter of 1971–2. But two national strikes, each extended for several weeks, with only two years separating them, greatly damaged the record as a whole, and the large number of unofficial strikes, some of them backed by union local officials, which occurred every year, showed that nationalization had brought nothing like a complete cure. Indeed, the first ten years of nationalization formed a period of increasing numbers of strikes, and there were more of them than there had been under private ownership. Most of them were very small, very short, and very localized (often only

[1] Chap. 4 (pp. 143–5) above for details.

[2] NCB *Ann. Rep.* 1947, 15–16, 20–2, and 24–7 and 1948, 34–40 and 51–8 describe the arrangements set up at the outset; MMC, *National Coal Board*, I, 258–64 describes them as they had become by 1982.

one shift on one face in one colliery), and though numerous, they were not nearly as numerous as the disputes which were settled peacefully through the conciliation arrangements or just by informal talking. But though they lost fewer man-days than the previous less frequent but bigger strikes, they were a running sore, interrupting the continuity of production and taking up a lot of management time, and occasionally one of these disputes flared up and spread rapidly. A maximum of tonnage lost through disputes was reached in 1955–7, and in 1957 78 per cent of all the recorded disputes in British industry and services were in the coal industry. Thereafter there was a huge reduction in the number of small unofficial strikes, far more than proportionate to the contraction in the size of the coal industry. The number of rather larger strikes fell more slowly and so, in consequence, did the output lost through disputes. Nevertheless, the period from 1958 to 1968 was one of great improvement and the coal industry ceased to be the outstandingly bad example of proneness to dispute. From 1969 there was a renewed increase in strike activity, though the number of separate strikes remained far lower than it used to be. The features were a larger scale in the unofficial strikes that took place and a greater readiness to contemplate official national disputes. Even in the years when there was no official national dispute the proportion of output lost through disputes was usually higher than in the more strike prone opening decade of nationalization. Yet the symptoms of industrial unrest were somewhat intermittent. The number of man-days lost through strikes, per thousand employees, was better in the coal industry than the average for all industries and services in 1971, 1973, 1975, 1977, and 1979, and was just about the same in 1980.[1]

Other characteristics, which may have some connection with the state of industrial relations, or at least with labour morale, show rather different trends. In the first years of nationalization, absence from work fell below the high rates of wartime, but in the fifties it began persistently to rise again, and the upward trend continued in the sixties. Absence, in fact, was greater than in the pre-war privately owned industry. The one favourable achievement in the sixties was that, even when total absences were at their new high levels, the rate of voluntary

[1] Chap. 5, section i, Chap. 6, section vii, Chap. 7, sections i, iv, and vii above. MMC, *National Coal Board*, I, 260–2 summarizes the contrasts in strike action between periods, and provides supporting figures (II, 188–9), though in detail only from 1969. There are figures providing comparisons with pre-nationalization experience in W. H. Sales, 'Human Relations and Welfare', in *Colliery Guardian*, *National Coal Board*, 1957, 102.

absence was falling. The seventies had a better record, but it was only near the end of the decade, after the incentive and bonus element had become established in the wage system, that the rate of absence fell sharply to lower levels than had been known at any time since before the Second World War.

Other features dependent on co-operation likewise show a mixed record. In the first decade the mineworkers were generally very helpful about Saturday working as long as it was needed, but they were often unhelpful about recruitment of additional workers, despite the national policies of the NUM and genuine official efforts to apply them. In particular, the repeated resistance to the employment of foreign workers, including refugees, was impossible to overcome in many areas. In the seventies, although after the two national strikes there were often very harmonious personal relations at the top between the NCB and the unions, it was sometimes very difficult to secure actions to execute in detail policies which the NUM had agreed to support. There was the protracted difficulty in moving to an incentive scheme, despite the agreement after the Wilberforce settlement to look for one; and there was the detailed resistance to colliery closures that were in accordance with the agreed strategy of the 1974 *Plan for Coal*. And there were the signs that such problems were not arising just because of traditional assertions of local independent views, but were symptoms of increasing division of outlook and purpose within the NUM. This was a division not confined to broad new issues of principle but extending to standard items of wage negotiation. A general increase in polarization of attitudes was noted.

Thus there were many ways in which the subsequent history of industrial relations fell short of the hopes attached to nationalization. In relation to the prior expectations the shortcomings call for more explanation than the successes, which were by no means absent. Especially, perhaps, the growth and persistence of numerous unofficial stoppages, which a complete and orderly operation of the agreed conciliation procedures would have abolished, demands comment. Any explanation must refer to the deep roots of traditional behaviour, complexities in the formal machinery of negotiation, and the influence of individual personalities, but it is bound also to include elements of speculation and uncertainty. Human beings do not cease to be inexplicably odd on occasion, even when they act with enough similarity to be presented as statistics.

Tradition was important, not least because it embraced very wide local variations. Lots of people reacted to a particular sort of situation in a particular way because this was the local norm, long familiar to everyone involved. There was no good reason to expect nationalization to break such established patterns of behaviour, at least not more than very gradually. To some extent it may even have reinforced them, because any vague expectations that somehow vesting date would bring a magical transformation were inevitably disappointed. To most working miners nationalization must have seemed to change little in their daily lives. They did the same things, in the same colliery, alongside the same companions, under the supervision of the same officials and managers. If they were soon doing them for rather more pay and with a shorter normal working week, this might be discounted because these benefits were regarded as no more than a response to past promises. In one respect, indeed, change for the worse might be assumed, because the ultimate employer was farther away, out of sight and apparently out of reach in London. So why should miners act any differently, if there was nothing that was both better and new? Indeed, as an expanding market made employment more secure, there was less risk in being rather more obstreperous than before.

It may have been just after nationalization that such disappointments were most obvious, but the situation did not change quickly, and if resort to unofficial strikes was partly traditional, by the fifties the tradition had some post-nationalization experience to draw on. Probably not before 1960 did the ultimate employer seem much less remote. But when markets were contracting and government policy was becoming less helpful, it was more obvious that everybody in the coal industry faced some common problems. Particularly in the years when Lord Robens was very publicly fighting a political battle for bigger sales of coal, even the highest level of management was seen to be not so remote as to be unconcerned with the job security which worried everyone.

Much in the habit of local action in disputes sprang from the conditions of a miner's work and life, which instilled a strong sense of mutual interdependence. Miners, especially those underground, are in some sense aware of being a race apart, whose members must rely on each other. They are physically cut off from all other industrial workers. Much of their work involves co-operation, either by working as a team or by finding a fair way of sharing out individual tasks. The safety of all can be dependent on the responsibility of each one and a readiness in emergency to go to the limit in giving aid. At the same time many mining jobs present a hard challenge to physical effort and judgement, and as the precise form of the challenge is always changing a little, it calls for individual initiative and adaptability. Such conditions encourage quick decisions, joint action, self-confidence, and a frequent belief that the rest of the world is so different as to be unaware of what is at stake and needs to be shown. Where miners live in communities of their own kind, without a large mixture of others whose working experience is different yet equally satisfying, home life tends to reinforce the attitudes developed at work.[1]

These characteristics have been one of the main sources of the strength of trades unionism in coalmining. They have equally strengthened many other forms of corporate or joint social action, often small and local. But even in relation to working matters, these have been of many different kinds, tending to vary from place to place. The joint deputation, or participation in some decision-making process in accordance with pre-arranged constitutional rules, were just as characteristic outlets for collective feeling as the combined walkout. But one outlet was favoured in one region, and a different outlet used much more frequently in another. Differences in this respect, which are very hard to explain, seem to have been established early in the growth of a coalfield and to have become built in to patterns of behaviour which, by example, exhortation, and absorption, were transmitted through successive generations. This is probably why it was unrealistic to expect nationalization to bring any swift changes. Though there were many local variations in both behaviour and its background, it seems as though poor industrial relations, with a propensity to go quickly into dispute (and often quickly out of it), were particularly common in coalfields which, between about 1880 and 1920, grew very fast and recruited workers by migration (mostly from outside the coal industry) either into isolated one industry communities, or into what were initially rather amorphous districts, lacking in amenities, where roots of responsibility were put down only slowly. Much of South Wales, South Yorkshire (especially the district around Doncaster), and Kent owed their coalfield development to this period and had a long subsequent history of troublesome labour relations, continuing after nationalization. In some of the districts where coal was rapidly developed earlier, in communities shared with other large industries, as, for instance, in Lancashire, West Yorkshire, and North Staffordshire, or where the main development was a little later, as in much of

[1] Cf. Sales, loc. cit., 105.

Nottinghamshire, things generally appear to have started better and remained better.

If the machinery of industrial relations sometimes seemed to reveal weaknesses, this also was mainly because of its involvement in local variations, a legacy of the past which had to be retained because there was so much local diversity and so many decisions had to be taken locally, and also because the miners had so many local loyalties that they had attached to their unions and would not readily forgo. At the official national level the relations between the NCB and the unions worked well most of the time. Divisions among the unions were not a very serious problem (though there were difficulties about the representation of supervisory staff and clerks), and there were no serious demarcation problems. The unions did not need to assail the employer with any actions arising from doubts about their own strength. Belief in the virtues of solidarity and the need for collective action was so general among miners that the NUM could rely on almost 100 per cent membership without ever needing to ask for union membership to be made a condition of employment. But there were problems with the ability of the unions, mainly the NUM, to get their agreed policies observed by all their own members. The chief reasons lay in the peculiar structure of the NUM and the reality of local union power which necessitated this peculiarity.

The NUM had existed only since the beginning of 1945 and still retained many characteristics of the MFGB, which was then dissolved. Trades unionism among coalminers had grown up locally and many of the small local unions had either merged with or affiliated to district unions, mostly on a county basis. Wages were settled on a district basis, so negotiation was a matter for district unions, though there were complications because the boundaries of union districts and wage ascertainment districts were not identical. The district unions were retained as 'areas' of the NUM and the smaller bodies as 'constituent associations'. The original agreement was for the NUM to have twenty areas and forty-one constituent associations, the latter figure reduced to thirty-six by the fusing of the six Scottish associations before the formal inauguration of the NUM. Of the twenty 'areas', sixteen were geographically defined and four were occupational. To the latter there were added COSA and a second 'power group'. From the geographical areas Bristol soon disappeared, and Somerset merged with South Wales some ten years before coalmining ceased in Somerset.[1] The main reasons for the

[1] Arnot 1961, Chap. X, describes the transformation of the MFGB into the NUM. The original

establishment of the NUM were to give better protection to members against the effects of competition among themselves and to move towards a national system of wages and conditions of employment. So the greatest practical change was the obligation imposed on area unions to abide by decisions of the NUM national executive on wage questions, and this was associated with the transfer of responsibility to the national executive for negotiating national agreements on hours and wages and related benefits such as sick pay, holiday pay, and pensions.

But this still left a lot of power and influence to the areas, and freedom to operate very diversely, which they did. Each area could register individually as a separate trade union. The areas had bigger funds than the national body and they were directly responsible for all the union non-industrial affairs. They, rather than the national body, were directly in touch with the local lodges or branches, which used the colliery as the normal unit of organization. The areas and the branches for many years also retained a large role in wage negotiation. Wages were still fixed on a district basis, and nationally negotiated changes had to be adapted to the rates established in each district. The piece-rates which prevailed underground had so many separate components which varied from pit to pit, and with changes in physical conditions from week to week, that local negotiations were continually necessary and could have a bigger influence than national agreements on an individual's pay packet. As long as this state of affairs continued, it was inevitable that the lodge officials (especially the secretary), who negotiated directly with the colliery manager, should have great influence. How individual that influence was depended both on personality and on the way relations between areas and lodges were conducted.

It used to be argued sometimes that the Yorkshire Area NUM was organized in a way that made it specially likely for local grievances to be quickly turned into local disputes. This was because the area was unusually large and had a highly centralized administration, and there was in nearly every one of the large collieries in the south of the county a full-time lodge secretary (rare in the rest of the industry), easily accessible to angry men and able to act on his own with no area official readily available to be consulted. It was an unusual combination of circumstances. The Durham Area was also large and centralized but had few collieries with a full-time lodge secretary. The only other region with many full-time lodge secretaries was the East Midlands, but the

structure of the NUM is explained in Baldwin 1955, 35–51. There is a helpful brief summary of structure and function in Handy 1981, 19–22.

NUM areas there were much smaller: there were four NUM areas within the one NCB East Midlands Division. Nevertheless, the argument in relation to unofficial strikes seems highly uncertain. Both the Scottish and South Wales NUM Areas had a decentralized administration and there were very few full-time lodge secretaries, but these areas were as prone to local stoppages as South Yorkshire.[1]

In any case, there were important matters affecting pay that were dealt with directly by the areas. Each of the seventeen wage districts had a conciliation board composed of representatives of the relevant area unions and constituent associations and representatives of the NCB—divisions before 1967 and areas from 1967 onward.[2]

Thus for most union matters it was natural for miners to look to the people and the organizations they could see doing what immediately concerned them: the lodges, the constituent associations, and the areas. As all these had a greater variety of activities than the national executive, long and varied traditions, and entrenched ideas about the proper way to do things, it was not surprising that they often followed their own line, without much regard to what the NUM nationally might have proclaimed as policy. There were other problems. One of the most troublesome in the early years of nationalization was the mistrust of the NUM by some members who were skilled specialists and contemplated breaking away. The series of disputes involving winding enginemen all sprang from such feelings. For the first five years of nationalization these disputes kept flaring up, being suppressed, and breaking out again in some slightly different form or some new place.[3] They seriously upset production several times, but after 1952 a more peaceful, though sometimes rather uneasy, existence was maintained for the winding enginemen within the union. Other things which detracted from the efficient conduct of industrial relations were matters of awkward administrative design rather than fundamentals. There were the continuing differences in the boundaries of wage districts, NUM areas, and NCB divisions and areas. There was the somewhat irrational departmental diffusion of personnel functions at NCB headquarters, and the lack of uniformity in personnel arrangements at NCB area level; but this weakness was dealt with when the recommendations of the Fleck Committee led to the creation of an Industrial Relations Department with comprehensive duties.

[1] Baldwin 1955, 45–9.
[2] Handy 1981, 23.
[3] Baldwin 1955, 57–62.

There was an important change in the influence of union organization as more and more mineworkers moved on to nationally negotiated daywages as a result of the three large wage structure agreements of 1955, 1966, and 1971.[1] After 1971 the role of union areas and lodges in relation to wages was greatly reduced. The long proclaimed union policy of national wages nationally negotiated had been attained. There were few local additions or modifications to take into account, though there was some recovery of function when incentive schemes were adopted and all the local norms, to which incentive rates were applied, had to be agreed.[2] But even this was much less complex than what had to be done under the old piece-rate system. There were still all the detailed local problems of working practice to be dealt with, all the regular consultative activities and participation in the colliery review procedure, which might involve discussion of colliery closure or expansion. These were important industrial matters, and the non-industrial work continued as before. But there had been a gain of functional authority in the centre at the expense of the rest, and since 1960 the area unions had contracted with the industry, some of them very drastically indeed. The power and influence of some area unions thus seemed somewhat diminished, but there had been little change of organization in reaction to this. It was perhaps a pointer suggesting that greater effect henceforward was to be obtainable through the NUM nationally rather than through the constituent areas. As the national executive was composed of senior officers from the area unions, this could mean that long standing differences between areas, which had been prominent in their own practices, were pressed more strongly than before into differences about policy and politics within the union at national level.

But any such change must depend also on personalities and is a reminder that personalities, and the ideas they propagate, interact with inherited traditions and administrative machinery as a third important group of influences on the conduct of industrial relations. In the first twenty years of nationalization the signs are that, in all that could be done centrally and nationally, the personal influences were strongly helpful, on balance, to the conduct of industrial relations. The long period of office of Arthur Horner and Will Paynter as successive general secretaries of the NUM meant that, to the end of 1968, negotiations and measures for the implementation of agreements were handled on the union side by men of ability and integrity, with an exceptional range of

[1] Chapter 5 (pp. 210–12) and Chap. 6 (pp. 292–6) above.

[2] Chap. 7 (pp. 371–3) above.

qualifications. Both were Communists from South Wales, which commended them to the left wing, yet both collaborated successfully with presidents who did not share their political views; both put the interests of mining and miners first and ideology far enough second to ensure that it did not interfere with a practicable settlement; and both believed that agreements, once made, should be kept, and strove to act on this belief.

On the employer's side there was a long period when the national board almost always included someone with a long practical experience of the affairs of the NUM. Of former union leaders there were Ebby Edwards, the board member for labour relations under Hyndley and, for a time, under Houldsworth; Sir James Bowman who, while chairman of the NCB, personally took the leading part in negotiations with the unions; and Sam Watson, for long the grey eminence in NUM national affairs as well as the leader of the Durham miners, whose advice was available to the NCB as a part-time board member after his retirement from union office. Such personal links were usefully educative to the NCB, as well as practically helpful. Initially there were too many board members who took an excessively idealistic view of the way to conduct industrial relations and who did not always distinguish clearly enough between generosity and weakness. In the early fifties industrial relations shared in some of the confusions of the time, when there were too many private understandings which left uncertainty about the exercise of authority. Both Edwards occasionally and Bowman more frequently could demonstrate the usefulness of plain and critical speaking as a means of calling back the unions to greater reasonableness; and Bowman knew the necessity of clarity in the responsibilities for the daily practice of industrial relations, as well as in the content of formal agreements, and saw that it was restored.

If some of the advantages arising from the efforts of these influential people were not wholly realized in practice, the reason is probably to be found mainly in the many effective local obstacles. However anxious Horner and Paynter may have been that agreements should be honoured and constitutional procedures followed, if particular groups acted contrariwise there were virtually no disciplinary sanctions that the NUM *would* apply, and very few that they even *could* apply. And one of the diplomatic limitations of even the most fair-minded union leaders was that they were reluctant to offer public approval of the application of discipline by the NCB. Neutrality on occasion was about the most that could be hoped for.[1]

[1] Cf. Baldwin 1955, 93–4.

To get over difficulties without the need for formal discipline could be helped by the personal qualities of countless individuals in informal contact, but it was too much to expect that the right qualities would always be displayed at the appropriate moment. It could also be the case that the direct experience of mining life and trade union activity, which was possessed by nearly all the NCB labour relations staff at division and area in the earlier years, was not used as advantageously as it could have been. This was because many of them felt that they were in a rather uncertain position. Nearly all the NCB labour relations officers had been recruited from the unions in the coal industry. Many were recruited directly, with union agreement, after nationalization. Others had been recruited during the war by the Ministry of Fuel and Power, and subsequently transferred to the NCB. Production staff of the NCB often felt that these men hampered schemes for more efficient output by recommending excessive concessions to the workers. The NUM expected them to act favourably in the interests of their members, but treated them as people cut off from the union. They were denied any prospect of ever returning to employment by the union and were restricted to only occasional formal contacts with union activities. So there was not the ease which might have enabled them to deploy their specialized experience to best effect, and they did not have specialized training in modern methods of personnel management which might have been a compensating influence.[1] There were limitations to effectiveness here, which worried the Fleck Committee, whose report led to a diminution in the extent to which NCB officials dealing with personnel were drawn from those with prior experience in the unions.[2]

Nevertheless, the settlement of the enormous number of disagreements which were raised formally and led to no strikes or acts of non-co-operation could not have been achieved without many displays of personal goodwill by all involved, and the goodwill was the more influential when it could be guided by familiarity with the experience out of which the problem had emerged. Even when attention is concentrated on possible reasons for the disappointment of some hopes, it is realistic to recall that there were also innumerable small successes and a few very large ones. It might well be claimed that the rationalization of the mineworkers' wage system was among the greatest and most skilful achievements in industrial relations in any industry in the third quarter of the twentieth century. Even though there were serious handicaps from

[1] Baldwin 1955, 29–34.
[2] Fleck Report 1955, 25–6.

having moved to a general system of daywages, the complexities and the indefensible disparities that had been removed, the irrationalities that had been replaced by ordered simplicity, the mystifying barriers of regional vocabulary that had given way to a generally understood naming of jobs, all indicated the transition from a world in which every element in pay was a potential point of dispute in any week, to one in which the grounds for conflict might, on occasion, be bitter but were clear and were reduced to very few. That was a notable gain. It was an achievement for which the initiative came from the NCB, and the detailed design of the new system was mainly the work of the industrial relations staff of the NCB, but it would have been impossible without the co-operation of the unions and their participation in long periods of tiring work.

The other outstanding achievement was the maintenance of general industrial peace during the great contraction of the coal industry in the sixties. It was, of course, aided by the generally high level of employment in the economy, which offered remunerative choices to displaced miners. But it depended on a huge effort to arrange alternative employment or compensation for workers from closing pits, and to ensure that they knew in good time what was happening, how each individual would be affected, and what options were available to him. This was an example of the function of industrial relations work in involving not just the specialists but everybody, in starting with decisions at the top, examining the range of possible consequences at every level, and assisting the maximum choice (within that range) on the basis of the fullest information given to those affected.

From the late sixties the context of industrial relations began to change. Many of those who had worked together from both the employer's and the union side to build up the system after nationalization, and to make it work, had retired or died. There was probably less that was approached with an awareness of shared experience, and less frequently was there a common standpoint for the review of current problems. There were deliberate attempts to make the common standpoint less likely by inducing more aggressive policies and a more extreme political outlook in the NUM. For this purpose, besides the effort to build up support for Lawrence Daly against Joe Gormley in the contest for general secretary, there was an attempt to change the dominant outlook and policy of an influential NUM area as a means of shifting the balance in the NUM.

The Yorkshire Area which, despite its high incidence of unofficial strikes, always had a president and general secretary devoted to caution and moderation, was from 1967 used for this purpose. The campaign was centred on two full-time union officials who were Communists and on the Barnsley Miners' Forum. The latter, which served as a weekly discussion centre for miners, with speakers drawn from left-wing mining unionists from all parts of the country, was formed by Arthur Scargill, who was at that time a face worker and union delegate from his own colliery, Woolley. The effectiveness of the campaign was assisted by the informal organizational features which already existed in the Yorkshire Area NUM. The area was divided into four sections, each with a representative panel (headed by a chairman and a secretary) which met to discuss specific issues and to mandate pit delegates on the Area Council. The Doncaster Panel was fairly tightly controlled by five representatives and had an established reputation for militant policies and for initiating strike movements. It played a significant role in the movement to change the political stance of the Yorkshire Area. Gradually, some of the leading participants in the Barnsley Miners' Forum also began to increase their influence on the Barnsley Panel, until they effectively controlled it, and the Doncaster and Barnsley Panels then coordinated their strategy. Even though the campaign drew nearly all its support from collieries in only two of the four Yorkshire sections, viz. Barnsley and Doncaster, it had a growing influence on the policies of the NUM Yorkshire Area Council. The way was being prepared for a thorough turnround of the area's policies and its role within the national executive, to be completed after the area president and general secretary departed from office. This came about in 1973.[1]

This campaign, and others in areas that already had more aggressive policies, had an effect on union demands and practice over the next few years, though the NUM nationally, after the more moderate presidential candidate had been elected in 1971, did not go nearly the whole way that the campaigners wanted. Equally influential was the participation of mineworkers in the two successful national strikes of 1972 and 1974. For several reasons this experience was almost bound to make part of the labour force more radical.

In the first place the strikes were *successful*, when most of the press and political speechmakers had said they were bound to fail, and the dire consequences for jobs and wages that had been forecast did not happen. So strikes and threats of strikes gained a new prestige, as a

[1] Allen 1981, 134–40, 145, 154–8. This is an account wholly favourable to the maximization of militancy.

means of getting what was seen as no more than fairness when negotiation and compromise could not.

Secondly, the government enhanced the prestige by proclaiming to the public that the strikes were political. To those who called the strikes and those who voted in favour of them they were simply and straightforward struggles for better wages, and this case is easy to understand and hard to refute by anyone who looks at the figures. Undoubtedly there were those who welcomed and encouraged the strikes still more because they realized that they could introduce the participants to hardships, criticisms, a new identification of opponents, triumphs in gaining particular objectives, and other experiences that might make them readier for a political push in a new direction. But those who gave the welcome did not, and could not on their own, provoke the strikes. They could only use them. The 1974 strike was likely to seem somewhat political because the miners' pay demands went well over the limits set by the government's incomes policy. But the thing that mattered to the miners was the erosion of their wages, not the particular source of the blockage to a wage rise. The motivation for the mass of miners was economic, not political. When the government called a general election on the issue, it then, and only then, became political. The effect was enhanced because these extraordinary tactics produced an outcome in which almost simultaneously the government lost the election and the miners won the strike. So the idea was instilled that the miners could not only win a strike but could defeat a government. This was a dangerous gift to present to those who wanted to turn industrial relations into a medium of political combat, which hitherto they had been unable to do. It was not less dangerous for being unsupported by the evidence, which suggested that the government had been defeated by itself rather than by the miners.

A third legacy came from the use of mobile mass pickets on a very large scale, applied not to collieries but to the premises of large users, suppliers, and distributors of coal—power stations, docks, fuel depots. The technique was to close premises by sheer weight of numbers, and if mere presence was not enough, to use the physical force of the numbers. It had been tried out in the large unofficial strike of 1969[1] by the group that was seeking to reorientate the Yorkshire Area.[2] In 1972 the national

[1] This was the strike over surface workers' hours.

[2] Cf. the interview given by Arthur Scargill to *New Left Review* in 1975, reprinted in *The Times*, 26 June 1984. Allen 1981, 158 notes that the president, vice-president, and general secretary of the Yorkshire Area were all opposed to this strike and excluded from its organization. The attitude of

executive of the NUM issued many written instructions requiring orderly, lawful, peaceful picketing, but they were widely disregarded and the union did little to enforce them.[1] (Peaceful, secondary picketing of the premises of customers of the coal industry was legal at this time and central to the tactics of the national executive.) In 1972 very large mobile groups of pickets were first deployed from Yorkshire and initially operated mainly in Yorkshire and at power stations and docks in East Anglia, but the tactics were quickly copied elsewhere and applied in many parts of the country. The symbolic climax was the closure of the Saltley coke depot of the West Midlands Gas Board,[2] the last large fuel depot operating normally in the Midlands. From 7 February 1972 mass pickets, at first brought from Yorkshire but soon reinforced from elsewhere, were active there. Estimates of numbers vary. It is suggested that by 10 February there were 2,000 miners, and on that day, when some other unions called a sympathy strike in Birmingham and its neighbourhood, an estimated 10,000 other trade unonists marched to Saltley. This crowd, confronted by only 1,000 police, forcibly closed the coke depot, which thenceforward allowed out coke only against permits approved by the miners. The principal organizer was Arthur Scargill, who later described this episode as producing 'the greatest day of my life', and who, immediately after the closure, addressed the crowd in a political speech, telling them this was the greatest victory of the working class in his lifetime. His later comment was: 'Here was the living proof that the working class had only to flex its muscles and it could bring governments, employers, society to a total standstill.'[3] This was politicization with a vengeance!

All these mass actions involved breaches of the law and bore little resemblance to the peaceful picketing which is one of the legal privileges given to trade unions. But not much was done about their illegality.

those who did organize it is suggested by the Yorkshire Area financial secretary, who told the delegates to the Area Council that if the rule book stood in the way of a strike, it should be thrown out of the window (Allen 1981, 154).

[1] Gormley 1982, 101–2 maintains that in a seven-week strike, with over 1,000 picket lines, there were no more than a dozen reported cases of violence and most of these were contrived for the benefit of television crews or freelance photographers. This scarcely fits the style of Scargill, loc. cit., who remarked, 'We had to declare *war* on them and the only way you could declare war was to attack the vulnerable points'; and, on the Saltley coke depot, not on the day of its final closure but on an earlier day, 'Battles raged outside that coke depot.' See also Allen 1981, 194. He maintains that 'in so far as the miners were determined to win their strike violence was implicit in it . . . there were provocative incidents and violence in all the coalfields.'

[2] Chap. 6 (pp. 308–9) above.

[3] Scargill, loc. cit.

There were arrests at Saltley, more than the miners experienced elsewhere, but they did not lead to charges and prosecutions commensurate with the scale and nature of what had happened; and this was a general feature of the reaction to the episodes of intimidation and violence.[1] The outcome of the strike would probably have been just the same if Saltley had been left in undisturbed peace, but the legacy of myth would have been very different, and the mass deployment of crowds of men at one lot of coal using or coal distributing premises after another, in quick succession, unquestionably speeded up the effects of the strike.[2] The conclusion that was encouraged by events and the way they were tackled was that the intimidation worked and, as long as it succeeded, would attract no reprisal. For some there was the corollary that if the intimidation was used widely and long enough, political as well as industrial results were possible. Prominence in organizing flying mass pickets brought some rewards of fame. In particular, it was very helpful to the career of Mr Scargill, who was also aided by the serious illness of the senior Yorkshire Area officials in 1972. He was elected compensation agent for the Yorkshire Area immediately after the strike, was acting area general secretary for a time later in the year, joined the NUM national executive in September 1972, and was elected Yorkshire Area president on the death of the incumbent in January 1973.[3]

The prestige allowed to mass picketing and intimidation did not give these practices an immediate and permanently enhanced role in trade unionism. They had little part in the miners' strike of 1974, for those most addicted to them could see that they would have been counterproductive if used during a general election campaign. It was a somewhat offsetting influence that this strike was won without such methods. But the government made its own different contribution to the politicization of industrial relations through the 1974 strike, and the experiences of 1972 were not forgotten.

Some of the influences of the experience of the miners' strikes showed themselves in the general outlook and attitudes of those conducting industrial relations quite as much as they did in the coal industry. Here the new were still very much countered by older

[1] Allen 1981, 194–7 writes of 'solidarity through force' and his phrase for policing is 'police retaliation', but his conclusion (195) is that, 'considering the scale and manner of the picketing, there were relatively few arrests.' To 21 Feb. the total number of arrests for offences connected with picketing was 263. Not all those arrested were miners.

[2] Allen 1981, 200 maintains that in 1972 the strike could not have been won without this type of picketing. But there is no proof, and the 1974 strike was won without it.

[3] Allen 1981, 224–6.

attitudes, the two staying alongside each other rather than mingling. But the situation was a good deal different from what it had been a few years earlier. Two major strikes had greatly weakened the NCB's finances, their standing with their customers, and their position *vis-à-vis* the government, just when they saw the opportunity to stop the long decline of the industry. They felt they could not risk further damage from the same source, and their deliberations over and over again show a cautious concern with likely union reactions which affected their decision-making. In fact, the institution of the tripartite examination in 1974 was a contribution to the maintenance of regular contact between the NCB and the NUM in the consideration of developments which were obviously of both fundamental importance and joint concern.[1] But it has already been seen that, although in some respects there was a very easy top level relationship between the NCB and the NUM, the former found it difficult to feel confident how far the latter would go in translating general professions into practical actions.

There were still many important matters to be dealt with between NCB and NUM areas, and it was still the case that different NUM areas saw things very differently from one another and sometimes from the national executive, whereas there was much more national unity of approach from the NCB than there had been in the days of the old divisional structure. But even in the NUM more had passed to the centre, where there were sharper differences than before. The president, Joe Gormley, was like his predecessors, by temperament and skill, a negotiator, anxious to extract a good price but always looking for a timely settlement, and he could usually judge what the members would support in a vote, and could usually obtain a majority in the national executive. But it was a narrow majority, and the demands of the opposition were both more divergent and more sharply presented than had been customary. There was less possibility of compromise and therefore more risk of lurches of policy between extremes, especially as those who had guided the more aggressive methods of 1972, and claimed credit for the success, had gained a lot of influence.

The conduct of industrial relations had thus moved into a context of increasing uncertainty. It was not made easier by the large rise of general unemployment in the late seventies and eighties, which increased the anxiety of miners to preserve their own island of apparent industrial security in an insecure world. And besides the creaking of the machinery, which had been devised in different times with a different

[1] Chap. 7 (pp. 353–4) above.

wage system, there were new personalities moved by different ideas and aims. The election in late 1981 of a new NUM president to succeed Joe Gormley at the end of March 1982 brought a large majority for Arthur Scargill, who had been his most vociferous adversary in the national executive. The new president had made his national reputation as a confronter rather than a negotiator, and had often expressed his belief in the use of industrial action as a weapon in class war. All these changes suggested that the uncertainty would only become greater and that a reversal of the trend was not attainable either by adjusting the administrative machinery or by practical measures of goodwill.

### iii. Administration of the National Coal Board

Discussion of the administrative structure of the National Coal Board, and the way the structure was used, was a favourite topic in political circles at the time when the legislation on nationalization was passing through Parliament, and for about three years afterwards.[1] From about 1950 outside interest in the subject appeared to evaporate, and for thirty years there was virtually no pressure from government or civil service for any change in organization, apart from the demand in the early seventies that the NCB should either have fewer non-mining activities or have a structure which put them more at arm's length. From time to time the NCB announced proposals for fairly large administrative changes and these would lead to some public discussion, but it never lasted long. In the early eighties the government began to take more interest, but it did not show in any but a piecemeal and superficial way. On the whole, the implication of public and governmental attitudes appears to have been either that by the fifties the basic administrative questions had been satisfactorily answered, or that those immediately involved could be relied on to work out and introduce the necessary adaptations as the need arose; but it could be that politicians, civil servants, and journalists were all easily bored by the subject.

In practice, initiatives for change in structure and practice came almost entirely from within the NCB. This was desirable not only as an indication of liveliness and adaptability within the undertaking, but also because there was a large amount of freedom of choice. The structural features prescribed by statute were few, though they were sufficiently basic to have a permanent influence for good or ill. The rest were developed and modified by the NCB themselves. It has become evident

[1] Chester 1975, 388–91, 493–5, 550–8, 1025–34.

as events have unfolded that there were periods of several years during which administrative history was a matter of very gradual evolution, and that there were a few occasions when large structural changes, with attendant innovations of practice, were made quickly.[1] The fundamental principles of managerial method and relationships were not greatly altered,[2] but the way in which they were interpreted varied between one period and another (most obviously during Houldsworth's chairmanship), and the formal structure was very different by the seventies from what it had been in the early years of nationalization.[3]

The virtues or shortcomings of the administration were thus to be attributed largely to the statutory elements in the machinery, as originally laid down and (to a fairly small extent) subsequently revised, and to the policies and judgement of the NCB. The coal industry was nationalized in haste. Two consequences of the haste emerged. The organization of the industry received rather little discussion during the drafting of the legislation (probably less than in the preparation of any other major nationalization bill),[4] and only the indispensable minimum provision for administrative structure was written into the legislation.

There was obviously a risk that the haste would cause unintended difficulties simply through oversight, but not much that was serious resulted in this way. Most of the omissions affected not the structure and methods of the organization, but the kind of business the NCB could conduct and the assets they could use. This was mainly because of rather careless drafting of the sections of CINA which were intended to restrict the NCB to the ownership and operation of assets in Great Britain.[5] For instance, the drafters of the Act had forgotten to make any reference to the Isle of Man (or the Channel Islands, though this raised no immediate problem), with the result that the Isle of Man coal depot of Lancashire Associated Collieries was left with no legal owner, and two ships whose function was to bring coal to the Isle of Man from Point of Ayr colliery in Flintshire could not legally be used. The bar on NCB activity overseas was so restrictively worded that the NCB could

[1] For the detailed history see Chap. 4, Chap. 5, section iii, Chap. 6, section iv, and Chap. 7, section ii above. It will be recalled that the divisional (or regional) structure was intended by the government from the outset, although, for convenience, it was not specified by statute.

[2] A detailed description of NCB organization and a discussion of the principles involved was given in NCB, *Ann. Rep.* 1948, 101–17. An updated account was given in *Ann. Rep.* 1977–8, 20–5, and some of the continuities were stressed.

[3] D. G. Brandrick, 'Organisation of the Coal Board', *Colliery Guardian*, Dec. 1975 and Jan. 1976.

[4] Chester 1975, 1025.

[5] CINA, ss. 63 (3) and 65 (2).

not hold patents overseas or appoint purchasing agents in foreign countries or form a sales office abroad or charter ships for foreign voyages. Although they clearly had authority to sell coal after it had been imported, it was even probably illegal for them to place import contracts themselves with foreign suppliers, as they were required to do almost from the start of their existence. All the restrictions illustrated in relation to foreign countries applied equally to Northern Ireland. By the spring of 1948 the NCB had drawn up forty points of detail on which they claimed that CINA needed amendment.[1] The government would not contemplate any new legislation as lengthy as that would be, but the more troublesome points were corrected in 1949, especially those relating to activities outside Great Britain.[2]

There was much more concern in government circles that haste and lack of thought in 1946 had led to an organization which was unsuitable for its purpose and caused an unsatisfactory quality of administration in the NCB. The doubts were extended both to the constitution of the national board laid down by statute[3] and to the character of the original appointments. There was particular criticism that all the members, except the chairman and deputy chairman, were appointed with a specific functional responsibility, although there was no statutory requirement for them to be, or not to be, functional. There was also no statutory requirement for board members to be full time, and a demand quickly arose that the experience of the national board should be broadened by the inclusion of part-time members; but it was felt that the statutory prescription of only eight members in addition to the chairman was too small a number to give scope for the appointment of part-time members.

Most of the early outside criticism of the administration of the NCB alleged the remoteness of national headquarters from those working in the coalfields, and an absence of decentralization. This is criticism which relates to the kind of administrative structure and practices that the NCB had set up below the national level, though there was often the apparent assumption that the deficiencies proceeded from the nature of the national board and the characteristics of those appointed to it. Much of the criticism from ministers was on similar lines, but Gaitskell at least, who was most closely concerned, was very much concerned about

[1] PRO, POWE 37/111.

[2] Coal Industry Act 1949, s. 2. Some small amendments of CINA were also included in Coal Industry (No. 2) Act 1949.

[3] CINA, s. 2.

administrative efficiency and the clarity of lines of managerial responsibility. He did not like functional boards, and argued that the presence of a board member at the head of a department gave the corresponding director on the divisional boards an excessive degree of authority as against the divisional board itself. This ought not to have been true if line management was properly defined and operated in accordance with the NCB's own description of it. But Gaitskell also claimed that if a board member, acting as a head of department, gave instructions to the staff of the same department at division or area, he was acting in parallel with the proper line of communication from national board to divisional board to area general manager, and causing confusion and dispersal of responsibility.[1] Again, this should not be the case with proper procedures, well understood. It should be clear enough (and it is certainly necessary) for instructions, provided they are for the implementation of policies already agreed, to pass between levels within departments, whereas the communication of policy needs to go between the boards. Gaitskell was probably right that confusion sometimes arose, but it was not because the national board had functional members. One forgotten relevant point is that the NCB operated for more than a year before there was a secretary to the board in post, and it is the secretary who would normally take charge of the central channel of communication. If things had not gone quite right in this respect, it is doubtful whether the first appointee was fully able to correct them.

There were, indeed, weaknesses in the early administration of the NCB, though they were not so serious as just after 1951, but their nature and causes were usually missed by the critics, most of whom were both inaccurate and unhelpful.[2] Gaitskell was probably premature in making major criticisms of the structure and operation of the national board as early as November 1947, almost as soon as he became Minister of Fuel and Power. Down to that time the main preoccupation of the board had had to be with the creation of a complete organizational system, something which could be done only from the top, and the simultaneous operation of the industry had been in a context of crisis inherited from the Ministry and the previous owners. There could hardly have been a

[1] Chester 1975, 551–3. See also Chap. 5 (pp. 181–3) above. Shinwell, the minister responsible for the choice of functional board appointments, also came to the conclusion that he had made a mistake (E. Shinwell, 'The Purpose of Nationalisation', in *Colliery Guardian*, *National Coal Board*, 1957, 2). The Fleck Report 1955, 37, discussed the arguments about the duality of loyalties when functional channels are used in addition to a central channel, and reached conclusions quite different from Gaitskell's.

[2] Chap. 5, section iii above.

more untypical period for testing the appropriateness of the new arrangements; and hindsight, which reveals the durability and soundness of the work of organization-building, suggests that the national board had functioned far better than the critics, including Gaitskell, maintained.

Such remedies as were attempted by the government were those concerning the constitution of the national board, which were included in the Coal Industry Act 1949. The innovations had a lasting effect, though not a very powerful impact. The rearrangement of the size of the national board was intended to ensure that there would always be a significant number of part-time members who could bring in experience and judgement from outside the coal industry. This arrangement persisted and was, on balance, helpful. The contribution of part-time members depended very little on constitutional arrangements and almost entirely on their personal qualities and the amount of effort they could give to their role. Normally the full-time members were left in a significant majority, and virtually all part-time members recognized from their experience that this was desirable, for they themselves had neither the detailed knowledge of the coal industry, nor the time, to be principal designers and supervisors of policy. Their role was to put up suggestions and advice. Only after 1980 was the cry occasionally going up, usually from politicians and with no proffered support from evidence, that what was really wanted for sound policy making was a board made up mainly of part-timers drawn from a wide range of businesses.

The other statutory innovation of 1949 was to permit, but not require, the appointment of two deputy chairmen, in the hope of having a broader range of full-time policy makers who combined authority with freedom from departmental distraction. Not much use has been made of this provision, perhaps because the initial trial, from 1951 to 1955, was a total failure. One of the deputy chairmen then appointed, Sir Eric Coates, came with no knowledge of the coal industry, failed to win the trust of colleagues, and could not match the qualities of the other, much respected, deputy chairman, Sir Walter Drummond, who carried many of the heavy burdens of a difficult period. Since then there has occasionally been a second deputy chairman, for instance to give Lord Robens a running-in period before he took up the chairmanship, or to provide a period of overlap when a deputy chairman was approaching retirement. The flexible possibilities were useful, but normally the board operated with only one deputy chairman and experienced no difficulty.

It might be contended that the statutory changes dealt with matters that were less basic than the criticisms that were so freely bandied about in 1948, however inaccurately directed many of these may have been. Remoteness and centralization, and their relation to speedy and sensible decision making, were indeed important subjects. A long chain from top to bottom of the organization, in marked contrast with conditions in many of the old colliery companies, was the inevitable result of the merging of almost the entire coal industry, which was the form that nationalization took. When the NCB in February 1948 were required to tell ministers what they had done about decentralization, they were justified in their conclusion:

It is not centralisation which the critics object to, but nationalisation; and they would not be satisfied unless the process of decentralisation was pushed to the point where the industry once more consisted of numbers of competing units uncoordinated and in the main ill-assorted and ill-equipped.[1]

It was politically impracticable to nationalize the coal industry on the basis of competing regional groups. The miners, whose support was being sought, would never have accepted it, for they regarded competition between coalfields as a source of weakness in their position and they had resolved to move towards national wages; and the problems of cross-subsidy and investment in high cost coalfields, as a contribution to the maximization of output, would have been enormous, even if the Coal Charges Account had been kept in being. So there had to be one national board, and since in manpower terms this was Europe's largest business unit, sheer size created the problem of the long chain of command.

There were other problems. The need for continuity of output at the highest attainable level meant that old familiar practices and institutional forms were kept, especially by the areas and at the collieries, even though they clashed to some extent with the new organization. The combination of an abundance of men who had some managerial experience with a desperate shortage of those who had an understanding of the way to run large undertakings posed a dilemma. Either too many men had to be put in posts beyond their ability, or the organization had to encompass a large number of relatively small sub-organizations that were within the scope of the available talent. In practice, both things happened to some extent.

The high degree of delegation, even at the cost of some managerial

[1] Memo. for Socialisation of Industries Committee (Ministerial), quoted in Chester 1975, 553.

laxity, was evident in the variety of styles in which different divisions and areas operated, and in the many ways in which the Fleck Report was able to illustrate the need to establish or recover central control over the execution and monitoring of national policy. These matters have been passed in review, and the inaccurate knowledge and slapdash diagnosis of most of the early critics is evident.[1] Nevertheless, some questions of lasting importance remain. In particular, it may be asked whether the structure could have been simplified, if not at the outset then at least a good deal earlier than it was.

Since 1955 the general trend within the NCB has been towards greater centralization. Various influences have been making this easier and more appropriate. The great reduction in the size of the coal industry, the narrowing of the range of coals demanded, the reduction in the number of customers accounting for most of the sales—all meant that there were fewer units to be administered, fewer decisions to be made, fewer occasions when local discretion was essential in order to get necessary action promptly. In addition, the great advances in the electronic processing of information made it much easier in all large organizations for the centre to keep abreast of what was happening in any subordinate formation, draw its own conclusions, and use them as the basis of its own programme of action. But despite these aids to centralization, the fundamental structure of NCB organization was not changed until 1967. There are some signs that particular elements in the structure may have been retained rather beyond their useful life, though not for very long.

The hierarchical character of the NCB organization, with its five or six tiers, tended inevitably to make the top seem farther from the bottom and to be a source of delay in the channels of decision making. But its existence was the penalty for the combination of huge size with multiplicity of operating units. The only bits that might have been treated as superfluous from the beginning were some of the formations between area and colliery. There were numerous variants, which probably resulted from the nature of the local supply of managers and the continuation of arrangements as near as possible to those familiar before nationalization. The Fleck Report rightly urged a tidying-up of these excrescences and irrationalities, particularly where there were both a group and a sub-area between the area and its collieries. They not only complicated the tiered structure, they muddled the line of administrative responsibility and so made inefficiency more likely.[2]

[1] Chap. 5, section iii above.
[2] Fleck Report 1955, 49–51.

Until the industry began to shrink in the late fifties, and both the urgency and volume of its investment programme were reduced, there was probably nothing else that could advantageously have been cut out. The divisions were necessary in order to try to get the right number and quality of senior staff to match each local situation, and to choose, justify, co-ordinate, and control the investment projects of the first *Plan for Coal*, even if the projects originated in the areas, as they usually did. This was perhaps even more important than relieving national headquarters of all but the largest policy decisions, and keeping fairly large decisions on finance and administration closer to the point of production. When it was evidently impracticable to press ahead with development investment in the way previously thought indispensable, and when decisions about the size of the industry had to be made centrally, with a close eye on what the government would accept, the role of the divisions was obviously being reduced. Such conditions were evident by 1960. And when, despite the disappearance of the need to give every latitude and encouragement for the securing of maximum output and additional development, the divisions could still not impose their authority in cases of disagreement with area general managers, as frequently happened, the value of divisions as a separate tier of organization began to look doubtful. Their abolition, when it came in 1967, was a few years overdue.

There were no other clear examples of avoidable delay in simplifying structure in practicable general ways, though some changes in the creation or rearrangement of particular departments look as though they would have been just as appropriate if they had happened earlier. After 1967 the policy was to try to keep organization under more continuous review and guide it into more evolutionary adaptation. This had obvious advantages, though it did not always avoid delay. The variations in area management to match detailed local needs, without sacrificing the common elements in structure and procedure that were needed for control and accountability, showed the advantages. But some of the mergers of areas came more slowly than the size of the business, and the increased means of exerting *informed* direction despite physical separation, would have suggested. There is also the possibility that the attempt to rely on gradual adaptation could cause a delay in recognizing that conditions had developed which might once again make more general changes appropriate. It may be that by 1981, with a further concentration of the market and growing difficulty in containing losses, more comprehensive reorganization had again become immediately desirable.

Internal discussions about this were in progress from then on, but as the characteristics and trends of market change, and also some of the forces making for greater concentration in other respects, had been showing several years before this, it might have been possible for action to have come earlier.[1]

Attitudes to the possibility of fairly comprehensive structural change are bound to be influenced by judgements of the value of what had been achieved when it was tried before. This is particularly so because on every such occasion there is upheaval in the working life of many people, and they need to be convinced that their discomfort is in a good cause. The NCB experienced three episodes of drastic reorganization. The first was the initial creation of the nationalized organization. This was often not put in the same category as the others, but for many of those engaged in the coal industry, it was by far the biggest upheaval of all. The durability of the new institutions and of the management principles then adopted was its justification. Later concern was not to overthrow these, but to modify them as circumstances changed and to alter them in detail to make them work better. The second episode was the sequel to the Fleck Report of 1955. The third was the change to a three-tier structure in 1967 and the follow-up of detailed changes over the next few years.

In most respects the changes after the Fleck Report were less disturbing than those in the other two periods. This was partly because one effect was to undo some recent changes that many had found very frustrating, and partly because the Committee found the basic structure of organization to be sound and did not seek its drastic amendment. The Fleck reforms were much more concerned to correct the details of organization and management in the interest of effectiveness. In so far as they tended towards upheaval, it was because the number of details needing improvement was so large. The effects of the Fleck Report were beneficial in many ways. The changes in departmental structure, the emphasis on better recruitment and training of managers, and the insistence on clear definition of lines of authority and accountability were invaluable. There were real improvements in the quality of administration and management as a result, though for a good many years managerial behaviour still did not always conform as closely as it should to the prescriptions about authority and accountability.

[1] Cf. Chap. 7 (pp. 349–50) above. NCB *Ann. Rep.* 1983–4, 9 refers briefly to the discussions from 1981 onwards, and the changes in structure and style which resulted (though only in part) from them, two to three years later.

The report itself included a warning about the limited short-run effect of its recommendations, and the Fleck enquiry had drawbacks, which were not all fully appreciated. It came rather late in the day in relation to some of the accumulated problems, and however much the authors of the report sought to ground their recommendations in basic principles, they had also to put many of them in the context of the controversial divagations of the Houldsworth regime. For these reasons some of their influence was likely to be reduced by at least partly successful resistance. People in managerial positions who had had plenty of time to entrench themselves in habits that were now condemned were not likely to change readily. Retirement and transfers between posts could be more effective than even a nominally implemented committee recommendation, but such sources of improvement worked rather slowly. The Fleck recommendations also met the unforeseen problem that their attempt to raise not only the quality but the quantity of administrative input came just before the onset of contraction in the coal industry made such a change harder to justify.

The contraction was a far bigger upheaval than the implementation of the Fleck recommendations. It established a new context in which it seemed possible that the Fleck reforms had not been drastic enough, particularly when it appeared that they had not solved some of the problems to which their principles had been specifically directed. This was probably most worrying in the continuing weakness of some of the divisions *vis-à-vis* their areas, to which reference has already been made. The Fleck Report stated that the divisions 'must co-ordinate the work of the Areas; must help, guide and stimulate the Areas in the light of the superior knowledge and experience which should be available at Division, and must supervise the Area managements'.[1] But even half a dozen years later, some Area General Managers continued very much to pursue their own line, whether or not this was what the division preferred. Some of the reasons for this are implicit in other passages in the report. Area General Managers had been left free from the beginning to organize their areas as they thought best, because the national board had not laid down any standard pattern. Area General Managers had also been given larger delegated powers for capital expenditure than were known in any other industry. Some of them regarded themselves as high grade mining engineers rather than as managers.[2] Those with the latter view were unlikely to believe that divisions had superior knowledge and experience, and men who had enjoyed so much latitude for so long, and

[1] Fleck Report 1955, 36. [2] Ibid., 42–4.

had often had high status and a fair amount of independence before nationalization, were not likely to change without compulsion.[1] Personal difficulties were supplemented by limitations in the application of strict managerial control, despite what the Fleck Report had said. For instance, the refusal until 1962 to give agreed estimates the status of firm budgets[2] made it easier for lower formations to avoid some of their accountability to higher formations.

It appeared in the early sixties that the upheaval produced by the Fleck Report had not been big enough. This was partly because some of the recommendations could not have quick results; but also because some of them had not been applied thoroughly enough, and some of the supplementary enquiries which the report insisted ought to be made and used as a basis for further detailed changes (notably in the staffing of each division and the devising of a common pattern of area organization) had not been carried out.[3] Moreover, the size and commercial environment of the industry had changed. So there was a suitable context for another attempt to design fundamental changes of structure, which Lord Robens at the beginning of 1964 asked the national board to investigate. The result was the Collins Report of 1965 and the change to a three-tier structure in 1967.

Such a change was urgently needed by then. The old structure was too elaborate for an undertaking with the reduced size and much more reduced number of separate operating units that characterized the NCB by the mid-sixties. With fewer tiers it was easier to be more strict in operating the lines of communication, delegation, and accountability. It was also practicable to revise the functional boundaries, location, and staffing of departments so as to serve the needs of a simpler structure more economically. These changes took time. The Collins Report, though basically sound, showed some reluctance to follow through its own logic. It left scope for quite a lot of vestigial survivals of divisional organization and did not go into much detail about the implications of

[1] Robens 1972, 109–14 discusses this clash of authority between divisions and areas with some vivid personal illustrations. He comments that some divisional chairmen 'contented themselves with a Post Office role, exercising judgment only to the extent that they occasionally decided not to press their Areas to get on with any aspect of the National Board's policies with which they disagreed'. This is almost certainly an exaggeration of a point that had genuine substance. The enforcement of authority and accountability was improving, though more gradually than was wanted.

[2] Chap. 6 (p. 266) above.

[3] Fleck Report 1955, 39, 42, and 92–3. It appears that from 1957 attention was diverted to short-term expedients to cope with falling demand, and long-term reforms not already in train were neglected.

its proposals for departmental organization and staffing. The report therefore needed a lot of tidying up before a practical scheme could be brought into operation, and a high proportion of the changes in detailed practice and departmental arrangements, in order to adapt efficiently to the new form of organization, did not come until after the Sheppard Report in 1968 and the proposals of the various working parties that followed up some of its specific points.[1] In this phase of reform there were the advantages of greater thoroughness, but some possible drawbacks from a rather gradual carrying through of the later adjustments. It was possible that the undertaking was still left with rather more administrative input than it should have needed. The MMC in its 1982 enquiry, without asserting that there could be no justification, was clearly worried that non-industrial (including management) staff had been increasing, relative to mining activity, in the seventies, though there was a reduction from 1980.[2] But the main reason for the increase was the need to implement the 1974 *Plan for Coal*.

Worries of this kind obviously have a bearing on questions about the efficiency of the administrative arrangements. That the administrative system was made to work without a lot of obvious creaking is evident. That it had weaknesses is also clear, both from the occasional large upheavals and from the reforms at particular points that were found desirable from time to time, though this is mostly evidence of weaknesses for which remedies were attempted. But whether the system in general worked at most times as near to the best practicable standards as could reasonably be expected nobody knows. The special enquiries, such as the Fleck Report, suggest that there were appreciable periods (including perhaps all the first ten years, but more particularly the second five) when it did not. Any additional answers depend very much on the subjective impressions of individuals who could spread their observations widely. Lord Robens has given his view that when he arrived in the early sixties the standard was not good enough, and many changes of attitude as well as procedure had to be achieved.[3] There are fewer suggestions of that sort for any later period, though there was no regular practice of reviewing all parts of the organization in accordance with the most objective criteria available. In any case, the form of the organization probably made, at most times, less difference than the way it was used.

[1] Chap. 6, section iv above.

[2] MMC, *National Coal Board* I, 304–5.

[3] Robens 1972, 112–19.

Stricter views were beginning to be applied in the seventies. In 1975 the NCB established a separate Audit Department when it was decided that the internal audit function should no longer be restricted to the correctness of accounting records, the prevention and detection of fraud, and the security of assets. From then on, one of the tasks of the new department was the audit of management controls with the aim of making them more effective and getting better value for money. This was not just a matter of checking retrospectively what had been done. The Audit Department was encouraged to look for places and procedures where significant improvements could be made. Other developments included the setting up of a Methods Evaluation Unit in headquarters Staff Department, mainly to carry out detailed measurement of clerical work. From 1978 to 1982 there was also, in conjunction with Hay Management Consultants Ltd., an evaluation of all managerial posts by means of job analysis. Nevertheless, there was some feeling that there were still gains to be made by extending the systematic review and measurement of non-industrial jobs and applying the additional knowledge to the achievement of economy in methods and in staffing levels.[1]

The effectiveness of the organization requires not only economy and smoothness in the regular routines of administration and management, but the existence of scope for policy initiatives to be put forward without being automatically stifled, and for them to be translated into action. The NCB organization always allowed scope of this kind, though it appears to have been achieved in different ways at different times. Whether there was enough it is impossible to say. The range and quality of policy making appears to have depended much more on the ability of individuals than on the nature of the administrative structure.

Initiatives requiring investment in mining projects came mostly from Area General Managers and, to some extent, from colliery managers. This remained so through all the changes of organizational structure. For many years Area General Managers also made and carried out many decisions of their own which added up to something like a policy for operating their areas. From the mid-sixties there were changes as the areas came to be directly supervised by national headquarters and the older Area General Managers, who had taken a high degree of independence for granted, retired or were removed, and were replaced by a generation who had known no employer but the NCB, and no procedures but those of the nationalized industry. They were still expected to take the forward look and devise the projects that would

[1] MMC, *National Coal Board* I, 289–90, 297, and 305.

keep the industry's output and costs at the required level, but all this was within a framework of policy originating higher up.

A somewhat similar shift took place about the same time in the source of initiatives on general policy. National headquarters was the normal place for these and they might be produced by either board members or officials, who all contributed. A notable feature was that headquarters departments, and individual specialists within them, used to take it for granted that one of their advisory functions was, without specific request, to take a forward look at likely policy needs and make proposals for meeting them. It was not just the officials who were departmental heads who acted in this way. Many of the proposals came from lower down; i.e. many policies originated in ideas generated and volunteered by middle management. The national board decided whether they should be developed and adopted. Here was one of the offsetting strengths of an organizational system which received some accusations of lack of clear definition and control. By 1970 the definition and control were established, but policy initiatives had then to start nearer the top, with the full-time board members and the officials heading the larger departments.

At least two important questions are suggested by these arrangements. One is the extent to which those who had to supply the initiatives were open to outside stimulus. Since, for so long, so many new proposals started with officials, this was bound up with the way these were recruited and trained. This connection did not lose its importance in the later years, when much more began at board level, for, from the later sixties, most board members had spent most of their careers as officials on the NCB staff. The NCB non-industrial staff in the early years had mostly had experience with a variety of companies, many of which had additional interests outside coal. But the staff became more and more completely home-bred as time went on. The general practice was to recruit school leavers and new graduates into the lower grades, and fill higher posts by internal transfers and promotions. Indeed, from 1951 the NCB had agreements with the unions representing non-industrial staff that vacancies above the levels for school and college leavers would be filled internally, as far as possible. So openness to external stimulus had to come from training policies, professional contacts, and the increasing number of formal relationships established with the coal-mining undertakings in other countries.

The MMC suggested that there might be lost opportunities of raising the level of efficiency by means of outside talent.[1] This may or may not

[1] Ibid., I, 291–2 and 305–6.

have been a justified fear. There can be little doubt that the general quality of managerial staff had been much improved since the early years of nationalization, but this is not proof that the staff could not have been more innovative. On the other hand the choice of staff was restricted because mines legislation ensured that there were so many posts that could be filled only by men with particular types of qualification. The majority of the staff recruited to management grades were engineers. But to ensure a broad outlook in engineers was one of the important tasks for staff policy, as it became increasingly rare for non-engineers to make a career in general management at any level in the NCB organization. There was a marked contrast with the first twenty years when the national and divisional boards had a preponderance of non-engineers. Here was a policy problem which remained unsolved in the seventies, though satisfactory results could be obtained in practice by careful choice of individuals to fill senior posts, and also because appropriate training and development of staff, whatever their first specialism, was a major role for the staff college.

A second question concerned the nature and purpose of the national board, and so came back to all the old arguments of the forties about functional and non-functional members. The clear-minded among those who denounced the functional appointments maintained that the board should devise policies and should set up an organization and procedures to ensure that the policies could be carried out, and the officials should be responsible for executive actions to carry out those policies. It was certainly desirable to make plain who was responsible for what. Both the Burrows and Fleck Reports insisted on this, and the Fleck Report sought to indicate how the functions of board members and departmental heads could be distinguished, even though the former could and should retain a clear responsibility for the existence and updating of a policy for the whole field of a department's work.[1]

But in practice, the line between policy and executive functions cannot always be sharp. On the one hand, the national board always chose and authorized policy, but for many years some of its choices were made from proposals which came from those whose functions were nominally executive and which board members might never have thought of for themselves. This did not necessarily weaken the role of either party. More obvious weakness appeared when settled policies were ignored elsewhere, either without the higher formations knowing, or (despite their knowledge) without sanctions against the transgressor.

[1] Fleck Report 1955, 12.

In such a situation there is obviously an executive failure, but, surely, there is also a policy failure; for is not a general rule for the enforcement of particular policies, and holding to account, an element of the policy function itself? 'Strategy' might be a better term than 'policy' for the board's basic function, but strategy in a vacuum is useless; it must include arrangements for reaching a particular set of results at a particular time, and that is something which must contain a significant executive element, i.e. a strategy must include means to make possible its own realization. So even a policy board did not, and could not, have the purity of role for which critics sometimes clamoured.

The NCB, though constrained by the political environment, shared many of the administrative needs of other large business undertakings. The Fleck Committee, convinced of the necessity of functional knowledge and responsibilities for a board and its members, tried to summarize what the requirements were.

> The men at the top must be sure that the undertaking is moving in the right direction. For the leaders to know when and how to step in and exercise the difficult role of decision, whether the purpose be to initiate new principles and methods or to end indecisive discussion or to reverse a trend in practice, calls for the highest qualities of industrial statesmanship.... We know from experience that in a large organisation it is usual for two or three men occupying top positions not to have their duties specified, but to be left to exercise leadership and to watch over the activities of the undertaking as a whole. Beyond that, however, it is essential that every activity in the organisation should be the particular concern of at any rate one Member of the Board.[1]

In short, someone or some group must 'run' the business, which includes not only starting new things but watching it as it runs and repeatedly stepping in to guide the way it is running. And what could be more 'executive' than stepping in to reverse a trend in practice?

At the time of the establishment of the NCB the term 'chief executive' for a top member of a company board was unfamiliar in British business, but it was normal to have a 'managing director', who might also be the chairman but very often was not. In later years the terms 'chief executive', who also might or might not be chairman, and 'executive directors' became more and more common. Their growing familiarity clearly indicated the character of essential roles in the direction of firms. The roles had to be performed in the NCB, as elsewhere, but there they were complicated by the additional need to gain political support and, sometimes, to work out modifications of industry strategy

[1] Ibid., 7–8.

in order to fit in with some overriding requirement of government policy. So there was a political role of informing, persuading, and, if necessary, opposing. In dealings with ministers it had to be discharged mainly by the NCB chairman, because it called for the maximum authority. In dealings with civil servants more of it could be performed by other board members and senior officials. Such requirements could cause some diversion from normal business methods of discharging the executive role of the board, but there were differences from time to time in the extent to which chairmen and their colleagues thought it appropriate to allow this to happen.

The NCB did not have a chief executive,[1] and this was not merely the absence of a name. It would be a mistake to think of the earlier chairmen regarding their function in that sort of way. Much of the executive role of the national board was performed collectively. In particular, the General Purposes Committee, from its inception in 1948, had a key position. The NCB chairman was always its regular chairman and, although this was not so at its inception, for most of its history it included all the full-time national board members, and they met with their senior officials present. It dealt with all major questions which were specific to individual issues and projects, though its proceedings were all reported to the national board and some of the questions were further considered and finally decided there. It was described in 1982 as 'the main executive Committee of the Board for day-to-day business', and had long served that purpose. Lord Robens, when he became chairman, thought the General Purposes Committee was not fulfilling its role effectively because some of its important business went earlier to other committees which, he maintained, pre-empted its decisions. He therefore sought to abolish the other committees, and succeeded.[2] While his complaints may have been justified by particular items, some of his colleagues doubted their general applicability. But his changes encouraged speed in decision-making by reducing the extent to which the same subject was passed from one place to another for discussion before it was disposed of, though this was a habit that crept back in the seventies.

The individual role in discharging the top executive function varied from one period to another, according to personal interests and temperament. The influence of the NCB chairman was crucial. Down to

[1] An office of the chief executive was established for the first time in 1983–4 and comprised both the chairman and deputy chairman, but this was in the context of a board differently composed (NCB, *Ann. Rep.* 1983–4, 9).

[2] Robens 1972, 114–17.

1960 it was commoner for the deputy chairman than the chairman to act as something recognizably like a managing director. Lord Robens clearly wanted something different. He commented in retrospect:

> Some of my Board Members appeared surprised that I wanted to be involved in fundamental business decisions. The impression in some quarters at Hobart House had apparently been that I was going to be content to be a political front-man whilst other people ran the business. That was an erroneous impression, especially as I was far from satisfied with the way they were running it.[1]

This was a lasting change, which fits in with the trend towards greater central direction.

It is an illustration of the strong influence which individuals could exert, and supports one of the basic contentions of the Fleck Committee:

> The prime necessity for the success of any large undertaking is the constant exercise of leadership at the highest level. . . . A soundly constructed organisation cannot, of itself, provide a solution to any . . . special problems. . . . The solution must be sought in better leadership.[2]

But a recognition that individually induced change also fitted in with a more general trend is a reminder of the constraints to which even the most energetic individual is subject. Chairmen with greater executive power taken into their own hands still had to rely heavily on their deputies and on other colleagues to be effectively executive in taking over major specific issues, which were sometimes very difficult. And there was still an important political role to be played, more than that of a mere front-man. The nationalized coal industry never was, and never could be, quite as similar to any other large business undertaking as the Fleck Committee implied.[3] Its political status, and ultimate political dependence, made it different. Organization, and some of the activities of those staffing the industry, had to be different because they had to meet the requirements not only of normal business operations but also of a superior authority whose understanding and demands were liable to change for reasons which, from a business standpoint, were accidental or arbitrary. The two sorts of requirement were not always wholly compatible with each other. 'Running the business' had to incorporate the task of making the best of an inescapable political relationship.

[1] Ibid., 115.
[2] Fleck Report 1955, 7 and 9.
[3] Ibid., 7.

### iv. Government and nationalized industry

The government necessarily had ultimate responsibility for any activity that was nationalized. It was conceivable for the responsibility to be exercised directly, by bringing the activity wholly within a department of state, but this sort of arrangement was regarded with little favour in the nineteen-thirties and forties.[1] When this was not done, and a nationalized activity was placed under a public corporation or board, specific provisions had to be made, and new conventions worked out, to enable the government's responsibility to be defined and exercised. There was scope for variety, and necessity for innovation, in the arrangements. Procedures applicable to government departments had become more or less standardized by a long history; those applicable to public boards had not.[2] The concept of a public board implied a large degree of autonomy, but there had to be a residual and overriding authority exercisable by the government, if nationalization was not to lose its meaning. It is helpful to consider what formal and informal provisions were made for this authority, how they were used, and what effect they had.

A basic question was the degree of detail which might be covered by governmental powers of control.[3] It did not necessarily have to be answered finally at the outset. Powers of control could be prescribed in a form sufficiently general to permit a variety of methods of application; but the presumption was that boards would be left free of interference in all matters of detail. It was, however, also widely accepted that the responsible minister must be given power to issue general directions to a nationalized board. This was largely new, though the Coal Act 1938 had given some ministerial powers of direction to the Coal Commission. The formal provisions in CINA for ministerial control were brief.[4] The main item was that:

> The Minister may, after consultation with the Board, give to the Board directions of a general character as to the exercise and performance by the Board of their functions in relation to matters appearing to the Minister to affect the national interest, and the Board shall give effect to any such directions.

In other words, the Minister could direct the board, but only on matters which were general and affected the national interest, and he could

[1] Chester 1975, 383–6 and 865–6 discusses the evolution of opinion on this subject.

[2] Ibid., 386–7.

[3] Ibid., 868–78 deals with the examination of this question by government and civil service before the enactment of the nationalization measures after the Second World War.

[4] Ibid., 888–901 for the governmental and parliamentary discussions which produced the final form of these provisions.

not direct the board without first consulting them. In addition, and more specifically, the board were required to settle in advance with the Minister their programmes involving substantial capital outlay, and their provisions for training, education, and research. These clauses, also, implied ministerial concern only with whole programmes and not with individual schemes. The board were also required to provide the Minister with returns, accounts, and information on such lines as he required.[1]

It is evident that ministerial authority to direct how the NCB should conduct their affairs was far from being unlimited. Ultimate power depended more on two other provisions. The chairman, deputy chairman, and members of the board were appointed by the Minister, who was empowered to make regulations for their tenure and vacation of office; and nearly all the capital, including working capital, was to be advanced (with statutory limits on the total) by the Minister, who had power (subject to Treasury approval) to issue directions about the operation of a reserve fund and the application of any surpluses.[2] Power to hire, fire, and finance gave the Minister as much controlling authority as he formally needed, though, of course, the circumstances at any particular time (and in some respects at all times) could impose political and other practical constraints on his ability to exercise the formal authority in the way he might prefer. Regulations about vacation of office which made it easy to fire board members might have been self-defeating. On the other hand, the appreciable statutory limitations on the kind of directions the Minister could issue gave the board an assurance of substantial freedom in management.[3] In practice, much depended on the extent to which informal requests by either party were met; and, in the earlier years, it was also relevant for some purposes that the Minister had wider powers under the Defence Regulations than he had under CINA. Under Defence Regulation 55 the Minister could make orders to provide for regulating or prohibiting the production, treatment, keeping, storage, movement, transport, distribution, disposal, acquisition, use, or consumption of any article. Power could not be much more

[1] CINA, s. 3.

[2] CINA, s. 2 (3), (5), and (7), s. 26 (2)(*b*), and s. 30.

[3] CINA, s. 4 (5) raised the theoretical possibility of significant interference with this freedom, but it proved to have no practical application. This sub-section provided that if a report from a Consumers' Council satisfied the Minister, after consultation with the Board, that a defect had been disclosed in the Board's general arrangements for the production, sale, or supply of coal, coke, or manufactured fuel, the Minister could give the Board any directions he thought requisite in order to remedy the defect.

comprehensive than this. In fact, such powers were not used to exercise authority by the Minister over the NCB, but they were used to some extent by the Minister for his own purposes and to enable the NCB to do certain things. Until 1958 opencast mining and the whole system of coal distribution operated under the powers conferred by Defence Regulations.[1]

Because the ministerial powers were set out in general terms there was sometimes doubt and argument about their precise extent, though this was not important in practice because doubtful cases could be circumvented by informal discussion and persuasion. In 1952 it was suggested that the Minister probably did have power to issue a direction on coal prices, provided it was not inconsistent with the statutory obligation of the NCB to pay their way. But this was irrelevant, as the Minister relied on the pre-nationalization 'gentlemen's agreement' for the purpose of influencing prices. In the same year the permanent secretary of the Ministry of Fuel and Power, Sir Donald Fergusson, said that the powers were there to ensure that the particularist duty or interest of the board could be subordinated to the general national interest as the Minister saw it. They were not intended to lead to the issue of general policy directions. He added:

> In practice, it is unlikely that it would ever be necessary for the Minister *actually* to issue a general direction except where the Board . . . were required to do something in the national interest which, however ready they may be to do it, would expose them to criticism as a commercial undertaking or possibly action in the courts.

This view fitted in with the general development of thinking and practice in all the nationalized industries. The Attorney-General, when asked for an opinion in 1949, inclined to a rather narrow interpretation of the extent of the statutory powers of ministerial direction, and ministers at that time adopted the line that when the government had formed a definite view on a general point of policy, they should normally try by informal discussion and persuasion to get it adopted by the boards of nationalized industries.[2]

The statutory provisions underwent only a small amount of change. The chief differences related to the financial powers. The section of CINA which dealt with advances by the Minister to the board for

[1] PRO, POWE 26/434 sets out the statutory powers of the Minister of Fuel and Power under emergency legislation, as they were in 1951.

[2] Chester 1975, 984–5, 998–9, and 1035–9.

capital purposes was repealed by the Coal Industry Act 1965. New provisions then enacted, and subsequent amendments in later Coal Industry Acts, redefined the NCB's powers to borrow both temporarily and otherwise, not only from the Minister but from other sources, including (from 1971) foreign sources. These changes, although they widened the sources of capital, did not reduce the strength of government control; rather the reverse. As the NCB's need for loans increased, they were bound to be more susceptible to government pressure. And although the only statutory restraints on foreign borrowing by the NCB were the requirement of the Secretary of State for Energy to consent and the Treasury to approve, the Bank of England in practice had almost unfettered discretion to regulate the timing, amount, and terms of such loans. The Bank, for instance, controlled the queuing order of would-be borrowers of foreign loans, and could object to the interest rate or to the particular currency of a proposed foreign loan. This is just one more reminder that statutory provisions did little more than loosely define the limits of the control of government over the NCB. To see how the relationship worked in practice, attention needs to be concentrated much more on informal arrangements, which showed far more fluctuation and development.

In the early years, despite the growing ministerial dislike of a functionally constructed national board, and ministerial and parliamentary criticism of the members' performance, relations between the NCB and government were generally fairly easy, though there were particular episodes that caused difficulty. The ease owed much to the views and behaviour of leading people. Fergusson had a very strong belief that ministers should not be involved in the detailed regulation of nationalized industries, and did his best to ensure that his ministers acted accordingly. Gaitskell, who was Parliamentary Secretary and then Minister, was an exceptionally fair-minded man. There were times when he was determined to change the NCB line on some particular point, but he did it by persuasion and persistence rather than by domination. Hyndley believed that it was both proper and sensible for the politicians to be kept continuously informed about the general policies of the NCB and the background to them, though not about detail. Street would carry the argument for the NCB as far as it would go, but retained his civil service view that if, in the end, his argument had not been accepted, the Minister's view must prevail. So there was an adequate sharing of knowledge, a fair measure of agreement about the differentiation of roles, and a willingness to seek an amicable settlement of differences.[1]

[1] In the first year or two it was customary for some or all of the Minister of Fuel and Power,

In these circumstances there was no need for general directions by the Minister, and none were issued.[1] The only formal directions by the Minister were those specifying (after discussion and agreement) the form of the accounts and the dates and amounts of the repayment of interest and principal on loans. But there were some conflicts about the interference of ministers over the heads of the national board. In 1947 ministers took initiatives on labour questions to a quite unexpected degree. Those members of the national board who foresaw the adverse financial consequences wanted to refuse agreement to the five-day week unless they had a ministerial direction, but their colleagues overruled them. Had things gone on in this way a clash would have been inevitable (and the whole system of industrial relations would have had to be reconsidered), but after 1947 ministers and civil servants for some twenty years left labour questions to the NCB and the unions without much interference. Other instances of ministerial pressure were on more isolated issues. Gaitskell prevented the NCB making an agreement with the coal merchants to establish fixed wholesale margins, and told them that he was prepared to issue a general direction which, in the event, they did not require him to do. He also persuaded the NCB to export more anthracite to Canada at £1.25 per ton below European prices. This, he claimed, was because of the overriding importance of increasing total exports to Canada, though it would have been at least equally accurate to note that he was covering up for his ministerial colleague who had promised the anthracite to the Canadian government without checking that it was available. He also intervened to require the NCB to agree terms with Peterlee Development Corporation for the sterilization of coal reserves to provide support for buildings in the new town.[2]

These last examples were claimed as interventions in support of general government policies, but they were getting near to the point at which a minister might be accused of meddling with detailed management. Had there been more of them, such an inference might well have

his permanent secretary, and the chairman and deputy chairman of the NCB to dine together each Monday. These occasions appear to have been used for the general exchange of views and to keep each other in the picture, not to float new proposals or to pre-arrange the settlement of outstanding questions.

[1] As already indicated, this was developing as normal ministerial practice. No general ministerial direction was issued to any nationalized industry until the incoming Conservative government gave one in 1951 to the Iron and Steel Corporation, and this was in the special circumstances of intended denationalization (Chester 1975, 1036).

[2] Chester 1975, 996–7.

been drawn; but in fact, the doubtful cases were rare enough for the distinction of roles to be maintained.

The one way in which ministers persistently overrode the NCB in such a way as to have an appreciable and lasting effect on their commercial management was by restriction of price changes. As has been seen, they did this not by means of statutory powers but by relying on the gentlemen's agreement. The insistence of ministers in prolonging this agreement was for years a source of irritation and frustration for the NCB. It was a significant restriction on commercial freedom to manage, and it had an effect on attainable financial results. It illustrates the possible conflict between the conditions for efficient, independent management and the aims of ministers. It was a general view in the post-war Labour governments 'that socialisation should have the constructive effect of enabling the Government through their broad control over the running and development of the industries to use them as effective instruments of national economic planning'.[1] Conservatives did not put it quite like that, but their practice for many years does not suggest a greatly different outlook in this respect. Arbitrary decisions on price changes and investment programmes were obvious ways of pursuing the aim.[2] Though they might be proclaimed as economic planning decisions, they could be related to purely short-term conditions of government finance, or to the search for a quick improvement in political popularity. Yet the timing of such decisions could be in direct conflict with the needs of the industries affected and could make it impossible for them to reach their planned financial and commercial targets. After the late fifties the NCB did not suffer much from the exercise of pricing powers in this way, because they were preoccupied with the effort to retain sales by keeping prices down. By the sixties they were conscious of more onerous exercises of government power, but until then the lack of freedom on pricing was, together with the total dependence on the Exchequer for capital, the chief political limitation on the NCB's freedom of operation.

There was, however, one other change worth noting. This resulted from the wish of backbenchers in Parliament to have more say in the affairs of the nationalized industries. Neither the statutory provisions

[1] Ibid., 998.

[2] It is noteworthy that, apart from the special case in anticipation of the denationalization of iron and steel, the first general ministerial direction to a nationalized industry was from a Conservative Minister of Transport to the British Transport Commission in 1952, to forbid the increase of railway fares outside London (ibid., 1036).

nor the preferences of ministers offered them much scope. The industries issued reports, which could be debated, but that was only once a year. MPs could ask questions, but both the Clerks of the House of Commons, in advising members about the acceptability of questions, and civil servants, in briefing their ministers and preparing parliamentary replies, were quite firm that ministers did not answer in Parliament about the day-to-day business of nationalized boards. Yet backbenchers believed that Parliament had a proper interest in anything nationalized, and that this interest ought to be satisfied. There were many fears that, if they were given more scope, MPs would seek a degree of accountability from the nationalized industries that was both in conflict with the statutes and inimical to good management. Herbert Morrison, perhaps the greatest influence on the form taken by the nationalization programme, thought there ought to be a body investigating and reporting on the efficiency of the nationalized industries, but he assumed that everyone in government and the industries would consider that a Parliamentary Select Committee was the most undesirable form which such a body could have. He was shocked to find he was mistaken. Some chairmen of nationalized boards felt that they were unreasonably criticized in Parliament because of the ignorance of MPs about their activities, and that a select committee might improve this state of affairs.

Eventually, late in 1951, a House of Commons Committee was set up to consider how the House was informed about the affairs of nationalized industries and what changes might be desirable in the methods by which it was informed.[1] The committee did not complete its work, but a similar committee in the following session reported in 1953. It recommended that a Committee of the House of Commons should be set up to examine the nationalized industries, for the purpose of informing Parliament about their aims, activities, and problems, and not with the object of controlling their work.[2] It was a result of this recommendation that reports on the various nationalized industries from a select committee of the House of Commons began to appear frequently from the later fifties onward. The NCB were first investigated in this way in 1957, with the report appearing in 1958.[3]

This development illustrates an important feature of the environment in which nationalized industries had to operate. They were subject to much fuller and more frequent enquiry, and had to prepare far more

[1] Chester 1975, 980–1001.

[2] *Report of SC on Nationalised Industries* (HC 235 of 1952–3), pp. vii and xii.

[3] HC 304 of 1957 and HC 187–I of 1958.

information for both public and governmental consumption, than businesses in the private sector. Their administrative resources had to be deployed with this requirement in mind. The division of effort between what was strictly strategic and managerial and what was politically informative or inescapable was less in favour of the former than it was in private business. On the whole, the pressures of enquiry from outside the industries increased as time went on. The external demands became more frequent and more detailed. The first enquiry into the NCB by the Select Committee on Nationalised Industries, for instance, was mainly directed to eliciting a lot of information that was fairly readily available within the NCB, and the putting of generally sympathetic questions about it. Ten years later, when the Select Committee reported on *Ministerial Control of the Nationalised Industries*,[1] its enquiries were far more elaborate and specialized and led to detailed proposals for reform. The same is true of the further enquiry into the coal industry which the Select Committee undertook in 1969.

The change in the way the relations between the NCB and the government were conducted was more marked from around 1960. The greatest immediate influence was the weakened economic position of the coal industry, which made it more vulnerable to political pressure while making it necessary for the NCB to approach the government with a greater number of specific requests (many of them requiring legislation), the granting or refusal of which provided additional means of exerting the pressure. In the later sixties the government's formulation of a national fuel policy, on which the NCB had no influence, created market conditions more extreme than would otherwise have existed and compelled the NCB to contract at an almost unmanageable rate. One consequence was that relations with the government tended to be diffused among more departments. Contraction, for instance, meant that much of the NCB's discussions had to be directly with the Department of Employment as well as with their own sponsoring department. Even the latter became more remote for several years, as the Ministry of Power was abolished and the NCB came under the aegis of a succession of vast, composite ministries until the Department of Energy was created. This further weakened the NCB's position, for the sponsoring department was not only the source of immediate control over a nationalized industry, but was also the medium through which the

[1] HC 371-I, II, and III of 1967-8. The Committee enquired into the coal industry both in 1966 and 1969, as well as including it in the large enquiry into ministerial control in 1967-8. This gives an indication of the burdensome frequency of enquiries.

needs of the industry were represented to other areas of government, of which the Treasury was particularly significant. There was less chance of successful representations if they had to be made through a medium with a reduced interest and a smaller fund of relevant specialism.

This change reinforced the effects of the exceptionally rapid turnover of ministers with whom the NCB had to deal and some reduction in civil service continuity in the relevant divisions. Moreover, the adoption of the government's fuel policy and the assumptions underlying it had spread the belief that coal was an industry with no future. It was probably hard for civil servants, or the ministers they briefed, to put forward with conviction any case presented on behalf of the NCB. Until the oil price revolution of 1973–4 showed the flaws in the basis of the existing fuel policy, the NCB had an exceptionally weak position in their dealings with the government.

There were also some more general influences which led the government to seek and exert greater authority over the nationalized industries. One was the sheer increase in the number and capitalization of the industries and in their financial liabilities to the government. This increased the concern of the Treasury with them and its determination to seek better returns from them. So there were efforts to exercise closer financial control, including the setting of financial objectives. Lord Robens was convinced that the influence of the Treasury had increased, with a consequent diminution in the influence of the Ministry of Power, in the twenty years after his first association with the coal industry, as Parliamentary Secretary to the Ministry of Fuel and Power from the end of 1947. As the Treasury hardly ever dealt directly with a nationalized industry, there was a feeling that the latter was placed under constraints without being able to make any contrary representations. Robens complained in 1967 that, as chairman of the NCB, he had only once been able to talk to Treasury representatives, and had had a flat refusal to a request to do so.[1]

Another factor was government concern for short-term regulation of the economy by means of an intermittent series of prices and incomes policies. The effects were heightened by the enthusiasm of the Labour governments of 1964–70 for national economic planning. Nationalized industries were obvious instruments to use in such schemes, on the plea of adapting their policies to the national interest as a whole. So they were liable to find themselves with specific restrictions and objectives which originated in neither their own policy making nor the controls of

[1] HC 371–II, 138 and 140.

their sponsoring departments. The latter, however, became for some purposes the media for imposing the policies, without publicly taking responsibility. The NCB complained of 'arm-twisting' by both ministers and civil servants, which caused them to act contrary to their own judgement on some matters in order not to be denied particular resources that they needed.[1]

It appears that by the late sixties, while the NCB were additionally oppressed by the weakness of their position, relations between the government and nationalized industries generally had deteriorated. They were not helped by the Treasury's public comment that the average quality of members of nationalized boards had not been high enough.[2] No doubt that was seen as a justification for a display of effortless superiority by government and the giving of reduced weight to the views of the industries. It is not very surprising that in 1968 the Select Committee on Nationalised Industries, after repeating the view that ministers should look after the wider public interest and oversee the efficiency of the industries, but should not intervene in management, reported that 'the practice has revealed an almost reversed situation'.[3] This statement was an exaggeration, but was a symptom of the muddle that had been created. Unfortunately, the Select Committee proposed remedies that could only have made matters worse. They included the creation of a Ministry of Nationalised Industries, in addition to the existing sponsoring departments, with a complex allocation of functions that would have guaranteed confusion and meddling.[4] It was easy for the government to reject such absurdities and propose very little change,[5] though it was conceded that nationalized industries ought to have the right to meet the central economic departments.[6] In practice, this did not make a lot of difference.

Changing conditions had some influence on subsequent relations. Both poor performance in national planning and a change of government lessened the urge to force nationalized industries into dominant general policies that did not necessarily suit them, though the constraints of prices and incomes policies recurred from time to time. The NCB's position *vis-à-vis* the government was strengthened because after 1973 coal was wanted again and the industry was no longer regarded as being in terminal illness. But it was also weakened because

[1] HC 371–II, 149–50.
[2] HC 371–I, 70.
[3] HC 371–I, 190.
[4] HC 3471–I, 193–4.
[5] *Ministerial Control of the Nationalised Industries*, Cmnd. 4027 (1969), esp. 3–5.
[6] Ibid. 6.

of the two national strikes and the financial damage they caused, and because of the regular receipt of government grants, most of which had to be negotiated annually. There were some other changes and improvements. The inflated amount of information which the Minister had been getting from the NCB was reduced and made more orderly,[1] though there was a tendency by the late seventies for the civil servants to take up a greater number of rather random questions *ad hoc* and demand to be told all about them. Government attempts to interfere with prices were lessened, partly because there were now more direct methods of financial control and partly because of the changed legal position after the UK entered the ECSC. But the national strikes brought a return of direct government involvement in labour relations, with serious implications for trade union attitudes, even though at other times the government kept aloof from detailed labour negotiations.

Some of the features of relations in the sixties left a permanent legacy, in particular the assumption of a wider gap in status between the personnel of government and of nationalized industries, with the corollary that the former felt entitled to range freely over the affairs of the industries without making very sharp distinctions between general policy and specific detail. The legacy showed itself in various ways: in the acceptance and answering of parliamentary questions which, in the fifties, would have been rejected as relating to detailed matters which were the concern of the board, not of the Minister; in the relegation of major items concerning the board to civil servants of lower rank than in the early years; even in the gradual change in the relativities of the salaries of the board members and of civil servants, to the detriment of the former.

How far these features were extended into purely political matters depended on personalities, ideologies, and circumstances. The NCB in the seventies and eighties generally experienced greater pressure from Conservative than Labour administrations, though the latter from time to time urged changes which the NCB thought detrimental.[2] The difference may have been because the former were less favourably committed

[1] Chap. 7 above (pp. 350–1).

[2] For instance, the NCB were urged to divert some of their coal exploration to Northern Ireland. (Private exploration did take place there but found lignite rather than coal as defined in CINA.) There was also some pressure to give the trade unions a greater say in managerial policies, including decisions on pit closures. All these pressures were successfully resisted. But the NCB noted in 1982 that 'relations with the Unions have become a progressively more important part of the Board's relations with the Government.' This tended to make the NCB's task in negotiating with the unions more difficult.

to the idea of nationalization, but not necessarily so, for the experience was the reverse of that in the sixties. Political pressure was probably at its height between 1979 and 1981.[1] It was difficult to resist then because it involved the pursuit of economic objectives defined mainly on abstract grounds with little attention to the empirical characteristics of the coal industry and its markets. A different balance in government–board relations re-emerged mainly because, within a couple of years, the empirical characteristics asserted themselves to a degree that even the most theoretically inclined had to take into account. But the relations continued to reflect some of the weaknesses of the NCB's position as a result of their financial difficulties. And in the late seventies and early eighties there was a particularly heavy exposure to different sorts of governmental or parliamentary enquiry. Inevitably there was a distraction of administrative resources and some inhibition as a result of the repeated need to try to demonstrate the rationale of management policies.

When it is recognized that the informal relationships were more influential than the statutory provisions, the rather wide fluctuations in the former suggest the persistence of an unsolved political and organizational problem. It is very doubtful whether any consensus was ever achieved about what the relative roles of department and board should be. Failure to agree about the end was coupled with uncertainty about the best means to achieve any particular definition of the end. There was probably never a time when ministers and nationalized boards did not each feel hampered by what they regarded as some usurpation of their territory by the other. In the late forties and early fifties such difficulties were to be expected because some of the recently nationalized undertakings, the NCB most notably of all, were on an unprecedented scale and had a novel political and constitutional status.[2] Some of the day-to-day business came to be conducted easily enough, especially when there were individuals in both organizations who had a chance to get used to each other and to an increasingly familiar range of subject matter. But, in the making and application of high level decisions, the passage of time and the cumulation of experience did not make things better. Indeed, they probably became, on the whole, more erratic.

[1] Cf. Chap. 7, section vii above.

[2] Chester 1975, 1038 rightly points out that such bodies as the Port of London Authority, Central Electricity Board, and London Passenger Transport Board 'were so different in scale and statutory arrangements and operated in such a different political climate as to be no guide to the handling of such bodies as the National Coal Board'. Ibid., 1035–49 has a detailed discussion of the factors influencing in the early years both the manner of exercising the government–board relationship and the ambiguity of the relationship itself.

In 1982 the NCB described to the MMC what they hoped for in relationships with governments. They wanted 'continuity in political attitudes to the industry; balance, orderliness and consistency in financial policy; and encouragement of a reasonable rate of change'. They expressed the view that the coal industry could not give the country the full economic benefit of which it was capable unless there were a strategy for the rest of the century which was supported by the government and the mining unions as much as by the NCB. And they added the perhaps wishful comment that 'the fact that the policy horizon of a particular Government may be shorter than that of the Board or the mining Unions need not prevent this'.

The choice of desirable characteristics was a list of things peculiarly difficult to attain in a political environment. One of the main sources of difficulty was the difference in the length of the policy horizon. The ability to take a long view has been one of the main foundations of success in the growth of business in the last few centuries. It is particularly important for such an industry as coal, in which investment schemes take years to complete. Politicians in government have to take a long view about some subjects, notably defence. They have sometimes been persuaded to include aspects of industry and trade among those subjects, but it is not easy unless the politicians can convince themselves that they are preparing a long-term future so productive that they can hope for some fairly rosy interim results. Usually they want quicker success. The life of a parliament is over in much less time than it needs to bring a large colliery into production. A politician in government office wants success while he is in office, and the habit of switching ministers around makes this an unhelpfully short period.

In such conditions it was not likely that government and nationalized board would act together in such a way that each would stimulate improved performance by the other, both in devising policy and carrying it out. Expectation in this respect was seldom upset. The merits and shortcomings of the NCB and of the government in dealing with energy policy depended in practice much more on themselves than on their mutual influence. At most times the government succeeded in keeping the NCB subject to an overriding energy policy. This meant very little in the first ten years, when the main aim of energy policy was to increase the output of coal as much as possible. It meant much more afterwards when one important aim was to reinforce the market forces that were compelling a contraction of the coal industry. Even in the mid-seventies, though pressure for contraction had ceased and the NCB had been able to reassert themselves a little as it was they, rather than the government, who had taken the first steps in formulating a revised energy policy with a bigger place for coal, the degree of general control was not much less in practice. This was because of the NCB's greater dependence on government money. The power of the purse was always the strongest means of control.

But within the scale of activities permitted by general government policies and by the financial provision (a significant qualification), the NCB continued to operate as they themselves judged best. Even if governments, from time to time, would have preferred this to be otherwise, they were not well placed to achieve their preference. This was mainly because they lacked the personnel with the specialized knowledge needed in order to make or to challenge managerial decisions. Unlike most of his ministerial colleagues, Gaitskell had foreseen this from the outset. He pointed out in 1949 that duplication of staff was undesirable and impracticable. The best available experts were already employed by the boards, and it could not be justified for the Ministry of Fuel and Power to appoint others of less standing in order to criticize board activities. Moreover, he pointed out that it was undesirable for ministers continually to try to teach boards how to do their own job better. That would undermine the sense of responsibility of the boards, make it harder to recruit the best industrialists for them, and run counter to the principle of decentralization of responsibility which it was hoped the nationalized industries would adopt.[1]

The effect of the concentration of technical expertise in the service of board rather than ministry proved to be lasting. Nearly twenty years later, when it had become required practice for all the NCB's largest capital projects to be submitted to the Ministry of Power, Lord Robens pointed out that they always came back unaltered, because 'they have not the technical people to indicate any changes'.[2] The practical effect in such cases was delay, which was a detraction from good management, not a stimulus to it. In some cases, especially in later years, lack of expertise did not deter ministers and civil servants from entering into long discussions of some specialist aspect of management policy and demanding detailed information about it. But this was usually a distracting rather than an effective intervention.

[1] Chester 1975, 995–6.

[2] Chester 1975, 1040–1; *Report of SC on Nationalised Industries*, 1967–8, II, 144. A new requirement for the NCB to submit to the Ministry details of major projects for the creation of new capacity had been adopted in 1963.

It would thus appear that, in practice, distinct roles for government and NCB were maintained because neither of the two parties could easily change the boundaries of the role, even if they sometimes wished to step beyond them. But the exercise of the government's role was felt by the NCB as more onerous from early in the sixties than it had been before. This was because the government's general policy became more constricting, rather than because of habitual extension of government action into more detailed matters. Occasional extensions of this kind occurred both earlier and later, without becoming systematic.

Whatever hopes and theories may have been put forward, it seems probable that the actual effect on efficiency of the kind of government–NCB relationship which was adopted was not far from neutral, though on balance it may have been slightly adverse. This, of course, is very different from saying that the effects of nationalization were nearly neutral, which they were not. The amount and prompt availability of government finance were sometimes very helpful in enabling valuable projects to be launched at the best possible time, and this may have been the most beneficial element in the relation; but it has been evident that there were also elements in the financial relationship which often handicapped the NCB.

There were also two specific sources of difficulty. One was the extra expense and effort required of NCB administration in providing additional information and dealing with repeated external investigations, instead of just getting on with running the industry. Related to this were some of the activities of the relevant government departments, and their cost. Despite the differentiation of roles, the attempt to display some sort of government supervision of the NCB did involve a certain amount of duplication and 'second-guessing', which wasted something in irrelevance without obviously producing any gains through tighter control. The most serious effect of this aspect of relations was on management attitudes, especially the tendency to inhibit the taking of risks.

The second problem arose from the combined effects in the seventies of the government's return to direct intervention in major labour disputes and its provision of extra money to deal with the financial losses which they left behind. This laid bare a relationship that had been conveniently obscured for a quarter of a century. It showed that, in the last resort, the financial terms of a settlement between employer and workers in the coal industry did not depend only on the revenue the employer could earn. Because it had been shown that they actually had depended (and not merely that they might depend) on what the government could be induced to pay out, the NCB's ability to discharge their managerial function in dealing with labour was weakened.

This last point might be regarded as an illustration of a problem inherent in national ownership, whereas other problems depend on treating nationalization in one way rather than another. The most pervasive difficulties perhaps sprang not from any choice of one preferred embodiment of authority, but from uncertainty and inconsistency about the objectives of a nationalized industry. There were clauses in CINA which were concerned with public service and others which had a commercial and financial purpose, but legislation gave no guidance to the resolution of conflicts between them. Government policy was not much clearer. The tendency was to make some *ad hoc* arrangement of a limited kind when an obvious contradiction appeared, as it did with some of the colliery closures of the late sixties, and hope for the best in everything else. A demonstration of public service and a campaign to get more value for less money can both appeal to politicians as ways of gaining, or at least avoiding the loss of, votes. Ministers were therefore prone to perpetuate the public service idea even while they were preaching that the nationalized industries should show a more commercial outlook. Yet most of the ideas and practices for reconciling the two to some extent came from the boards rather than the government. The NCB certainly evolved a style of management, and framed their annual objectives, to express a concept of their obligations a good deal different from that of a private sector undertaking. The commercial aspect of their obligations was more restricted than that of a private sector firm, but their composite approach did not prevent them coming close to commercial requirements for most of their existence.[1] It seems unlikely that the commercial result could have come as close if its attainment had been more directly left to ministers, tempted to manipulate 'their' industries in response to short-term fluctuations in finance and politics.

But it may be misleading to attribute continuing imperfections in government–industry relations either to the choice of particular types of relation or to uncertainty of aim. It could well be that there were problems not completely soluble in any way, not least because there were justifiable interests involved that were not wholly reconcilable with each other. It is also probable that the most fundamental difficulties did not arise directly from nationalization, though they were indirectly modified by it.

[1] Cf. section i of this chapter.

In the first place, the most intransigent difficulties of the coal industry were those of resources, especially human resources, and attitudes moulded into very definite forms over many generations, and hard to change. Every ambiguity of the relation between economics and welfare was firmly built in to the daily activity of the coal industry at a myriad points. It was difficult to devise *any* system of organization and direction that would not reflect the ambiguities rather than remove them.

Secondly, there were difficulties inherent in the scale of the undertaking, however it was owned and however the government sought to control it. The Reid Report had made clear that, for technical and economic reasons, a large increase in the scale of colliery undertakings was essential. The political circumstances in which nationalization was enacted made a nationwide undertaking inevitable. But Gaitskell, for one, recognized that such large scale had disadvantages as well as advantages, and that 'when, in addition to large scale, you also have public accountability these disadvantages are accentuated'. He pointed out the problems of careful (if necessarily slow) decision making and control, because a bad mistake by so large a national undertaking could have such widely disastrous consequences. He concluded that it remained to be seen whether large-scale organization under public control could be made to work efficiently in all cases. He also expressed doubt whether the decentralization, for which people asked as a necessary element in NCB organization, was compatible with the conception of control by ministers and parliament.[1] This was very perceptive comment in 1949, fully justified by subsequent experience, which indicated that there were, indeed, incompatibilities that were not total but could not be eliminated.

Finally, what the government could and would do about coal depended even more on the fact that it was a completely basic industry than on its being nationalized. The later experience of, for example, Rolls-Royce and British Leyland showed how such economic significance limited and guided governmental choices, irrespective of current industrial ownership. The relations of the government with the NCB had already provided many illustrations. Coal was so essential that the government had to ensure that the conditions of its adequate supply were continuously present. But that also meant that the government had to refrain from forms of control that might very seriously hamper the NCB in their task of producing the coal. For a few years in the sixties conditions changed because it was judged that most of the coal output

[1] Chester 1975, 1033.

was becoming dispensable. More restrictive government policies therefore seemed practicable and were applied. But the assumptions underlying them proved to be partly illusory, so they had to be relaxed. Relations between the government and the NCB did not return to what they had been in the fifties, but they continued to involve a slightly ambiguous mutual accommodation, in which the varying strength of the pressures depended less on anybody's concepts of the purposes of nationalization than on the degree to which coal, at the costs at which it could be produced, was judged to be essential.

CHAPTER 12

# Reflections, Explanations, and Unanswered Questions

In any history of the nationalized coal industry over three and a half decades some themes present themselves repeatedly for emphasis. Likewise, the questions on the subject which retain public interest tend to be few, large, and repetitive, though their detailed form may vary over time as different answers come to be assumed. Such subjects invite further reflection, not only because they are recurrent, but also because their connections extend far enough to give them some general significance. Probably the questions most often put in public relate to nationalization itself, and particularly to the outcome of an assumed choice between opposites, nationalization and private ownership. There is a desire to know in what ways and to what extent nationalization 'succeeded' or 'failed', and whether it stood as a barrier to alternatives which promised either better results or more troublesome shortcomings.

Many of those who ask such questions would want to relate the answers to narrower and more specific questions about costs and revenues. They might judge the success or failure of nationalization by asking what the nationalized industry cost the public and also, if they remembered to be fair, what the public got in return. But to go immediately to these criteria is probably too simple. It is necessary also to consider the nature and effects of government policy, especially in relation to the supply and use of energy, but to some extent over a wider economic and social field. The success or failure of the industry has to be relative to the objectives set for it and the context created for its operations. These influences may be decisive and they are derived mainly from government policy, which at different times may be sensible or not. It is only after such influences have been considered in association with what was done within the industry that one can approach the broad questions of success or failure. These questions cannot be answered solely in simple accounting terms, though the accounting results are weighty evidence. There needs to be some rather wider estimate of the benefits which society derived from the coal industry and of the costs which the latter imposed on the public. Neither have been fully evaluated, nor can they be in precise monetary terms, and this is one reason why some of the most fundamental questions are likely to receive, at best, only incomplete answers. But after such enquiry there is a basis for approximate general judgement, especially as some of the significant questions are not quantitative, but relate to the characteristics of institutional arrangements.

When general questions about the success of nationalization are put they are often presented in expectation of a particular type of answer. After the Second World War, the nationalization of the coal industry, while far from universally admired, was not resisted with passion, and was supported by a large body of opinion which regarded it as a panacea for past and current ills, the fulfilment of long-held hopes, and a promise of future efficiency and harmony. By 1980 there was much political utterance, not unchallenged but remaining highly confident, to the effect that all nationalization is inherently bad, its tendency to inefficiency amply demonstrated, and its replacement overdue. It would not do to take these movements of opinion as evidence of objective fact. A little pragmatism would not come amiss. The evidence indicates fairly clearly that the coal industry was nationalized, and remained nationalized, because such little alternative as existed was judged likely to do less good, or more harm, to the nation's coal supply. This applies most clearly in 1946, but in this respect circumstances did not obviously change thereafter.

The beliefs of the Labour Party, newly come to power, and the aspirations of the mining unions, who were among its most powerful supporters, provide part of the explanation of the nationalization of the coal industry. But the measure had much wider approval, or (at the very least) acquiescence, because coal was wanted in increasing quantity for economic recovery and renewal, and much of the industry was in such a sorry state that, under existing arrangements, there seemed little prospect of getting what was needed. Drastic reorganization was necessary for technical efficiency, tension between mineowners and miners had long been exceptionally high, and the industry was unattractive to the additional labour that was being sought. It was not an enormous step from the large inescapable minimum of rearrangement in ownership and control to complete nationalization. The Mining Association had proposed a scheme of reorganization under private ownership, but it came very late. Twenty years earlier it might have been politically acceptable and industrially helpful, but things had changed since then, and even as late as 1944 the plans of the Mining Association had many admitted gaps affecting practical arrangements.[1] In the political conditions after the 1945 general election they had no chance of adoption. In any case, experience immediately afterwards showed that the coal industry was in even worse physical condition than had been thought and needed larger injections of capital than the private owners had been in a position to provide. If the industry had been reorganized under private ownership in 1945 or 1946, any government would almost certainly have had to attempt further reorganization, and provide more financial assistance, within four years.

Thus nationalization of the coal industry was, to a substantial degree, a response to the perceived shortcomings of private ownership. It was also a measure to stave off an anticipated continuing shortage of output which would have had a most deleterious effect on the whole national economy. In fact, of course, a short-term failure of this kind was already happening when the NCB took over and might fairly have been attributed more to the errors of government control than to any current omissions by the private owners, though their past limitations were relevant. But the threat of long-term failure *was* lifted, in ways which were more practicable for the NCB than for the former owners. There was this early element of justification by the outcome.

Once nationalization was an accomplished fact, there was no opportunity to undo it. The former private firms had either gone out of business or regrouped to pursue other activities, sometimes related, sometimes not. They would not have wanted their colliery undertakings back again. These would usually no longer have fitted into their restructured businesses, and collieries were becoming increasingly demanding of capital on which it was difficult to get an adequate return into the hands of the owner, as, first, prices were held down by the government and, later, the problems of a fluctuating and contracting market had to be faced. For similar reasons the coal industry could have no attractions to new private entrants, except on an extremely selective basis which was impracticable on any but the smallest scale, because of the complex interrelatedness of the main elements in the industry.

In continuing practical terms, choice lay among different forms of nationalization. The abundance of criticism, especially in the early days, indicates doubt whether the best form had been chosen, but nobody came up with an alternative that was both radically different

[1] Lee 1954, 169–203 describes the proposals in detail. See also vol. 4 of the present series of histories.

and immediately workable. Pre-nationalization discussion of the options had paid most attention to the possibility of direct operation by a government department, and to operation by an elected or appointed or mixed council which could include trade union representatives.[1] The former was an alternative to the public board or corporation, the latter a variant of it. As the criticisms of the NCB alleged an excess of bureaucracy and hierarchical influence, it looks as though the fault was seen as too much resemblance to a government department. Direct operation by a department would have posed greater difficulties of avoiding administrative delay and clumsiness, and would have provoked louder cries of denunciation. Elements of worker control were not practicable in the forties because the unions wanted to keep free of the possibility of conflicting allegiances which a share in managerial responsibility might create. They were not easy to introduce later, partly because the same view continued to be widely held, and partly because of fear that those seeking worker participation would burden management with a constant struggle to resist sectionally biased and anti-commercial policies.

There were those who, after nationalization, advocated a change to a system of common public ownership of a number of largely separate regional units which, in greater or less degree, might compete with each other, subject to some overriding national form of control. A scale and structure for these units which might have made them satisfactorily manageable were not worked out, but this might have been regarded as a problem of detail which could have been solved. But, in the circumstances of the time, there were greater difficulties. One was that the industry was intended to offer greater attractions to the recruitment and co-operation of labour, and the NUM were seeking the reduction and removal of regional competition, which they regarded as a source of pressure keeping down their members' wages. In this respect attitudes did not change. Another problem was the increasingly national character of the market and the need to supply any part of it from any coalfield. A third was that unless competition had been financially tempered to the point where it did not justify the name, it would have driven some regional units out of business at a time when their coal output was regarded as indispensable. There were arguments in favour of being rid of such backward producers instead of relying on them, but to have adopted such arguments would have involved a change in the nature and objectives of national fuel policy. From about 1960 considerations of

[1] Chester 1975, 383–7.

this kind no longer presented themselves in the same way, though the wide range of production costs kept them relevant. But by then there were new problems as some authority had to juggle with the regional balance of closures, and the effects of contraction were so drastic that the industry could hardly cope additionally with even greater reorganization than the NCB themselves undertook.

Another recipe for the nationalized industries generally, which became fashionable around 1980, though there were hints of it in the preceding ten years, was to split them between the existing public corporations and private owners by hiving off profitable parts for sale in the private sector. Many of the propositions on this theme, put forward in political speeches and the press, were the progeny of waffle out of inexperience and of wishful thinking out of ignorance. The line of argument appeared to be that all nationalization could be assumed to be inefficient and socially undesirable, so anything that a nationalized undertaking operated profitably would be operated even more profitably by private owners, who should be given the opportunity to take it over. As a generalization it was untested and demanded a contribution of blind faith. It was an unwelcome counter proposition that successful operation within an existing nationalized undertaking was evidence in favour of leaving an activity where it was. There were, indeed, activities that were no longer central to the business of a nationalized board and could be usefully hived off. Just so the NCB had got rid of most of their brick undertakings for a good price in 1973, and a governmental shove in that direction was beneficial. There were also some main businesses that were running so smoothly that new owners could probably take them over without seriously disturbing their mode and profitability of operation, though it is hard to see the point of disturbance for such a neutral result. But in most cases the search for a new combination of public and private ownership needed detailed empirical enquiry if its results were not to be derisorily small or commercially disappointing.

The coal industry always retained some private operation through its licensed mines and licensed opencast sites. These undertakings were, on the whole, commercially successful, as one would expect, since the operators could pick and choose where and whether they would operate, but their performance gave little guidance to the relative merits of public and private ownership. It is often overlooked that, whoever owns or operates a colliery, there are constraints on its management, staffing, and working practices which are imposed by mining legislation, and that these are less onerous for small mines, statutorily defined, than for the rest. After nationalization the NCB left the small mines to private operators, as far as they were willing to take them on, and operated the others themselves. Licensed opencast sites were limited in size by statute and had outputs roughly comparable to those of licensed mines. Private mining was in some cases carried on by operators as ancillary to other businesses of their own, and they did not have comparable overheads to meet the need for continuous large-scale opening up of new reserves. The small size of the mines lessened the risks of great variability of mining conditions, which can add greatly to costs, and it also reduced the cost of meeting statutory standards. In all these ways the chances of making private mines profitable were improved.

The NCB in their earlier years were left to operate numerous rather small mines, particularly in Scotland and South Wales, because either these were just above the statutory definition of 'small mine' or they did not suit any surviving private undertaking, and they often operated at a loss until they were closed in the sixties or earlier. But in the few cases where they retained small mines taken over elsewhere in more favourable circumstances, the NCB were able to operate them as profitably as private firms. Whether private operators could have done as well as the NCB with somewhat larger mines was not tested (and there was not much demand for a test), and this was a receding possibility from about 1970 because of technological trends. The costs of the new mechanized equipment of the fifties and sixties could often be justified by the improved performance from a single face small mine, whereas the advanced technology and the elaborate electronics of the next phase raised overheads to a level too high to be borne even by medium size mines.

Opencast mining was rather a different matter because the NCB and their civil engineering contractors evolved their own form of public and private sector association under nationalized industry control.[1] It was the complementary achievements of these associates that turned a loss-making activity into something so profitable in the seventies and eighties that some ideologues assumed it must be ripe for privatization; but it was not the experienced private contractors who clamoured for this. The resources of the NCB had played too large a part for them to be dispensed with. It was the NCB who maintained the constant supply of sites necessary for a continuous national programme, who increased and eked out a national reserve of shallow coal so that it could be continuously worked for, probably, seventy or more years when it had been

[1] Chap. 8 above.

deemed to be exhaustible in twenty, and who maintained the balance between restraint and determination when confronting the demands of environmentalists; and who bore the costs of these achievements. There were private licensed operators, a few of whom by about 1980 were demanding greater scope and complaining (against the evidence) that the NCB operated their licensing policy restrictively.[1] Nothing indicated that they were doing the job more cheaply or efficiently than it was done by the joint operation of the NCB and their contractors, especially as the licensed operators were not maintaining a balanced national programme and were rather inclined to cream off a little here and a little there.

The truth was that there was very little recent empirical evidence from which to make a general estimate of the comparative potential merits of public and private ownership of coal mines. The most relevant was still that from 1946 and before. It showed, at best, that a small number of firms had been able to run a small part of the coal industry very competently. But the past evidence about private ownership and operation was, as a whole, not reassuring.

It would not do, however, simply to conclude from all this that nationalization, in the form that was adopted, was shown by events to have been, at worst, the least unsatisfactory solution to the problems of the coal industry. There was, for instance, much scope for argument about the objectives and financial structure of the NCB. Probably most of the criticism of the NCB in later years consisted of complaints about financial and commercial results. These are quite proper subjects for judgement, but if they are to be used as the main criteria of success, it is fair to say that they should have been laid down at the outset as the principal objectives; and they were not. The NCB were left to supply all the coal the public needed, on terms that met the public interest, and to look after the welfare of their employees, and, over a period of good and bad years, they were to avoid making losses. This was not the right prescription to prepare an undertaking to face judgement on strictly commercial grounds. Its inappropriateness was increased by the uncertain limits of the government's supervisory powers and the consequent variations and unpredictability in the way they were exercised. However much attention was given to making managerial decisions in accordance with commercial criteria, the context provided was such that the commercial criteria were sometimes not adequately available or were heavily intermingled with others. The NCB were not in a position to behave exactly like a commercial undertaking.[2]

[1] Chap. 8 (pp. 469–70) above.

[2] Cf. Chap. 11 above, esp. pp. 579–82, 634–5, and 645.

In some ways the problems arising from the ambiguities became worse as the industry contracted. For instance, the requirements to provide for welfare, which are additional to those affecting any employer through general legislation, permitted departures from the norm (with some open-ended financial implications) in the division of responsibilities for social policy between the NCB and the government. While the government made exceptional financial provisions for some needs of present and former NCB employees, the NCB carried on some of their 'commercial' activities with an exceptional regard for limiting social damage that might have been regarded as a responsibility of government. There is a case for arguing that too much of the problem of economic and social adjustment in some areas, such as South Wales, was carried by the NCB and the BSC, and by their accounts, instead of by the government directly.

It seems probable that, though the consequences were not foreseen, some of the ambiguous features of the form of nationalization were deliberately chosen, because they reflected some confusion about the purposes of nationalization. Nothing could be more evidently 'commercial' than running the country's biggest industry, but the NCB must, first and foremost, render a public service and conform to government's general economic plans. They must not be overtly profit-making because the pursuit of profit was one of the things which had rendered the private coalowners objectionable, but neither must they be unprofitable, and they must be efficient. It has been suggested that the company form was deliberately adapted with an extension of the developing separation of ownership and management.

> If it was not essential to good management for the directors to own all or most of the capital could there not be bodies without equity capital, i.e. without shareholders, but managed by a Board of Directors? It would then be largely a question of who appointed the Directors and on what terms. In so far as the arguments for nationalisation were arguments against the profit motive, a company without equity shareholders met the demand.[1]

R. H. Tawney was called in aid, with his claim that managers, liberated from the profit motive, would not make money but would make service.[2] All his specific examples were services and thus begged the question how well such managers would make goods.

Any institution operating under the sort of conditions just described was bound to have some aspects of performance disturbed by a conflict

[1] Chester 1975, 384. [2] Ibid., 385.

of aims. To have to serve several worlds simultaneously ensures that less than the best will be made of some of them. But it is not impossible to carry on tolerably and even to seek the best commercial results practicable and to pursue efficiency in operations, though all is done under some handicap. This, in fact, is more or less what the NCB did. In such conditions, however, it is more difficult to test the level of performance. Profit and loss are only partly relevant, but some of the possible criteria are qualitative and may be partly subjective, and others may be quantitatively defined but of restricted use because of the difficulty of supplying relevant data that are equally precise.[1]

If the purpose had been to have a coal industry run on strictly commercial lines and judged mainly by financial results, a different adaptation of company form would have been appropriate. Something nearer to what the NCB did with their non-mining activities, when they were required to operate them at arm's length and without direct access to government loans, might have suggested itself. That is to say, there might have been an undertaking endowed by the government from the outset with adequate capital, of which a substantial proportion would be equity, and instructed to aim for a surplus, to be used to finance internally all or most of future investment needs, with any additional surplus put into reserve for coal industry needs or (if and when profit became politically respectable) used to pay dividends to the equity owner, i.e. the government. The government would have left it to operate on its own commercial judgement, but would have intervened, like any large majority shareholder, to change personnel, or methods, or structure, or objectives, if management performance were seen to be seriously deficient. Nothing like this was attempted, and such an arrangement would at most times have been assumed to give less government influence than nationalization was intended to secure. But something of the sort was needed for strictly commercial operation. There were times when governments sought to prescribe a more commercial outlook, with some specific financial target other than mere break-even on revenue account, though they were never for long able to retain such a target for coal. But an appreciable change in objectives calls eventually for some change in structure, which was not attempted.

[1] MMC, *National Coal Board*, I, 124–5 discusses in a fairly narrow way the need for the NCB to devise additional measures of performance; its suggestions about methods of investment appraisal (I, 199–209) also have some relevance. Cf. also R. Davis, 'Making sense of sensible accounting', *The Times*, 9 Oct. 1984, which considers in a wider way the indicators that are needed to supplement the published accounts of nationalized industries.

It seems to have been the case that governments did not clearly think out what they wanted from nationalization; that, nevertheless, they wanted a degree of control, however uncertain they were of its specific uses, which was bound to limit the commercial autonomy of a nationalized board; and that from time to time they varied both the requirements which they sought from boards and the way they expected them to be fulfilled. The NCB were probably less drastically disturbed by these conditions than some other nationalized boards. But the conditions were reflected both in their own permanent structure and in the context in which they operated. Inevitably, both what they attempted and what they achieved were modified to some degree.

It might well seem, from these considerations, that the very fact of nationalization ensured that there were some respects in which an industry's ability to be thoroughly commercial was restricted. There was another aspect of operations in the coal industry which national ownership was explicitly intended to change. Relations between employees and private owners had caused so much unrest for so long that it was widely believed that no remedy was possible without a substitution of public for private ownership. The assumption was that the new employer would be welcomed, and attract more co-operation from the mineworkers, just because he had got rid of the old, and still more because he was more generous about pay, conditions, and consultation. Above all, it was hoped that the workers would identify with the employer, and share with him a sense of common interest, because the employer represented the whole nation.

Although for more than twenty years the NCB had better official relations with the unions than the private owners had done, the outcome never matched the expectation. Early indications of better co-operation from the men at the collieries began to recede after the first few years. They did not revive until the sixties, when there was a great reduction in the number of unofficial disputes. But from the end of the sixties a sense of a sharp opposition of interests was more frequently apparent, as in the days of private ownership. Indeed, industrial relations tended to be not only more disturbed but more politicized.

For this disappointing experience plenty of reasons have been found as the sequence of events has been described and discussed. Although it was quickly evident—and perhaps should have been anticipated—that a nationalized institution is too large and distant for a worker to identify himself closely with it, and that working life is initially too little changed for nationalization to be perceived as a transforming novelty, not all the reasons were attributable to the nature and actions of the NCB. The inheritance of deep-rooted patterns of behaviour was probably more influential, and their strength was too great for them to be overthrown by a change in the ownership and management of the industry. The most it was ever reasonable to hope for was that national ownership, merely as one of many social influences, might contribute to a gradual modification of old attitudes.

The later disturbance and politicization of labour relations was also in part the result of external forces. Some trade unionists took up the doctrines of class warfare and were willing or anxious to see miners developed into a weapon in the war, in which any employer, public or private, was treated as a class enemy. Those who acted on these doctrines got no general or consistent response, but they helped to ensure that industrial relations in the coal industry were a political matter for everyone, whatever his outlook. To the confirmation and continuance of this state of affairs the government made a large contribution, especially by the mishandling of the 1973–4 dispute. It was then that the government taught a disastrously lasting lesson that a coal strike could be a major political event when it had not been intended as one.

But there was also an important influence from the NCB, not by way of their policy on industrial relations but through their increased financial dependence. The NCB could not meet the full cost of running the industry after the settlement of the 1972 strike, so the government made the necessary money available. Until then settlements which raised labour costs had to be financed by the NCB, and there was tacit recognition that the employer's ability to pay set a limit in negotiations. After 1972 the limit was no longer recognized, because it had been shown that it could be breached if the government would let it be. In this respect it seemed that a positive disadvantage of nationalized status had emerged. The original expectation that industrial conflict would be dissolved by social and political harmony had been almost reversed. The disputes of party politics, which became sharper in the late seventies than they had been for more than twenty years, spilled over into industrial relations, and the restraint of employees was diminished by experience of access to the public purse.[1]

Though there may have been little alternative to the creation and retention of a nationalized undertaking to run the coal industry, it is evident

[1] Chap. 11, section ii above for fuller examination of influences on industrial relations, and for a consideration of the same points in a rather different context. Cf. also. Chap. 11, pp. 644–5 above.

that the form of what was created and the effects of nationalization depended to a large extent on the objectives and decisions of government, including its inconsistencies. The influence of government policy went beyond the effects of direct measures to determine the characteristics of the NCB and to conduct relations with them. In particular, the government's general policy for the supply and use of energy had a decisive influence on the coal industry. At all times it modified the nature of the market for coal, sometimes in one direction, sometimes in another. By this means it altered both the magnitude of the NCB's needs for finance and other resources, and their opportunity to meet those needs out of the revenue derived from sales.

In the first ten years there were two features of energy policy which had a powerful effect not only on the current but on the whole subsequent history of the coal industry. The first was the decision—perhaps hardly a decision but rather something taken for granted—that all energy needs should be met, to the maximum degree possible, by coal. As nearly full employment and an expanding economy could be expected to increase the consumption of energy, this meant that the output of coal must be increased as much and as fast as possible. For several years the NCB were put under repeated pressure to try to enlarge their plans for additional output. In retrospect this decision looks like a mistake, though a perfectly understandable one which anyone and everyone at the time would have made. The example of continental western Europe, where there was a greater and earlier turning not only to hydroelectric power, for which some regions had much more potential than Britain, but also to fuel oil, and the subsequent example of Japan which, a little later, resumed industrialization with an increasing reliance on oil—both suggest the immediate and long-term advantages of a greater diversity of fuel supplies. Even in the UK it turned out in practice that a significant proportion of the *increase* in energy consumption (though not the total) had to be met by oil,[1] and it would not have been very difficult or costly to add to the proportion.

The adverse effects of this policy on the British coal industry can readily be seen. Many collieries which had no chance of ceasing to make large losses had to be kept in production, even while their remaining reserves had been reduced to small amounts. The need to keep them going absorbed capital and skill which could have been better used in other ways. There were years of effort to try to expand the total labour force in what was a seller's market for labour, and the effect was to

[1] Chap. 5 (p. 230) above.

enlarge the surplus which would need to be shed in a few years' time. Efforts to speed up recruitment encouraged the building of more one-industry mining communities, which only a few years earlier had been generally deplored as something not to be copied again. The whole investment policy to enlarge productive capacity was pushed to the limit, with some indications of overstrain: there were at any one time too many uncompleted projects, and there was rather too much willingness to embark (especially in Scotland and South Wales) on schemes more attractive for additions to physical output than for a satisfactory financial return. The sixties would have been difficult in any case, but the difficulties were enhanced by the belated culmination of the drive for maximum capacity. Yet, without the benefit of hindsight, any other policy was almost inconceivable. The UK after the Second World War was a country with a serious balance of payments problem and large indigenous reserves of coal. It would have seemed absurd to worsen the balance by importing some fuel and by neglecting to make the most of the native resources as a foundation for industrial production and exports. It would have seemed even worse when the rest of Europe was crying out for fuel, including British coal whenever it was available. But the period of European energy shortage turned out to be much briefer than had been expected, and in the not very long run the British choice showed a slight disadvantage.

The other most influential early feature of fuel policy was the holding of coal prices around average production cost rather than at the international competitive level, which was higher. This policy had the general agreement of the NCB, though they were aggrieved by the delays in the approval of price increases to match costs. It was a defensible policy because, by keeping down the cost of inputs, it could be claimed to be assisting the recovery of the economy and the competitiveness of exports. It was also argued at the time that it lessened the temptation to seek greater wage increases. But it failed to encourage the economic use of coal, and fuel efficiency was one of the weak points of British industry. So it helped to raise demand for coal to the level which imposed some strain on the creation of adequate production capacity. It also confirmed and reinforced the policy of keeping the NCB dependent on the Exchequer for development finance, and thus demonstrated that a nationalized coal industry was not to be as strictly commercial as industries in the private sector, just as had been intended from the beginning.

A changed fuel policy in the sixties developed piecemeal, but new opportunities (especially the discovery of North Sea gas) and the pursuit of national economic planning encouraged the government to quantify and formalize it. This was done in the White Paper of 1967. The main forces working against coal were not those of policy but those of the competitive market: the price of oil undercut the average price of British (and much other) coal in many uses. Nevertheless, government policy, which had to provide for the future as well as the present, unwisely pushed the market further against coal and sent the wrong signals to other fuel consumers. This was done by decisions about public consumption of fuel, especially in power stations, by the allocation of investment funds among the energy industries, and by public statements and forecasts about the future level of supply and consumption of different fuels. The message was clearly conveyed that the coal industry would continuously run down and coal would never again be the principal fuel, but only a modest supplement to more important sources of energy. In so far as government money was provided for new purposes in the coal industry, it was to offer help and relief for people to cease working in the industry, rather than to increase the amount of coal producing capacity that was highly efficient.

The principal sign of unwisdom in this policy was the assumption of permanence in the current market conditions for oil. The UK in the fifties had turned too quickly to atomic energy and too little to oil. In the sixties, while continuing (with rather small results for the time being) to pursue unproven techniques of nuclear power generation, it jumped on the oil bandwagon and apparently hoped to ride it for ever. There were widespread illusions about oil which were hard to resist because they were held and acted on worldwide. The rate of consumption in proportion to known reserves ought to have raised doubts as to whether current prices covered all long-term costs, including depletion costs. Prices were certainly far below those that might be maintained if the market were ever to pass more into the hands of a politically supported cartel. The risks inherent in these conditions were brushed aside until they suddenly became realities. Cheap energy was seen as an established and continuing basis of economic expansion, which should not be compromised and which the UK had the chance to reinforce by the rapid depletion of North Sea gas reserves at low prices.

It was rational to plan for a large reduction in the use and output of coal. It was not rational to carry such a policy so far and so fast as to treat most of the coal industry as dispensable in the not very distant future. Coal offered the security of greater abundance than oil or natural gas. Probably somewhere between a third and a half of coal output was fully competitive on costs with oil, and more could have been made so with suitable investment. What was needed in the mid-sixties was a continuous but more gradual closure of high cost collieries, so that the contraction of the labour force and other resources could be kept under control; investment in some replacement capacity on the most cost reducing lines (it would have been sensible to pursue the discoveries in the Selby coalfield straight away instead of waiting for a decade); and some assurance of a somewhat larger continuing market. The latter could have been helped by the choice of coal rather than other fuels for some new power stations (even though most of the existing power station market was retained by coal), and by the publication of a general policy which did not help to persuade industrial users to switch away from coal because it was the fuel that would be in steadily diminishing supply.

It would not have been easy to adopt such a policy. In the very short term it was not the cheapest. Other countries were not following these lines and the UK did not want to be at an international disadvantage by having higher energy costs. But, for a ten-year review, it would have been preferable to what was done, and it would have lessened the debilitating spread of pessimism among those engaged in the coal industry. In 1974, after the unthought of had become reality in the price of oil, a programme very much of this type was approved as the new policy for coal. But although this was regarded as the opening of a hopeful prospect for coal, it was in important respects too late. There was not a big enough proportion of low cost capacity in existence in the coal industry, and, because of the interruption of investment, there was no possibility of having it for a good many years; the composition and morale of the labour force had been damaged, with adverse effects on industrial relations and costs; and much of the damage to the industrial market for coal had already been done. In these circumstances there was bound to be a gap of several years in which the development of demand and the ability of the coal industry to respond to it in terms of output and cost were uncertain. While that gap existed there was a strong possibility that the apparent requirements of government fuel policy would become uncertain, and that consequently the justification for a renewed larger share for coal would also become uncertain.

This, indeed, is roughly what happened. The principal signs of government commitment to an increased share for coal in energy supply and use were two: a willingness from 1973 to support the operation of the coal industry by grants, and the commitment from 1974 to the restructuring of the industry outlined in *Plan for Coal*. The provision of government capital for long-term investment was consistently maintained, though in 1980 and 1981 it was briefly under threat from the proposed progressive reduction of the external financing limit, within which new borrowings as well as grants had to be accommodated. But the amount and terms of grants were not at all consistent. It does not seem to have been clearly thought out what the justification for grants was, nor why they should have been repeatedly needed.

For a short time from 1973 the purpose of grants was obviously to enable the NCB to meet the higher costs resulting from the Wilberforce settlement without having to return to now unwanted policies of rapid contraction. By the end of 1975 the general rise in energy prices had taken care of that particular difficulty. From then on it appeared that grants were part of the process of smoothing the transition to the day when a physically reconstructed coal industry would be operating at higher output (in response to higher demand) and competitively lower costs. This was always assumed to be a protracted transition, continuing well into the eighties; but when even the approach to its completion failed to become evident, the purpose of financial assistance became less clear. Justification in terms of similar but larger targets, attached to more distant dates, was less convincing. It was not altogether surprising that there should have been an attempt to change the policy. In 1980 there was a new programme for the phasing out of all grants in aid of production and the financing of a bigger proportion of investment out of revenue.[1] This might have been fine for an institution designed and instructed from the outset to operate on strict commercial lines. But it was a drastic new instruction, unaccompanied by a significant change in physical targets or institutional design, and issued at a time of serious economic recession which had particularly severe effects on the demand for energy. As it took no account of either the historical or the current context in which it was to be applied, the new policy inevitably fell apart very quickly.

In fact, after the settlement of the 1974 strike and until 1981–2, the NCB depended hardly at all on government grants (other than social grants) for anything except, indirectly, some usually substantial relief to the mounting sum of interest charges for new and still incomplete capital projects. It was mainly because the long-term reconstruction programme was increasing interest payments so fast, though also because of rising social costs, that at the end of the seventies government

[1] NCB *Ann. Rep.* 1979–80, 2, 5–6, and 8.

grants were steadily rising in proportion to NCB turnover. But the need for government grants to support current operations hardly arose until 1981–2;[1] i.e. the significant change came *after* the abortive attempt to reverse government policy. From 1981–2 NCB finances deteriorated seriously in the face of the current recession. But until then the major emerging questions concerned the extent to which the reconstruction programme remained appropriate in the light of contemporary results and future prospects, and the suitability of the methods of financing it. These were not the questions to which the immediate changes of government policy were principally addressed, though they were among the current preoccupations of planning within the NCB.

To look back over thirty-five years of nationalized operation suggests that, in setting the framework of policy for the coal industry, governments and their civil service advisers had an undistinguished record for clarity and consistency, though it is easy to see the plausibility of some of the assumptions which look to have been dubious. It all suggests that it was probably a good thing that the relation of politicians to the running of the industry was no closer than it was. But this, in turn, raises the question whether the people who did run the industry performed any better.

The answer to this question is that, given the somewhat mixed objectives prescribed for them and the fluctuating constraints of policy, they probably did at most times, at any rate in what may be regarded as the regular business of keeping the industry going week-by-week and year-by-year, supplying what the market wanted within uneven but generally defensible rates of progress in efficiency. The results of current operations have already been examined. They suggest that in both physical and financial terms they were fairly well in line with the required objectives at most times.[2] The main exceptions were in the great strike period of 1972–4 and in the difficult market from the end of the seventies. It was a respectable, if imperfect, record, even by international standards, especially when the physical difficulties of operation are taken into account. The British coalfields are very far from being the most geologically consistent and easily worked in the world.

But the favourable record has to be qualified in various ways. One is the extent to which, from the sixties onward, deep mining, which was the principal activity, lagged behind the other activities in financial

[1] Chap. 7, section vi, pp. 403–6 above for an analysis of government grants in relation to NCB financial results.

[2] Chap. 11, section i above.

performance. Another was the failure at some critical times to achieve increases in labour productivity, especially in the early fifties when increased output was being desperately sought, and for much of the seventies when more competitive pricing was wanted in order to keep alive the prospect of re-expanding the industry. The delays in technological progress at some times (even though they were offset by spectacular advances at other times) were also relevant, especially in the early years of nationalization and to some extent in the seventies, when the diffusion of the new phase of heavy duty equipment and of electronics was slow. Most serious, probably, was the slowness in carrying out major programmes for the reconstruction of the industry in the fifties and the later seventies, so that new market opportunities were ebbing away before the projects intended to take advantage of them had borne much fruit. No doubt, all these shortcomings were to some extent interrelated. But it appears that the NCB were somewhat less successful in the execution of long-term plans than with the regular requirements of continuous operation. This was partly the result of handicaps imposed by government. The call for reconstruction came first in the late forties after years of neglect, but was not maintained, and the painfully acquired expertise was dissipated again. When the call for reconstruction was repeated in 1974 too little of this expertise was still readily at hand. Even when all the necessary qualifications have been made, it still looks as though the immediate responsibility for running the coal industry was better discharged than that for framing the policies which set the context.

So it is possible to come back to the joint contribution of the various parties in the running of the coal industry and ask what they had made of it through thirty-five years of nationalization; and, as the industry had been made a public enterprise, operated by a body charged to serve the public interest, to ask what the public had derived from it. The general public benefit had been the removal (overwhelmingly dependent on coal at first, less so thereafter) of the prospect of energy shortage, which in the forties had appeared as a long-term threat to economic progress. It was a contribution most obvious in coal output but, in relation to long-term security, just as much demonstrated in the maintenance and increase of known reserves, not least impressive the reserves of shallow coal that made possible the long continuance of opencasting, with a helpful effect on the level of average costs. But, even though the coal industry had at most times come close to achieving the objectives laid down for it, the assumption that nationalization would bring cheaper coal had, over the long period, been disproved by experience. Natural conditions, and the necessity for coalmining to consume some of its own productive capacity as a condition of maintaining current output, both mean that costs of production always tend to rise, and too little allowance for this had been made in public expectations. The tendency could be offset by continual improvements in technology. Many such improvements were made and for short periods they brought reductions in the real cost of coal to consumers, but not for long enough to reverse the general trend. Much of the improvement in efficiency, achieved mainly by new technology and higher capital expenditure, appears to have gone, especially in the first ten years and in the seventies, to improve the lot of the workers rather than to lower costs. It was not just a matter of higher wages but of better working conditions and greater safety (which were among the industry's most urgent needs) and non-wage benefits such as pensions, sick pay, and longer holidays. Such improvements were proclaimed by politicians as among the purposes of nationalization, and they were achieved to a substantial degree. Indeed, the great improvement in safety and health at work was a humane triumph, worthy of pride. But many of the public seem to have felt that coal consumers got less benefit than they had a right to expect.

Complaints about the coal industry were probably most widely expressed in the early years of nationalization and in the seventies and eighties. Apart from the somewhat whipped-up campaign about the alleged bureaucratic inefficiencies of the NCB, the main early complaints were about the quality and suitability of the coal supplied to different types of consumer. This was a genuine disservice which was only gradually removed by better provision for coal preparation and greater care in the allocation system.

The later complaints were mainly about cost and the environment. That costs of fuel grew in the long term more than the average of prices could not be denied and the difference was likely to be resented. On the other hand, coal was not the fuel subject to the largest price increases, and its better (or, if one prefers, 'less bad') performance had a helpful influence on electricity prices. By the end of the seventies complaints about cost to the consumer were being coupled with complaints about cost to the taxpayer, but the latter were somewhat misconceived. The strikes of 1972 and 1974 did throw some significant burdens on the taxpayer, but otherwise, until 1981–2, government payments related mainly to investment and were not, taking one year with another, in any appreciable way mitigating high production costs by subsidy.

Complaints about the environment concerned both the disamenity of opencasting and the alleged ecological damage caused by the products of the combustion of coal. The usually effective counter to the former was the combination of the cost saving effects of opencasting with the quality of land restoration. The latter charge was harder to deal with because of the total confidence of the objectors in very incomplete research and their unreadiness to face the question why coal should be thought to cause much more ecological damage when much less coal was being burned. It was a charge which could have brought a new threat to the sales prospects for coal, had it not been that the objectors complained just as much about ecological damage from the combustion of oil and the generation of nuclear power. If all were sinners, perhaps the market would forgive coal as readily as the others.

This is perhaps a not very exciting catalogue of likely public benefits and costs. But the objectives built into the nationalization programme, seeking a course between profit and loss, and between commercial achievement, workers' welfare, and public interest, were not designed for exciting results. It might be concluded that, on the whole, the public got what the public, through legislation, had asked for. In any case, judgement was being based on a few central but simply stated results. The real evidence of accomplishment, or its lack, was hidden behind the results, in the way in which constantly changing conditions were tackled in order to achieve any sort of result. By the standards of the great majority of economic activities, the coal industry is of exceptional complexity; complexity of physical environment which differs not only from location to location but quickly from one week to another in the same location; complexity of human relationships, conducted in a unique setting; all intermingled with repeated innovations of technique and organization which are necessary in order to cope with the other complexities. Output and economy in such conditions depend on a high degree of adaptability, and when they show shortcomings it is not easy to tell whether adaptability has declined or the complexities have so changed as to be less manageable.

There is, indeed, another way of looking over the history of the coal industry, in which nationalization and government policy and their effects on performance are only large elements in a much more broadly constituted total. Seen in this way, the course of the industry looks much more remarkable. The period from the forties to the eighties then becomes, in the magnitude of its contrasts, the most extraordinary in the entire history of the coal industry, and one realizes that by the seventies there were dominant characteristics of the industry, especially its reduced size, which thirty years earlier were inconceivable. It was the period in which the use of coal in the UK reached its maximum, in 1956, and output was painfully built up to match it. It was also the period when, soon afterwards, coal underwent a greater and faster contraction than was experienced by any other major British industry at any time: in ten years output fell by more than a quarter, the number of collieries and their labour force by more than half. Indeed, the labour force was reduced by over a third in only five years. All this meant that employment at collieries was dropping by an average of more than 30,000 a year for a decade. Yet all this was happening at a time when the most drastic technological changes for at least a century, perhaps the most drastic in all the history of the industry, were being rapidly spread in the collieries, and Britain became the world's leader in the techniques of cutting and loading coal underground. Some other countries with better mining conditions did even better by copying British techniques, but in British mines in the decade of great contraction overall OMS rose by nearly 60 per cent. The course of events shows the coal industry and its members receiving their most restrictive treatment and their lowest valuation when they were giving their best performance.

Rarely have managers and workforce been called upon to adapt in such numbers to such sweeping changes in so short a time. It was a severe test of human nature and it was not over yet, for, after the headlong downward rush, everyone in the coal industry was besought to stabilize it, improve its quality further, and prepare more gradually to re-expand its output (though not its labour force), and, somewhat belatedly, was offered new inducements.

The transformation of the coal industry in the sixties was its most influential experience in the first thirty-five years of nationalization. It was made easier by the relative abundance of employment opportunities in other industries, an aid not present when substantial, though less severe, adaptations of a not dissimilar kind were needed a dozen and more years later. The degree to which the coal industry modernized its technological outlook and the degree to which the labour force achieved redeployment and accepted new practices were remarkable; but the adaptations were not complete. There was some lasting damage. Resources of skill and loyalty were lost to the coal industry and could not readily be recovered when they were wanted again. The steep decline in sales and output meant that miners derived much less material reward for co-operating with new technology than they otherwise would have done. So resentments were stored up, and a resolve to secure a better share in the future. The national strikes of 1972 and 1974, which, more than anything else, were reactions to the loss of relative pay and status that had accumulated in the sixties, increased and prolonged the damage. They also spread some of it to the consuming public, many of whom cast a warier eye on the coal industry and wondered about its reliability as a source of supply: a change of outlook which threw a shadow over expectations of a reviving market.

Though the rate of change may have slowed down after the sixties, the qualitative variety of necessary adaptation was as common as before, and the predictability of output, costs, and sales was no less difficult even if trends had changed. But the resilience of the coal industry in the face of new conditions appears to have been somewhat reduced, probably by the effects of what happened in the sixties. It may also have been affected by what had not happened. The size and methods of operation of the coal industry had changed fundamentally, but not the structure of its responsible institutions. The organization of the NCB had changed a lot in detail, but much remained as it was, and the unions had changed their organization much less. It was unlikely that activities and institutions could be matched to each other in the same way as before.

In terms of sales and the rewards of effort and enterprise, the seventies were a time of hope aroused and hope deferred. The arousal seems to have clouded judgement a little about the length and hardness of the road before the hope could be attained. The speed with which the NCB got back into financial surplus after the traumas of 1972 and 1974 demonstrated the new ability to defeat oil in straight competition by exploiting the latter's rising costs, and encouraged a belief in coal's steady recovery of markets for fuel. But the signs were already there that energy saving and structural changes in the economy, to the detriment of large fuel using industries, were altering the market, and that natural gas and cheaper foreign coal were undefeated competitors as significant as oil. On the other side, there were symptoms of insufficient adaptability in parts of the British coal industry, particularly the great difficulty in resuming a rise in labour productivity and the slow progress in the restructuring of the industry; and, for some years, there was the apparent political willingness to protect those who resisted adaptation against those who sought it. When general economic recession set in, the difficulties of the struggle for output and sales, implicit in such features, were exacerbated, at least for a short term which could be severe enough to weaken the resources for seizing the opportunities of the long term.

Analysis of the state of the coal industry as it entered the nineteen-eighties has shown how fragile it was.[1] Though its spectacular contraction of the sixties had been stemmed, the industry had never been able to throw off all the damage it had then sustained, and its impressive rise in efficiency then had also been seriously checked. It had to survive in a world that was much more competitive, politically as well as economically, than that before 1960. It sought not mere survival, but some eventual growth, which could come only by demonstrating to the public that there was a necessary product competitively available permanently and regularly. Events in the recent past had from time to time induced some scepticism about the fulfilment of one or other of these requirements. In 1980 the doubts on cost and security of supply, especially the latter, were not particularly severe, but were familiar enough to be easily revived. The coal industry had recovered a leading position in the British supply of energy. It still had abundant reserves of unworked fuel, great resources of capital and skill, and a substantial proportion of highly efficient mines. But there were many great uncertainties in a rapidly changing and depressed economic environment, there was the prospect of intense competition in energy supply for years ahead, and there were various uncured weaknesses and inefficiencies in the coal industry itself. For all these reasons the phase of coal industry history which came down to the early nineteen-eighties had to have, as its closing symbol, a large question mark.

[1] Chap. 7, section ix above.

APPENDIX I

# Statistics

Except for matters to which attention is drawn in the notes, the figures in the appendix tables are on a consistent basis. In a few cases, however, they differ slightly from some figures cited in the text. There are two reasons. (1) Most of the appendix figures are rounded to the nearest integer or one decimal place. If the annual figures are totalled for a fairly extended period there are sometimes slight consequential inaccuracies in the result, whereas there are occasional period totals in the text which are exact. (2) Some figures in the text relate to rather different definitions from those applicable to the appendix tables. Thus the figures of disputes tonnage lost which are given in Chapter 5 (but not those in Chapter 6) include estimates for tonnage lost by restrictions of effort as well as actual stoppages and are therefore higher than the figures in Table A2. In Chapters 6 and 7 there are some small differences between the totals given for outstanding loans and those given in Table A3, and these differences relate to loans raised by subsidiaries. In Chapter 7 there are figures of average price of coal which relate to all NCB output and therefore differ from those in Table A3, which relate only to deep mined output. In Chapter 9, Table 9.1 shows a slightly lower total of operating profits than would be derived from Table A3, partly because of errors due to rounding of annual figures in the latter, but partly because the former excludes profits from associated companies and the latter includes them.

*Note to all Appendix Tables*

— means zero.

... means not applicable (Table A2) or not available (Table A3).

Table A1. *Coal output, imports, consumption, exports, stocks, and manpower*

| | 1947 | 1948 | 1949 | 1950 | 1951 | 1952 | 1953 | 1954 | 1955 |
|---|---|---|---|---|---|---|---|---|---|
| Number of NCB producing collieries at year end | 958 | 940 | 912 | 901 | 896 | 880 | 875 | 867 | 850 |
| Output (million tonnes)[a] | | | | | | | | | |
| NCB deep mined | 187.5 | 197.8 | 204.0 | 205.6 | 212.9 | 214.1 | 213.3 | 214.9 | 211.3 |
| Opencast | 10.4 | 11.9 | 12.6 | 12.4 | 11.2 | 12.3 | 11.9 | 10.3 | 11.6 |
| Licensed | 2.1 | 1.8 | 1.6 | 1.5 | 1.5 | 2.0 | 1.9 | 2.1 | 2.3 |
| TOTAL | 200.0 | 211.5 | 218.2 | 219.6 | 225.6 | 228.4 | 227.1 | 227.2 | 225.2 |
| Imports (million tonnes) | 0.7 | 0.1 | — | — | 1.2 | 0.3 | 0.6 | 3.0 | 11.8 |
| Inland consumption (million tonnes) | 187.5 | 197.0 | 198.9 | 205.9 | 211.1 | 210.1 | 211.1 | 217.2 | 218.7 |
| Exports (million tonnes) | 5.3 | 16.4 | 19.5 | 17.2 | 11.6 | 15.1 | 16.9 | 16.3 | 14.1 |
| Stocks at year end (million tonnes) | | | | | | | | | |
| NCB stock | 1.4 | 2.6 | 2.6 | 1.7 | 1.4 | 4.6 | 1.9 | 1.1 | 2.2 |
| Consumer stock[b] | 16.7 | 14.8 | 14.9 | 12.6 | 16.5 | 16.4 | 17.7 | 15.7 | 18.6 |
| TOTAL[c] | 18.1 | 17.5 | 17.6 | 14.3 | 18.0 | 21.0 | 19.6 | 17.0 | 20.8 |
| Average colliery manpower (thousands)[d] | 703.9 | 716.5 | 712.5 | 690.8 | 692.6 | 709.7 | 711.5 | 701.8 | 698.7 |
| Average age (years) | . . . | 39.7 | 40.1 | 40.2 | 40.5 | 40.2 | 40.3 | 40.3 | 40.4 |
| Output per man-year (tonnes)[d] | 267 | 276 | 287 | 298 | 307 | 302 | 300 | 306 | 302 |

| | 1956 | 1957 | 1958 | 1959 | 1960 | 1961 | 1962 | 1963–4 (15 months) | 1964–5 |
|---|---|---|---|---|---|---|---|---|---|
| Number of NCB producing collieries at year end | 840 | 822 | 793 | 737 | 698 | 669 | 616 | 576 | 534 |
| Output (million tonnes)[a] | | | | | | | | | |
| NCB deep mined | 210.7 | 210.8 | 202.2 | 195.7 | 186.8 | 182.7 | 190.7 | 241.3 | 186.7 |
| Opencast | 12.3 | 13.8 | 14.5 | 11.0 | 7.7 | 8.6 | 7.3 | 7.0 | 6.5 |
| Licensed | 2.6 | 2.6 | 2.6 | 2.7 | 2.2 | 2.2 | 2.5 | 3.2 | 2.3 |
| TOTAL | 225.6 | 227.2 | 219.3 | 209.4 | 196.7 | 193.6 | 200.6 | 251.5 | 195.6 |
| Imports (million tonnes) | 5.3 | 2.9 | 0.8 | 0.1 | — | — | — | — | — |
| Inland consumption (million tonnes) | 221.0 | 216.3 | 205.6 | 192.4 | 199.9 | 194.9 | 194.3 | 253.6 | 190.3 |
| Exports (million tonnes) | 9.9 | 8.0 | 5.0 | 4.4 | 5.6 | 5.8 | 4.9 | 9.7 | 5.4 |
| Stocks at year end (million tonnes) | | | | | | | | | |
| NCB stock | 2.9 | 8.7 | 20.0 | 36.3 | 29.7 | 21.7 | 25.8 | 18.7 | 20.0 |
| Consumer stock[b] | 18.3 | 19.1 | 17.8 | 14.6 | 13.6 | 15.9 | 15.3 | 12.2 | 12.6 |
| TOTAL[c] | 21.2 | 27.7 | 37.8 | 50.9 | 43.3 | 37.6 | 41.1 | 30.9 | 32.6 |
| Average colliery manpower (thousands)[d] | 697.4 | 703.8 | 692.7 | 658.2 | 602.1 | 570.5 | 550.9 | 517.0 | 491.0 |
| Average age (years) | 40.4 | 40.5 | 40.7 | 41.3 | 41.6 | 42.1 | 42.4 | 42.7 | 42.7 |
| Output per man-year (tonnes)[d] | 302 | 300 | 292 | 298 | 310 | 320 | 346 | 368 | 379 |

Table A1 (*cont.*)

| | 1965–6 | 1966–7 | 1967–8 | 1968–9 | 1969–70 | 1970–1 | 1971–2 | 1972–3 | 1973–4 |
|---|---|---|---|---|---|---|---|---|---|
| Number of NCB producing collieries at year end | 483 | 438 | 376 | 317 | 299 | 292 | 289 | 281 | 259 |
| Output (million tonnes)[a] | | | | | | | | | |
| NCB deep mined | 177.0 | 167.3 | 165.4 | 155.6 | 142.1 | 135.5 | 111.0 | 129.1 | 98.7 |
| Opencast | 6.9 | 6.8 | 6.8 | 6.4 | 6.3 | 8.1 | 10.1 | 10.1 | 9.0 |
| Licensed | 1.8 | 1.6 | 1.3 | 1.2 | 1.3 | 1.1 | 1.1 | 1.3 | 1.1 |
| TOTAL | 185.7 | 175.8 | 173.6 | 163.2 | 149.8 | 144.7 | 122.3 | 140.5 | 108.8 |
| Imports (million tonnes) | — | — | — | — | — | 1.2 | 5.3 | 3.4 | 2.0 |
| Inland consumption (million tonnes) | 184.0 | 172.6 | 168.1 | 167.6 | 161.7 | 150.7 | 128.4 | 130.2 | 121.2 |
| Exports (million tonnes) | 3.7 | 2.5 | 2.0 | 3.1 | 3.6 | 3.0 | 2.1 | 2.3 | 2.1 |
| Stocks at year end (million tonnes) | | | | | | | | | |
| NCB stock | 18.8 | 20.6 | 27.0 | 25.3 | 14.6 | 6.3 | 7.4 | 12.1 | 8.5 |
| Consumer stock[b] | 13.2 | 14.6 | 14.5 | 11.6 | 10.1 | 11.7 | 10.3 | 18.9 | 10.7 |
| TOTAL[c] | 32.0 | 35.2 | 41.5 | 36.9 | 24.8 | 18.0 | 17.7 | 31.0 | 19.1 |
| Average colliery manpower (thousands)[d] | 455.7 | 419.4 | 391.9 | 336.3 | 305.1 | 287.2 | 281.5 | 268.0 | 252.0 |
| Average age (years) | 43.3 | 43.5 | 43.7 | 43.8 | 43.9 | 43.9 | 43.7 | 43.7 | 44.2 |
| Output per man-year (tonnes)[d] | 387 | 396 | 421 | 461 | 464 | 471 | 393 | 480 | 390 |

| | 1974–5 | 1975–6 | 1976–7 | 1977–8 | 1978–9 | 1979–80 | 1980–1 | 1981–2 | 1982–3 |
|---|---|---|---|---|---|---|---|---|---|
| Number of NCB producing collieries at year end | 246 | 241 | 238 | 231 | 223 | 219 | 211 | 200 | 191 |
| Output (million tonnes)[a] | | | | | | | | | |
| NCB deep mined | 116.9 | 114.5 | 108.5 | 106.3 | 105.5 | 109.3 | 110.3 | 108.9 | 104.9 |
| Opencast | 9.2 | 10.4 | 11.4 | 13.6 | 13.5 | 13.0 | 15.3 | 14.3 | 14.7 |
| Licensed | 1.0 | 0.9 | 0.9 | 1.0 | 0.9 | 1.0 | 1.1 | 1.1 | 1.2 |
| TOTAL | 127.2 | 125.8 | 120.8 | 120.9 | 119.9 | 123.3 | 126.6 | 124.3 | 120.9 |
| Imports (million tonnes) | 4.0 | 4.8 | 2.4 | 2.7 | 2.1 | 5.1 | 7.3 | 4.2 | 3.4 |
| Inland consumption (million tonnes) | 127.2 | 122.2 | 124.6 | 121.6 | 122.5 | 128.4 | 120.3 | 117.0 | 110.4 |
| Exports (million tonnes) | 2.1 | 1.4 | 1.4 | 1.8 | 2.1 | 2.5 | 4.7 | 9.4 | 7.1 |
| Stocks at year end (million tonnes) | | | | | | | | | |
| NCB stock | 5.6 | 11.0 | 9.6 | 10.3 | 14.1 | 12.0 | 20.9 | 24.9 | 24.9 |
| Consumer stock[b] | 16.1 | 18.9 | 18.5 | 19.5 | 14.7 | 15.8 | 17.5 | 18.6 | 28.3 |
| TOTAL[c] | 21.6 | 29.9 | 28.1 | 29.8 | 28.8 | 27.7 | 38.4 | 43.5 | 53.3 |
| Average colliery manpower (thousands)[d] | 246.0 | 247.1 | 242.0 | 240.5 | 234.9 | 232.5 | 229.8 | 218.5 | 207.6 |
| Average age (years) | 43.6 | 43.2 | 42.8 | 41.1 | 40.6 | 39.6 | 39.4 | 39.1 | 38.7 |
| Output per man-year (tonnes)[d] | 474 | 462 | 448 | 441 | 448 | 470 | 479 | 497 | 504 |

*[See p. 676 for Notes]*

*Notes*: [a] Because the figures are given to only one decimal place, the constituent items in some years do not sum exactly to the total. 'NCB deep mined' includes NCB tip and capital coal. For this reason the figures differ slightly for most of the period from those printed contemporaneously in Annual Reports, where tip and capital coal were included in the figure for 'Licensed and other'. In this table licensed opencast output appears under 'Licensed' and not under 'Opencast', the opposite of the practice adopted in the Annual Reports for 1963–4 to 1971–2 inclusive.

[b] From 1970–1 onwards the figures for 'consumer stock' relate to a narrower definition and exclude stocks held by what had by then become rather small classes of coal user. For earlier years all stocks are included, as was done in the contemporaneous Annual Reports. These earlier figures therefore differ from the retrospective statistics published in recent Annual Reports, which use the narrower definition back to 1955.

[c] Rounding off to one decimal place causes the two components not to sum exactly to the total in a few years. The figures in the table cannot be balanced in order to show the origin of the year-to-year changes in total stocks. This is probably due in part to leads and lags in the reporting of some constituent items, but also arises because the output figures do not include non-vested coal, which enters into consumption and stocks, and in peak years reached 3 million tonnes.

[d] The figures for average colliery manpower, and therefore for output per man-year, from 1954 onwards were affected by a change of definition. There was a further and slighter effect from 1979–80 on the figures of output per man-year as a result of a change in the statistical treatment of capital coal. All the manpower figures for 1963–4 relate to the financial year, not the 15-month period.

*Source*: NCB.

Table A2. *Labour statistics, NCB mines*

| | 1947 | 1948 | 1949 | 1950 | 1951 | 1952 | 1953 | 1954 | 1955 |
|---|---|---|---|---|---|---|---|---|---|
| Average colliery manpower (thousands) | 703.9 | 716.5 | 712.5 | 690.8 | 692.6 | 709.7 | 711.5 | 701.8 | 698.7 |
| Output per manshift (tonnes)[a] | | | | | | | | | |
| Face | 2.97 | 3.02 | 3.14 | 3.24 | 3.30 | 3.26 | 3.25 | 3.33 | 3.34 |
| Production | ... | ... | ... | ... | ... | ... | ... | ... | ... |
| Overall | 1.09 | 1.13 | 1.19 | 1.23 | 1.24 | 1.23 | 1.25 | 1.26 | 1.25 |
| Disputes tonnage lost (million tonnes) | 1.7 | 1.0 | 1.4 | 1.0 | 1.0 | 1.7 | 1.1 | 1.5 | 3.3 |
| Accidents per 100,000 manshifts[b] | | | | | | | | | |
| Fatal | 0.34 | 0.25 | 0.25 | 0.28 | 0.27 | 0.23 | 0.21 | 0.21 | 0.24 |
| Reportable injuries[c] | 1.36 | 1.29 | 1.31 | 1.17 | 1.10 | 1.17 | 1.10 | 1.06 | 1.08 |
| Average cash earnings per week, £ | 6.65 | 7.85 | 8.37 | 8.73 | 9.75 | 10.78 | 11.10 | 11.75 | 12.46 |
| Average value of allowances in kind per week, £ | 0.29 | 0.34 | 0.35 | 0.38 | 0.43 | 0.47 | 0.50 | 0.54 | 0.59 |

Table A2. (*cont.*)

| | 1956 | 1957 | 1958 | 1959 | 1960 | 1961 | 1962 | 1963–4 | 1964–5 |
|---|---|---|---|---|---|---|---|---|---|
| Average colliery manpower (thousands) | 697.4 | 703.8 | 692.7 | 658.2 | 602.1 | 570.5 | 550.9 | 517.0 | 491.0 |
| Output per manshift (tonnes)[a] | | | | | | | | | |
| Face | 3.40 | 3.43 | 3.59 | 3.81 | 4.04 | 4.24 | 4.62 | 5.03 | 5.26 |
| Production | ... | ... | ... | ... | ... | ... | ... | ... | ... |
| Overall | 1.26 | 1.26 | 1.30 | 1.37 | 1.42 | 1.47 | 1.59 | 1.70 | 1.77 |
| Disputes tonnage lost (million tonnes) | 2.1 | 1.8 | 1.5 | 1.0 | 1.6 | 2.1 | 1.1 | 1.4 | 1.3 |
| Accidents per 100,000 manshifts[b] | | | | | | | | | |
| Fatal | 0.19 | 0.22 | 0.19 | 0.23 | 0.22 | 0.18 | 0.20 | 0.21 | 0.16 |
| Reportable injuries[c] | 1.03 | 1.10 | 1.24 | 1.36 | 1.12 | 1.11 | 1.22 | 1.14 | 1.16 |
| Average cash earnings per week, £ | 13.55 | 14.52 | 14.21 | 14.19 | 14.70 | 15.60 | 16.22 | 17.20 | 17.97 |
| Average value of allowances in kind per week, £ | 0.68 | 0.73 | 0.80 | 0.83 | 0.88 | 0.96 | 1.05 | 1.10 | 1.16 |

| | 1965–6 | 1966–7 | 1967–8 | 1968–9 | 1969–70 | 1970–1 | 1971–2 | 1972–3 | 1973–4 |
|---|---|---|---|---|---|---|---|---|---|
| Average colliery manpower (thousands) | 455.7 | 419.4 | 391.9 | 336.3 | 305.1 | 287.2 | 281.5 | 268.0 | 252.0 |
| Output per manshift (tonnes)[a] | | | | | | | | | |
| Face | 5.57 | 5.78 | 6.06 | 6.73 | 7.00 | 7.30 | 7.05 | 7.56 | 7.30 |
| Production | ... | ... | ... | ... | ... | ... | ... | ... | ... |
| Overall | 1.83 | 1.86 | 1.98 | 2.16 | 2.20 | 2.24 | 2.13 | 2.33 | 2.15 |
| Disputes tonnage lost (million tonnes) | 1.2 | 1.7 | 0.4 | 0.3 | 2.9 | 3.1 | 26.7 | 0.6 | 21.3 |
| Accidents per 100,000 manshifts[b] | | | | | | | | | |
| Fatal | 0.22 | 0.15 | 0.15 | 0.15 | 0.12 | 0.15 | 0.10 | 0.13 | 0.12 |
| Reportable injuries[c] | 1.16 | 1.07 | 1.13 | 1.06 | 1.01 | 0.95 | 1.07 | 0.99 | 0.96 |
| Average cash earnings per week, £ | 19.14 | 20.15 | 21.13 | 22.37 | 23.82 | 27.07 | 30.93 | 35.67 | 37.53 |
| Average value of allowances in kind per week, £ | 1.22 | 1.35 | 1.41 | 1.48 | 1.63 | 1.88 | 2.08 | 2.52 | 2.56 |

Table A2. (*cont.*)

| | 1974–5 | 1975–6 | 1976–7 | 1977–8 | 1978–9 | 1979–80 | 1980–1 | 1981–2 | 1982–3 |
|---|---|---|---|---|---|---|---|---|---|
| Average colliery manpower (thousands) | 246.0 | 247.1 | 242.0 | 240.5 | 234.9 | 232.5 | 229.8 | 218.5 | 207.6 |
| Output per manshift (tonnes)[a] | | | | | | | | | |
| Face | 7.89 | 7.89 | 7.75 | 7.91 | 8.53 | ... | ... | ... | ... |
| Production | ... | ... | ... | ... | ... | 8.88 | 9.09 | 9.56 | 10.10 |
| Overall | 2.29 | 2.28 | 2.21 | 2.19 | 2.24 | 2.31 | 2.32 | 2.40 | 2.44 |
| Disputes tonnage lost (million tonnes) | 0.4 | 0.5 | 1.1 | 0.8 | 1.5 | 0.9 | 1.5 | 1.1 | 3.6 |
| Accidents per 100,000 manshifts[b] | | | | | | | | | |
| Fatal | 0.10 | 0.11 | 0.07 | 0.09 | 0.13 | 0.06 | 0.07 | 0.07 | 0.09 |
| Reportable injuries[c] | 1.07 | 1.01 | 0.98 | 1.00 | 0.93 | 0.87 | 1.16 | 1.69 | 1.69 |
| Average cash earnings per week, £ | 57.52 | 74.00 | 76.10 | 84.10 | 101.76 | 119.15 | 140.20 | 156.36 | 166.07 |
| Average value of allowances in kind per week, £ | 3.01 | 4.30 | 5.13 | 6.02 | 6.54 | 7.87 | 9.88 | 10.85 | 11.75 |

[*See opposite page for Notes*]

*Notes*: [a] Changes in the definition of manpower, introduced in 1954, slightly increased the figures of OMS. Changes in job definition below ground in 1958 caused a further increase in the figures of face OMS, but did not alter the overall figure. A new set of definitions of OMS was introduced in 1979 and the ratio previously described as face OMS ceased to be used. 'Face OMS' took into account not only shifts at production faces and at any face heading directly associated with a production face, but also all other non-capital drivage shifts associated with a face heading. The latter were not taken into account in 'production OMS', which therefore tended to be a little higher, but more accurately monitored performance at the primary centre of production.

[b] The accident figures for 1947 and 1948, unlike all the other figures in the table, include licensed mines.

[c] From 1959 'minor' reportable injuries were removed from this category and the figures are therefore lower than they would have been if they were strictly comparable to those for 1947 to 1958. For instance, the total of reportable injuries in 1959 was 1,662, but on the previous definition it would have been 2,025. New regulations for reporting accidents applied to all industries from 1 January 1981 and had the effect of greatly increasing the figures. They affected the figures for the last quarter of 1980–1, but did not have their full effect on the annual figure until 1981–2.

*Source*: NCB.

Table A3. *NCB financial statistics*
(£ million except where otherwise stated)

| | 1946[a] | 1947 | 1948 | 1949 | 1950 | 1951 | 1952 | 1953 | 1954 | 1955 |
|---|---|---|---|---|---|---|---|---|---|---|
| Turnover[b] | — | 371 | 457 | 478 | 481 | 541 | 636 | 682 | 720 | 809 |
| Profit (loss) on trading[c] | — | (8) | 18 | 31 | 26 | 19 | 8 | 23 | 16 | 2 |
| Surplus (deficit)[c] | (0.2) | (23) | 2 | 10 | 8 | (2) | (8) | 1 | (4) | (20) |
| Assets employed at year end[d] | | | | | | | | | | |
| Fixed assets | — | 361 | ... | ... | ... | ... | ... | 449 | 510 | 564 |
| Total assets | — | 408 | ... | ... | ... | ... | ... | 473 | 519 | 585 |
| Loans outstanding at year end[d] | — | 429 | ... | ... | ... | ... | ... | 483 | 533 | 618 |
| Colliery proceeds per tonne[e] (£) | — | 1.98 | 2.32 | 2.36 | 2.35 | 2.52 | 2.82 | 3.01 | 3.12 | 3.35 |
| Colliery costs per tonne[e] (£) | — | 2.03 | 2.24 | 2.21 | 2.23 | 2.42 | 2.79 | 2.91 | 3.04 | 3.31 |
| Annual rate of inflation[f] (%) | ... | ... | 4.9 | 3.5 | 3.2 | 12.0 | 6.4 | 1.0 | 4.0 | 5.8 |

| | 1956 | 1957 | 1958 | 1959 | 1960 | 1961 | 1962 | 1963–4 (15 months) | 1964–5 |
|---|---|---|---|---|---|---|---|---|---|
| Turnover[b] | 888 | 937 | 922 | 851 | 834 | 863 | 900 | 1,189 | 893 |
| Profit (loss) on trading[c] | 40 | 20 | 19 | 13 | 20 | 29 | 45 | 73 | 43 |
| Surplus (deficit)[c] | 13 | (5) | (4) | (24) | (21) | (15) | 1 | — | 1 |
| Assets employed at year end[d] | | | | | | | | | |
| Fixed assets | 615 | 669 | 720 | 774 | 794 | 817 | 822 | 831 | 842 |
| Total assets | 612 | 692 | 811 | 889 | 910 | 876 | 884 | 853 | 886 |
| Loans outstanding at year end[d] | 633 | 717 | 839 | 941 | 974 | 958 | 963 | 928 | 960 |
| Colliery proceeds per tonne[e] (£) | 3.79 | 4.03 | 4.18 | 4.10 | 4.23 | 4.47 | 4.52 | 4.55 | 4.52 |
| Colliery costs per tonne[e] (£) | 3.66 | 4.00 | 4.13 | 4.05 | 4.15 | 4.35 | 4.29 | 4.27 | 4.35 |
| Annual rate of inflation[f] (%) | 3.0 | 4.6 | 1.8 | 0.0 | 1.8 | 4.4 | 2.6 | 1.4 | 4.5 |

Table A3 (*cont.*)

| | 1965–6 | 1966–7 | 1967–8 | 1968–9 | 1969–70 | 1970–1 | 1971–2 | 1972–3 | 1973–4 |
|---|---|---|---|---|---|---|---|---|---|
| Turnover[b] | 863 | 863 | 867 | 855 | 852 | 905 | 841 | 1,033 | 914 |
| Profit (loss) on trading[c] | — | 29 | 35 | 29 | 9 | 35 | (118) | (39) | (99) |
| Surplus (deficit)[c] | (25) | — | — | (9) | (26) | 1 | (157) | (84) | (131) |
| Assets employed at year end[d] | | | | | | | | | |
| Fixed assets | 775 | 749 | 642 | 636 | 630 | 633 | 642 | 370 | 384 |
| Total assets | 796 | 782 | 735 | 722 | 662 | 667 | 630 | 473 | 513 |
| Loans outstanding at year end[d] | 575 | 613 | 695 | 698 | 668 | 675 | 692 | 442 | 480 |
| Colliery proceeds per tonne[e] (£) | 4.51 | 4.95 | 4.85 | 4.83 | 5.04 | 5.75 | 6.40 | 6.87 | 7.06 |
| Colliery costs per tonne[e] (£) | 4.56 | 4.84 | 4.71 | 4.75 | 5.12 | 5.71 | 7.61 | 7.55 | 9.84 |
| Annual rate of inflation[f] (%) | 4.3 | 3.5 | 3.4 | 6.3 | 5.1 | 8.8 | 7.6 | 8.2 | 13.5 |

| | 1974–5 | 1975–6 | 1976–7 | 1977–8 | 1978–9 | 1979–80 | 1980–1 | 1981–2 | 1982–3 |
|---|---|---|---|---|---|---|---|---|---|
| Turnover[b] | 1,590 | 1,982 | 2,427 | 2,733 | 2,989 | 3,740 | 4,186 | 4,727 | 4,932 |
| Profit (loss) on trading[c] | 40 | 52 | 110 | 109 | 121 | 28 | 69 | (84) | (123) |
| Surplus (deficit)[c] | — | 5 | 27 | 21 | (19) | — | (58) | — | (111) |
| Assets employed at year end[d] | | | | | | | | | |
| Fixed assets | 459 | 527 | 701 | 1,001 | 1,336 | 1,796 | 2,333 | 2,788 | 3,224 |
| Total assets | 499 | 733 | 1,003 | 1,337 | 1,733 | 2,128 | 2,656 | 3,553 | 3,712 |
| Loans outstanding at year end[d] | 466 | 694 | 935 | 1,175 | 1,590 | 1,984 | 2,569 | 3,470 | 3,740 |
| Colliery proceeds per tonne[e] (£) | 11.11 | 15.55 | 18.72 | 21.34 | 23.06 | 27.38 | 32.83 | 35.59 | 37.91 |
| Colliery costs per tonne[e] (£) | 12.08 | 16.20 | 19.03 | 22.08 | 24.87 | 29.38 | 34.88 | 37.81 | 40.94 |
| Annual rate of inflation[f] | 21.2 | 21.2 | 16.7 | 9.1 | 9.8 | 19.8 | 12.6 | 10.4 | 4.6 |

*Notes:* [a] The NCB did not trade until vesting date (1 January 1947) but existed from July 1946 and necessarily incurred expenditure which produced a small deficit. This, unlike all the other figures, is shown to one decimal place because the scale of expenditure was so small.

[b] Turnover excludes sales between constituent parts of the NCB organization. In the figures for each year to 1962 inclusive, variations in stocks of products were incorporated, but these were excluded from the figures for all subsequent years.

[c] Trading profit from 1973–4 to 1978–9, and the surplus or deficit in all years from 1973–4 onwards, were affected by the inclusion of government grants. The treatment of these differed from time to time. For the detailed effect on the figures see the discussion and tables in Chapter 7, above (pp. 403–6).

[d] Because of the very long delays in the valuation of ancillary assets acquired at nationalization, no reasonably complete figure of assets could be produced until the end of 1953, by which time some of the original assets had been disposed of or heavily depreciated. As payment for the assets, whenever their valuation was settled, had to be made from loans provided by the Minister of Fuel and Power, it was similarly uncertain at all times

## Notes to Table A3 (*cont.*)

until the end of 1953 what was the true liability for loans. Hence the absence of figures for the earlier years. Ten years after the event a calculation was made, in the light of subsequent valuations, of what the true capital position would have been in 1947. The relevant figures are included in the table.

The figures are affected by the write-offs and capital reconstructions of 1965 and 1973. For the detailed effects see the discussion in Chapter 6, section v.

The peculiarity that in a few years the value of fixed assets exceeds that of total assets arises because of the subtraction of substantial amounts of deferred liabilities.

[e] The figures exclude opencast coal, for which proceeds usually differed a little and costs, for most of the period, very substantially.

Every price index for coal as a single commodity has significant defects, because the prices of different categories of coal, for periods of several years, sometimes followed very different trends, and because most coals were priced according to standard formulae which allowed for various characteristics of quality, so that price changes partly resulted from changes in average quality. The annual averages of proceeds per tonne provide a useful indicator of the course of current prices of coal, which is, in effect, reweighted each year in accordance with the proportions of different qualities produced in the year. It is, however, an indicator with its own crudities (e.g. in not separating the proceeds of the small proportion of coal exported), and, like any price index of coal in general, should not be treated as a minutely exact indicator of price changes.

[f] The percentage increase in the official retail price index for the twelve months ending with the last month of the financial year, i.e. December up to and including 1962, and March for each year from 1964 to 1983. No figure is given for 1947 because the retail price index began only in June of that year, when it replaced the old cost of living index. The two indices together suggest that inflation in the year to December 1947 was of the order of 3 per cent.

*General note*: There were various relatively small retrospective revisions of financial data. For this reason some of the figures in this table differ slightly from those printed in contemporaneous Annual Reports. It should also be noted that the following NCB financial years covered 53 weeks: 1958, 1967–8, 1972–3, 1978–9.

*Source*: NCB and (for last line only) Dept. of Employment, *Retail Prices Indices 1914-1983*.

APPENDIX 2

# Principal Office Holders 1946–1982

I. NATIONAL COAL BOARD

*Chairmen*

| | |
|---|---|
| Lord Hyndley | 1946–51 |
| Sir Hubert Houldsworth | 1951–56 |
| J. (later Sir James) Bowman | 1956–61 |
| A. (later Lord) Robens | 1961–71 |
| D. J. (later Sir Derek) Ezra | 1971–82 |

*Deputy Chairmen*

| | |
|---|---|
| Sir Arthur Street | 1946–51 |
| Sir Eric Coates | 1951–55 |
| W. J. (later Sir Walter) Drummond | 1951–55 |
| J. Bowman | 1955–56 |
| J. (later Sir Joseph) Latham | 1956–60 |
| E. H. (later Sir Humphrey) Browne | 1960–67 |
| A. Robens | 1960–61 |
| D. J. Ezra | 1967–71 |
| W. V. Sheppard | 1971–75 |
| N. Siddall | 1973–82 |

II. NATIONAL UNION OF MINEWORKERS

*Presidents*

| | |
|---|---|
| Sir William Lawther | 1945–54 |
| W. E. Jones | 1954–60 |
| S. (later Sir Sidney) Ford | 1961–71 |
| J. Gormley | 1971–82 |

*General Secretaries*

| | |
|---|---|
| A. L. Horner | 1946–59 |
| W. Paynter | 1959–68 |
| L. Daly | 1968–84 |

III. GOVERNMENT: SENIOR MINISTERS WITH OVERSIGHT OF THE COAL INDUSTRY

*Ministers of Fuel and Power*

| | |
|---|---|
| E. Shinwell | 1945–47 |

| | |
|---|---|
| H. T. N. Gaitskell | 1947–50 |
| P. J. Noel-Baker | 1950–51 |
| G. Lloyd | 1951–55 |
| A. Jones | 1955–57 |

*Ministers of Power*

| | |
|---|---|
| Lord Mills | 1957–59 |
| R. Wood | 1959–63 |
| F. J. Erroll | 1963–64 |
| F. Lee | 1964–66 |
| R. Marsh | 1966–68 |
| R. J. Gunter | 1968 |
| R. Mason | 1968–69 |

*Ministers of Technology*

| | |
|---|---|
| A. W. Benn | 1969–70 |
| A. G. F. Rippon | 1970 |

*Secretaries of State for Trade and Industry*

| | |
|---|---|
| J. Davies | 1970–72 |
| P. E. Walker | 1972–74 |

*Secretaries of State for Energy*

| | |
|---|---|
| Lord Carrington | 1974 |
| E. Varley | 1974–75 |
| A. W. Benn | 1975–79 |
| D. Howell | 1979–81 |
| N. Lawson | 1981–83 |

# Select Bibliography

## A. The principal source and its citation

This book has been written with the very abundant papers of the National Coal Board as its central, but by no means its only, source. Much of it has been written very close in time to the events described. For this reason many papers have been consulted before they have become archival and, in general, the author has drawn more on administrative than archival materials. The papers of the National Coal Board, unlike those of other nationalized industries, are public records. The Public Record Office retains the main series of documents and makes them publicly available in conformity with its normal rules. Many of the sources of this book will, however, not be public for many years and, when they are released, they will be classified with different identification marks from those used while they were current material. It would, therefore, be inappropriate and often impracticable to print full documentary references to the NCB papers, or to attempt a detailed classification in a bibliographical note.

Various series of papers for the earlier years of the NCB are, however, already open to readers at the PRO. Details are provided in BENSON, J., NEVILLE, R. G., and THOMPSON, C. H., *Bibliography of the British Coal Industry* (Oxford, 1981), entry 6073, p. 592. Most of the important events and decisions are recorded and dated in the minutes of the national board (which the PRO files in the series COAL 21), even if the main discussion took place elsewhere. So, as these documents are publicly available, references to them are printed in the footnotes of the present book for the first ten years of the NCB's existence. The citations are given simply by the number and date of the meeting of the national board and do not include any PRO reference.

Readers requiring a detailed bibliography will find it, for all except the last few years, in Benson, Neville, and Thompson, op. cit. The present note makes no attempt to be comprehensive and does not include all the works that have been consulted. It merely records, in broad classifications, a number of primary and secondary sources which have been helpful in the preparation of this book. Many of them are included in Benson, Neville, and Thompson, but some which are very recent or rather specialized will supplement that bibliography.

## B. Other Manuscript Sources

*Public Record Office, London*
Cabinet: files on 1947 fuel crisis
CAB 128/6, 128/9, 129/14, 129/16, 129/17, 130/16, 134/62.

Ministry of (Fuel and) Power

| | |
|---|---|
| POWE 1 | Miners' Welfare Commission |
| POWE 9 | Funding of departmental experience |
| POWE 10 | Establishments Division |
| POWE 16 | Coal Division: early correspondence and papers |
| POWE 17 | Emergency series |
| POWE 18 | Fuel and lighting |
| POWE 19 | House coal and Domestic Coal Consumers' Council |
| POWE 20 | Labour and labour relations |
| POWE 21 | Prices |
| POWE 26 | 'A' files |
| POWE 28 | Statistics and General Division (28/195–208 and 28/232–3 deal with Committee on National Fuel Policy (1952) and action on its report). |
| POWE 35, 36, and 42 | Valuation and compensation for nationalization |
| POWE 37 | Post-nationalization 'B' files. (This is the most wide-ranging and useful series.) |
| POWE 40 | Coal Division: Opencast Coal |
| POWE 41 | International matters |
| POWE 43 | Miscellaneous financial topics |

Some files in the POWE series, which have been used for this book, remain closed under the 30-year rule. References to these files have not been given in the footnotes.

## C. **Official Publications** (in chronological order from each source)

### (a) *Government*

#### (i) *Annuals*

*Annual Reports of Chief Inspector of Mines*, 1947–74 (issued by the Ministry or Department currently supervising the coal industry (see Appendix 2) ).

Department of Energy (other Ministries and Departments before 1975), *Digest of UK Energy Statistics*, 1969–.

Health and Safety Executive, *Health and Safety: Mines*, 1975–.

#### (ii) *Other publications*

Scottish Home Department, *Scottish Coalfields: Report of the Scottish Coalfields Committee*, 1944. Cmd. 6575.

Ministry of Fuel and Power, *Regional Survey Reports: Northumberland and Cumberland Coalfields; Durham Coalfield; North Western Coalfields; North Eastern Coalfield; North Midland Coalfield; Coalfields of the Midland Region; Kent Coalfield*; 1945. *South Wales Coalfield; Forest of Dean Coalfield; Bristol and Somerset Coalfield*; 1946.

Ministry of Fuel and Power, *Coal Mining: Report of the Technical Advisory Committee* (Reid Report), 1945. Cmd. 6610.

*Report of the Committee on National Policy for the Use of Fuel and Power Resources* (Ridley Report), 1952. Cmd. 8647.
*Agreement concerning the relations between the United Kingdom of Great Britain and Northern Ireland and the European Coal and Steel Community*, 1954. Cmd. 9346.
*Report of SC on Nationalised Industries*, 1958. HC 187–I of 1958.
*The Financial and Economic Obligations of the Nationalised Industries*, 1961. Cmnd. 1337.
*Fuel Policy*, 1965. Cmnd. 2798.
*The Finances of the Coal Industry*, 1965. Cmnd. 2805.
NBPI, Report no. 12, *Coal Prices*, 1966. Cmnd. 2919.
*Second Report of SC on Nationalised Industries: Gas, Electricity and Coal Industries*, 1966. HC 77 of 1965–6.
*Nationalised Industries: A Review of Economic and Financial Objectives*, 1967. Cmnd. 3437.
Ministry of Power, *Fuel Policy*, 1967. Cmnd. 3438.
*Report of the Tribunal appointed to inquire into the Disaster at Aberfan on October 21st 1966*, 1967. HL 316, HC 553 of 1966–7.
*First Report of SC on Nationalised Industries*, 1967–8. 3 vols. HC 371 of 1967–8.
*Ministerial Control of the Nationalised Industries: the Government's reply to the First Report from the SC on Nationalised Industries session 1967-8*, 1969. Cmnd. 4027.
*Report of SC on Nationalised Industries*, 1969. HC 471 of 1969–70.
SC on Nationalised Industries, *National Coal Board: Observations by the Minister of Technology*, 1970. Cmnd. 4323.
NBPI, Report no. 138, *Coal Prices*, 1970. Cmnd. 4255.
NBPI, Report no. 153, *Second Report on Coal Prices*, 1970. Cmnd. 4455; with *Supplements nos. 1 and 2*, 1971. Cmnd. 4455–I and 4455–II.
Department of Employment and Productivity, *Ryhope: a pit closes*, 1970.
Department of Employment, *Report of a Court of Inquiry into a Dispute between the National Coal Board and the National Union of Mineworkers under the chairmanship of the Rt. Hon Lord Wilberforce*, 1972. Cmnd. 4903.
Pay Board, *Special Report: Relative Pay of Mineworkers*, 1974. Cmnd. 5567.
Department of Energy, *Coal Industry Examination: Interim Report* and *Final Report*, 1974.
Department of Energy, *Coal for the Future: Progress with "Plan for Coal" and prospects to the year 2000*, n.d. [1977].
Department of Energy, *Energy Policy Review*, 1977.
Department of Energy, *Energy Policy: A Consultative Document*, 1978. Cmnd. 7101.
Department of Energy, *Coal Technology: Future Developments in Conversion, Utilisation and Unconventional Mining in the United Kingdom*, n.d. [1978].

House of Commons, *Second Report from the Energy Committee: Pit Closures*, 1982. HC 135 of 1982–3.
Monopolies and Mergers Commission, *National Coal Board: a report on efficiency and costs*, 2 vols., 1983. Cmnd. 8920.

(b) *National Coal Board*

(i) *Annuals and serials*

*Annual Report and Accounts*, 1946–.
*Annual Report of Research on Medical etc. Problems*, 1952–61.
*Medical Service Annual Report*, 1957–61, 1971/2–.
*Medical Service and Research on Medical and Allied Problems Annual Report*, 1962–70/1.
*MRDE Annual Report* 1969/70–.
*MRDE Projects* 1974/5–.
*Opencast News* (irregular, twice or three times a year), 1973–7.

(ii) *Other publications*

*Plan for Coal*, 1950.
*The Bolsover Story*, 1951.
*Report of the Advisory Committee on Organisation* (Fleck Report), 1955.
*Investing in Coal: Progress and Prospects under the Plan for Coal*, 1956.
*Revised Plan for Coal*, 1959.
*National Power Loading Agreement*, 1966 (jointly with NUM).
*Land Use Aspects of Mining–Opencast*: written evidence for the Commission on Energy and the Environment, 1979.
*Opencast Coal Mining in Britain*, 1979.
*Fluidised Bed Combustion*, n.d. [1979].
*New Mines for Old*, n.d. [1979].
*High Technology in Coal*, n.d. [1982].

(c) *Coal Industry Social Welfare Organisation*

*Annual Report*, 1952–.

(d) *Safety in Mines Research Establishment*

*An Introduction to the Safety in Mines Research Establishment*, 1952; revised edn. 1957.

(e) *Conseil d'Association* (ECSC and UK)

Rapport annuel, 1956–72.

## D. Other Printed Works

This section includes both articles in periodicals and books.

(i) *Annuals and serials*

*Colliery Guardian*, continuous since 1858. Of particular value for reference is the long series of articles, 'Structure of the National Coal

Board', in successive issues from October 1975 to June 1978. A few other specialist articles are separately specified below.
*Colliery Year Book and Coal Trades Directory*, annual to 1964 (ceased publication).
*Guide to the Coalfields*, annual since 1948.

(ii) *Other publications*

*ACHIEVEMENT* magazine, *Coal's New Burning Challenge* (articles reprinted in book form) (London, n.d. [1981]).
AGNEW, D., *Bevin Boy* (London, 1947).
ALLEN, V. L., *The Militancy of British Miners* (Shipley, 1981).
ANON., 'Arthur Street', *Public Administration*, Winter 1951.
ARGUILE, R. T., 'Reclamation: five industrial sites in the East Midlands', *Journal of the Institution of Municipal Engineers*, vol. 98, no. 6, 1971.
ARGUILE, R. T., 'Opencast Coal Mining in Britain: The First 32 Years', *Colliery Guardian*, February 1975.
ARGUILE, R. T., 'Reclamation—a success story', *Chartered Municipal Engineer*, vol. 105, no. 6, 1978.
ARNOT, R. P., *The Miners in Crisis and War* (London, 1961).
ARNOT, R. P., *The Miners: One Union, One Industry* (London, 1979).
BALDWIN, G. B., *Beyond Nationalization: The Labor Problems of British Coal* (Cambridge, Mass., 1955).
BRENT-JONES, E., 'Methods and Costs of Land Restoration', *Quarry Managers' Journal*, October 1971.
CARVEL, J. L., *Fifty Years of Machine Mining Progress 1899-1949: Anderson Boyes Ltd.* (Motherwell, 1949).
CASELEY, J. R., *Opencast Mining: Some Facts and Figures* (London, 1959).
CHESTER, D. N. (Sir Norman), *The Nationalisation of British Industry 1945-1951* (London, 1975).
COLE, G. D. H., *The National Coal Board*, revised edn. (London, 1949).
COLE, M. I., *Miners and the Board* (London, 1949).
*COLLIERY GUARDIAN*, *National Coal Board: the first ten years* (London, 1957).
COSTAIN MINING LTD., *Westfield Open Pit* (London, n.d. [?1975]).
COURT, W. H. B., *Coal (History of the Second World War: UK Civil Series)* (London, 1951).
DAVISON, D. J., 'Opencast Coal Mining: The Future and the Environment', *Colliery Guardian*, May 1975.
DELL, C. C., 'The Principles of Froth Flotation of Coal', *Mine and Quarry*, June 1976.
EZRA, D. J., 'The British Coal Industry and the European Economic Community', *BACM News Letter*, June 1970.
EZRA, D. J., *Coal and Energy* (London, 1978).
EZRA, D. J., *European Community Coal Policy* (Energy Commission Paper no. 14) (London, 1978).

FAY, S. and YOUNG, H., *The Fall of Heath* (London, 1976).

FINDLAY, R. V., 'Volks- und energiewirtschaft: Der Beitritt Grossbritanniens zum Europäischen Gemeinschaft aus der Sicht des britischen Kohlenbergbaus', *Glückauf*, 108, nr. 2 (1972).

GORMLEY, J., *Battered Cherub* (London, 1982).

GRIFFIN, A. R., *The British Coalmining Industry* (Buxton, 1977).

GRIFFIN, A. R., *Coalmining* (London, 1971).

GRIMSHAW, W. A., *The Role of Parliament in the Grimshaw Affair* (privately printed, 1977).

HANDY, L. J., 'British Coalminers' Strike in Perspective', *Economic and Political Weekly*, 4 March 1972.

HANDY, L. J., *Wages Policy in the British Coalmining Industry: a study of national wage bargaining* (Cambridge, 1981).

HANNAH, L., *Engineers, Managers and Politicians: the first fifteen years of nationalised electricity supply in Britain* (London, 1982).

HARLOW, C., *Innovation and Productivity under Nationalisation: The First Thirty Years* (London, 1977).

HAYNES, W. W., *Nationalization in Practice: The British Coal Industry* (London, 1953).

HORNER, A., *Incorrigible Rebel* (London, 1960).

HOUSE, J. W. and KNIGHT, E. M., *Pit Closure and the Community* (Newcastle upon Tyne, 1967).

HOWARD, C. W., 'British Coal and the Common Market', *Coal Merchant and Shipper*, June 1971.

HUGHES, J. and MOORE, R., *A Special Case? Social Justice and the Miners* (London, 1972).

JACOBSEN, M., RAE, S., WALTON, W. H., and ROGAN, J. M., 'New Dust Standards for British Coal Mines', *Nature*, vol. 27, 1 August 1970.

JENSEN, W. G., *Energy in Europe 1945-80* (London, 1967).

KELF-COHEN, R., *British Nationalisation 1945-1973* (London, 1973).

KIRBY, M. W., *The British Coalmining Industry 1870-1946* (London, 1977).

KIRK, G. (ed.), *Schumacher on Energy: Speeches and Writings of E. F. Schumacher* (London, 1982).

KNIGHT, E. M., *Men Leaving Mining: West Cumberland 1966-67* (Newcastle upon Tyne, 1968).

LEE, W. A., *Thirty Years in Coal* (London 1954).

LISTER, L., *Europe's Coal and Steel Community* (New York, 1960).

MANCHESTER JOINT RESEARCH COUNCIL, *Economic Aspects of Fuel and Power in British Industry* (Manchester, 1960).

MOFFAT, A., *My Life with the Miners* (London, 1965).

PAVITT, K. (ed.), *Technical Innovation and British Economic Performance* (London, 1980).

PAYNTER, W., *Outlook for Mining* (London, n.d. [1958]).

PAYNTER, W., *My Generation* (London, 1972).
PEP, *A Fuel Policy for Britain* (London, n.d. [1966] ).
POSNER, M. V., *Fuel Policy: a study in applied economics* (London, 1973).
REID, C. C., 'The Problem of Coal', *The Times*, 22, 23, 24 November 1948.
REID, G. L., ALLEN, K., and HARRIS, D. J., *The Nationalized Fuel Industries* (London, 1973).
ROBENS, LORD, *Ten Year Stint* (London, 1972).
ROGOW, A. A., *The Labour Government and British Industry 1945-1951* (Oxford, 1955).
ROUND, C. L., 'Opencast Coal Mining. Production Techniques and Marketing', *Colliery Guardian*, April 1975.
SEWELL, J., *Colliery Closure and Social Change* (Cardiff, 1975).
SHEPHERD, R. and WITHERS, A. G., *Mechanised Cutting and Loading of Coal* (London, 1960).
SHEPHERD, W. G., *Economic Performance under Public Ownership: British Fuel and Power* (New Haven, 1965).
SHINWELL, E., *Conflict without Malice* (London, 1955).
STREET, A., *Why Have Hobart House?* (London, 1950).
SUTCLIFFE, R. J. and E. D., *Richard Sutcliffe: the pioneer of underground belt conveying*, 3rd edn. (privately printed, 1955).
TOMLINSON, R. C. (ed.), *OR comes of age: a review of the work of the Operational Research Branch of the National Coal Board 1948-1969* (London, 1971).
TOWN, S. W., *After the Mines: changing opportunities in a South Wales Valley* (Cardiff, 1978).
TOWNSEND, J. F., *Innovation in Coal Mining Machinery. "The Anderton Shearer Loader"–the Role of the NCB and the Supply Industry in its Development* (University of Sussex Science Policy Research Unit, Occasional Paper Series, no. 3, 1976).
TOWNSHEND-ROSE, H., *The British Coal Industry* (London, 1951).
VARIOUS AUTHORS (VIELVOYE, R. and others) for NCB, *Coal: Technology for Britain's Future* (London, 1976).
WARD WALTERS, L., 'Opencast Mining, 1942–1948', *Opencast Journal*, no. 3, November 1948.
WILLIAMS, P. M., *Hugh Gaitskell: A Political Biography* (London, 1979).

# Index